Counterterrorism and Cybersecurity

Newton Lee

Counterterrorism and Cybersecurity

Total Information Awareness

Second Edition

 Springer

Newton Lee
Newton Lee Laboratories, LLC
Tujunga, CA
USA

ISBN 978-3-319-37460-4   ISBN 978-3-319-17244-6   (eBook)
DOI 10.1007/978-3-319-17244-6

Springer Cham Heidelberg New York Dordrecht London
© Springer International Publishing Switzerland 2012, 2015
Softcover reprint of the hardcover 2nd edition 2015

Printed on acid-free paper

Springer International Publishing AG Switzerland is part of Springer Science+Business Media
(www.springer.com)

To peace, love, and freedom

Contents

About the Author

Newton Lee is CEO of Newton Lee Laboratories LLC, president of the Institute for Education, Research, and Scholarships, adjunct professor at Woodbury University's School of Media, Culture & Design, and editor-in-chief of ACM Computers in Entertainment. Previously, he was a computer scientist at AT&T Bell Laboratories, senior producer and engineer at The Walt Disney Company, and research staff member at the Institute for Defense Analyses. He was founder of Disney Online Technology Forum, creator of Bell Labs' first-ever commercial AI tool, and inventor of the world's first annotated multimedia OPAC for the US National Agricultural Library.

Lee graduated Summa Cum Laude from Virginia Tech with a B.S. and M.S. degree in Computer Science, and he earned a perfect GPA from Vincennes University with an A.S. degree in Electrical Engineering and an honorary doctorate in Computer Science. He is the coauthor of *Disney Stories: Getting to Digital*; author of the Total Information Awareness book series including *Facebook Nation* and *Counterterrorism and Cybersecurity*; coauthor/editor of the Digital Da Vinci book series including *Computers in Music* and *Computers in the Arts and Sciences*; and editor-in-chief of the *Encyclopedia of Computer Graphics and Games*.

Additional contributors are as follows:

Ray Balut, chief information security officer in the healthcare industry.

Darren Manners, senior security engineer at SyCom Technologies and Black Hat speaker.

Andy Marken, president of Marken Communications Inc.

Emily Peed, founder of Developer's Club—IT Girls 2.0 and DEF CON speaker.

Philip Polstra, associate professor at Bloomsburg University and DEF CON speaker.

Jean C. Stanford, lecturer at Georgetown University and DEF CON speaker.

About the Book

From 9/11 to Charlie Hebdo along with Sony-pocalypse and DARPA's $2 million Cyber Grand Challenge, this book examines counterterrorism and cybersecurity history, strategies, and technologies from a thought-provoking approach that encompasses personal experiences, investigative journalism, historical and current events, ideas from thought leaders, and the make-believe of Hollywood such as 24, *Homeland*, and *The Americans*.

President Barack Obama said in his 2015 State of the Union address, "We are making sure our government integrates intelligence to combat cyber threats, just as we have done to combat terrorism." Demystifying Total Information Awareness along with Edward Snowden's NSA leaks, the author expounds on the US intelligence community, artificial intelligence, data mining, social media, cyber attacks and prevention, causes and cures for terrorism, and long-standing issues of war and peace.

This book offers practical advice for businesses, governments, and individuals to better secure the world and protect cyberspace. It quotes former US Navy Admiral and NATO's Supreme Allied Commander James Stavridis: "Instead of building walls to create security, we need to build bridges." The author also provides a glimpse into the future of Plan X and Generation Z, along with an ominous prediction from security advisor Marc Goodman at TEDGlobal 2012 that "if you control the code, you control the world."

This expanded edition also includes new chapters and commentaries by healthcare chief information security officer Ray Balut, security engineer and Black Hat speaker Darren Manners, communications expert Andy Marken, and DEF CON speakers Emily Peed, Philip Polstra, and Jean C. Stanford.

Part I
Counterterrorism History: Then and Now

Chapter 1
September 11 Attacks

We're a nation that is adjusting to a new type of war. This isn't a conventional war that we're waging. Ours is a campaign that will have to reflect the new enemy.... The new war is not only against the evildoers themselves; the new war is against those who harbor them and finance them and feed them.... We will need patience and determination in order to succeed.
— President George W. Bush (September 11, 2001)

Our company around the world will continue to operate in this sometimes violent world in which we live, offering products that reach to the higher and more positive side of the human equation.
— Disney CEO Michael Eisner (September 11, 2001)

He risked his life to stop a tyrant, then gave his life trying to help build a better Libya. The world needs more Chris Stevenses.
— U.S. Secretary of State Hillary Clinton (September 12, 2012)

We cannot have a society in which some dictators someplace can start imposing censorship here in the United States because if somebody is able to intimidate us out of releasing a satirical movie, imagine what they start doing once they see a documentary that they don't like or news reports that they don't like. That's not who we are. That's not what America is about.
— President Barack Obama (December 19, 2014)

1.1 September 11, 2001

I was waking up in the sunny California morning on September 11, 2001. Instead of music playing on my radio alarm clock, I was hearing fragments of news broadcast about airplanes crashing into the Pentagon and the twin towers of the World Trade Center. I thought I was dreaming about waking up in an alternative universe.

© Springer International Publishing Switzerland 2015
N. Lee, *Counterterrorism and Cybersecurity*, DOI 10.1007/978-3-319-17244-6_1

Christopher Nolan, writer-director of *Inception* (2010), once said that "ever since he was a youngster, he was intrigued by the way he would wake up and then, while he fell back into a lighter sleep, hold on to the awareness that he was in fact dreaming. Then there was the even more fascinating feeling that he could study the place and tilt the events of the dream" [1].

Similarly, whenever I was having a nightmare, I would either semiconsciously alter the chain of events to achieve a less sinister outcome, or I would force myself to wake up and escape to reality or to a new dream state about waking up.

However, as I awoke to the radio news broadcast on the morning of September 11, I realized that it was not a lucid dream. New York City and the Pentagon were under attack. The U.S. airspace was shut down. The alternative universe was the present reality.

I went to work that morning in a state of shock and disbelief. As I entered the lobby of Disney Online, I saw a television on a portable TV cart adjacent to the reception desk. Several people were standing in front of the television. No one spoke a word. We watched the replays of an airplane crashing into the South Tower of the World Trade Center, people jumping out of the broken windows, and the collapse of the North and South Towers. It was surreal and somber.

The Disney Online office was uncharacteristically quiet that morning. I tried my best to focus on work, but my mind kept wandering off to the unfolding disasters in the East Coast and my reminiscence of the year 1984 in Virginia.

In the summer of 1984, Virginia Tech professor Dr. Timothy Lindquist introduced me to Dr. John F. Kramer at the Institute for Defense Analyses (IDA), a nonprofit think tank serving the U.S. Department of Defense (DoD) and the Executive Office of the President [2]. A year prior, I received a surprise letter from the White House signed by President Ronald Reagan, thanking me for my support. Partially motivated by the letter, I accepted the internship offer at IDA and became a research staff member. My summer project was to assist DoD in drafting the Military Standard Common APSE (Ada Programming Support Environment) Interface Set (CAIS) [3].

My winter project was to design a counterterrorism software program for a multi-agency joint research effort involving the Defense Advanced Research Projects Agency (DARPA), National Security Agency (NSA), and Federal Bureau of Investigation (FBI). The FBI was investigating the deadly terrorist attacks against the U.S. embassy and military barrack in Beirut, Lebanon. As a co-pioneer of artificial intelligence applications in counterterrorism, I helped develop a natural language parser and machine learning program to digest news and articles in search of potential terrorist threats around the globe.

I left IDA for Bell Laboratories in 1985. However, the IDA counterterrorism project came across my mind in December 1988 when the New York-bound Pan Am Flight 103 exploded from a terrorist bomb over Lockerbie, Scotland, killing all 259 aboard and 11 on the ground [4]. The passengers included 35 Syracuse University students returning home for a Christmas break. Their surviving families were devastated by the Lockerbie bombing.

International terrorism arrived on American soil in February 1993 when a truck bomb was detonated in a public parking garage beneath the World Trade Center. "It felt like an airplane hit the building," said Bruce Pomper, a 34-year-old broker working at the Twin Towers [5]. Unbeknownst to anyone, the eyewitness' statement turned out to be an eerie prophecy. The World Trade Center bombing in 1993 and the Oklahoma City bombing in 1995 were totally eclipsed by the terrorist attacks on September 11, 2001 when the hijacked planes hit the Twin Towers and the Pentagon.

I wanted to know exactly what had happened, who was responsible, and why. I recalled one very smart computer programmer to whom Disney Online made a job offer a few years ago, but he turned it down and instead took a much higher paying job at a Wall Street firm in World Trade Center. I hoped he evacuated safely from the Twin Towers.

I checked ABCNews.com, CNN.com, and MSNBC.com, but they returned a "Server Busy" error message in the morning. The Google homepage displayed the messages: "If you are looking for news, you will find the most current information on TV or radio. Many online news services are not available, because of extremely high demand. Below are links to news sites, including cached copies as they appeared earlier today."

At 10:08 AM Pacific Time, I received an email from Steve Wadsworth, president of Walt Disney Internet Group (WDIG):

Subject: Today
Author: Steve Wadsworth
Date: Tuesday, September 11, 2001 10:08 AM

Team -

As I'm sure you are well aware, there has been a series of tragic and frightening terrorist attacks in the United States today. I want to make sure you all know that the personal safety and security of employees is our first concern. If for any reason you have any concerns or personal issues to attend to, please do so. Coming to or staying at work is discretionary.

We are currently assessing the situation at all WDIG facilities. At this point, simply as precautionary measures:

- The 34th Street building in New York is being evacuated.
- Smith Tower in Seattle has been closed by the landlord.
- The Westin building in Seattle is only assessable with a security ID.

Should the situation change for these or any other WDIG facilities, you will be alerted.

In the meantime, please know that your safety and that of your loved ones is our top priority.

- Steve

I felt like the world was getting smaller that day. Ironically, I was working on Disney's "Small World Paint" application with my colleagues Alejandro Gomez and Perigil Ilacas. But my mind kept pondering about a million things, and I ended up spending most of the day reading online news.

1.2 Disney's Responses to the 9/11 Attacks

On the morning of 9/11, Disneyland and Disney California Adventure remained closed for the entire day. The Disneyland closure marked only the second time it has locked its gates due to a national tragedy. The first time was November 23, 1963, in honor of President John F. Kennedy [6].

After the second plane hit World Trade Center on September 11, The Walt Disney Company announced the closure of its theme parks worldwide. To shut down the Disney theme parks in Florida, a "human wall procedure" was conducted at Magic Kingdom, Disney's Hollywood Studios, Animal Kingdom, and Epcot:

> Once the guests were forced to the streets of the park because all the rides were closed, all the cast members were instructed to hold hands and basically form a human wall and gently (without touching anyone) walk towards the hub of the park and eventually towards Main Street. That way we could basically force the guests out of the park. Disney Security obviously followed each human wall and made sure no one got past it [7].

The Disney resort hotels stayed open to provide free accommodations for the stranded guests who were unable to leave since the U.S. had grounded all civilian air traffic for two days.

At 7:24 PM Pacific Time on September 11, Disney CEO Michael Eisner sent out an email to all employees:

> Subject: 09/11/01 Today's events
> Author: Eisner, Michael
> Date: Tuesday, September 11, 2001 7:24 PM
>
> Dear Fellow Cast Members:
>
> Today's catastrophic act of violence to the World Trade Center, Pentagon, and four commercial planes is such a calamity that no words to our cast can express the sorrow or outrage that we feel. Let me offer my sincere condolences to those who lost a family member or friend or knew someone who was injured.
>
> During this difficult time, all of our employees whose jobs require them to rise to occasions such as these in fact rose with dignity and strength. As I write this, ABC News continues to perform an outstanding job in keeping the nation informed about these deeply disturbing events, as usual with professional, informative and meticulous coverage. WABC-TV in New York, under extreme pressure including the loss of its broadcast antenna, operated with honor, as did ABC Radio.
>
> Our parks closed for the day without incident and cast members assisted hotel guests in Florida and California, many of whom were obviously unable to get home. The Disney Stores closed as well for the day as did our Broadway shows in New York and on the road. Everybody acted with stoic determination to maintain Disney operations in efficient and caring ways.
>
> I want to thank all of you – who are understandably upset, normally confused about our complicated world and tolerably angry – for being calm and calming to our guests. Finally let me say our company around the world will continue to operate in this sometimes violent world in which we live, offering products that reach to the higher and more positive side of the human equation.
>
> Michael

Security tables were set up overnight at all Disney theme parks. The "crash course in flying" joke was eliminated from The Jungle Cruise ride. The Walt Disney Company Foundation established DisneyHAND: Survivor Relief Fund to provide $5 million corporate gift and $700,000 employee donations to organizations providing assistance to victims and their families. Each of the victims' families was issued a check for $1,000. Disney VoluntEARS, including trained medical personnel from Health Services and the Industrial Engineering department, helped man six of the Central Florida Blood Bank branches.

The 9/11 terrorist attacks had a big impact on all Americans and Disney employees. Ken Goldstein, General Manager of Disney Online, said at a weekly senior staff meeting, "This tragedy reminds us that the most important things in our lives are our families and friends." My colleague Eric Haseltine left his position as Executive Vice President of R&D at Walt Disney Imagineering to join the National Security Agency (NSA) as Director of Research in the following year.

1.3 FBI Warning of Hollywood as a Terrorist Target

Nine days after 9/11 on September 20, the Federal Bureau of Investigation (FBI) informed The Walt Disney Company about the possibility that a Hollywood studio could be a terrorist target. Disney president Bob Iger sent out an alarming email to all employees at 6:33 PM Pacific time:

Subject: 09/20/01 Important Security Notice
Author: Iger, Bob
Date: Thursday, September 20, 2001 6:33 PM

Dear Fellow Cast Members:

Today we, along with other studios, received information from the F.B.I. that a Hollywood studio could become a target for terrorist activity.

Your safety and security are our top priority and as such we are taking extraordinary precautions to ensure your safety. As you have doubtless already seen, we have already increased security at our facilities, and further steps will now be taken. We are also working diligently with the F.B.I. and local law enforcement.

We urge your patience and cooperation at this time. If you do have any concerns or notice anything suspicious, please contact security at 8228-3220 (818-560-3220).

Again, please know that our goal is to continue to provide a safe environment for you. Thank you again for your patience and understanding.

Bob

Thankfully, the uncorroborated terrorist threats against Hollywood did not materialize. Nevertheless, Disney had increased security measures at all of its facilities in Burbank, Glendale, and North Hollywood. All employees were required to wear their Company I.D. while at work. Street parking adjacent to the Disney headquarters was prohibited. Visitor parking was moved to surface lots, and all incoming packages were x-rayed. Security tables at Disney theme parks were deemed a necessity in spite of inconveniencing the visitors.

Terrorists often opt for soft targets in order to strike fear into the heart of civilians. However, it would be political suicide for a terrorist group to target children, schools, and hospitals. During an assault against a Yemeni Ministry of Defense compound in December 2013, an al-Qaeda fighter mistakenly attacked a hospital. Qassim Al-Raimi, head of AQAP (al-Qaeda in the Arabian Peninsula), did not wait long to publicly apologize for the deadly attack [8].

On September 27, 2001, President George W. Bush became Disney's newest cheerleader. In his speech to airline employees at O'Hare International Airport in Chicago, Bush remarked, "When they struck, they wanted to create an atmosphere of fear. And one of the great goals of this nation's war is to restore public confidence in the airline industry. It's to tell the traveling public: Get on board. Do your business around the country. Fly and enjoy America's great destination spots. Get down to Disney World in Florida. Take your families and enjoy life, the way we want it to be enjoyed" [9].

Nonetheless, America and its allies were forced to tighten security at all high-profile locations since 9/11. International travelers have been annoyed by long waits, pat-downs, and full-body scanners at airports. In a latest controversial move, tourists arriving in Israel after April 2013 may be asked to show their personal emails in addition to the travel documents [10].

1.4 Hollywood Realism on Terrorism

Oscar Wilde wrote in his 1889 essay *The Decay of Lying* that "Life imitates Art far more than Art imitates Life" [11]. Using a commercial airliner as a weapon was not entirely a new idea. In the 1996 film *Executive Decision* starring Kurt Russell, Halle Berry, and Steven Seagal, terrorists seize control of a Boeing 747 in their plan to attack Washington D.C. and cause mass causalities [12].

Premiered on November 6, 2001, the counterterrorism drama *24* set a television precedent for portraying the first African-American president (season 1 through 5) and the first female president (season 7 and 8). Dennis Haysbert, who played the U.S. president, predicted that his role in *24* would benefit Barack Obama in the 2008 presidential election [13]. U.S. Supreme Court Justice Clarence Thomas, Homeland Security Secretary Michael Chertoff, Defense Secretary Donald Rumsfeld, and Vice President Dick Cheney were all huge fans of *24* [14]. Because of the 9/11 attacks, the premiere was postponed for one week and the footage of an airplane explosion was edited out of the show [15].

President Barack Obama is reportedly a big fan of *Homeland*, a Showtime counterterrorism television series starring Claire Danes as CIA agent Carrie Mathison. Iranian-born actor Navid Negahban played terrorist Abu Nazir in *Homeland*. Negahban said, "The president is watching. It is fantastic that he is a big fan…. But at the same time I think I will be nervous when he sends me an invitation to the White House" [16].

In fact, the U.S. intelligence community has been seeking advice from Hollywood on handling terrorist attacks [17]. In 1999, the U.S. Army established the Institute for Creative Technologies (ICT) at the University of Southern California (USC) to tap into the creative talents of Hollywood and the game industry [18].

Sometimes there can be a blurred distinction between make-believe Hollywood and real life events. In the 2011 film *God Bless America* starring Joel Murray and Tara Lynne Barr, the two lead characters shot several rude teenagers in a movie theater for talking loudly and using their cell phones during the show [19]. In July 2012, a gunman killed 12 people and injured 58 inside a movie theater in Aurora, Colorado during a midnight screening of the film *The Dark Knight Rises* [20].

Abu Jandal, al-Qaeda terrorist and former bodyguard of Osama bin Laden, was captured and held in prison before September 11, 2001. After the 9/11 attacks, the FBI interrogated Jandal to determine the identities of the hijackers and the role of bin Laden. Jandal was shown a Yemeni news magazine with photographs of the airplanes crashing into the Twin Towers and people jumping a hundred stories off the buildings. He could hardly believe his eyes and said that they looked like a "Hollywood production" [21].

1.5 A New Day of Infamy and America at War

In the early afternoon on September 11, 2001, U.S. Senator John McCain of Arizona spoke on CNN and Fox News. McCain said, "I'm not in disagreement with others [members of Congress] who said that it is an act of war" [22]. On the night of September 11, President George W. Bush addressed the nation. Bush said, "We will make no distinction between the terrorists who committed these acts and those who harbor them…. America and our friends and allies join with all those who want peace and security in the world; and we stand together to win the war against terrorism" [23].

On September 12, newspapers around the world headlined the 9/11 attacks. Some compared the national tragedy to the attack on Pearl Harbor and some declared America at war (see Fig. 1.1):

"New day of infamy" — *Boston Globe, Albuquerque Journal,* and *The Charleston Gazette.*

"America's Bloodiest Day: This is the second Pearl Harbor" — *Honolulu Advertiser.*

"It's War" — *New York Daily News.*

"Act of War" — *Portland Press Herald, New York Post, USA Today,* and *Sun Journal.*

"War at home" — *The Dallas Morning News.*

"War on America" — *The Daily Telegraph* (UK).

"Pearl Harbor im Jahr 2001" — *Die Welt* (Germany, Der Kommentar).

Fig. 1.1 Newspaper headlines on September 11, 2001

On September 23, National Security Adviser Condoleezza Rice told reporters that the U.S. government had "very good evidence of links" between Osama bin Laden's operatives "and what happened on September 11" [24]. Secretary of State Colin Powell remarked that the evidence also came from the investigation of the USS *Cole* bombing in Yemen. The Bush administration persuaded most of the world, including some Muslim nations, that a military response was justified.

On October 7, the United States and its allies began the invasion of Afghanistan under "Operation Enduring Freedom" in an attempt to capture Osama bin Laden and to force the Taliban government to hand him over to the U.S. [25]

The invasion further heightened security at high-profile American businesses nationwide and overseas. On October 8, the Disney corporate office in Hong Kong issued an email to all Disney employees in Asia:

Subject: Escalating Tensions
Author: Corporate-Broadcast AP at DISNEY-TV-HK
Date: 10/8/01 9:58 AM

Dear Team Asia,

By now, you know that war against terrorism has reached Afghanistan today with the retaliatory attacks by US and British forces.

It is of course expected that retaliatory attacks by the terrorists will follow across the globe. The back-and-forth has begun. In addition, as has been reported over the past month, this battle will likely go on for some time.

We are of course in contact with the American Consulate in Hong Kong and local officials to make sure we stay updated on any security issues or news related to us locally as this battle grows. In addition, we are in constant communication with Burbank. In all cases, there is a heightened sense of alert and security in place now.

The main thing that each of us can do is to stay alert and pay attention to the situation. As has been the plan, we will move forward with business the way we do each and every day. Now, more than ever, it's key for each division and person within Disney to work closely together for the mutual benefit of the company. Let's help each other out whenever possible. Safety at home, in the office and during travel is our key priority while we move forward with our business — our being a family entertainment company.

Please let me know if you have any question or concerns. We'll keep you posted as we get more information.

Thank you,

Jon

On October 26, President George W. Bush signed into law the USA PATRIOT (Uniting and Strengthening America by Providing Appropriate Tools Required to Intercept and Obstruct Terrorism) Act of 2001 [26]. The Patriot Act expanded the power of law enforcement agencies to gather intelligence within the United States via surveillance of communications, financial transactions, and other activities.

On November 13, President Bush issued a military order for detention, treatment, and trial of certain non-citizens in the war against terrorism [27]. The Bush Administration established in 2002 the controversial detainment and interrogation facility at the Guantánamo Bay Naval Base in southeastern Cuba where 779 people had been detained [28].

Although Operation Enduring Freedom in Afghanistan (OEF-A) failed to locate Osama bin Laden, it toppled the Taliban government who provided sanctuary to bin Laden and his "al-Qaeda" ("The Base") network. Osama bin Laden had evaded capture for nearly 10 years before he was killed by U.S. Navy SEALs in Abbottabad, Pakistan on May 2, 2011 [29]. The war in Afghanistan continued after bin Laden's death; and it has become the longest war in American history, surpassing Iraq (8 years and 7 months) and Vietnam (8 years and 5 months) [30].

1.6 September 11, 2012

The intense hunt for terrorists over the past decade has diminished the al-Qaeda network, but the threat of terrorism is far from gone. On September 11, 2012, terrorists attacked the U.S. Consulate in Benghazi, Libya during the massive protests across the Muslim world over an anti-Islam film by Egyptian-American Mark Basseley Youssef [31]. Among the casualties were U.S. ambassador J. Christopher Stevens, ex-Navy SEALs Glen Doherty and Tyrone Woods, and an online gaming maven Sean Smith [32].

London-based think tank Quilliam believed that al-Qaeda "came to avenge the death of Abu Yahya al-Libi, al-Qaeda's second in command killed a few months ago [by a U.S. drone strike in Pakistan]" [33]. The U.S. had obtained information that Ansar al-Sharia, al-Qaeda in the Islamic Maghreb, and the Egypt based Muhammad Jamal group were likely responsible for the terrorist attack [34]. Former CIA director David Petraeus initially gave the impression that the attack arose out of a spontaneous demonstration against an anti-Islam film, but he later testified in Congress that it was an act of terrorism committed by al-Qaeda linked militants [35].

Ambassador J. Christopher Stevens, a former Peace Corps volunteer, was well-regarded among Libyans for his support of democratic transition in Libya against dictator Muammar Gaddafi's regime. "In the early days of the Libyan revolution, I asked Chris to be our envoy to the rebel opposition," said U.S. Secretary of State Hillary Clinton. "He arrived on a cargo ship in the port of Benghazi and began building our relationships with Libya's revolutionaries.... He risked his life to stop a tyrant, then gave his life trying to help build a better Libya. The world needs more Chris Stevenses" [36].

In September 2001, President George W. Bush described the long, drawn-out war on terror: "We're a nation that is adjusting to a new type of war. This isn't a conventional war that we're waging. Ours is a campaign that will have to reflect the new enemy.... These are people who strike and hide, people who know no borders... The new war is not only against the evildoers themselves; the new war is against those who harbor them and finance them and feed them. We will need patience and determination in order to succeed" [9].

1.7 Sony-pocalypse: Invoking 9/11 Attacks in Cyber Terrorism

"The world will be full of fear. Remember the 11th of September 2001. We recommend you to keep yourself distant from the places at that time. (If your house is nearby, you'd better leave.)" These were the words of the cybercriminals who called themselves the Guardians of Peace (GOP).

Enterprise security analyst Adrian Sanabria coined the term "Sony-pocalypse" to describe the most destructive cyber attack reported to date against a company on U.S. soil [37]. In November 2014, the GOP had stolen 100 terabytes of data [38], destroyed 75 % of corporate computer servers, and crippled the company's data centers [39]. Among the stolen data were five feature films, executive emails, business contracts, company budgets, employee personal data, salary information, medical records, and celebrity secrets [40]. *The New York Times* reported that "administrators hauled out old machines that allowed them to cut physical payroll checks in lieu of electronic direct deposit" [39].

On December 5, the GOP threatened Sony employees and their families via an email written in broken English: "Removing Sony Pictures on earth is a very tiny work for our group which is a worldwide organization.... Please sign your name to object the false of the company at the email address below if you don't want to suffer damage. If you don't, not only you but your family will be in danger" [41].

Sony did not budge. On December 8, the GOP condemned the controversial satire comedy *The Interview* about a hapless TV host recruited by the CIA to assassinate North Korean dictator Kim Jong-un. The group demanded Sony to "stop immediately showing the movie of terrorism which can break regional peace and cause the War!" [42]

On December 10, Sony fought back by launching denial of service (DoS) attacks on websites that were leaking the stolen data, as well as spreading bogus files to the torrent network to fool downloaders [43].

Cyber warfare escalated to cyber terrorism on December 16 when the GOP invoked the 9/11 terror attacks in a warning to moviegoers who might consider watching *The Interview* [44]. In response to the physical terrorism threat, the U.S. Department of Homeland Security issued a statement saying that "at this time there is no credible intelligence to indicate an active plot against movie theaters within the United States" [45].

Nevertheless, the damage was done: Three largest theater chains — AMC, Regal, and Cinemark — along with the majority of 3,000 cinemas called off the screenings of *The Interview* in fear of physical violence and collateral damages to other movies shown in the same theaters [46].

On December 17, Sony gave in and canceled the Christmas Day theatrical release for *The Interview*. "We are deeply saddened at this brazen effort to suppress the distribution of a movie, and in the process do damage to our company, our employees, and the American public," Sony said in a statement. "We stand by our filmmakers and their right to free expression and are extremely disappointed by this outcome" [47].

President Barack Obama strongly disapproved of Sony's decision to cancel the film release. "I am sympathetic to the concerns that they face. Having said all that, yes, I think they made a mistake," Obama said at a news conference on December 19. "Let's not get into that way of doing business. We cannot have a society in which some dictators someplace can start imposing censorship here in the United States because if somebody is able to intimidate us out of releasing a satirical movie, imagine what they start doing once they see a documentary that they don't like or news reports that they don't like. That's not who we are. That's not what America is about" [48].

The FBI pinned blame on North Korea for the Sony-pocalypse [49] and the GOP replied with a mocking message: "The result of investigation by FBI is so excellent that you might have seen what we were doing with your own eye. We congratulate you [sic] success. FBI is the BEST in the world. You will find the gift for FBI at the following address. https://www.youtube.com/watch?v=hiRacdl02w4 Enjoy!" [50]

Even if North Korea was actually responsible for the massive cyber attack against Sony, President Obama did not consider it "an act of war." In an interview on CNN's *State of the Union* that aired on December 21, Obama told news anchor Candy Crowley, "We've got very clear criteria as to what it means for a state to sponsor terrorism. And we don't make those judgments just based on the news of the day. We look systematically at what's been done and based on those facts, we'll make those determinations in the future…. It was an act of cyber vandalism that was very costly. We take it very seriously and we will respond proportionally. If we set a precedent in which a dictator in another country can disrupt through cyber, a company's distribution chain or its products, and as a consequence we start censoring ourselves, that's a problem. And it's a problem not just for the entertainment industry; it's a problem for the news industry…. So the key here is not to suggest that Sony was a bad actor. It's making a broader point that all of us have to adapt to the possibility of cyber attacks, we have to do a lot more to guard against them" [51].

On December 22, New York's Treehouse Theater announced that it would present a live reading of the movie's script. "In the wake of recent events surrounding the controversial film 'The Interview,' the feeling that a threat to free speech has been imposed is inescapable and terrifying," reads the theater's announcement for the event. "In response to this, three comedians have acquired a draft of the script for the banned film and are producing a live-read at the Treehouse Theater on Saturday, December 27th at 7 pm. Featuring some of the finest performers from the Upright Citizen's Brigade theater, this is an opportunity for people to come together in the name of free speech, in defiance of all who have threatened it" [52].

On December 23, with the support of about 300 independently-owned theaters across the country, Sony reversed its decision to cancel the theatrical release. Seth Rogen who co-directed and co-starred in *The Interview* sent out a victory tweet: "The people have spoken! Freedom has prevailed! Sony didn't give up! The Interview will be shown at theaters willing to play it on Xmas day!" (See Fig. 1.2).

Although most critics considered *The Interview* "a pretty terrible movie" [53], the premiere draw many sell-out audiences on Christmas Day [54]. Sony made Hollywood history by releasing the movie simultaneously in theaters and online via streaming on YouTube, Google Play, Microsoft Xbox, and the dedicated website SeeTheInterview.com — a strategy known as same-day-and-date release [55].

Fig. 1.2   Seth Rogen's Victory Tweet on December 23, 2014

For the opening weekend, the film barely earned $3 million in ticket sales from 331 screens but it pulled in over $15 million from online purchases and rentals, making it the Sony Pictures #1 online film of all time [56]. By January 2015, *The Interview* made $31 million in online and video-on-demand revenues and $5 million in limited theatrical release [57]. Although the financial damage to Sony was estimated to be around $100 million [58], Sony CEO Kazuo Hirari downplayed the financial impact of the cyber attacks on the company at a news conference during the 2015 Consumer Electronics Show (CES) in Las Vegas, Nevada [59]. In February 2015, Sony gave an official figure for damages from the massive hack: $15 million [60] and Amy Pascal resigned from her co-chair job at Sony Pictures Entertainment [61].

It is a cosmic irony that Japan was the aggressor in the surprise attack on Pearl Harbor in December 1941 and now Sony became the victim of a massive cyber attack 73 years later in November 2014. More importantly, however, the Sony saga has brought to light a hidden sinister: Cyber terrorism has reached a whole new level by combining cyber attacks and threats of physical terrorism.

Kim Zetter, senior staff reporter at *Wired* magazine, expressed her concern: "Even if members of GOP lack the means or intent to pull off a terrorist attack on their own, they've now created an open invitation for opportunistic attackers to do so in their name — in essence, escalating their crimes and influence to a level no other hackers[1] have achieved to date" [62].

The Sony-pocalypse serves as a wakeup call for everyone to take cybersecurity seriously. President Obama considered the Sony saga "cyber vandalism" [51] whereas one senior Sony employee called it "a terrorist attack" [63]. I believe that "cyber terrorism" is a more appropriate term for the Sony saga. Cyber attacks and terrorist threats are a lethal combination that can only be resolved by aligning conscientious counterterrorism policies with cybersecurity technologies.

Bibliography

1. **Boucher, Geoff.** 'Inception' breaks into dreams. [Online] Los Angeles Times, April 4, 2010. http://articles.latimes.com/2010/apr/04/entertainment/la-ca-inception4-2010apr04.
2. **Institute for Defense Analyses.** IDA's History and Mission. [Online] [Cited: August 29, 2012.] https://www.ida.org/aboutus/historyandmission.php.
3. **Ada Joint Program Office.** Military Standard Common APSE (Ada Programming Support Environment) Interface Set (CAIS). [Online] Defense Technical Information Center, 1985. http://books.google.com/books/about/Military_Standard_Common_APSE_Ada_Progra.html ?id=EjEYOAAACAAJ.
4. **The Guardian staff and agencies.** Lockerbie bombing – timeline. [Online] The Guardian, May 20, 2012. http://www.guardian.co.uk/uk/2012/may/20/time-line-lockerbie-bombing-megrahi.
5. **BBC.** World Trade Center bomb terrorises New York. [Online] BBC, February 26, 1993. http://news.bbc.co.uk/onthisday/hi/dates/stories/february/26/newsid_2516000/2516469.stm.
6. **Rivera, Heather Hust.** Disneyland Resort Remembers. [Online] Disney.com, November 23, 2009. http://disneyparks.disney.go.com/blog/2009/11/disneyland-resort-remembers/.
7. **Hill, Jim.** What Was It Like at Walt Disney World on 9/11. [Online] Huffington Post, September 7, 2011. http://www.huffingtonpost.com/jim-hill/what-was-it-like-at-walt-_b_952645.html.
8. **Basil, Yousuf and Shoichet, Catherine E.** Al Qaeda: We're sorry about Yemen hospital attack. [Online] CNN, December 22, 2013. http://www.cnn.com/2013/12/22/world/meast/yemen-al-qaeda-apology/index.html.
9. **Office of the Press Secretary.** At O'Hare, President Says "Get On Board". [Online] The White House, September 27, 2001. http://georgewbush-whitehouse.archives.gov/news/releases/2001/09/20010927-1.html.
10. **Sidner, Sara.** Israeli security allowed to seek check of tourists' e-mail. [Online] CNN, April 25, 2013. http://www.cnn.com/2013/04/25/travel/israel-travelers-email/index.html.
11. **Wilde, Oscar.** The Decay of Lying. [Online] The Nineteenth Century, January 1889. http://cogweb.ucla.edu/Abstracts/Wilde_1889.html.
12. **IMDb.** Executive Decision. [Online] IMDb, March 15, 1996. http://www.imdb.com/title/tt0116253/.

---

[1]It is unfortunate that the media has given the term "hacker" a negative connotation. The truth is that hackers create new ideas and invent new products that we use every day [66]. Take Facebook as an example, its headquarters' street address is named 1 Hacker Way in Menlo Park, California [64] and its company blog states that "hacking is core to how we build at Facebook" [65]. Positive news about hackers is hard to come by. For instance, a team of hackers at the Lunar Orbiter Image Recovery Project (LOIRP) was able to recover some 2,000 lost lunar photos for NASA [67].

13. **Reynolds, Simon.** Haysbert: '24' president helped Obama. [Online] Digital Spy, July 2, 2008. http://www.digitalspy.com/celebrity/news/a106486/haysbert-24-president-helped-obama.html.
14. **Tapper, Jack.** Conservative Lovefest for '24'. [Online] ABC, June 23, 2006. http://abcnews.go.com/Nightline/story?id=2112624&page=1&singlePage=true.
15. **24 Spoilers.** Official 24 Season 1 Trailer . [Online] 24 Spoilers, January 28, 2011. http://www.24spoilers.com/2011/01/28/official-24-season-1-trailer/.
16. **Kelly, Suzanne.** Mistaken for a terrorist: Homeland star only plays one on TV. [Online] CNN, October 5, 2012. http://security.blogs.cnn.com/2012/10/05/mistaken-for-a-terrorist-homeland-star-only-plays-one-on-tv/.
17. **BBC News.** Army turns to Hollywood for advice. [Online] BBC News, October 8, 2001. http://news.bbc.co.uk/2/hi/entertainment/1586468.stm.
18. **USC ICT.** ICT Overview. [Online] University of Southern California. [Cited: December 26, 2012.] http://ict.usc.edu/about/.
19. **IMDb.** God Bless America. [Online] IMDb, May 31, 2012. http://www.imdb.com/title/tt1912398/.
20. **Frosch, Dan and Johnson, Kirk.** Gunman Kills 12 in Colorado, Reviving Gun Debate. [Online] The New York Times, July 20, 2012. http://www.nytimes.com/2012/07/21/us/shooting-at-colorado-theater-showing-batman-movie.html?pagewanted=all.
21. **Wright, Lawrence.** The Agent. Did the C.I.A. stop an F.B.I. detective from preventing 9/11? [Online] The New Yorker, July 10, 2006. http://www.newyorker.com/archive/2006/07/10/060710fa_fact_wright?currentPage=all.
22. **Fox News.** 9/11/01 Fox News Brit Hume interviews Senator John McCain. [Online] YouTube, September 11, 2001. http://www.youtube.com/watch?v=5ZCY8Rthpok.
23. **Bush, George W.** George W. Bush The Night of 9-11-01. [Online] YouTube, September 11, 2001. http://www.youtube.com/watch?v=XbqCquDl4k4.
24. **Perlez, Jane and Weiner, Tim.** U.S. to Publish Terror Evidence on bin Laden. [Online] The New York Times, September 24, 2001. http://www.nytimes.com/2001/09/24/international/24DIPL.html?pagewanted=all.
25. **CNN.** Bush announces opening of attacks. [Online] CNN, October 7, 2001. http://articles.cnn.com/2001-10-07/us/ret.attack.bush_1_qaeda-targets-al-kandahar.
26. **Bush, George W.** President Bush Signs Anti-Terrorism Bill. [Online] PBS Newshour, October 26, 2001. http://www.pbs.org/newshour/updates/terrorism/july-dec01/bush_terrorismbill.html.
27. **Office of the Press Secretary.** President Issues Military Order . [Online] The White House, November 13, 2001. http://georgewbush-whitehouse.archives.gov/news/releases/2001/11/20011113-27.html.
28. **Yasui, Hiromi.** Guantánamo Bay Naval Base (Cuba). [Online] The New York Times, November 30, 2012. http://topics.nytimes.com/top/news/national/usstatesterritoriesandpossessions/guantanamobaynavalbasecuba/index.html.
29. **Phillips, Macon.** Osama Bin Laden Dead. [Online] The White House Blog, May 2, 2011. http://www.whitehouse.gov/blog/2011/05/02/osama-bin-laden-dead.
30. **Dermody, William.** The Longest War. [Online] USA Today. http://www.usatoday.com/news/afghanistan-ten-years-of-war/index.html.
31. **Reuters.** Terrorists killed U.S. ambassador to Libya: Panetta. [Online] Chicago Tribune, September 27, 2012. http://www.chicagotribune.com/news/sns-rt-us-libya-usa-investigationbre88q1jw-20120927,0,3904573.story.
32. **Smith, Matt.** Ex-SEALs, online gaming maven among Benghazi dead. [Online] CNN, September 15, 2012. http://www.cnn.com/2012/09/14/us/benghazi-victims/index.html.
33. **Quilliam.** The Attack on the US Consulate Was A Planned Terrorist Assault Against US and Libyan Interests. [Online] Quilliam, September 12, 2012. http://www.quilliamfoundation.org/press-releases/the-attack-on-the-us-consulate-was-a-planned-terrorist-assault-against-us-and-libyan-interests/.

34. **Hosenball, Mark.** Congress to continue probes of Benghazi attacks. [Online] Chicago Tribune, November 7, 2012. http://www.chicagotribune.com/news/politics/sns-rt-us-usa-campaign-benghazibre8a62cg-20121107,0,1506034.story.

35. **CNN Wire Staff.** Ex-CIA chief Petraeus testifies Benghazi attack was al Qaeda-linked terrorism. [Online] CNN, November 16, 2012. http://www.cnn.com/2012/11/16/politics/benghazi-hearings/index.html.

36. **Pearson, Michael.** Slain ambassador died 'trying to help build a better Libya'. [Online] CNN, September 15, 2012. http://www.cnn.com/2012/09/12/world/africa/libya-us-ambassador-killed-profile/index.html.

37. **Pagliery, Jose.** 'Sony-pocalypse': Why the Sony hack is one of the worst hacks ever. [Online] CNNMoney, December 5, 2014. http://money.cnn.com/2014/12/04/technology/security/sony-hack/.

38. **Estes, Adam Clark.** The Sony Pictures Hack Was Even Worse Than Everyone Thought. [Online] GIZMODO, December 3, 2014. http://gizmodo.com/the-sony-pictures-hack-exposed-budgets-layoffs-and-3-1665739357/1666122168/+ace.

39. **Cieply, Michael and Barnes, Brooks.** Sony Cyberattack, First a Nuisance, Swiftly Grew Into a Firestorm. [Online] The New York Times, December 30, 2014. http://www.nytimes.com/2014/12/31/business/media/sony-attack-first-a-nuisance-swiftly-grew-into-a-firestorm.html.

40. **Zetter, Kim.** Sony Hackers Threaten to Release a Huge 'Christmas Gift' of Secrets. [Online] Wired, December 15, 2014. http://www.wired.com/2014/12/sony-hack-part-deux/.

41. **Zeitlin, Matthew.** The Sony Hackers Are Now Threatening Employees And Their Families. [Online] BuzzFeed News, December 5, 2014. http://www.buzzfeed.com/matthewzeitlin/the-sony-hackers-are-now-threatening-employees-and-their-fam#.myKPdwyZP0.

42. **Pagliery, Jose.** Message to Sony: Don't show 'movie of terrorism'. [Online] CNNMoney, December 8, 2014. http://money.cnn.com/2014/12/08/technology/security/sony-hackers/.

43. **Chmielewski, Dawn and Hesseldahl, Arik.** Sony Pictures Tries to Disrupt Downloads of Its Stolen Files. [Online] < re/code >, December 10, 2014. http://recode.net/2014/12/10/sony-pictures-tries-to-disrupt-downloads-of-its-stolen-files/.

44. **Rushe, Dominic.** Hackers who targeted Sony invoke 9/11 attacks in warning to moviegoers . [Online] The Guardian, December 17, 2014. http://www.theguardian.com/film/2014/dec/16/employees-sue-failure-guard-personal-data-leaked-hackers.

45. **Rosenblatt, Seth.** DHS finds no evidence for attack on theaters showing 'The Interview'. [Online] CNet, December 16, 2014. http://www.cnet.com/news/dhs-finds-no-evidence-for-attack-on-theaters-showing-the-interview/.

46. **Kastrenakes, Jacob.** The Sony hackers won: The Interview just disappeared from America's biggest theater chains. [Online] The Verge, December 17, 2014. http://www.theverge.com/2014/12/17/7411155/us-biggest-movie-theaters-wont-show-the-interview.

47. **Lang, Brent.** Sony Cancels Theatrical Release for 'The Interview' on Christmas. [Online] Variety, December 17, 2014. http://variety.com/2014/film/news/sony-cancels-theatrical-release-for-the-interview-on-christmas-1201382032/.

48. **Perez, Evan, Sciutto, Jim and Diamond, Jeremy.** Obama: Sony 'made a mistake'. [Online] CNN, December 19, 2014. http://www.cnn.com/2014/12/19/politics/fbi-north-korea-responsible-sony/.

49. **Altman, Alex and Miller, Zeke J.** FBI Accuses North Korea in Sony Hack. [Online] TIME Magazine, December 19, 2014. http://time.com/3642161/sony-hack-north-korea-the-interview-fbi/.

50. **Schroeder, Stan.** Sony hackers apparently mock the FBI; Anonymous threatens more hacks. [Online] Mashable, December 21, 2014. http://mashable.com/2014/12/21/sony-hackers-apparently-call-the-fbi-idiots-anonymous-threatens-more-hacks/.

51. **Bradner, Eric.** Obama: North Korea's hack not war, but 'cybervandalism'. [Online] CNN, December 21, 2014. http://www.cnn.com/2014/12/21/politics/obama-north-koreas-hack-not-war-but-cyber-vandalism/index.html.

52. **Weinstein, Shelli.** New York's Treehouse Theater Plans Live Read of 'The Interview'. [Online] Variety, December 22, 2014. http://variety.com/2014/film/news/new-yorks-treehouse-theater-plans-live-read-of-the-interview-1201385490/.
53. **O'Brien, Chris.** In tragic new twist, critics say 'The Interview' is actually a pretty terrible movie. [Online] VentureBeat, December 26, 2014. http://venturebeat.com/2014/12/26/in-new-tragic-twist-critics-say-the-interview-is-actually-a-pretty-terrible-movie/.
54. **Cohen, Luc and Avila, Alicia.** 'The Interview' Draws Big Crowds In Opening Weekend. [Online] The Huffington Post, December 26, 2014. http://www.huffingtonpost.com/2014/12/26/the-interview-opening-weekend-sold-out_n_6381802.html.
55. **Stelter, Brian.** Sony streams 'The Interview' online and makes Hollywood history. [Online] CNNMoney, December 24, 2014. http://money.cnn.com/2014/12/24/media/interview-digital-release/index.html.
56. **D'Orazio, Dante.** The Interview makes over $15 million online, beating movie theaters . [Online] The Verge, December 28, 2014. http://www.theverge.com/2014/12/28/7458287/the-interview-opening-weekend-box-office-online-numbers.
57. **McNary, Dave and Lang, Brent.** 'The Interview' Online, VOD Sales Reach $31 Million. [Online] Variety, Jaunary 6, 2015. http://variety.com/2015/film/news/the-interview-online-sales-reach-31-million-1201394022/.
58. **Richwine, Lisa.** Cyber attack could cost Sony studio as much as $100 million. [Online] Reuters, December 9, 2014. http://www.reuters.com/article/2014/12/09/sony-cybersecurity-costs-idUSL1N0TT1YO20141209.
59. **Ando, Ritsuko.** Sony CEO sees no major financial impact from cyber attack. [Online] Reuters, January 6, 2015. http://www.reuters.com/article/2015/01/07/us-sony-cybersecurity-idUSKBN0KF1ZW20150107.
60. **The Associated Press.** Sony says massive hack cost the company $15 million. [Online] Mashable, February 4, 2015. http://mashable.com/2015/02/04/sony-hack-cost-15-million/.
61. **Zeitchik, Steven and Faughnder, Ryan.** Amy Pascal's exit hints at new script for Sony. [Online] Los Angeles Times, February 5, 2015. http://www.latimes.com/entertainment/envelope/cotown/la-et-ct-sony-amy-pascal-exit-20150206-story.html.
62. **Zetter, Kim.** The Evidence That North Korea Hacked Sony Is Flimsy. [Online] Wired Magazine, December 17, 2014. http://www.wired.com/2014/12/evidence-of-north-korea-hack-is-thin/.
63. **Stelter, Brian and Pallotta, Frank.** Sony zeroing in on North Korea as it investigates cyberattack. [Online] CNNMoney, December 3, 2014. http://money.cnn.com/2014/12/03/media/sony-north-korea-cyberattack/index.html?iid=EL.
64. **Facebook.** Company Info. [Online] Facebook Newsroom. [Cited: August 16, 2014.] http://newsroom.fb.com/company-info/.
65. **Alves, David.** Announcing Facebook's 2012 Hacker Cup. [Online] Facebook, January 4, 2012. https://www.facebook.com/notes/facebook-engineering/announcing-facebooks-2012-hacker-cup/10150468260528920.
66. **Lee, Newton.** Facebook Nation: Total Information Awareness (Second Edition). [Online] Springer Science + Business Media, November 2014. http://www.amazon.com/Facebook-Nation-Total-Information-Awareness/dp/1493917390/.
67. **Bierend, Doug.** The hackers who recovered NASA's lost lunar photos. [Online] CNN, July 30, 2014. http://www.cnn.com/2014/07/30/world/hackers-nasa-lunar-photos/index.html.

Chapter 2
U.S. Intelligence Community

Secrecy stifles oversight, accountability, and information sharing.
— The 9/11 Commission (July 22, 2004)

The situation was, and remains, too risky to allow someone to experiment with amateurish, Hollywood style interrogation methods - that in reality - taints sources, risks outcomes, ignores the end game, and diminishes our moral high ground in a battle that is impossible to win without first capturing the hearts and minds around the world. It was one of the worst and most harmful decisions made in our efforts against al-Qaeda.
— Former FBI Agent Ali Soufan (May 13, 2009)

In solving intelligence problems, including diversity of thought is essential.
— Letitia "Tish" Long, Director of the National Geospatial-Intelligence Agency (2012)

You know enough to know what's not true, but you can't necessarily connect all of the dots to know what is true. ... The most effective propaganda is a mixture of truths, half truths, and lies.
— American author Richard Thieme at DEF CON 22 (August 8, 2014)

2.1 "Need to Know" — Truths, Half Truths, and Lies

"You can't handle the truth!" exclaimed Col. Nathan R. Jessup played by Jack Nicholson in the 1992 legal drama *A Few Good Men*. "Son, we live in a world that has walls, and those walls have to be guarded by men with guns. ... You don't want the truth, because deep down in places you don't talk about at parties, you *want* me on that wall. You *need* me on that wall" [1].

On August 8, 2014 at DEF CON 22 in Las Vegas, American author Richard Thieme recalled a conversation with his friend from the National Security Agency (NSA) who

© Springer International Publishing Switzerland 2015
N. Lee, *Counterterrorism and Cybersecurity*, DOI 10.1007/978-3-319-17244-6_2

spoke of the difficulty in knowing the truth: "You know enough to know what's not true, but you can't necessarily connect all of the dots to know what is true" [2].

A commentator on technology and culture, Thieme gave his talk at DEF CON for the 19th years and is widely considered to be a "father figure" in hacker conventions worldwide. He told a story about U.S. Army General Alexander Haig who served as the U.S. Secretary of State under President Ronald Reagan from 1981 to 1982: "In a small Italian newspaper where a piece of really great investigative reporting revealed that the KGB during the Cold War was sponsoring terrorism all over the world and supporting groups that were antithetical to the United States and its intentions. That small story caught the attention of an author and journalist who wrote a piece about it for *The New York Times* and also wrote a book about it. That came to the attention of Alexander Haig who was Secretary of State; and he became very, very alarmed and he held a press conference in which he demanded that the CIA explore this revelation in order to counter the nefarious and insidious work of the KGB in this way. William Casey, director of CIA, said 'we'll get right on it.' Six months later the result came back: 'We've explored that, you don't have to worry about that particular thing.' Why? Because the story in the Italian newspaper was a CIA plant in the first place. In other words, it was just propaganda to smear the Soviets, but it was picked up by our own journalist who didn't know any better and couldn't, turned into a book which went to the Secretary of State and then became an alarming consideration, and the CIA could not tell him the truth" [2].

The U.S. Secretary of State is the head of the State Department responsible for overall direction, coordination, and supervision of activities of the U.S. government overseas. Yet CIA director William Casey had kept U.S. Army General Alexander Haig in the dark for most of his career as Secretary of State. The "need to know" syndrome had reached epidemic proportions within the U.S. government, all the way to the top including Presidents of the United States (i.e. giving the Presidents the benefit of the doubt).

In March 1987 during the Iran-Contra affair, President Ronald Reagan said in a televised press conference that "he was not aware of a two-year secret campaign organized by key White House aides, including two advisers he saw nearly every day, to ship millions of dollars in arms to Nicaraguan contra rebels" and that "he was not aware of the alleged diversion of funds from U.S. arms sales to Iran to the rebels" [3].

In October 2013 amid Edward Snowden's NSA leaks, President Barack Obama was reportedly unaware of the United States spying on its ally leaders: "The National Security Agency ended a program used to spy on German Chancellor Angela Merkel and a number of other world leaders after an internal Obama administration review started this summer revealed to the White House the existence of the operation, U.S. officials said … The account suggests President Barack Obama went nearly five years without knowing his own spies were bugging the phones of world leaders. Officials said the NSA has so many eavesdropping operations under way that it wouldn't have been practical to brief him on all of them" [4].

"The most effective propaganda," Thieme said, "is a mixture of truths, half truths, and lies" [2].

2.2 FBI Ten Most Wanted Fugitive: Osama Bin Laden

In August 1996, *Al Quds Al Arabi* in London published Osama bin Laden's fatwā entitled "Declaration of War against the Americans Occupying the Land of the Two Holy Places" [5]. The land referred to Saudi Arabia and the two holy places Mecca and Medina. In February 1998, bin Laden issued a second fatwā declaring a holy war or jihad against America and Israel [6]. In August 1998, al-Qaeda bombed the U.S. embassies in Kenya and Tanzania simultaneously, killing 224 people [7]. In December 1998, Director of Central Intelligence's (DCI) Counterterrorist Center (CTC) reported to President Bill Clinton that "Bin Ladin and his allies are preparing for attacks in the US, including an aircraft hijacking" [8].

In June 1999, two years before 9/11, Osama bin Laden (aka Usama Bin Laden) was placed on the FBI's Ten Most Wanted Fugitives list [9]. (See Figure 2.1). Bin Laden was wanted for the murder of U.S. nationals outside the U.S., in connection with the August 1998 bombings of the U.S. Embassies.

After the September 11 attacks, the U.S. Department of State offered a reward of up to $25 million for information leading directly to the apprehension or conviction of Osama bin Laden. An additional $2 million was offered through a program developed and funded by the Airline Pilots Association and the Air Transport Association. However, the revised FBI poster in November 2001 inexplicably did not mention 9/11 in bin Laden's list of crimes. Instead, it simply stated, "In addition, Bin Laden is a suspect in other terrorist attacks throughout the world."

Bin Laden initially denied any involvement in 9/11. In a statement issued to the Arabic satellite channel Al Jazeera, bin Laden said, "The U.S. government has consistently blamed me for being behind every occasion its enemies attack it. I would like to assure the world that I did not plan the recent attacks, which seems to have been planned by people for personal reasons. I have been living in the Islamic emirate of Afghanistan and following its leaders' rules. The current leader does not allow me to exercise such operations" [10].

To contradict bin Laden's denial, the Bush administration released in December 2001 a video tape showing bin Laden bragging about the 9/11 attacks: "I was thinking that the fire from the gas in the plane would melt the iron structure of the building and collapse the area where the plane hit and all the floors above it only. This is all that we had hoped for" [11].

Nonetheless, Saudi Defense Minister Prince Sultan bin Abdul Aziz questioned, "Who stands behind this terrorism and who carried out this complicated and carefully planned terrorist operation? … Are bin Laden and his supporters the only ones behind what happened or is there another power with advanced technical expertise that acted with them?" [12].

In an audiotape surfaced in May 2006, bin Laden claimed responsibility for 9/11: "The truth is that he [Zacarias Moussaoui] has no connection whatsoever with the events of September 11th, and I am certain of what I say, because I was responsible for entrusting the 19 brothers — Allah have mercy upon them — with those raids, and I did not assign brother Zacarias to be with them on that mission" [13].

FBI TEN MOST WANTED FUGITIVE

Murder of U.S. Nationals Outside the United States; Conspiracy to Murder U.S.
Nationals Outside the United States; Attack on a Federal Facility Resulting in Death

USAMA BIN LADEN

Date of Photograph
Unknown

Aliases:
Usama Bin Muhammad Bin Ladin, Shaykh Usama Bin Ladin, the Prince, the Emir, Abu Abdallah, Mujahid Shaykh, Hajj,
the Director

DESCRIPTION

Date(s) of Birth Used:	1957	**Hair:**	Brown
Place of Birth:	Saudi Arabia	**Eyes:**	Brown
Height:	6' 4" to 6' 6"	**Complexion:**	Olive
Weight:	Approximately 160 pounds	**Sex:**	Male
Build:	Thin	**Nationality:**	Saudi Arabian
Occupation:	Unknown		
Scars and Marks:	None known		
Remarks:	Bin Laden is the leader of a terrorist organization known as Al-Qaeda, "The Base". He is left-handed and walks with a cane.		

CAUTION

Usama Bin Laden is wanted in connection with the August 7, 1998, bombings of the United States Embassies in Dar es
Salaam, Tanzania, and Nairobi, Kenya. These attacks killed over 200 people. In addition, Bin Laden is a suspect in other
terrorist attacks throughout the world.

REWARD

The Rewards For Justice Program, United States Department of State, is offering a reward of up to $25 million for
information leading directly to the apprehension or conviction of Usama Bin Laden. An additional $2 million is being
offered through a program developed and funded by the Airline Pilots Association and the Air Transport Association.

CONSIDERED ARMED AND EXTREMELY DANGEROUS

If you have any information concerning this person, please contact your local FBI office or the nearest American Embassy
or Consulate.
June 1999 Poster Revised November 2001

Fig. 2.1 FBI ten most wanted fugitive — Usama Bin Laden (November 2001, courtesy of the
Federal Bureau of Investigation)

The initial denials and subsequent admissions might have suggested that bin
Laden eventually decided to go for this once-in-a-lifetime opportunity and take
credit for the actions of his top lieutenants in order to maintain a unified support
of jihad against America. Although the authenticity of the bin Laden "confes-
sion" audio and video tapes is subject to debate, the al-Qaeda terrorist organization has been inextricably linked to the 9/11 attacks, thanks to the FBI's relentless
investigations.

2.3 An American Hero Born in Lebanon

Back in March 2000, the U.S. Central Intelligence Agency (CIA) learned that two military trained al-Qaeda terrorists, Khalid al-Mihdhar and Nawaf al-Hazmi, had entered the United States [14]. However, the CIA did not inform the FBI or the U.S. State Department until August 2001, a year and half later [15]. By then, it was too late for the FBI to track down al-Mihdhar and al-Hazmi who, along with three other terrorists, hijacked American Airlines Flight 77 that crashed into the Pentagon on September 11, killing 189 people [16].

"When [Ali] Soufan realized that the CIA had known for more than a year and a half that two of the hijackers were in the country he ran into the bathroom and threw up," wrote Pulitzer Prize-winning author Lawrence Wright in the July 2006 issue of *The New Yorker* [17].

Lebanese-American Ali Soufan is a former FBI agent who might have prevented 9/11 if the CIA and the FBI had been cooperative with one another. Soufan's close friend and colleague John O'Neill was killed in the attacks. O'Neill was the former head of the FBI National Security Division. He died less than a month after he started his new job as head of security at the World Trade Center.

Back in 1998, O'Neill drafted Soufan into the FBI's I-49 squad to investigate al-Qaeda in connection with the U.S. embassy bombings in Kenya and Tanzania [7]. Soufan was one of eight FBI agents who spoke Arabic fluently. The FBI collected evidence linking the bombings to Osama bin Laden, and placed him on the FBI's Ten Most Wanted Fugitives list [9]. The FBI also discovered a phone number in Yemen that functioned as a virtual switchboard for al-Qaeda's global organization. However, the CIA had jurisdiction over monitoring the overseas phone conversations, and they refused to share the intercepted messages with the FBI.

The CIA followed al-Qaeda terrorist Khalid al-Mihdhar to Dubai where they learned that he had a U.S. visa. Nevertheless, the CIA did not alert the FBI or the U.S. State Department to put al-Mihdhar on the terrorist watch list. The CIA had hoped to recruit al-Mihdhar to infiltrate the al-Qaeda organization. However, the CIA realized in March 2000 that al-Mihdhar and his accomplice Nawaf al-Hazmi had entered the U.S. legally, and the spy agency failed to inform the FBI and the U.S. State Department that had jurisdiction inside the United States.

In October 2000, a fiberglass fishing boat containing plastic explosives blasted open a hole in USS *Cole* in Yemen, killing 17 American sailors [18]. O'Neill and Soufan went to Yemen for the investigation. The FBI identified the one-legged al-Qaeda lieutenant Khallad who orchestrated the *Cole* attack. Despite numerous requests from the FBI, the CIA withheld information about Khallad and his meeting in Malaysia with the future 9/11 hijackers.

O'Neill retired from the FBI in August 2001 and took a job as head of security at the World Trade Center. Meanwhile, the FBI learned that al-Mihdhar and al-Hazmi were in the U.S. but the agency was unable to track them down.

Less than a month later on September 11, al-Mihdhar and al-Hazmi hijacked American Airlines Flight 77 that crashed into the Pentagon, killing a total of 189 people [16]. Other 9/11 hijackers piloted American Airlines Flight 11 and United

Airlines Flight 175 that crashed into the North Tower and South Tower of the World Trade Center, killing O'Neill and a total of 2,762 people [19]. Including the crash of United Airlines Flight 93 after its 40 passengers revolted against the four hijackers [20], the total death toll on September 11, 2001 was 2,995 people.

On September 12, the FBI ordered its agents to identify the 9/11 hijackers "by any means necessary." Soufan was tasked to interrogate al-Qaeda terrorist Fahd al-Quso in Yemen, the only lead that the FBI had at the time. The CIA handed Soufan the surveillance photos of 20 al-Qaeda terrorists and a complete report on the Malaysia meeting of the suspected hijackers. When Soufan learned that the CIA had known for more than a year and a half that two al-Qaeda terrorists were living in the U.S., the shock made him physically ill.

Unlike Jack Bauer (played by Kiefer Sutherland) in the counterterrorism TV drama *24*, Soufan did not torture or threaten the suspects in order to extract information from them and exact revenge for the death of his close friend and colleague O'Neill. Instead, Soufan's arsenal of interrogation included American bottle water, sugarless wafers, theological debates, and a history book of America in the Arabic language.

Soufan was able to persuade Fahd al-Quso and disarm Abu Jandal (aka Nasser al-Bahri) who was trained in counter-interrogation techniques. Al-Quso and Jandal eventually identified many 9/11 hijackers from the photos of known al-Qaeda terrorists, including Mohammed Atta who was the lead hijacker. Working on the 9/11 case for days and nights with almost no sleep, his coworkers referred to Soufan as "an American hero."

For every hero recognized by the news media, there are many more unsung heroes who work steadfastly and risk their lives to keep the world safer. In 2009, for example, seven CIA officers were killed by a Jordanian double agent when he detonated his suicide bomb vest at their secret meeting in Khost, Afghanistan [21] .

2.4 The FBI-CIA Bureaucratic Rivalries

Former *Newsweek* investigative correspondent Michael Isikoff said, "The CIA knew who they were, they knew that they were suspected al-Qaeda operatives, they failed to alert the INS [Immigration and Naturalization Service], the State Department, the Customs Service, agencies who could have kept them out of the country. And, perhaps more importantly, they failed to alert the FBI, which could have tracked them while they were in the country" [14].

Although the Ali Soufan story placed the blame squarely on the CIA, Michael Scheuer, former chief of the Osama bin Laden unit at the CIA Counterterrorism Center, accused the FBI instead. Scheuer wrote in the July 4, 2006 issue of *The Washington Times* [22]:

> FBI officers sat in the unit I first commanded and then served in and they read the same information I did. If the data did not get to FBI headquarters it is because the FBI then lacked, and still lacks, a useable computer system. The FBI did not know the September

11 hijackers were here because Judge Louis Freeh [5th FBI Director] and Robert Mueller [6th FBI Director] have failed to provide their officers computers that allow them to talk securely to their headquarters and other intelligence community elements. ... In my own experience, Mr. O'Neill was interested only in furthering his career and disguising the rank incompetence of senior FBI leaders. He once told me that he and the FBI would oppose an operation to capture bin Laden and take him to a third country for incarceration. When I asked why, he replied, "Why should the FBI help to capture bin Laden if the bureau won't get credit among Americans for his arrest and conviction?"

Contrary to Scheuer's negative assessment of O'Neill who was killed in the 9/11 attacks, the FBI I-49 squad intercepted phone calls between bin Laden and his al-Qaeda operatives by installing two antennae in the Pacific and the Indian Ocean, as well as a satellite phone booth with a hidden camera in Kandahar, Afghanistan [17].

The CIA apparently did not get along with the National Security Agency (NSA) either. A military officer from the 1998 Middle East signals-intelligence operation team told Seymour Hersh of *The New Yorker* that he was unable to discuss the activity with representatives of the CIA and the NSA at the same time. "I used to meet with one in a safe house in Virginia, break for lunch, and then meet with the other," the officer said. "They wouldn't be in the same room" [23].

A senior manager at a U.S. intelligence agency disclosed that the major intelligence centers were so consumed by internecine warfare that the professional analysts find it difficult to do their jobs. "They're all fighting among each other," said the senior manager. "There's no concentration on issues" [23].

To make matter worse, the U.S. Justice Department in 1995 established a policy known as "the Wall" or "Chinese Wall," that hindered the exchange of foreign intelligence information between FBI agents and criminal investigators. U.S. Attorney Mary Jo White voiced her dissonance in a memorandum faxed to Deputy Attorney General Jamie Gorelick in December 1995. White argued that the Justice Department and the FBI were structured and operating in a way that did not make maximum legitimate use of all law enforcement and intelligence avenues to prevent terrorism and prosecute terrorists. White asserted that the Justice Department was building "unnecessary and counterproductive walls that inhibit rather than promote our ultimate objectives" and that "we must face the reality that the way we are proceeding now is inherently and in actuality very dangerous" [24]. Using the Chinese Wall policy as an excuse, the FBI withheld intelligence from the White House, and the CIA refused to share intelligence with the FBI.

In his 1998 book *Secrecy: The American Experience*, U.S. Senator Daniel Patrick Moynihan wrote that "Departments and agencies hoard information, and the government becomes a kind of market. Secrets become organizational assets, never to be shared save in exchange for another organization's assets. ... Too much of the information was secret, not sufficiently open to critique by persons outside government. Within the confines of the intelligence community, too great attention was paid to hoarding information, defending boundaries, securing budgets, and other matters of corporate survival. ... The system costs can be enormous. In the void created by absent or withheld information, decisions are either made poorly or not at all" [25].

2.5 Operational Failures of the U.S. Intelligence Community

After 9/11, CIA director George Tenet and CIA Counterterrorist Center director Cofer Black testified to Congress that the CIA had divulged information about future hijacker Khalid al-Mihdhar's to the FBI in a timely manner. However, the 9/11 Commission concluded that their statements were false [26].

The 9/11 Commission Report identified several major operational failures — opportunities that the U.S. intelligence community and governmental agencies could have exploited to prevent 9/11 [27]. The failures included:

Central Intelligence Agency (CIA)

- Not watchlisting future hijackers Khalid al-Mihdhar and Nawaf al-Hazmi, not trailing them after they traveled to Bangkok, and not informing the FBI about one future hijacker's U.S. visa or his companion's travel to the United States.
- Not sharing information linking individuals in the *Cole* attack to al-Mihdhar.

Federal Bureau of Investigation (FBI)

- Not taking adequate steps in time to find al-Mihdhar or al-Hazmi in the United States.
- Not linking the arrest of Zacarias Moussaoui, described as interested in flight training for the purpose of using an airplane in a terrorist act, to the heightened indications of attack.
- Not expanding no-fly lists to include names from terrorist watchlists.

U.S. State Department and Immigration and Naturalization Service (INS)

- Not discovering false statements on visa applications.
- Not recognizing passports manipulated in a fraudulent manner.
- Not realizing false statements to border officials to gain entry into the United States.

Federal Aviation Administration (FAA)

- Not making use of the U.S. TIPOFF terrorist watchlist where two of the hijackers were listed.
- Not searching airline passengers identified by the Computer-Assisted Passenger Prescreening System (CAPPS).
- Not hardening aircraft cockpit doors or taking other measures to prepare for the possibility of suicide hijackings.

Furthermore, a 2001 Congressional report disclosed that the NSA was faced with "profound needle-in-the-haystack challenges." *The New York Times* revealed in 2002 that there were 200 million pieces of intelligence in a regular workday, and less than one percent of it was ever decoded, translated, or processed [28].

Despite the establishment of the Terrorist Screening Center (TSC) in 2003, U.S. agencies handling the terrorist watch lists have continued to "work from at least 12 different, sometimes incompatible, often uncoordinated and technologically archaic databases" [29]. The TSC has been tasked to consolidate the State Department's TIPOFF, Homeland Security's No-fly list, and FBI's Interpol terrorism watch list [30].

For spy agencies like the CIA and military intelligence organizations, *The Wall Street Journal* revealed in 2009 that there are hundreds of databases each and most of them are not linked up [31]. Analysts often have to query individual databases separately and analyze the data through the conventional "pen and paper" method.

In short, finger-pointing, interagency rivalries, personal animosity, operational inefficiency, outmoded government regulations, and intentional withholding of vital information had resulted in a colossal failure of the U.S. intelligence community to protect and serve the American public.

2.6 Unity of Counterterrorism Effort Across U.S. Government

Most U.S. military officers serving on the front lines overseas have heard "One Team, One Fight" — everyone working together to accomplish the mission [32]. Aesop's fable *The Four Oxen and the Lion* reminds us that "united we stand, divided we fall." To effectively fight against terrorism, the 9/11 Commission called for unity of effort in five areas of the U.S. government [27]:

1. **National Counterterrorism Center (NCTC)** unifying strategic intelligence and operational planning against Islamist terrorists across the foreign-domestic divide. Placed in the Executive Office of the President, the NCTC would:

   a. Build on the existing Terrorist Threat Integration Center, and would replace it and other terrorism "fusion centers" within the government.
   b. Become the authoritative knowledge bank with information collected from both inside and outside the United States.
   c. Perform joint operational planning with existing agencies and track implementation of plans.
   d. Influence the leadership and the budgets of the counterterrorism operating arms of the CIA, the FBI, the departments of Defense, and Homeland Security.
   e. Follow the policy direction of the President and the National Security Council.

2. **National Intelligence Director (NID)** unifying the U.S. intelligence community. Located in the Executive Office of the President, the NID would:

   a. Oversee national intelligence centers (e.g. CIA, DIA [Defense Intelligence Agency], NSA, and FBI) that combine experts from all the collection disciplines against common targets such as counterterrorism or nuclear proliferation.
   b. Oversee the agencies that contribute to the national intelligence program including setting common standards for personnel and information technology.

    c. Have authority over three intelligence deputies:

- Director of the CIA (for foreign intelligence)
- Under Secretary of Defense for Intelligence (for defense intelligence)
- Executive Assistant Director for Intelligence at the FBI or Under Secretary of Homeland Security for Information Analysis and Infrastructure Protection (for homeland intelligence)

3. **Network-based Information Sharing System** unifying the many participants in the counterterrorism efforts and their knowledge. Transcending traditional governmental boundaries, the decentralized network-based information system would replace the system of "need to know" by a system of "need to share." The 9/11 Commission acknowledged that "secrecy stifles oversight, accountability, and information sharing." Legal and policy issues must be resolved in order for the new information sharing system to be used effectively.

4. **U.S. Congress** unifying and strengthening congressional oversight to improve quality and accountability. For intelligence oversight, Congress would combine authorizing and appropriating intelligence committees to empower their oversight function. To minimize national security risks during transitions between administrations, Congress would create a permanent standing committee for homeland security in each chamber.

5. **U.S. Government** strengthening the FBI and homeland defenders. The FBI would establish a specialized and integrated national security workforce consisting of agents, analysts, linguists, and surveillance specialists, in order to develop a deep expertise in intelligence and national security. The Department of Defense's Northern Command would regularly access the adequacy of strategies and planning to defend against military threats to the homeland. The Department of Homeland Security would regularly assess the types of threats the country faces, in order to determine the readiness of the U.S. government to respond to those threats.

Figure 2.2 shows the post-9/11 organizational chart for the U.S. intelligence community recommended by the 911 Commission to achieve a unity of effort in managing intelligence. The Executive Office of the President includes the President of the United States (POTUS), the National Intelligence Director (NID), and the National Counterterrorism Center (NCTC). The three NID deputies are the CIA Director (foreign intelligence), the Under Secretary of Defense for Intelligence (defense intelligence), and the FBI Intelligence Director (homeland intelligence).

The "new" Open Source Agency in Figure 2.2 became a reality in 2005 when Douglas Naquin was appointed the director of the Open Source Center (OSC). Replacing the former Foreign Broadcast Information Service (FBIS) head by Naquin, the OSC collects information available from "the Internet, databases, press, radio, television, video, geospatial data, photos and commercial imagery" [33]. According to the OSC website, "OpenSource.gov provides information on foreign political, military, economic, and technical issues beyond the usual media from an ever expanding universe of open sources. Our website contains sources from more than 160 countries, in more than 80 languages and hosts content from several commercial providers, as well as content from OSC partners" [34].

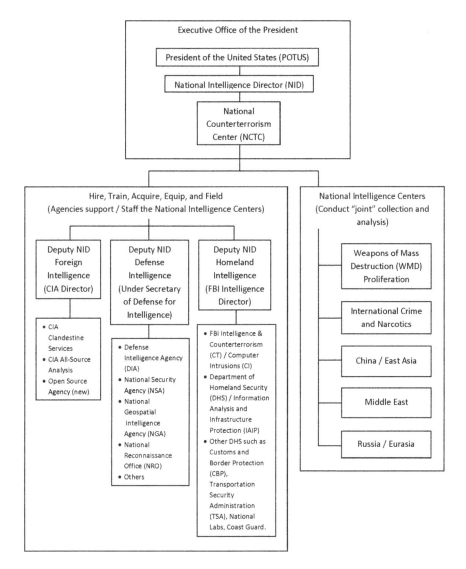

Fig. 2.2 Unity of effort in managing intelligence

2.7 Transition from Need-to-Know to Need-to-Share and Need-to-Provide

Based on the 9/11 Commission's recommendations, Congress passed the Intelligence Reform and Terrorism Prevention Act in 2004 and established the Office of the Director of National Intelligence (ODNI) [35]. Former Ambassador John Negroponte served as the first Director of National Intelligence (DNI) in April 2005.

Although the CIA had objected to the creation of the ODNI, the CIA issued a statement in 2006 admitting the necessity of interagency cooperation: "Based on rigorous internal and external reviews of its shortcomings and successes before and after 9/11, the CIA has improved its processing and sharing of intelligence. CIA's focus is on learning and even closer cooperation with partners inside and outside government, not on public finger pointing, which does not serve the American people well" [36].

Former NSA director and vice admiral John Michael McConnell became the second Director of National Intelligence in February 2007. He implemented an aggressive "100 Day Plan for Integration and Collaboration" in order to address and overcome barriers — legal, policy, technology, process, and culture — in the U.S. intelligence community [37]. The 100 Day Plan focused on six enterprise integration priorities:

1. Create a Culture of Collaboration
2. Foster Collection and Analytic Transformation
3. Build Acquisition Excellence and Technology Leadership
4. Modernize Business Practices
5. Accelerate Information Sharing
6. Clarify and Align DNI's Authorities

McConnell's 100 Day Plan was followed by his "500 Day Plan for Integration and Collaboration" in October 2007 that "continues to build the foundation to enable the IC to work as a single, integrated enterprise" [38].

As of 2015, the U.S. intelligence community is a coalition of 17 agencies and organizations headed by the Office of the Director of National Intelligence (ODNI) [39] (See Figure 2.3):

1. ODNI
2. Air Force Intelligence, Surveillance, and Reconnaissance (AF ISR)
3. Army Intelligence (G-2)
4. Central Intelligence Agency (CIA)
5. Coast Guard Intelligence
6. Defense Intelligence Agency (DIA)
7. Department of Energy (DOE)
8. Department of Homeland Security (DHS)
9. Department of State: Bureau of Intelligence and Research (INR)
10. Department of the Treasury: Office of Intelligence and Analysis (OIA)
11. Drug Enforcement Administration (DEA): Office of National Security Intelligence (ONSI)
12. Federal Bureau of Investigation (FBI): National Security Branch (NSB)
13. Marine Corps Intelligence
14. National Geospatial-Intelligence Agency (NGA)
15. National Reconnaissance Office (NRO)
16. National Security Agency (NSA) /Central Security Service (CSS)
17. Navy Intelligence (ONI)

Fig. 2.3 U.S. intelligence community (Courtesy of the Office of the Director of National Intelligence)

The intelligence community employs more than 100,000 people including private contractors. The annual budget exceeds $50 billion, more than the U.S. government spends on energy, scientific research, and the federal court and prison systems [40].

Letitia "Tish" Long, Director of the National Geospatial-Intelligence Agency (NGA), described the shift across the post-9/11 intelligence community as the transition from a need-to-know atmosphere to a need-to-share and need-to-provide culture. "In solving intelligence problems, including diversity of thought is essential," said Long in a 2012 interview. "[In] the Osama bin Laden operation, the intelligence community witnessed the true value of merging many thoughts and perspectives, and we must continue to replicate this kind of integration across the enterprise in the future" [41]. The NGA was responsible for the analysis of satellite imagery of the bin Laden compound in Pakistan [42].

2.8 Informed Interrogation Approach

At the 2012 Aspen Institute Security Conference, Admiral William McRaven touted the bin Laden raid as one of the "great intelligence operations in history" [43]. However, the FBI and CIA have been at odds with each other's interrogation techniques. FBI agent Ali Soufan who firmly believes in "knowledge and empathy" has accused CIA's "enhanced interrogation" methods for being harsh, ineffective, and borderline torture. Senate Intelligence Committee Chairman Dianne Feinstein and Senate Armed Services Committee Chairman Carl Levin released their findings in April 2012 that seriously questioned the effectiveness of CIA's coercive interrogation methods [44]. Ali Soufan testified before Congress in May 2009 [45]:

> The Informed Interrogation Approach is based on leveraging our knowledge of the detainee's culture and mindset, together with using information we already know about him.... This Informed Interrogation Approach is in sharp contrast with the harsh interrogation approach introduced by outside contractors and forced upon CIA officials to use.... The [enhanced interrogation] approach applies a force continuum, each time using harsher and harsher techniques until the detainee submits.... There are many problems with this technique. A major problem is that it is ineffective. Al-Qaeda terrorists are trained to resist torture. As shocking as these techniques are to us, the al-Qaeda training prepares them for much worse — the torture they would expect to receive if caught by dictatorships for example.... A second major problem with this technique is that evidence gained from it is unreliable. There is no way to know whether the detainee is being truthful, or just speaking to either mitigate his discomfort or to deliberately provide false information....
>
> Another disastrous consequence of the use of the harsh techniques was that it reintroduced the 'Chinese Wall' between the CIA and FBI — similar to the wall that prevented us from working together to stop 9/11.... The situation was, and remains, too risky to allow someone to experiment with amateurish, Hollywood style interrogation methods — that in reality — taints sources, risks outcomes, ignores the end game, and diminishes our moral high ground in a battle that is impossible to win without first capturing the hearts and minds around the world. It was one of the worst and most harmful decisions made in our efforts against al-Qaeda.

U.S. senators Dianne Feinstein (Senate Intelligence Committee Chairman) and Carl Levin (Senate Armed Services Committee Chairman) released a joint statement in April 2012 saying that "We are deeply troubled by the claims of the CIA's former Deputy Director of Operations Jose Rodriguez regarding the effectiveness of the CIA's coercive interrogation techniques.... We are also troubled by Rodriguez's statements justifying the destruction of video tapes documenting the use of coercive interrogation techniques as 'just getting rid of some ugly visuals.' His decision to order the destruction of the tapes was in violation of instructions from CIA and White House lawyers, illustrates a blatant disregard for the law, and unnecessarily caused damage to the CIA's reputation" [46].

The 2012 movie *Zero Dark Thirty* dramatized the hunt for bin Laden by the U.S. intelligence community. Acting CIA director Michael Morell issued a statement to agency employees on December 21, 2012 saying that *Zero Dark Thirty* is not a historically accurate film [47]. U.S. senators Dianne Feinstein, Carl Levin, and John McCain wrote in a letter to Sony Pictures Entertainment that "*Zero Dark Thirty* is factually inaccurate, and we believe that you have an obligation to state that the role of torture in the hunt for Usama Bin Laden is not based on the facts, but rather part of the film's fictional narrative.... The filmmakers and your production studio are perpetuating the myth that torture is effective. You have a social and moral obligation to get the facts right" [48].

CNN national security analyst Peter Bergen also remarked that "the compelling story told in the film captures a lot that is true about the search for al-Qaeda's leader but also distorts the story in ways that could give its likely audience of millions of Americans the misleading picture that coercive interrogation techniques used by the CIA on al-Qaeda detainees — such as waterboarding, physical abuse and sleep deprivation — were essential to finding bin Laden" [49].

2.9 U.S. Fusion Centers

To encourage information sharing among governmental agencies, seventy-seven fusion centers have been set up in various cities across the U.S. "Fusion centers have stepped up to meet an urgent need in the last decade," Homeland Security and Governmental Affairs Committee Chairman Joe Lieberman said in October 2012. "Without fusion centers, we would not be able to connect the dots. Fusion centers have been essential to breaking down the information silos and communications barriers that kept the government from detecting the most horrific terrorist attack on this country — even though federal, state, and local officials each held valuable pieces of the puzzle" [50]. Lieberman cited examples of counterterrorism cases:

- "Raleigh Jihad" case — This case from 2009 involved Daniel Patrick Boyd and six others who planned to attack Marine Base Quantico. The North Carolina fusion center partnered with the local FBI Joint Terrorism Task Force on this investigation.
- Rezwan Ferdaus — In 2010, homegrown violent Islamist extremist Ferdaus was arrested in Boston for planning to attack the Pentagon and the Capitol with explosives attached to remote control small planes. The Massachusetts state fusion center was credited with making a "significant contribution" to the investigation.
- Seattle military recruiting center plot — In 2011, two homegrown violent Islamist extremists were arrested in Seattle for planning to attack a military recruiting center. The initial lead in this case came from a Seattle Police Department informant, and the investigation was jointly coordinated by the FBI and state and local agencies at the Washington state fusion center.

However, U.S. Senators Carl Levin and Tom Coburn issued a report in October 2012 of their two-year investigation that uncovered the ineffectiveness and inefficiency of the state and local fusion centers around the country [51]:

Sharing terrorism-related information between state, local and Federal officials is crucial to protecting the United States from another terrorist attack. Achieving this objective was the motivation for Congress and the White House to invest hundreds of millions of tax-payer dollars over the last nine years in support of dozens of state and local fusion centers across the United States.

The Subcommittee investigation found that DHS [Department of Homeland Security]-assigned detailees to the fusion centers forwarded "intelligence" of uneven quality — oftentimes shoddy, rarely timely, sometimes endangering citizens' civil liberties and Privacy Act protections, occasionally taken from already-published public sources, and more often than not unrelated to terrorism.

The findings of both the 2010 and 2011 assessments contradict public statements by DHS officials who have described fusion centers as "one of the centerpieces of our counter-terrorism strategy," and "a major force multiplier in the counterterrorism enterprise." The Subcommittee investigation found that the fusion centers often produced irrelevant, use-less or inappropriate intelligence reporting to DHS, and many produced no intelligence reporting whatsoever.

Despite some success stories cited by Joe Lieberman, the Senate investigative report clearly reveals that the U.S. fusion centers are in dire need of revamp. Reorganization is a common practice in successful businesses, the U.S. government should learn from the private sector.

2.10   International Collaboration on Counterterrorism

During the investigation of the Boston Marathon bombing suspect Tamerlan Tsarnaev, U.S. Senator Saxby Chambliss cited an apparent lack of information-sharing between the federal law enforcement and intelligence agencies. Senator Lindsey Graham concurred, "It was clear to me that the homeland security shop had information about the travel to Russia, the FBI did not, and they're not talking to each other and they're going back to the pre-9/11 problems here" [52].

Had there been better international collaboration on counterterrorism, the Boston Marathon bombings on April 15, 2013 would have most likely been averted. In March 2011, Russian intelligence warned the FBI about Tamerlan Tsarnaev for pos-sible terrorist activities, but the FBI cleared Tsarnaev in its own investigation. Six months later in September 2011, the Russian security agency FSB sent a similar warning letter to the CIA about Tsarnaev becoming radicalized, but refused to offer more information. "There were no details, no examples, no threads to pull," said a senior U.S. official. "Because of the rather light nature of the information we did go back to them and asked can you tell us more. We never heard back…. There's still a lot of suspicion [between U.S. and Russian intelligence operatives]. I am not sure they would share their source information with us" [53].

In September 2014, Russia Direct released a new memo explaining why U.S.-Russia joint counterterrorism collaboration has failed to materialize in the post-Cold War era. Reporter Igor Rozin wrote, "After 9/11 and the Beslan terrorist attacks in 2004, the United States and Russia have shown signs that cooperation in a global fight against terrorism is possible. Yet, the past two years have witnessed a number of notable setbacks, culminating with the current crisis in Ukraine.... Cooperation is now more important than ever, as the Islamic State of Iraq and the Greater Syria (ISIS) continues to threaten both the West and Russia with further terrorist attacks.... Clearly, after the end of the Cold War, international terrorism has become the 'very quintessence of other global threats'" [54].

Earlier on September 22, 2011, U.S. Secretary of State Hillary Clinton and Turkish Foreign Minister Ahmet Davutoglu launched the Global Counterterrorism Forum (GCTF) in New York [55]:

> The GCTF's overarching and long-term goal is to reduce the vulnerability of people everywhere to terrorism by effectively preventing, combating, and prosecuting terrorist acts and countering incitement and recruitment to terrorism.
>
> It provides a venue for national counterterrorism (CT) officials and practitioners to meet with their counterparts from key countries in different regions to share CT experiences, expertise, strategies, capacity needs, and capacity-building programs. It prioritizes civilian capacity building in areas such as rule of law, border management, and countering violent extremism.
>
> A core part of the Forum's mission is to support the implementation of the UN Global Counter-Terrorism Strategy around the globe and it places particular emphasis on working closely with the United Nations and other relevant multilateral bodies. The GCTF maintains an inclusive, even-handed, and transparent approach to its work while continuing to be an informal, action-oriented, and flexible platform committed to ensuring that it attracts the most capable and experienced CT policymakers and experts to the table.

The main focus areas of the Global Counterterrorism Forum are:

1. Addressing the "foreign terrorist fighters" problem.
2. Promoting criminal justice responses to terrorism grounded in human rights and the rule of law.
3. Supporting victims of terrorism.
4. Taking action against kidnapping for ransom and other sources of terrorism funding.
5. Supporting multi-sectoral approaches to countering violent extremism, including community engagement and community-oriented policing.
6. Rehabilitating and reintegrating violent extremist offenders.
7. Supporting efforts to address instability in the Sahel and other key regions.
8. Inspiring and supporting new international centers and initiatives to address critical challenges.
9. Developing a worldwide network of civilian CT practitioners.
10. Catalyzing implementation of the United Nations framework for countering terrorism.

2.11   Hard Lessons from Pearl Harbor and 9/11

In an unclassified document released on September 2007, Tom Johnson, head of the Division of History and Publications at the NSA, analyzed the surprise Japanese attack on Pearl Harbor in 1941. Johnson wrote in *Cryptologic Quarterly* [56]:

Interservice and intraservice rivalry always loses. If the coordination between Army and Navy was bad in Washington and in Hawaii, it was even worse within the Navy itself. Admiral Turner unquestionably harmed the defense effort through overzealous aggrandizement and turf quarrels. It was inexcusable then — it is inexcusable today. And it can be seen everywhere one goes in the Federal Government.

There is a delicate balance between the requirements of secrecy and the needs of the customer. At Pearl Harbor this balance was not properly struck. Information was kept from field commanders on whose shoulders the administration had placed a great deal of responsibility. Information did not flow because we feared losing the source. It remained bottled up in Washington, serving as small talk for intelligence professionals, State Department officials, and a limited number of operational staff planners. It is not easy to achieve a balance, but it must be done, constantly, in thousands of daily decisions over disclosure and dissemination. We face the same decisions today, in far greater quantity, though with no greater consequence. We weren't smart about it then. Are we now?

After the surprise 9/11 attacks in similar magnitude to the attack on Pearl Harbor, the U.S. intelligence community has been moving in the right direction from "need-to-know" to "need-to-share" and "need-to-provide," but there is still plenty of room for improvement.

On September 11, 2012, terrorists attacked the U.S. Consulate in Benghazi, Libya during the massive protests across the Muslim world over an anti-Islam film [57]. U.S. ambassador J. Christopher Stevens and three other Americans were killed.

Fran Townsend, CNN national security analyst and former homeland security advisor to President George W. Bush, reported on the progress of the FBI investigation 15 days after the attack. Townsend said on CNN's "Anderson Cooper 360°," "They've gotten as far as Tripoli now, but they've never gotten to Benghazi," said Townsend. "They had difficulty, and we understand there was some bureaucratic infighting between the FBI and Justice Department on the one hand, and the State Department on the other, and so it took them longer than they would have liked to get into the country. They've now gotten there. But they still are unable to get permission to go to Benghazi" [58]. The FBI finally arrived at the crime scene on October 3, three weeks after the deadly attack on the U.S. Consulate [59].

On October 10, former regional security officer Eric Nordstrom testified before a congressional hearing that his superiors denied his request for additional security for the U.S. Consulate in Benghazi months before the terrorist attack. Nordstorm told the House Oversight Committee, "You know what makes it most frustrating about this assignment? It's not the hardships. It's not the gunfire. It's not the threats. It's dealing and fighting against the people, programs, and personnel who are supposed to be supporting me…. For me, the Taliban is on the inside of the building" [60].

"Taliban on the inside of the State Department" is a dire metaphor. Bureaucratic red tape continues to hamper the effectiveness and efficiency of the U.S. intelligence community. The U.S. government needs to learn from successful private businesses that run an effective and efficient operation in serving their customers and outwitting their competitors. Like running any successful business, the government cannot afford to make serious mistakes. Bruce Riedel, former CIA officer and chair of Strategic Policy Review of Afghanistan and Pakistan, voiced his opinion in *The Daily Beast*, "In fighting terror, our team has to stay lucky 100 percent of the time. Al-Qaeda needs to be lucky only once" [61].

Bibliography

1. **IMDb.** A Few Good Men. [Online] IMDb, December 11, 1992. http://www.imdb.com/title/tt0104257/.
2. **Thieme, Richard.** DEF CON 22 - Richard Thieme - The Only Way to Tell the Truth is in Fiction. [Online] DEFCONConference, December 21, 2014. https://www.youtube.com/watch?v=EdsJulQdUcg.
3. **Lama, George de.** Reagan: In Dark On Contra Deal. [Online] Chicago Tribune, March 20, 1987. http://articles.chicagotribune.com/1987-03-20/news/8701210915_1_contra-deal-arms-sales-press-conference.
4. **Gorman, Siobhan and Entous, Adam.** Obama Unaware as U.S. Spied on World Leaders: Officials . [Online] The Wall Street Journal, October 28, 2013. http://www.wsj.com/articles/SB10001424052702304470504579162110180138036.
5. **Public Broadcasting Service.** Bin Laden's Fatwa. [Online] PBS Newshour, August 23, 1996. http://www.pbs.org/newshour/updates/military/july-dec96/fatwa_1996.html.
6. **PBS.** Al Qaeda's Fatwa. [Online] PBS Newshour, February 23, 1998. http://www.pbs.org/newshour/terrorism/international/fatwa_1998.html.
7. **Farnsworth, Elizabeth.** African Embassy Bombings. [Online] PBS Newshour, August 7, 1998. http://www.pbs.org/newshour/bb/africa/embassy_bombing/news_8-7-98.html.
8. **DCI Counterterrorist Center.** Bin Ladin Preparing to Hijack US Aircraft and Other Attacks. [Online] Central Intelligence Agency, December 4, 1998. http://www.foia.cia.gov/sites/default/files/document_conversions/89801/DOC_0001110635.pdf.
9. **The FBI.** FBI Ten Most Wanted Fugitive. [Online] Federal Bureau of Investigation, June 1999. http://www.fbi.gov/wanted/topten/usama-bin-laden.
10. **CNN.** Bin Laden says he wasn't behind attacks. [Online] CNN, September 16, 2001. http://articles.cnn.com/2001-09-16/us/inv.binladen.denial_1_ bin- laden- taliban-supreme-leader-mullah-mohammed-omar.
11. **—.** Bin Laden on tape: Attacks 'all that we had hoped for'. [Online] CNN, December 13, 2001. http://articles.cnn.com/2001-12-13/us/ret.bin.laden.videotape_1_government-translation-bin-laden-talks-osama-bin.
12. **Dudney, Robert S.** Verbatim Special: War on Terror. [Online] AIR FORCE Magazine, December 2001. http://www.airforce-magazine.com/MagazineArchive/Documents/2001/December%202001/1201verb.pdf.
13. **ABC News.** Transcript of the Alleged Bin Laden Tape. [Online] ABC News, May 23, 2006. http://abcnews.go.com/International/Terrorism/story?id=1995630.
14. **Koch, Kathleen.** White House downplays Newsweek report. [Online] CNN, June 3, 2002. http://articles.cnn.com/2002-06-03/politics/white.house.newsweek_1_bin-laden-operatives-qaeda-bush-officials.
15. **Ensor, David.** Sources: CIA warned FBI about hijacker. [Online] CNN, June 4, 2002. http://articles.cnn.com/2002-06-03/us/cia.fbi.hijackers_1_almihdhar-and-nawaf-alhazmi-cia-analysts-al-qaeda-meeting.
16. **The Guardian.** Suspected Hijackers. [Online] The Guardian. http://www.guardian.co.uk/september11/suspectedhijackers/0,,605011,00.html.
17. **Wright, Lawrence.** The Agent. Did the C.I.A. stop an F.B.I. detective from preventing 9/11? [Online] The New Yorker, July 10, 2006. http://www.newyorker.com/archive/2006/07/10/060710fa_fact_wright?currentPage=all.

18. **The FBI.** The USS Cole Bombing. [Online] Federal Bureau of Investigation. http://www.fbi. gov/about-us/history/famous-cases/uss-cole.

19. **CNN.** New York reduces 9/11 death toll by 40. [Online] CNN, October 29, 2003. http:// articles.cnn.com/2003-10-29/us/wtc.deaths_1_death-toll-world-trade-center-names.

20. **Roddy, Dennis B., et al.** Flight 93: Forty lives, one destiny. [Online] Pittsburgh Post-Gazette, October 28, 2001. http://old.post-gazette.com/headlines/20011028flt93mainstoryp7.asp.

21. **Benson, Pam.** Yemen plot exposes new world of U.S. spying. [Online] CNN, May 11, 2012. http://security.blogs.cnn.com/2012/05/11/yemen-plot-exposes-new-world-of-us-spying/.

22. **Scheuer, Michael F.** Bill and Dick, Osama and Sandy. [Online] The Washington Times, July 4, 2006. http://www.washingtontimes.com/news/2006/jul/4/20060704-110004-4280r/.

23. **Hersh, Seymour M.** What Went Wrong: The C.I.A. and the failure of American intelligence. [Online] The New Yorker, October 8, 2001. http://www.newyorker.com/archive/2001/10/08/011008fa_FACT?currentPage=all.

24. **Office of the Inspector General.** A Review of the FBI's Handling of Intelligence Information Prior to the September 11 Attacks. [Online] U.S. Justice Department, November 2004. http://www.justice.gov/oig/special/0506/chapter2.htm.

25. **Moynihan, Daniel Patrick.** Secrecy: The American Experience. [Online] Yale University Press, 1998. http://www.nytimes.com/books/first/m/moynihan-secrecy.html.

26. **The 9/11 Commission.** The System was Blinking Red. [Online] National Commission on Terrorist Attacks Upon the United States, July 22, 2004. http://www.9-11commission.gov/report/911Report_Ch8.htm.

27. **The 9-11 Commission.** The 9/11 Commission Report: Final Report of the National Commission on Terrorist Attacks Upon the United States (9/11 Report). [Online] U.S. Congress, July 22, 2004. http://www.9-11commission.gov/report/index.htm.

28. **Bamford, James.** War of Secrets; Eyes in the Sky, Ears to the Wall, and Still Wanting. [Online] The New York Times, September 8, 2002. http://www.nytimes.com/2002/09/08/weekinreview/war-of-secrets-eyes-in-the-sky-ears-to-the-wall-and-still-wanting.html?pagewanted=all.

29. **Block, Robert, Fields, Gary and Wrighton, Jo.** U.S. 'Terror' List Still Lacking . [Online] The Wall Street Journal, January 2, 2004. http://online.wsj.com/article/0,,SB1073005342680 21800,00.html.

30. **Investigation, Federal Bureau of.** Terrorist Screening Center. [Online] FBI. [Cited: November 23, 2012.] http://www.fbi.gov/about-us/nsb/tsc.

31. **Gorman, Siobhan.** How Team of Geeks Cracked Spy Trade . [Online] The Wall Street Journal, September 4, 2009. http://online.wsj.com/article/SB125200842406984303.html.

32. **Rudy, Charles J.** One team, one fight. [Online] U.S. Air Force (Kandahar Airfield), March 16, 2011. http://www.kdab.afcent.af.mil/news/story.asp?id=123246976.

33. **GeekyRoom's Chief Editor.** CIA's Open Source Center Analyze Twitter in Real Time. [Online] GeekyRoom, November 5, 2011. http://geekyroom.com/2011/11/05/cia-open-source-center-analyze-twitter-in-real-time/.

34. **Open Source Center.** Open Source Center: Information to Intelligence. [Online] U.S. Government. https://www.opensource.gov/.

35. **108th Congress.** Intelligence Reform and Terrorism Prevention Act of 2004. [Online] National Counterterrorism Center, December 17, 2004. http://www.nctc.gov/docs/pl108_458.pdf.

36. **Rose, Derek.** Cia Hid Key Info On 9/11 . [Online] New York Daily News, July 3, 2006. http://articles.nydailynews.com/2006-07-03/news/18345695_1_two-al-qaeda-cia-cole-bombing.

37. **Office of the Director of National Intelligence.** 100 Day Plan for Integration and Collaboration. [Online] Defense Technical Information Center (DTIC), September 2007. http://www.dtic.mil/dtic/tr/fulltext/u2/a471965.pdf.

38. **Director of National Intelligence.** United States Intelligence Community 500 Day Plan Integration and Collaboration. [Online] Defense Technical Information Center (DTIC), October 10, 2007. http://www.dtic.mil/dtic/tr/fulltext/u2/a473641.pdf.

39. **Office of the Director of National Intelligence (ODNI).** Office of the Director of National Intelligence: Leading Intelligence Integration. [Online] U.S. Government. http://www.dni.gov/.

40. **Wright, Lawrence.** The Spymaster. Can Mike McConnell fix America's intelligence community? [Online] The New Yorker, January 21, 2008. http://www.newyorker.com/reporting/2008/01/21/080121fa_fact_wright?currentPage=all .

41. **Young, Denise.** Letitia Long: A Global Vision. Alumna leads intelligence agency in new era of collaboration. [Online] Virginia Tech Magazine, Spring 2012. http://www.vtmag.vt.edu/spring12/letitia-long.html.

42. **Hawkins, Arielle.** Bin Laden raid nets one intel employee big bonus. [Online] CNN, May 18, 2012. http://security.blogs.cnn.com/2012/05/18/bin-laden-raid-nets-one-intel-employee-big-bonus/.

43. **Crawford, Jamie.** McRaven on bin Laden raid: One of history's "great intelligence operations". [Online] CNN, July 26, 2012. http://security.blogs.cnn.com/2012/07/26/u-s-special-ops-commander-discusses-role-in-serious-and-humorous-tones/.

44. **Feinsten, Dianne and Levin, Carl.** Feinstein, Levin Statement on CIA's Coercive Interrogation Techniques. [Online] Dianne Feinstein: U.S. Senator for California, April 30, 2012. http://www.feinstein.senate.gov/public/index.cfm/2012/4/feinstein-levin-statement-on-cia-s-coercive-interrogation-techniques.

45. **Soufan, Ali.** Testimony of Ali Soufan. [Online] United States Senate Committee on the Judiciary, May 13, 2009. http://www.judiciary.senate.gov/hearings/testimony.cfm?id=e655f9e2809e5476862f735da14945e6&wit_id=e655f9e2809e5476862f735da14945e6-1-2.

46. **Feinstein, Dianne and Levin, Carl.** Feinstein, Levin Statement on CIA's Coercive Interrogation Techniques. [Online] United States Senate, April 30, 2012. http://www.feinstein.senate.gov/public/index.cfm/press-releases?ID=f3271910-3fad-40a5-9d98-93450e0090aa.

47. **Benson, Pam.** CIA challenges accuracy of 'Zero Dark Thirty'. [Online] CNN, December 21, 2012. http://security.blogs.cnn.com/2012/12/21/cia-challenges-accuracy-of-zero-dark-thirty/.

48. **Feinstein, Dianne; Levin, Carl; McCain John.** Feinstein Releases Statement on 'Zero Dark Thirty'. [Online] United States Senate, December 19, 2012. http://www.feinstein.senate.gov/public/index.cfm/press-releases?ID=b5946751-2054-404a-89b7-b81e1271efc9.

49. **Bergen, Peter.** 'Zero Dark Thirty': Did torture really net bin Laden? [Online] CNN, December 10, 2012. http://www.cnn.com/2012/12/10/opinion/bergen-zero-dark-thirty/index.html.

50. **Lieberman, Joe.** Fusion Centers Add Value to Federal Government Counterterrorism Efforts. [Online] U.S. Senate Committee on Homeland Security and Governmental Affairs, October 3, 2012. http://www.hsgac.senate.gov/media/fusion-centers-add-value-to-federal-government-counterterrorism-efforts.

51. **Levin, Carl and Coburn, Tom.** Federal Support for and Involvement in State and Local Fusion Centers. [Online] U.S. Senate Committee on Homeland Security and Governmental Affairs, October 3, 2012. http://www.hsgac.senate.gov/download/?id=49139e81-1dd7-4788-a3bb-d6e7d97dde04.

52. **Cratty, Carol and Cohen, Tom.** Bombing suspect not on terror watch list. [Online] CNN, April 23, 2013. http://www.cnn.com/2013/04/23/politics/boston-bombings-fbi.

53. **Johns, Joe, et al.** Russia asked U.S. twice to investigate Tamerlan Tsarnaev, official says. [Online] CNN, April 24, 2013. http://security.blogs.cnn.com/2013/04/24/russia-asked-u-s-twice-to-investigate-tamerlan-tsarnaev-official-says/.

54. **Rozin, Igor.** The rise and fall of US-Russian counter-terrorism collaboration. [Online] Russia Direct, September 3, 2014. http://www.russia-direct.org/russian-media/rise-and-fall-us-russian-counter-terrorism-collaboration.

55. **Global Counterterrorism Forum.** About the GCTF. [Online] Global Counterterrorism Forum. [Cited: February 5, 2015.] https://www.thegctf.org/web/guest/mission.

56. **Johnson, Tom.** What Every Crytologist Should Know about Pearl Harbor. [Online] Cryptologic Quarterly, September 27, 2007. http://www.nsa.gov/public_info/_files/cryptologic_quarterly/pearlharbor.pdf.

57. **Reuters.** Terrorists killed U.S. ambassador to Libya: Panetta. [Online] Chicago Tribune, September 27, 2012. http://www.chicagotribune.com/news/sns-rt-us-libya-usa-investigation-bre88q1jw-20120927,0,3904573.story.

58. **CNN Wire Staff.** Sources: 15 days after Benghazi attack, FBI still investigating from afar. [Online] CNN, September 26, 2012. http://www.cnn.com/2012/09/26/world/africa/libya-investigation/index.html.
59. **Starr, Barbara.** FBI visits site of attack in Libya. [Online] CNN, October 4, 2012. http://www.cnn.com/2012/10/04/world/africa/libya-fbi-benghazi/index.html.
60. **CNN Wire Staff.** U.S. official says superiors worked against effort to boost Benghazi security. [Online] CNN, October 11, 2012. http://www.cnn.com/2012/10/10/politics/congress-libya-attack/index.html.
61. **Riedel, Bruce.** A Stubborn Terror. [Online] The Daily Beast, September 10, 2012. http://www.thedailybeast.com/newsweek/2012/09/09/a-stubborn-terror.html.

Part II
Counterterrorism Strategies: Causes and Cures, War and Peace

Chapter 3
Understanding Terrorism

You can kill a man but you can't kill an idea.
— Civil rights activist and U.S. Army sergeant Medgar Evers

Morality intervenes not when we pretend we have no enemies but when we try to understand them, to put ourselves in their situation. ... Trying to understand other people means destroying the stereotype without denying or ignoring the otherness.
— Italian philosopher Umberto Eco

The single story creates stereotypes, and the problem with stereotypes is not that they are untrue, but that they are incomplete. They make one story become the only story.
— Nigerian novelist Chimamanda Ngozi Adichie at TEDGlobal 2009 (July 23, 2009)

I study the future of crime and terrorism. And quite frankly I'm afraid. ... We consistently underestimate what criminals and terrorists can do.
— Global security advisor and futurist Marc Goodman at TEDGlobal 2012 (June 28, 2012)

In the past, women used to tell their children, "Go to bed or I will call your father. Now they say, "Go to bed or I will call the plane." That is a golden ticket you give al Qaeda to use against you.
— Former U.S. exchange student Farea Al-Muslimi (May 16, 2013)

I also expressed my concerns that drone attacks are fueling terrorism. ... If we refocus efforts on education it will make a big impact.
— Nobel Peace Laureate Malala Yousafzai (October 11, 2013)

© Springer International Publishing Switzerland 2015
N. Lee, *Counterterrorism and Cybersecurity*, DOI 10.1007/978-3-319-17244-6_3

[The Americans] is a show that asks us to give a human face
to people we might consider our enemies, to understand the
underlying humanity, complexity, and conflicts of people on
both sides of any particular divide.
— Richard Thomas who plays FBI agent Frank Gaad in
The Americans (TV Series, 2014)

3.1 "Give a Human Face to People We Consider Our Enemies" — *The Americans*

Academy Award-winning actress Angelina Jolie was in Japan on 9/11, and she recalled, "It was this other strange thing for me, because it was obviously a country where if you would have turned the clock back, it was an enemy. And now, on that date, for me as an American, they were my allies, my friends, taking care of me, giving me sympathy for my country. I became immediately conscious of how things shift, how the picture of the enemy shifts. I don't have any answers, but to be aware of all these things as it's coming down — it's not as simple as, 'Well, this is the bad guy'" [1].

Italian philosopher Umberto Eco once said, "I would argue that morality intervenes not when we pretend we have no enemies but when we try to understand them, to put ourselves in their situation. … Trying to understand other people means destroying the stereotype without denying or ignoring the otherness" [2].

Created and produced by former CIA officer Joe Weisberg, *The Americans* is a TV show that "asks us to give a human face to people we might consider our enemies, to understand the underlying humanity, complexity, and conflicts of people on both sides of any particular divide," said Richard Thomas who plays FBI Agent Frank Gaad in the show.

Drugs and terrorism are both mind-altering and deadly. Long before President George W. Bush declared the global war on terror in 2001, President Richard Nixon declared drug abuse "public enemy number one" in 1971. Today, the United States spends about $51 billion annually on the war on drugs with no end in sight [3]. The black market for illicit drugs follows the same fundamental concepts of economics: supply and demand. One may recall the fiasco of the 1920-1933 Prohibition in the United States that gave organized crime a major boost. Similarly, there will always be drug trafficking for as long as there are affluent customers paying high prices for illegal substances. "The billionaires I know, almost without exception, use hallucinogens on a regular basis," said Silicon Valley investor Tim Ferriss. "[They're] trying to be very disruptive and look at the problems in the world … and ask completely new questions" [4].

Give a human face to drug users and understand why they need narcotics is the only chance that we have to win the war on drugs [5].

After the 9/11 attacks in 2001, President Bush described the long, drawn-out war on terror: "We're a nation that is adjusting to a new type of war. This isn't a conventional war that we're waging. Ours is a campaign that will have to reflect the new enemy. … These are people who strike and hide, people who know no borders … The new war is not only against the evildoers themselves; the new war is against those who harbor them and finance them and feed them. We will need patience and determination in order to succeed" [6]. After 13 years and over $1.5 trillion spent in Iraq [7] and Afghanistan [8], the war on terror has managed to weaken but not wipe out the al Qaeda terrorist organization [9].

In 2014, ISIS (Islamic State of Iraq and Syria) or ISIL (Islamic State of Iraq and the Levant) became America's number 1 enemy who is more brutal than al Qaeda. ISIS has beheaded innocent civilians and challenged President Barack Obama to "bring it on" in the "Flames of War" video that looks like a Hollywood trailer. In February 2015, President Obama asked U.S. Congress to authorize the use of American military forces in the war against ISIS [10]. Analyst Laith Alkhouri remarked that ISIS "appears to be more relentless than ever, not only expanding in territory but also raising the bar in its confrontation with the world's top superpower" [11].

Give a human face to terrorists and understand why they resort to violence is the only chance that we have to win the war on terror.

During the Porte de Vincennes hostage crisis on January 9, 2015 following the Charlie Hebdo shooting in Paris two days earlier, a father named Teddy tried to pick up his young son from a nearby elementary school and he said, "It's like a war. I don't know how I will explain this to my 5-year-old son" [12]. On the same day, a senior U.S. intelligence official told CNN: "This is in perpetuity what we're dealing with. It's like the war on drugs. This isn't going to stop" [13]. Hacktivist group Anonymous posted a YouTube video stating that "We, Anonymous around the world, have decided to declare war on you the terrorists… we are going to shut down your accounts on all social networks" [14].

While there are a number of ways to combat terrorism, attempting to cure the symptoms without tackling the root causes is like waging a losing war on drugs. In the aftermath of Charlie Hebdo massacre, Middle East correspondent Patrick Cockburn told *The Guardian*, "An explanation of the roots of some ghastly atrocity such as Charlie Hebdo [is seen] to dilute the degree of evil involved and justify the actions of the perpetrators. This response is understandable but it is not very useful. Asking who is to directly to blame for Charlie Hebdo and explaining why it happened are two different things" [15].

Stephen Yates, former deputy assistant to the vice president for national security affairs in the George W. Bush administration, had asked the following three pointed questions [16]:

1. Who is the enemy, what does it seek, and by what means?
2. Are we at war or is the current threat simply another manifestation of violent crime?
3. Are today's terrorists predominantly disaffected individuals in search of the means to act out, or is there an organized movement seeking out disaffected individuals to use for its own violent purposes?

3.2 Bravery and Cowardice

One of my all-time favorite television shows was Bill Maher's *Politically Incorrect* on ABC between 1997 and 2002. On September 11, 2001, former assistant U.S. attorney Barbara Olson was traveling to a *Politically Incorrect* taping in Los Angeles. She was a passenger on American Airlines Flight 77 that crashed into the Pentagon. Maher left a panel chair empty for a week in honor of her memory.

On the September 17 show, Maher's guest Dinesh D'Souza quarreled with President George W. Bush's characterization of the terrorists as cowards. D'Souza argued, "Look at what they [the 9/11 terrorists] did. You have a whole bunch of guys who were willing to give their life; none of them backed out. All of them slammed themselves into pieces of concrete. These are warriors." Maher concurred, "We have been the cowards, lobbing cruise missiles from 2,000 miles away. That's cowardly. Staying in the airplane when it hits the building, say what you want about it, it's not cowardly" [17].

It was shocking, as Maher and his guests often upset the status quo by offering their opposing viewpoints on controversial topics, and sparking heated debates on the show. Maher later apologized and clarified that his "cowardly" comments were directed not at American soldiers, but at "the government, the elected officials, [and] the people who want to put up a giant missile shield, when plainly that's not where the threat is from" [18].

American musician and activist Henry Rollins once said, "A coward hides behind freedom. A brave person stands in front of freedom and defends it for others." Murdering innocent people who are defenseless is a cowardly act. Standing up to difficult life circumstances and oppressive regimes is bravery. Nevertheless, Maher challenged his millions of viewers, including myself, to dig deeper into the problem of terrorism, rather than fixating on a simplistic good guy versus bad guy paradigm.

In the 1988 novel *The Silence of the Lambs* by Thomas Harris, cannibalistic serial killer Hannibal Lecter told FBI agent-in-training Clarice Starling, "Nothing happened to me. I happened." Like the cold-blooded murderer Lecter, terrorists do not commit evil deeds simply because of some mental disorders or self-centered vendetta. A memorable dialogue between Lecter and Starling follows:

Lecter: "First principles, Clarice. Simplicity. Read Marcus Aurelius. Of each particular thing ask: what is it in itself? What is its nature? What does he do, this man you seek?"

Starling: "He kills women..."

Lecter: "No. That is incidental. What is the first and principal thing he does? What needs does he serve by killing?"

To combat terrorism, we need to understand terrorism. German journalist and author Juergen Todenhoefer traveled deep into ISIS territory and interviewed fighters, terrorists, prisoners, and civilians in Iraq and Syria. In June 2014, some 300 ISIS fighters defeated more than 20,000 Iraqi soldiers in just four days to take control of Mosul, Iraq's second largest city. A young fighter told Todenhoefer how ISIS won the battle: "Well, we didn't attack them all at once, we hit their front lines hard, also using suicide attacks. Then the others fled very quickly. We fight for Allah, they fight for money and other things that they do not really believe in" [19]. French soldier and military theorist Ferdinand Foch in World War I said, "The most powerful weapon on earth is the human soul on fire."

3.3 Drones Kill Terrorists, not Terrorism

Unmanned Aerial Vehicles (UAVs), Unmanned Aerial Systems (UAS), or simply drones have become the counterterrorism weapon of choice. In 2001, the U.S. Department of Defense had fewer than 50 military drones. In 2012, the number of armed and reconnaissance drones swelled to around 7,500 [20]. The Defense Advanced Research Projects Agency (DARPA) has developed a miniature hummingbird-size drone and the Air Force Research Laboratory has plans for a pigeon-sized UAV that can recharge while perching on power lines [21].

The drone program began in the Bush administration to target al-Qaeda leaders in the tribal regions of Pakistan, and it has expanded extensively under President Barack Obama [22]. *The Economist* reported in October 2011 that "under Barack Obama, the frequency of drone strikes on terrorists in Pakistan's tribal areas has risen tenfold, from one every 40 days during George Bush's presidency to one every four" [23].

In January 2012 during a YouTube-Google forum, President Obama defended the use of drones: "I think we have to be judicious in how we use drones. But understand that probably our ability to respect the sovereignty of other countries and to limit our incursions into somebody else's territory is enhanced by the fact that we are able to (execute a) pinpoint strike on al-Qaeda operatives in a place where the capacities of that military in that country may not be able to get them" [24].

John Brennan, President Obama's assistant for homeland security and counterterrorism, further explained that "in full accordance with the law — and in order to prevent terrorist attacks on the United States and to save American lives — the United States government conducts targeted strikes against specific al-Qaeda terrorists, sometimes using remotely piloted aircraft, often referred to publicly as drones" [25].

"The drone strikes are not just important in terms of eliminating the leadership of al-Qaeda," said Peter Bergen, director of the national security studies program at the New American Foundation. "They are also important in terms of preventing people from training in the tribal region and making that very difficult because you are always looking over your shoulder for a drone attack" [26].

When American-born al-Qaeda terrorist Anwar al-Awlaki was killed by a drone strike in Yemen in September 2011, a senior U.S. military official said, "It's critically important to send an important message to the surviving leaders and foot soldiers in the Qaeda affiliate. It sets a sense of doom for the rest of them. Getting Awlaki, given his tight operational security, increases the sense of fear. It's hard for them to attack when they're trying to protect their own back side" [27].

However, what is the unintended message that the U.S. is sending to the innocent people residing in their home countries? Many Pakistanis consider U.S. drone strikes a violation of sovereignty and a cause of unacceptable civilian casualties.

In his May 2013 speech at the National Defense University in Washington D.C., President Obama asserted, "Before any strike is taken, there must be near-certainty that no civilians will be killed or injured — the highest standard we can set" [28]. In reality, however, collateral damage is very difficult to avoid. In December 2013, six months after Obama's reassuring speech, U.S. intelligence mistook a wedding convoy in Yemen for al-Qaeda militants. The CIA drones targeted two of the 11 vehicles, killing 14 and injuring 22 others in the wedding party [29].

According to data compiled by the New America Foundation, 394 CIA [Central Intelligence Agency] drone strikes in Pakistan between mid-2004 and early 2015 have killed an estimated 2,217 to 3,598 people, including 1,760 to 2,957 confirmed militants and 258 to 307 innocent civilians [30]. The Amnesty International reported that a strike in October 2012 killed a 68-year-old woman, Mamana Bibi, as she picked vegetables with her grandchildren, a number of whom were injured in the attack [28]. Many civilian reactions to the drone strikes have echoed the following sentiments [31]:

1. 28-year-old Pakistani Mohammad Rehman Khan lost his father, three brothers, and a nephew in South Waziristan on the Afghan border. After Obama's reelection in 2012, Khan said, "The same person who attacked my home has gotten reelected. … When the Sandy hurricane came, I thought that Allah would wipe away America. America just wants to take over the world."
2. Haji Abdul Jabar who lost his 23-year-old son said, "Whenever he has a chance, Obama will bite Muslims like a snake. Look at how many people he has killed with drone attacks."
3. Warshameen Jaan Haji who lost his wife said, "Any American, whether Obama or Mitt Romney, is cruel. I lost my wife in the drone attack and my children are injured. Whatever happens, it will be bad for Muslims."

Widespread resentment, anger, and hatred toward Americans will not help the U.S. win the war on terror. On the contrary, as *Reuters* reported in November 2012, "anger over the unmanned aircraft may have helped the Taliban gain recruits, complicating efforts to stabilize the unruly border region between Pakistan and Afghanistan. That could also hinder Obama's plan to withdraw U.S. troops from Afghanis" [31].

Indeed, NATO's International Security Assistance Force (ISAF) in Afghanistan released dismal data in September 2012 showing that the conditions in Afghanistan are mostly worse than before the U.S. troop surge began two years prior in 2010 [32]. President Barack Obama had high hopes when he outlined his plan for an Afghanistan troop surge in 2009 [33], but the surge ended in 2012 with little fanfare [34].

In April 2013, CNN's Nic Robertson reported that the Taliban had been successfully recruiting young boys aged 8 to 18 by weaning them on tales of the U.S. drone strikes that killed women and children. British lawyer Ben Emmerson, the U.N. Special Rapporteur on drones, told Robertson after visiting Pakistan, "The consequence of drone strikes has been to radicalize an entirely new generation. … Through the use of drones you may win the immediate battle you are waging against this particular faction or that particular faction … but you are losing the war in the longer term. … If it is lawful for the U.S. to drone al Qaeda associates wherever they find them, then it is also lawful for al Qaeda to target U.S. military or infrastructure wherever [militants] find them" [35].

Retired U.S. Air Force General Michael Hayden who was NSA director (1999-2005) and CIA director (2006-2009) questioned if the CIA should be targeting terrorists overseas. Hayden wrote an opinion piece for *CNN* in April 2013 on the heels of Robertson's article: "…I had breakfast with Dave Petraeus. As we were leaving the table, I suggested that the CIA had never looked more like its wartime predecessor, OSS, than it did right then. That had made America safer, but I reminded the general that the CIA was not the OSS" [36].

Temporary safety is not the same as long-term security. A false sense of security is like the calm before the storm.

To raise public awareness of drone attacks, British artist James Bridle launched Dronestagram on Instagram in October 2012 to show Google Earth images of the locations of drone strikes [37]. Bridle told *CNN*, "We have gotten better at immediacy and intimacy online. Perhaps we can be better at empathy too" [38].

Farea Al-Muslimi, a former U.S. exchange student who grew up in a remote village in Western Yemen, talked to CNN's Jessica Gutteridge in May 2013 about a U.S. drone strike in his hometown. "In the past, women used to tell their children, 'go to bed or I will call your father,'" said Al-Muslimi. "Now they say, 'go to bed or I will call the plane.' That is a golden ticket you give al Qaeda to use against you" [39]. Yemeni Nobel Peace Laureate Tawakkol Karman weighed in, "The killing conducted by unmanned planes in Yemen is outside the law and worse than the terrorist activities of individuals and groups" [40].

Pakistani activist Malala Yousafzai and President Barack Obama are both Nobel Peace Laureate, but they share different views on counterterrorism strategies. Yousafzai wrote after her meeting with Obama and first lady Michelle at the White House on October 11, 2013: "I also expressed my concerns that drone attacks are fueling terrorism. Innocent victims are killed in these acts, and they lead to resentment among the Pakistani people. If we refocus efforts on education it will make a big impact" [41].

Pakistani attorney Yasser Latif Hamdani concurred, "First of all, hanging terrorists and defeating the Taliban militarily is tantamount to alleviating the symptom while not addressing the underlying cause, i.e. the mindset that tolerates, if not outrightly celebrates, Talibanisation of the country. To this end the government needs to take several intermediary steps, foremost of which will be curriculum reform. … Curriculum reform should ensure that young Pakistanis look at themselves as human beings and Pakistanis before they see themselves through the prism of their religious or linguistic identity. Without the inculcation of these basic principles, any victory on the battlefield will be a short lived one. … Real de-radicalisation begins with educating the people in universal human values instead of limited religious ones" [42].

Although Iraq's prime minister Haider al-Abadi welcomed the airstrikes against ISIS targets in Syria, he cautioned the U.S. and its Arab allies in September 2014: "I hope they do it right, and they don't do it their own way. I hope we're not going to see a crush of ISIS and the rise of another terrorist element instead of them" [43].

American civil rights activist and U.S. Army Sergeant Medgar Evers once said that "you can kill a man, but you can't kill an idea" [44]. Malala Yousafzai told CNN's Christiane Amanpour that "they can kill me... but it does not mean that they can kill my cause" [45]. Their statements can apply to good as well as evil. Drones are powerful assassination weapons that offer short-term fixes. Drones kill terrorists, but not the idea of terrorism in the long term.

3.4 War on Terror (Overseas Contingency Operation)

The Obama administration in 2009 renamed the war on terror campaign to "Overseas Contingency Operation." The phrase Global War on Terror "was enormously unfortunate because I think it pulled together disparate organizations and insurgencies," said John A. Nagl, former Army officer and now president of the Center for a New American Security. "Our strategy should be to divide and conquer rather than make of enemies more than they are. We are facing a number of different insurgencies around the globe — some have local causes, some of them are transnational. Viewing them all through one lens distorts the picture and magnifies the enemy" [46].

In order to win the war on terror or to be successful in the Overseas Contingency Operation, we need to understand terrorism so that we can find ways to end it. The 9/11 Commission Report stated, "The problem is that al-Qaeda represents an ideological movement, not a finite group of people. It initiates and inspires, even if it no longer directs. In this way it has transformed itself into a decentralized force. … Killing or capturing him [Osama bin Laden], while extremely important, would not end terror. His message of inspiration to a new generation of terrorists would continue" [47]. Wayne Murphy, deputy director for analysis and production at the National Security Agency (NSA) concurred, "In the end, I don't know if the benefits of getting bin Laden would balance out. And I don't know if it buys us anything. Think about what we just went through with Saddam Hussein" [48].

In fact, after the bombings of the American embassies in East Africa in 2006, al-Qaeda conspirator Abu Jandal told an FBI agent, "Can you imagine how many joined bin Laden after the embassy bombings? Hundreds came and asked to be martyrs" [49].

Nevertheless, General Stanley McChrystal and Michael Scheuer represented the majority opinion that bin Laden was a top priority in the war on terror. General McChrystal, commander of U.S. Forces Afghanistan (USFOR-A), testified in Congress, "I don't think that we can finally defeat al-Qaeda until he's captured or killed. I believe he is an iconic figure at this point, whose survival emboldens al-Qaeda as a franchising organization across the world" [50].

Michael Scheuer, a 22-year veteran with the CIA, created and served as the chief of the agency's Osama bin Laden unit at the U.S. Counterterrorist Center. Scheuer openly blamed President Bill Clinton's refusal to authorize the CIA to kidnap or assassinate bin Laden [51].

On May 2, 2011, Osama bin Laden was finally found and killed by U.S. Navy SEALs in Abbottabad, Pakistan [52]. By the time of his death, bin Laden was no longer an effective leader but simply a figurehead of the al-Qaeda terrorist organization. He hid in an unguarded compound with no intrusion alarm and no escape route; he was basically a sitting duck waiting to die a martyr or a coward depending on one's point of view.

"Jihad against America will not stop with the death of Osama," proclaimed Fazal Mohammad Baraich, a Muslim cleric. "Osama bin Laden is a shaheed (martyr). The blood of Osama will give birth to thousands of other Osamas" [53].

American-born al-Qaeda terrorist Anwar al-Awlaki was one of "the other Osamas." Dubbed the "bin Laden of the Internet," al-Awlaki had a blog, a Facebook page, and YouTube videos [54]. He had evaded capture for years until September 2011 when he was killed by a CIA drone strike in Yemen. Once again, he was considered a martyr by the Islamic extremists.

"The death of Sheik Anwar al-Awlaki will merely motivate the Muslim youth to struggle harder against the enemies of Islam and Muslims," said Anjem Choudhry, an Islamic scholar in London. "I would say his death has made him more popular" [27].

In spite of bin Laden and al-Awlaki's death, the war on terror continues without an end in sight. Ibrahim Al-Asiri has now replaced bin Laden and al-Awlaki as the world's most dangerous terrorist who is the mastermind behind printer cartridge bombs, underwear bombs, the attempted assassination of a top Saudi counterterrorism official, and the failed attack on a Detroit-bound plane [55].

Tom Engelhardt of tomdispatch.com wrote, "Unless we set aside the special ops assaults and the drone wars and take a chance, unless we're willing to follow the example of all those nonviolent demonstrators across the Greater Middle East and begin a genuine and speedy withdrawal from the Af/Pak theater of operations, Osama bin Laden will never die" [56].

3.5 A Stubborn Terror

In the TV series *Homeland*, Abu Nazir told U.S. Marine Sergeant Nicholas Brody, "Why kill a man when you can kill an idea?" [57]. It was a real irony because Nazir played a wanted terrorist leader who converted Brody into an undercover operative for al-Qaeda. Brody was turned after a U.S. drone strike killed Nazir's young child whom Brody grew to love like a son.

"The American public is underestimating the Islamic fundamentalist groups, and terrorism and extremism," said Amrullah Saleh, head of the National Directorate of Security, Afghanistan's domestic intelligence agency [58].

Bruce Riedel, former CIA officer and current senior fellow at the Brookings Institution, chaired the strategic review of policy toward Afghanistan and Pakistan in 2009 at President Barack Obama's request. On September 10, 2012, a day before the terrorist attack at the U.S. Consulate in Benghazi, Libya, Bruce Riedel published a chilling report titled "A Stubborn Terror" in *Newsweek/The Daily Beast* [59]:

> Eleven years after 9/11, al-Qaeda is fighting back. Despite a focused and concerted American-led global effort—despite the blows inflicted on it by drones, SEALS, and spies—the terror group is thriving in the Arab world, thanks to the revolutions that swept across it in the last 18 months. And the group remains intent on striking inside America and Europe. The al-Qaeda core in Pakistan has suffered the most from the vigorous blows orchestrated by the Obama administration. The loss of Osama bin Laden eliminated its most charismatic leader, and the drones have killed many of his most able lieutenants. But even with all these blows, bin Laden's successor, Ayman al-Zawahiri, is still orchestrating a global terror network and communicating with its followers.

Specifically, Riedel pointed out that:

1. The Taliban and Lashkar-e-Taiba, al-Qaeda's allies in Pakistan, have a global network with terrorist cells in the U.S., England, and the Persian Gulf.
2. Al-Qaeda is multiplying in the Arabian Peninsula, especially Yemen.
3. Al-Qaeda carries out waves of bombings every month in Iraq.
4. Al-Qaeda and other Islamist extremists has taken over half of Mali in North Africa.
5. A new al-Qaeda franchise has emerged in Egypt's Sinai Peninsula.
6. Al-Qaeda operation is fast growing in Syria.
7. Al-Qaeda dispatched Chechen terrorists to Spain in 2012 to attack Gibraltar.
8. Al-Qaeda has planned to attack New York, Chicago, and Detroit since 2009.

U.S. Secretary of State Hillary Clinton testified at the Benghazi hearings before Congress in January 2013: "The Arab Spring has ushered in a time when al-Qaeda is on the rise. The world in many ways is even more dangerous because we lack a central command [in al-Qaeda] and have instead these nodes that are scattered throughout North Africa and other places" [60]. For example, Islamist militants, including al-Qaeda in the Islamic Maghreb (AQIM), attacked a convoy of oil workers in Algeria and held captive dozens of hostages from the United States, United Kingdom, France, Japan, Norway, and other countries [61].

Newt Gingrich, former Speaker of the U.S. House of Representatives, admitted in September 2014 on CNN that "13 years after the September 11 attacks, the United States does not have an effective strategy for dealing with radical Islamists and their deep commitment to waging war against us and against our civilization" [62].

Besides international terrorism, the U.S. is also facing domestic terrorism by militant right-wing, left-wing, anarchist, and special interest groups. According to Southern Poverty Law Center (SPLC), the number of conspiracy-minded antigovernment "Patriot" groups reached an all-time high of 1,360 in 2012, up 7% from 2011 [63].

Years before the 9/11 attacks, former U.S. Army platoon leader Timothy McVeigh killed 168 people including 19 children in the Oklahoma City bombing in April 1995 [64]. It was the worst act of homegrown domestic terrorism in U.S.

history. During the 1996 Summer Olympics, U.S. Army veteran Eric Rudolph planted a pipe bomb at the Centennial Olympic Park in Atlanta, Georgia, killing 2 and injuring 112 people [65]. In 1996 alone, the Federal Bureau of Investigation (FBI) thwarted five planned acts of domestic terrorism directed against local law enforcement officials in Montana, an FBI facility in West Virginia, communications and transportation infrastructure, and banking facilities in Washington State.

A week after the Boston Marathon bombings in April 2013, CNN's LZ Granderson opined that violent gangsters should be treated as domestic terrorists: "Last week, millions watched as an entire city was shut down to look for one guy. Every major news station was covering the pursuit of one guy. We all know the face and relatives of this one guy. And it's all because he is an alleged terrorist. But more Americans were murdered in the south and west sides of Chicago than there were U.S. servicemen killed in Afghanistan last year, and yet for some reason we don't view those neighborhoods as terrorized" [66].

The FBI said, "As long as violence is viewed by some as a viable means to attain political and social goals, extremists will engage in terrorism" [65]. Marc Goodman, global security advisor and futurist, painted a bleak future as he told the audience at TEDGlobal 2012 in Edinburgh: "I study the future of crime and terrorism. And quite frankly I'm afraid. … We consistently underestimate what criminals and terrorists can do" [67].

In December 2012, Dr. Mathew Burrows of the National Intelligence Council (NIC) published a 166-page report, "Global Trends 2030: Alternative Worlds," in which he expressed a mixed outlook on the future of terrorism [68]:

> [Positive:] Several circumstances are ending the current Islamist phase of terrorism, which suggest that as with other terrorist waves—the Anarchists in the 1880s and 90s, the post-war anti-colonial terrorist movements, the New Left in 1970s—the recent religious wave is receding and could end by 2030.

> [Negative:] Taking a global perspective, future terrorists could come from many different religions, including Christianity and Hinduism. Right-wing and left-wing ideological groups—some of the oldest users of terrorist tactics— also will pose threats. … The worst-case outcome on nuclear proliferation … terrorists or extremist elements acquiring WMD (Weapon of Mass Destruction) material.

> [Conclusion:] Terrorism is unlikely to die completely, however, because it has no single cause. The traditional use of the term "root cause" for understanding what drives terrorism is misleading. Rather, some experts point to the analogy of a forest fire: a mixture of conditions—such as dry heat, a spark, and wind—that lead to terrorism.

3.6 Economic and Psychological Warfare

In the December 2012 NIC report "Global Trends 2030: Alternative Worlds," Burrows wrote that "terrorists … would focus less on causing mass casualties and more on creating widespread economic and financial disruptions" [68]. Al-Qaeda, has in fact, been employing the economic warfare tactic for over a decade while keeping their own expenses under control by recording every single transaction as small as $0.60 for a piece of cake that one of their fighters ate [69].

In a 2004 videotape sent to Alijazeera, Osama bin Laden spoke of "having experience in using guerrilla warfare and the War of Attrition to fight tyrannical superpowers" as al-Qaeda "bled Russia for 10 years, until it went bankrupt and was forced to withdraw in defeat" and so al-Qaeda is "continuing this policy in bleeding America to the point of bankruptcy" [70].

In October 2005, Abu Mus'ab al-Najadi, a Saudi supporter and member of al-Qaeda authored a seven-page document titled "Al-Qaeda's Battle is Economic not Military" in which he clarified bin Laden's strategy [71]:

> Usually, wars are based on military strength and victory belongs to those who are militarily superior on the battlefield … But our war with America is fundamentally different, for the first priority is defeating it economically. For that, anything that negatively affects its economy is considered for us a step in the right direction on the path to victory. Military defeats do not greatly effect how we measure total victory, but these defeats indirectly affect the economy which can be demonstrated by the breaching of the confidence of capitalists and investors in this nation's ability to safeguard their various trade and dealings.

> This reveals the importance of the blessed September 11th attacks, which is not that it killed a large number of infidels, but what is more important, the economic effect that this strike achieved. … I will not be exaggerating if I say that striking the Pentagon was purely symbolic and had no noticeable effect on the course of the battle.

The U.S. seems to have fallen into al-Qaeda's trap, as Chris Hedges, former foreign correspondent for *The New York Times*, wrote in February 2011 [72]:

> We are wasting $700 million a day to pay for the wars in Iraq and Afghanistan, while our teachers, firefighters and police lose their jobs, while we slash basic assistance programs for the poor, children and the elderly, while we turn our backs on the some 3 million people being pushed from their homes by foreclosures and bank repossessions and while we do nothing to help the one in six American workers who cannot find work. … [These wars] have turned our nation into an isolated pariah, fueling the very terrorism we seek to defeat. And they cannot be won. The sooner we leave Iraq and Afghanistan the sooner we will save others and finally save ourselves.

The war in Iraq officially ended on December 18, 2011 after a costly, nearly nine-year military engagement. Almost 4,500 U.S. troops had been killed in Iraq since 2003 [73]. The war in Afghanistan began on October 7, 2001 and it has become the longest war in American history, surpassing Iraq (8 years and 7 months) and Vietnam (8 years and 5 months) [74].

Retired Lt. Gen. Robert Gard and Brig. Gen. John Johns wrote a CNN article in December 2012, "In the last decade, America fought two expensive wars and Congress has yet to pay for them; that policy has contributed to our precarious economic position. … Cutting Pentagon spending recognizes that national security is more than military power. The United States is stronger with a strong economy, sustainable jobs, investment in education, renewal of our infrastructure and a sensible energy strategy. Continuing to waste money when our nation should have other priorities is bad policy and bad for security" [75].

In December 2012, U.S. Treasury Secretary Timothy Geithner said that the federal government had already reached the debt ceiling of nearly $16.4 trillion [76]. In October 2014, Todd Harrison of the Center for Strategic and Budgetary

Assessments estimated that the annual bill for U.S. military operations would range from $4 billion to $22 billion [77].

The war on terrorism is largely funded by U.S. taxpayers, but who is funding the terrorists? Professor Michael Soussan of New York University and Elizabeth Weingarten of New America explained, "The money that funded the rise of ISIS and other terrorist organizations came from countries that we now claim are allies, but which embrace religious ideologies that are similar in theory, if not always as openly brutal in practice (or at least on YouTube videos) as ISIS. Wahhabi and Salafi ideologies on the Sunni side, and the Iranian Ayatollahs' own brand of systematic suppression of female liberties, all prevail in the states that gave birth to extremist groups. By aiding states that were actively supporting groups that merged with ISIS, the United States seems to have favored the short-term alignment of interests over the long-term alignment of principles" [78].

To make matters worse, two main rebel groups — Harakat Hazm and the Syrian Revolutionary Front (SRF) — that received heavy weapons and military training from the United States to fight ISIS had surrendered to al Qaeda in November 2014 [79]. Syria's moderate rebels were no match for the well-funded jihadists. According to Todd Harrison, a senior fellow at the Center for Strategic and Budgetary Assessments, the United States ended up sending $30,000-bombs to destroy 41 Humvees captured by ISIS forces. With the cost of about $250,000 per armored vehicle, the total equipment loss was at least $10 million [80].

Also in November 2014, ISIS announced its plan to mint its own currency in gold, silver, and copper. Jimmy Gurule, a former U.S. Treasury undersecretary, told CNN's Jake Tapper, "Unlike its predecessor al Qaeda, which raised money principally from external sources ... ISIS is principally, primarily self-funded. This is the wealthiest terrorist organization that the world has ever known, and so with that kind of money it's hard to understand — what's the potential? What could they do with that? The difficulty, of course, with that kind of money is you can't just put that money in shoe boxes and place it under your mattress. It has to enter into the financial system at some point in time. So I think the Treasury needs to be focusing on banks — banks in Qatar for example, and in Kuwait — that may be the recipients and handling money for ISIS" [81].

British American broadcast journalist Ted Koppel wrote an eye-opening opinion piece in *The Wall Street Journal* in August 2013, "Terrorism, after all, is designed to produce overreaction. It is the means by which the weak induce the powerful to inflict damage upon themselves — and al Qaeda and groups like it are surely counting on that as the centerpiece of their strategy. It appears to be working. ... Will terrorists kill innocent civilians in the years to come? Of course. ... The challenge that confronts us is how we will live with that threat. We have created an economy of fear, an industry of fear, a national psychology of fear. Al Qaeda could never have achieved that on its own. We have inflicted it on ourselves" [82].

3.7 Inside the Minds of Terrorists and Their Sympathizers

Jeffrey Swanson, a professor in psychiatry and behavioral sciences at Duke University's School of Medicine, debunked a common misconception about terrorists being social misfits afflicted with psychological disorders. Swanson explained that "based on the best available scientific evidence on the link between violence and mental illness in populations, most violence is not caused by a major psychiatric condition like schizophrenia, bipolar disorder, or depression" [83]. Terrorists are different from those sociopathic shooters at Columbine High School, Virginia Tech, Aurora Colorado Century 16 movie theater, or Sandy Hook Elementary School.

On the contrary, Colonel Philip G. Wasielewski reported in *Joint Force Quarterly* that "an analysis of over 150 al-Qaeda terrorists displayed a norm of middle- to upper-class, highly educated, married, middle-aged men" [84]. Women are also appearing in increasing numbers, and have been significant actors in rebel groups such as the Tamil Tigers in Sri Lanka [85]. An 18-year-old woman was arrested at Stansted Airport near London on suspicion of terrorism offenses [86]; a young schoolteacher joined the Khansa'a Brigade — an all-female police for ISIS in Syria [87]; and France's most wanted woman, Hayat Boumedienne, was an accomplice to the Charlie Hebdo shooting in Paris [88].

Even more sadly, young adolescents and children are being manipulated by adults to join armed conflicts in war and terrorism as child soldiers [89]. Kids as young as eight have been used as bombers in Pakistan [90]. A 13-year-old Nigerian girl was coerced by her father to become a suicide bomber [91]. In February 2015, FBI counter terrorism chief Michael Steinbach told *CNN* that ISIS had been targeting and recruiting American children as young as 15 in the United States [92].

Why are innocent children being drawn into violence? Ed Husain, senior fellow for Middle Eastern studies at the Council on Foreign Relations (CFR), wrote an opinion piece on *CNN* back in November 2012: "Al-Jazeera Arabic gives prominence to the popular Egyptian Muslim Brotherhood cleric Yusuf al-Qaradawi, who has repeatedly called suicide bombings against Israelis not terrorism, but 'martyrdom.' He argues that since Israelis all serve in the military, they are not civilians. Even children, he despicably argues, are not innocent. They would grow up to serve in the military. Qaradawi is not alone. I can name tens of Muslim clerics, important formulators of public opinion in a region dominated by religion, that will readily condemn acts of terrorism against the West, but will fall silent when it comes to condemning Hamas or Islamic Jihad. Put simply, support for violent resistance against Israel among Arab and Muslim-majority countries — including allies of the United States such as Qatar, Saudi Arabia, Egypt, Turkey, Tunisia — remains popular" [93].

Peter Bergen, CNN's national security analyst and vice president at the New America Foundation, reported in September 2014 that according to the British Office of Security and Counter-Terrorism, the total number of Westerners who have fought in Syria is between 2,620 and 2,870, which included 700 fighters from France; 400 from Germany; between 300 and 500 from Belgium; 130 from the Netherlands; over 100 from Denmark; 100 from Austria; 80 from Sweden; between 50 and 100 from Spain, and about 100 from Canada [94].

Another 100 Americans have fought in Syria, become suicide bombers, or been arrested before they could get there [95]. U.S.-born Omar Hammami — who turned into an Islamist terrorist — wrote in his autobiography, "The real fear that the Americans feel when they see an American in Somalia talking about Jihad, is not how skillful he is at sneaking back across the borders with nuclear weapons. The Americans fear that their cultural barrier has been broken and now Jihad has become a normal career choice for any youthful American Muslim" [96]. The FBI has been working around the clock to combat terrorists' recruitment of American youths [92].

To understand terrorism is to comprehend the minds of terrorists and their sympathizers. They want their voice heard so badly at the expense of innocent lives including men, women, and children of all nationalities. Terrorists and their sympathizers are often self-righteous and motivated by ideologies, politics, religions, and vengeance, either singly or in combination:

1. **Osama bin Laden** was once considered an American ally during the Soviet war in Afghanistan between December 1979 and February 1989. Not only did bin Laden and his jihad fighters receive American and Saudi funding, some analysts believe that bin Laden himself had security training from the CIA [97]. In January 1991, the U.S. led a coalition force in Operation Desert Storm against Iraq in response to Iraq's invasion of Kuwait [98]. Bin Laden became furious, especially about the continuing U.S. military presence in his birthplace Saudi Arabia long after the Persian Gulf War ended in February 1991. In his 1996 fatwā "Declaration of War against the Americans Occupying the Land of the Two Holy Places," bin Laden wrote [99]:

   Terrorizing you, while you are carrying arms on our land, is a legitimate and morally demanded duty. ….. our problem will be how to restrain our youths to wait for their turn in fighting and in operations. … They stood up tall to defend the religion; at the time when the government misled the prominent scholars and tricked them into issuing fatwās of opening the land of the two Holy Places for the Christians armies and handing the Al-Aqsa Mosque to the Zionists. The youths hold you responsible for all of the killings and evictions of the Muslims and the violation of the sanctities, carried out by your Zionist brothers in Lebanon; you openly supplied them with arms and finance. More than 600,000 Iraqi children have died due to lack of food and medicine and as a result of the unjustifiable aggression (sanction) imposed on Iraq and its nation. The children of Iraq are our children. You, the USA, together with the Saudi regime are responsible for the shedding of the blood of these innocent children.

   As the Twin Towers were collapsing on September 11, 2001 in Manhattan, New York City, many Palestinians at refugee camps in Lebanon cheered the attacks, and children in east Jerusalem were seen celebrating news of the atrocity, according to BBC News [100].

2. **Timothy McVeigh**, former U.S. Army platoon leader who once served in the Persian Gulf, bombed the Murrah Federal Building in Oklahoma City in 1995 on the second anniversary of the Waco siege on April 19 [64]. To justify his criminal act that caused the death of 168 people including 19 children,

McVeigh authored "An Essay on Hypocrisy" from a federal maximum-security prison and he wrote a letter to *Fox News* in April 2001 [101]:

I chose to bomb a federal building because such an action served more purposes than other options. Foremost, the bombing was a retaliatory strike; a counter attack, for the cumulative raids (and subsequent violence and damage) that federal agents had participated in over the preceding years (including, but not limited to, Waco.) ... our government - like the Chinese - was deploying tanks against its own citizens. ... Bombing the Murrah Federal Building was morally and strategically equivalent to the U.S. hitting a government building in Serbia, Iraq, or other nations.

3. **Eric Rudolph**, a U.S. Army veteran and self-described Catholic, conducted a series of terrorist acts across southeastern U.S. between 1996 and 1998. He confessed to the 1996 Centennial Olympic Park bombing in Atlanta as well as bombings at a gay nightclub and two abortion clinics [102]. Michael Barkun, professor emeritus of political science at Syracuse University, called Rudolph a "Christian terrorist" [103]. In his April 2005 statement [104] and letters to his mother from behind bars [105], Rudolph offered his reasons:

In the summer of 1996, the world converged upon Atlanta for the Olympic Games. Under the protection and auspices of the regime in Washington millions of people came to celebrate the ideals of global socialism. ... even though the purpose of the Olympics is to promote these despicable ideals, the purpose of the attack on July 27th was to confound, anger and embarrass the Washington government in the eyes of the world for its abominable sanctioning of abortion on demand.

However wrongheaded my tactical decision to resort to violence may have been, morally speaking my actions were justified. ... I don't believe that a government that sanctions, protects and promotes the murder of 50 million unborn children in the heinous practice of abortion has the moral authority to judge a man for murder.

4. **Anders Behring Breivik**, a Norwegian native and self-proclaimed Christian, bombed government buildings in Oslo and shot 85 people dead at a youth camp held by the ruling Labor party on Utoeya Island in July 2011 [106]. Before carrying out the hideous acts, Breivik posted on Facebook to his 7,000+ friends a massive 1,516-page manifesto titled "2083: A European Declaration of Independence" [107], in which he plagiarized the Unabomber manifesto "Industrial Society and Its Future" by American terrorist Ted Kaczynski [108]. Breivik espoused a right-wing, anti-Islam, and anti-immigrant ideology [109]:

Let's end the stupid support for the Palestinians that the Eurabians have encouraged, and start supporting our cultural cousin, Israel. ... I believe Europe should strive for: A cultural conservative approach where monoculturalism, moral, the nuclear family, a free market, support for Israel and our Christian cousins of the east, law and order and Christendom itself must be central aspects (unlike now). Islam must be re-classified as a political ideology and the Quran and the Hadith banned as the genocidal political tools they are.

5. **Anwar al-Awlaki**, an American-born al-Qaeda cleric, was one of the most wanted terrorists. After a CIA drone strike killed al-Awlaki in Yemen in September 2011, his widow **Aminah**, a white European Muslim convert, vowed revenge on the United States. Aminah said, "I would be making a martyr operation, but Sheikh Basir al-Wuhayshi, the emir of AQAP [al-Qaeda in the Arabian

Peninsula], said that the sisters so far [can]not carry out operations because it will mean a lot of problems for them … so I cannot perform operation. … I want to be killed the same way as my husband was … Insha'Allah" [110].

Aminah is one among many terrorist sympathizers. Um Saad, a middle-aged woman in the Sunni district of Khadra in west Baghdad, lost her husband and two of her sons in a series of unfortunate events during "the surge" in Iraq in 2007. Saad exclaimed, "[The Americans] are monsters and devils wearing human clothes. One day I will put on an explosive belt under my clothes and then blow myself up among the Americans. I will get revenge against them for my husband and sons and I will go to paradise [111]."

6. 19 and 26-year-old brothers **Dzhokhar** and **Tamerlan Tsarnaev** carried out the Boston Marathon bombings on April 15, 2013 that killed an 8-year-old boy, a Boston University grad student, and a restaurant manager as well as wounded more than 260 people [112]. Citing the U.S. wars in Afghanistan and Iraq as motivating factors [113], the brothers were self-radicalized via the Internet [114]. Tamerlan Tsarnaev created a YouTube channel in August 2012 and uploaded videos from militant preachers. In one of the videos, Lebanese-Australian Feiz Muhammad who is an open sympathizer with al-Qaeda said, "The war on terrorism is nothing but a war on Islam and on Muslims" [115].

   Dzhokhar Tsarnaev was on the cover of the August 1st, 2013 issue of *Rolling Stone* with the cover story entitled "The Bomber: How a Popular, Promising Student Was Failed by His Family, Fell into Radical Islam and Became a Monster" [116]. The 19-year-old University of Massachusetts Dartmouth sophomore justified the atrocity in his writing: "The bombings were in retribution for the U.S. crimes in places like Iraq andAfghanistan [and] that the victims of the Boston bombing were collateral damage, in the same way innocent victims have been collateral damage in U.S. wars around the world. Summing up, that when you attack one Muslim you attack all Muslims" [117].

7. 32 and 34-year-old brothers **Chérif** and **Saïd Kouachi** were born, raised, and radicalized in Paris, France [118]. They were upset by images of American soldiers humiliating Muslims at the Abu Ghraib prison. In 2005, the younger brother was arrested by the French police on the day of his departure flight to join the jihadists in Iraq. While in prison, he met and became an acolyte of a top al-Qaeda operative. "He made him feel important, he listened to him, recognized him as an individual … Chérif Kouachi was fragile, looking for a family … he didn't have a family he could turn to for support," said an anonymous source. French anthropologist Dounia Bouzar explained, "A fundamentalist discourse is more easily latched on to by people who don't feel important … It's about transferring a feeling of malaise into a feeling of being all powerful. They become someone important."

   On January 7, 2015 after a decade of radicalization and terrorist training, Chérif and Saïd Kouachi forced their way into the offices of the French satirical weekly magazine *Charlie Hebdo* in Paris, killing 12 people including magazine editor Stéphane "Charb" Charbonnier, 4 cartoonists, 2 columnists, and a copy editor

[119]. However, they spared the life of magazine writer Sigolène Vinson, who later recounted her terrifying encounter with Saïd Kouachi: "I looked at him. He had big dark eyes, a gentle look. I felt he was slightly troubled, like he was searching for my name. He said 'don't be afraid, calm down. I won't kill you. You're a woman, we don't kill women. But think about what you do, what you do is bad. I'm sparing you and because I've spared you, you will read the Quran'" [120].

An amateur video of the assailants' gunfight with the police showed the men shouting, "We have avenged the Prophet Muhammad. We have killed Charlie Hebdo!" [121]. Vincent Ollivier, Chérif Kouachi's lawyer in his 2005 arrest, told a French newspaper, "He was a lost kid who was scared to death. I will never know if the person he became is the result of his time in jail, or else the result of a hardening of his commitment" [122].

Although the atrocity was condemned by most heads of state including Palestinian Authority President Mahmoud Abbas, protestors in Algeria, Jordan, Pakistan, Somalia, Syria, Turkey, and other countries demonstrated against *Charlie Hebdo* [123]. More than 45 churches along with a Christian school and orphanage were burned down by violent protestors in Niger [124]. Turkey's Deputy Prime Minister Yalcin Akdogan wrote on Twitter, "Those who are publishing figures referring to our supreme Prophet are those who disregard the sacred. [Such a move is] open incitement and provocation" [125]. A Muslim woman also tweeted, "Still confused by how supporting publishing material 1.6 billion peaceful people find personally offensive can bring anyone joy. 1.6 billion Muslims want to express solidarity for a crime, and yet you are telling them that to do so they must accept ridicule. No" [126].

3.8 Questioning Terrorism and Destroying Stereotypes

Nigerian novelist Chimamanda Ngozi Adichie told the audience at TEDGlobal 2009: "The single story creates stereotypes, and the problem with stereotypes is not that they are untrue, but that they are incomplete. They make one story become the only story" [127].

Paul R. Pillar, former national intelligence officer for the Near East and South Asia, reported that there has been a global decline of state-sponsored international terrorism since the end of the Cold War [128]. However, MIT Professor Emeritus Noam Chomsky opines that the United States is a leading terrorist state. In a November 2001 interview by David Barsamian, Chomsky cited Nicaragua as an example: In 1984, the U.S. was "the only country that was condemned for international terrorism by the World Court and that rejected a Security Council resolution calling on states to observe international law" [129].

Going back in history to the 1950's, *The New York Times* published in April 2000 a CIA document written in 1954 on "Operation Ajax" (aka TPAJAX) — a successful plot to overthrow the democratically-elected Iranian Prime Minister Mohammad Mossadegh. The document shows that "Iranians working for the CIA and posing as Communists harassed religious leaders and staged the bombing of one cleric's home in a campaign to turn the country's Islamic religious community against Mossadegh's government" [130].

Chomsky also blamed the Reagan administration for the 1985 Beirut car bombing near the home of Mohammad Hussein Fadlallah in a failed assassination attempt that instead killed about 60 and injured more than 200 civilians [131]. If the allegation was true, the administration would be in direct violation of President Reagan's Executive Order 12333 stating that "no person employed by or acting on behalf of the United States Government shall engage in or conspire to engage in assassination" [132].

Two years prior in 1983, the Beirut barracks bombing killed 220 U.S. Marines, 21 Americans, and 58 French paratroopers [133]. The United States categorized the bombing an act of terrorism towards off-duty American servicemen, but the Hezbollah regarded the Americans and Frenchmen as enemy combatants stationing in Beirut.

Fast forward to the post 9/11 era, one might argue that CIA drone attacks are state-sponsored assassinations. However, the U.S. Congress approved the Authorization for the Use of Military Force (AUMF), giving the President the authority to use "all necessary and appropriate force against those nations, organizations, or persons he determines planned, authorized, committed, or aided the terrorist attacks that occurred on September 11, 2001, or harbored such organizations or persons, in order to prevent any future acts of international terrorism against the United States by such nations, organizations or persons" [134].

Collateral damages from CIA drone strikes are pale in comparison to economic sanctions. In May 1996, CBS reporter and former White House correspondent Lesley Stahl asked U.S. Secretary of State Madeleine Albright on *60 Minutes*, "We have heard that half a million children have died [during the U.N. sanctions against Iraq]. I mean, that's more children than died in Hiroshima. And — and you know, is the price worth it?" Albright replied, "I think this is a very hard choice, but the price — we think the price is worth it" [135]. In her autobiography *Madam Secretary: A Memoir*, Albright regretted making that statement and she wrote, "My reply had been a terrible mistake… Nothing matters more than the lives of innocent people" [136].

After the 9/11 terrorist attacks, *The New York Times* reported in September 2001 that the Bush administration demanded Pakistan to cut off fuel supplies and eliminate truck convoys that provide much of the food and medicines to Afghanistan's civilian population [137].

Chomsky considered the U.S. sanctions and many CIA covert operations to be equivalent to state-sponsored terrorism: "The U.S. is officially committed to what is called 'low-intensity warfare.' That's the official doctrine. If you read the definition of low-intensity conflict in Army manuals and compare it with official definitions of 'terrorism' in Army manuals, or the U.S. Code, you find they're almost the same. Terrorism is the use of coercive means aimed at civilian populations in an effort to achieve political, religious, or other aims" [129].

HBO's *Real Time with Bill Maher* host wrote in his blog on November 30, 2012: "We utilize the best means at our disposal to go into foreign lands and blow up the people we consider the bad guys even if that means collateral damage in the form of civilian casualties. When someone does that exact same thing to us, don't we call it 'terrorism'?" [138].

In July 2014, North Korea filed a complaint with the United Nation over Seth Rogen and James Franco's upcoming movie *The Interview* about a hapless TV host recruited by the CIA to assassinate North Korean dictator Kim Jong-un. "To allow the production and distribution of such a film on the assassination of an incumbent head of a sovereign state should be regarded as the most undisguised sponsoring of terrorism as well as an act of war," North Korean U.N. Ambassador Ja Song Nam wrote in the letter to the United Nation [139].

The saga is somewhat reminiscent of CIA's psychological warfare during the height of the Cold War in 1954 when the CIA secretly funded the film version of George Orwell's *Animal Farm* as propaganda against communism and Joseph Stalin [140]. One should not underestimate the power of persuasion by well-crafted propaganda films. D.W. Griffith's *The Birth of a Nation* in 1915 helped to resurrect Ku Klux Klan in Georgia; and Leni Riefenstahl's award-winning *Triumph of the Will* in 1935 helped fuel the rise of Nazism in Germany.

In January 2015, a Jewish ultra-orthodox newspaper Hamevaser in Israel digitally removed Germany's female chancellor Angela Merkel from a photo of the Paris unity march against terrorism after the Charlie Hebdo attacks [141]. Media manipulation is commonplace. For people who do not look outside their comfort zone, they will never learn the whole truth. If scientists had not challenged the status quo, we would not have enjoyed modern medicine and technological innovations today. As Dave Skylark (played by James Franco) told Aaron Rapaport (played by Seth Rogen) in the 2014 action comedy *The Interview*, "Maybe the media is manipulating you."

In the words of novelist Chimamanda Ngozi Adichie, our lives and our cultures are composed of many overlapping stories, and if we hear only a single story about another person or country, we risk a critical misunderstanding. Antonin Scalia is the longest-serving U.S. Supreme Court Justice appointed by President Ronald Reagan in 1986. In a 2013 interview with *New York Magazine*, Scalia said that he ditched *The Washington Post* and *The New York Times* because they just "went too far for me. I couldn't handle it anymore. It was the treatment of almost any conservative issue. It was slanted and often nasty. And, you know, why should I get upset every morning? I don't think I'm the only one" [142]. He is certainly not the only one. Too many people prefer to stay inside their own comfort zones with a one-sided liberal or conservative sentiment.

Bibliography

1. **Junod, Tom.** Angelina Jolie Dies for Our Sins. [Online] Esquire, July 20, 2010. http://www.esquire.com/women/women-we-love/angelina-jolie-interview-pics-0707.
2. **Eco, Umberto.** Inventing the Enemy: Essays. [Online] Houghton Mifflin Harcourt, September 4, 2012. http://books.google.com/books?id=mFbYcy8uhCUC&pg=PA18&lpg=PA18.

3. **Drug Policy Alliance.** Drug War Statistics. [Online] Drug Policy Alliance. [Cited: December 31, 2014.] http://www.drugpolicy.org/drug-war-statistics.

4. **Fink, Erica.** When Silicon Valley takes LSD. [Online] CNN, January 25, 2015. http://money.cnn.com/2015/01/25/technology/lsd-psychedelics-silicon-valley/index.html.

5. **National Institute on Drug Abuse.** DrugFacts: Understanding Drug Abuse and Addiction. [Online] National Institute on Drug Abuse, November 2012. http://www.drugabuse.gov/publications/drugfacts/understanding-drug-abuse-addiction.

6. **Office of the Press Secretary.** At O'Hare, President Says "Get On Board". [Online] The White House, September 27, 2001. http://georgewbush-whitehouse.archives.gov/news/releases/2001/09/20010927-1.html.

7. **Childress, Sarah, Wexler, Evan and Rockwood, Bill.** The Iraq War: How We Spent $800 Billion (and Counting). [Online] PBS, March 18, 2013. http://www.pbs.org/wgbh/pages/frontline/iraq-war-on-terror/the-iraq-war-how-we-spent-800-billion-and-counting/.

8. **National Priorities Project.** Cost of National Security. [Online] National Priorities Project. [Cited: January 1, 2015.] https://www.nationalpriorities.org/cost-of/.

9. **Miller, Greg and Whitlock, Craig.** U.S. weakens al-Qaeda groups around the world but hasn't wiped any out. [Online] The Washington Post, September 11, 2014. http://www.washingtonpost.com/world/national-security/us-weakens-al-qaeda-groups-around-the-world-but-hasnt-wiped-any-out/2014/09/11/3c28d626-39bb-11e4-8601-97ba88884ffd_story.html.

10. **Acosta, Jim and Diamond, Jeremy.** Obama ISIS fight request sent to Congress. [Online] CNN, February 11, 2015. http://www.cnn.com/2015/02/11/politics/isis-aumf-white-house-congress/index.html.

11. **Schmidt, Michael S.** Islamic State Issues Video Challenge to Obama. [Online] The New York Times, September 17, 2014. http://www.nytimes.com/2014/09/17/world/middleeast/isis-issues-video-riposte-to-obama.html.

12. France: Raids kill 3 suspects, including 2 wanted in Charlie Hebdo attack. [Online] CNN, January 9, 2015. http://www.cnn.com/2015/01/09/europe/charlie-hebdo-paris-shooting/index.html.

13. **Brown, Pamela.** U.S. official on terror attacks: 'This isn't going to stop'. [Online] CNN, January 9, 2015. http://www.cnn.com/2015/01/09/us/france-attacks-u-s-/index.html.

14. **Thompson, Mark.** Anonymous declares war over Charlie Hebdo attack. [Online] CNN, January 9, 2015. http://money.cnn.com/2015/01/09/technology/anonymous-charlie-hebdo-terrorists/index.html.

15. **Shariatmadari, David.** Patrick Cockburn: 'An effective terrorist attack requires the complicity of governments'. [Online] The Guardian, January 24, 2015. http://www.theguardian.com/culture/2015/jan/24/patrick-cockburn-the-rise-of-the-islamic-state-books-interview-isis.

16. **Yates, Stephen.** Back to basics on terrorism. [Online] CNN, May 7, 2013. http://globalpublicsquare.blogs.cnn.com/2013/05/07/so-who-is-our-enemy/.

17. **Bohlen, Celestine.** In New War on Terrorism, Words Are Weapons, Too. [Online] The New York Times, September 29, 2001. http://www.nytimes.com/2001/09/29/arts/think-tank-in-new-war-on-terrorism-words-are-weapons-too.html.

18. **ABC News.** Maher Apologizes for 'Cowards' Remark. [Online] ABC News, September 20, 2001. http://abcnews.go.com/Entertainment/story?id=102318&page=1.

19. **Pleitgen, Frederik.** Author's journey inside ISIS: They're 'more dangerous than people realize'. [Online] CNN, January 4, 2015. http://www.cnn.com/2014/12/22/world/meast/inside-isis-juergen-todenhoefer/index.html.

20. **Bergen, Peter and Rowland, Jennifer.** A dangerous new world of drones. [Online] CNN, October 8, 2012. http://www.cnn.com/2012/10/01/opinion/bergen-world-of-drones/index.html.

21. **Weinberger, Sharon.** Pentagon's Tiny New Spy Drone Mimics Hummingbird. [Online] AOL News, February 18, 2011. http://www.aolnews.com/2011/02/18/pentagons-tiny-new-spy-drone-mimics-hummingbird/.

22. **Levine, Adam.** Obama admits to Pakistan drone strikes. [Online] CNN, January 30, 2012. http://security.blogs.cnn.com/2012/01/30/obama-admits-to-pakistan-drone-strikes/.

23. **The Economist.** Flight of the drones: Why the future of air power belongs to unmanned systems. [Online] The Economist, October 8, 2011. http://www.economist.com/node/21531433.

24. **Jackson, David.** Obama defends drone strikes. [Online] USA Today, January 31, 2012. http://content.usatoday.com/communities/theoval/post/2012/01/obama-defends-drone-strikes/1.

25. **Kelly, Suzanne.** Deadly drones and the classified conundrum. [Online] CNN, May 23, 2012. http://security.blogs.cnn.com/2012/05/23/deadly-drones-and-the-classified-conundrum/.

26. **Benson, Pam.** Yemen plot exposes new world of U.S. spying. [Online] CNN, May 11, 2012. http://security.blogs.cnn.com/2012/05/11/yemen-plot-exposes-new-world-of-us-spying/.

27. **Mazzetti, Mark, Schmitt, Eric and Worth, Robert F.** Two-Year Manhunt Led to Killing of Awlaki in Yemen. [Online] The New York Times, September 30, 2011. http://www.nytimes.com/2011/10/01/world/middleeast/anwar-al-awlaki-is-killed-in-yemen.html?pagewanted=all.

28. **Bergen, Peter and Rowland, Jennifer.** Did Obama keep his drone promises? [Online] CNN, October 25, 2013. http://www.cnn.com/2013/10/25/opinion/bergen-drone-promises/index.html.

29. **Almasmari, Hakim.** Yemen says U.S. drone struck a wedding convoy, killing 14. [Online] CNN, December 14, 2013. http://www.cnn.com/2013/12/12/world/meast/yemen-u-s-drone-wedding/.

30. **New America Foundation.** Drone Wars Pakistan: Analysis. [Online] New America Foundation. [Cited: January 18, 2015.] http://securitydata.newamerica.net/drones/pakistan/analysis.

31. **Fabi, Randy and Chowdhry, Aisha.** Obama victory infuriates Pakistani drone victims. [Online] Chicago Tribune, November 8, 2012. http://articles.chicagotribune.com/2012-11-08/news/sns-rt-us-usa-campaign-pakistanbre8a70a0-20121107_1_drone-strikes-drone-program-pakistani-taliban.

32. **Ackerman, Spencer.** Military's Own Report Card Gives Afghan Surge an F. [Online] Wired, September 27, 2012. http://www.wired.com/dangerroom/2012/09/surge-report-card/.

33. **Obama, Barack.** Obama Outlines Plan for Afghanistan Troop Surge. [Online] PBS Newshour, December 1, 2009. http://www.pbs.org/newshour/updates/military/july-dec09/obamapeech_12-01.html.

34. **Nordland, Rod.** Troop 'Surge' in Afghanistan Ends With Mixed Results. [Online] The New York Times, September 21, 2012. http://www.nytimes.com/2012/09/22/world/asia/us-troop-surge-in-afghanistan-ends.html?pagewanted=all.

35. **Robertson, Nic.** In Swat Valley, U.S. drone strikes radicalizing a new generation. [Online] CNN, April 15, 2013. http://www.cnn.com/2013/04/14/world/asia/pakistan-swat-valley-school/index.html.

36. **Hayden, Michael.** Should CIA be targeting terrorists? [Online] CNN, April 18, 2013. http://www.cnn.com/2013/04/18/opinion/hayden-cia-role/index.html.

37. **Bridle, James.** Dronestagram: The drone's-eye view. [Online] Instagram, October 2012. http://instagram.com/dronestagram.

38. **Boyette, Chris.** Dronestagram uses social media to highlight drone strikes. [Online] CNN, February 15, 2013. http://www.cnn.com/2013/02/14/tech/dronestagram/index.html.

39. **Gutteridge, Jessica.** 'You have given al Qaeda a golden ticket'. [Online] CNN, May 16, 2013. http://globalpublicsquare.blogs.cnn.com/2013/05/16/you-have-given-al-qaeda-a-golden-ticket/?hpt=hp_bn2.

40. **Almasmari, Hakim, Jamjoom, Mohammed and Brumfield, Ben.** More suspected al Qaeda militants killed as drone strikes intensify in Yemen. [Online] CNN, August 8, 2013. http://www.cnn.com/2013/08/08/world/meast/yemen-drone-strike/index.html.

41. **CNN Political Unit.** Malala confronts Obama. [Online] CNN, October 11, 2013. http://politicalticker.blogs.cnn.com/2013/10/11/obamas-meet-with-malala/.

42. **Hamdani, Yasser Latif.** The litmus test . [Online] Daily Times, February 12, 2015. http://www.dailytimes.com.pk/opinion/12-Jan-2015/the-litmus-test.

43. **Botelho, Greg.** Iraqi leader: U.S.-led military campaign against ISIS welcome if 'they do it right'. [Online] CNN, September 23, 2014. http://www.cnn.com/2014/09/23/world/meast/iraq-prime-minister/index.html.

44. **Arlington National Cemetery.** Medgar Wiley Evers. [Online] Arlington National Cemetery Website, October 11, 2009. http://www.arlingtoncemetery.net/mwevers.htm.

45. **Amanpour, Christiane.** The Bravest Girl in the World. [Online] CNN, October 11, 2013. http://amanpour.blogs.cnn.com/2013/10/11/the-bravest-girl-in-the-world/.

46. **Wilson, Scott and Kamen, Al.** 'Global War On Terror' Is Given New Name. [Online] March 25, 2009. http://www.washingtonpost.com/wp-dyn/content/article/2009/03/24/AR2009032402818.html.

47. **The 9-11 Commission.** The 9/11 Commission Report: Final Report of the National Commission on Terrorist Attacks Upon the United States (9/11 Report). [Online] U.S. Congress, July 22, 2004. http://www.9-11commission.gov/report/index.htm.

48. **Wright, Lawrence.** The Spymaster. Can Mike McConnell fix America's intelligence community? [Online] The New Yorker, January 21, 2008. http://www.newyorker.com/reporting/2008/01/21/080121fa_fact_wright?currentPage=all.

49. —. The Agent. Did the C.I.A. stop an F.B.I. detective from preventing 9/11? [Online] The New Yorker, July 10, 2006. http://www.newyorker.com/archive/2006/07/10/060710fa_fact_wright?currentPage=all.

50. **BBC.** Gen McChrystal: Bin Laden is key to al-Qaeda defeat. [Online] BBC News, December 9, 2009. http://news.bbc.co.uk/2/hi/americas/8402138.stm.

51. **Scheuer, Michael F.** Bill and Dick, Osama and Sandy. [Online] The Washington Times, July 4, 2006. http://www.washingtontimes.com/news/2006/jul/4/20060704-110004-4280r/.

52. **Phillips, Macon.** Osama Bin Laden Dead. [Online] The White House Blog, May 2, 2011. http://www.whitehouse.gov/blog/2011/05/02/osama-bin-laden-dead.

53. **msnbc.com staff and news service reports.** Protesters condemn 'brutal killing' of bin Laden. [Online] NBCNews.com, May 6, 2011. http://www.msnbc.msn.com/id/42927020/ns/world_news-death_of_bin_laden/t/protesters-condemn-brutal-killing-bin-laden/.

54. **Madhani, Aamer.** Cleric al-Awlaki dubbed 'bin Laden of the Internet'. [Online] USA Today, August 24, 2010. http://usatoday30.usatoday.com/news/nation/2010-08-25-1A_Awlaki25_CV_N.htm.

55. **Calabresi, Massimo.** The World's Most Dangerous Terrorist. [Online] TIME Magazine, August 5, 2013. http://content.time.com/time/magazine/article/0,9171,2148173,00.html.

56. **Engelhardt, Tom.** Osama Bin Laden's American Legacy: It's Time to Stop Celebrating and Go Back to Kansas . [Online] The Huffington Post, May 5, 2011. http://www.huffingtonpost.com/tom-engelhardt/osama-bin-ladens-american_b_858182.html.

57. **Haglund, David and Thomas, June.** The Riveting Season Finale of Homeland. [Online] Slate, December 19, 2011. http://www.slate.com/blogs/browbeat/2011/12/19/homeland_season_finale_discussing_the_dramatic_end_to_the_showtime_show_s_first_season_.html.

58. **CBS News.** Ex-CIA Operative Comes Out of the Shadows. [Online] CBS News, August 2, 2010. http://www.cbsnews.com/8301-18560_162-6014887.html?pageNum=2.

59. **Riedel, Bruce.** A Stubborn Terror. [Online] Newsweek/The Daily Beast, September 10, 2012. http://www.thedailybeast.com/newsweek/2012/09/09/a-stubborn-terror.html.

60. **Dougherty, Jill and Cohen, Tom.** Clinton takes on Benghazi critics, warns of more security threats. [Online] CNN, January 24, 2013. http://www.cnn.com/2013/01/23/politics/clinton-benghazi/index.html.

61. **Fleishman, Jeffrey.** Algeria hostages reportedly escape captors; some may have been slain. [Online] Los Angeles Times, January 17, 2013. http://www.latimes.com/news/nationworld/world/la-20-foreign-hostages-escape-islamist-captors-in-algeria-20130117,0,7004240.story.

62. **Gingrich, Newt.** We need to think outside the box on ISIS. [Online] CNN, September 4, 2014. http://www.cnn.com/2014/09/03/opinion/gingrich-isis-obama-strategy/index.html.

63. **Potok, Mark.** The Year in Hate and Extremism. Intelligence Report, Issue Number 149. [Online] Southern Poverty Law Center, Spring 2013. http://www.splcenter.org/home/2013/spring/the-year-in-hate-and-extremism.

64. **Federal Bureau of Investigation.** Terror Hits Home: The Oklahoma City Bombing. [Online] Federal Bureau of Investigation. [Cited: December 1, 2012.] http://www.fbi.gov/about-us/history/famous-cases/oklahoma-city-bombing.

65. **Counterterrorism Threat Assessment and Warning Unit, National Security Division.** Terrorism in the United States 1996. [Online] Federal Bureau of Investigation, 1996. http://www.fbi.gov/stats-services/publications/terror_96.pdf.

66. **Granderson, LZ.** Treat Chicago gangs as terrorists. [Online] CNN, April 24, 2013. http://www.cnn.com/2013/04/24/opinion/granderson-chicago-terror/index.html.

67. **Goodman, Marc.** The technological future of crime: Marc Goodman at TEDGlobal 2012. [Online] TED, June 28, 2012. http://blog.ted.com/2012/06/28/the-technological-future-of-crime-marc-goodman-at-tedglobal-2012/.

68. **National Intelligence Council.** Global Trends 2030: Alternative Worlds. [Online] U.S. National Intelligence Council, December 2012. http://www.dni.gov/files/documents/Global Trends_2030.pdf.

69. **Callimachi, Rukmini.** $0.60 for cake: Al-Qaida records every expense . [Online] Associated Press, December 30, 2013. http://bigstory.ap.org/article/060-cake-al-qaida-records-every-expense.

70. **Bin Laden, Osama.** Full transcript of bin Ladin's speech. [Online] Aljazeera, November 1, 2004. http://www.aljazeera.com/archive/2004/11/200849163336457223.html.

71. **Salama, Sammy and Wheeler, David.** From the Horse's Mouth: Unraveling Al-Qa`ida's Target Selection Calculus. [Online] James Martin Center for Nonproliferation Studies (CNS), April 17, 2007. http://cns.miis.edu/stories/070417.htm.

72. **Hedges, Chris.** No Other Way Out. [Online] Common Dreams, February 28, 2011. http://www.commondreams.org/view/2011/02/28-0.

73. **Cutler, David.** Timeline: Invasion, surge, withdrawal; U.S. forces in Iraq. [Online] Reuters, December 18, 2011. http://www.reuters.com/article/2011/12/18/us-iraq-usa-pullout-idUSTRE7BH08E20111218.

74. **Dermody, William.** The Longest War. [Online] USA Today. http://www.usatoday.com/news/afghanistan-ten-years-of-war/index.html.

75. **Gard, Robert G. and Johns, John.** Generals: Get real and cut Pentagon spending. [Online] CNN, December 12, 2012. http://www.cnn.com/2012/12/12/opinion/gard-johns-military-spending/index.html.

76. **CNN Political Unit.** Treasury Department rules out $1 trillion coin. [Online] CNN, January 12, 2013. http://politicalticker.blogs.cnn.com/2013/01/12/treasury-department-rules-out-1-trillion-coin/.

77. **Bilmes, Linda J.** Fighting the Islamic State — how much will it cost? [Online] The Boston Globe, October 8, 2014. http://www.bostonglobe.com/opinion/2014/10/07/fighting-islamic-state-how-much-will-cost/xub6sT2eWP1k67t1HWBsFL/story.html.

78. **Soussan, Michael and Weingarten, Elizabeth.** What really scares terrorists. [Online] CNN, December 26, 2014. http://www.cnn.com/2014/12/26/opinion/soussan-weingarten-gender-equality/index.html.

79. **Sherlock, Ruth.** Syrian rebels armed and trained by US surrender to al-Qaeda. [Online] The Telegraph, November 2, 2014. http://www.telegraph.co.uk/news/worldnews/middleeast/syria/11203825/Syrian-rebels-armed-and-trained-by-US-surrender-to-al-Qaeda.html.

80. **Alesci, Cristina and Trafecante, Kate.** One cost of war: U.S. blowing up its own Humvees. [Online] CNNMoney, September 25, 2014. http://money.cnn.com/2014/09/25/news/f22-raptor-humvee/index.html.

81. **Alkhshali, Hamdi and Ford, Dana.** ISIS announces new currency. [Online] CNN, November 14, 2014. http://www.cnn.com/2014/11/13/world/meast/isis-currency/index.html.

82. **Koppel, Ted.** Ted Koppel: America's Chronic Overreaction to Terrorism. [Online] The Wall Street Journal, August 6, 2013. http://www.wsj.com/articles/SB10001424127887324653004578650462392053732.

83. **Swanson, Jeffrey.** Looking into the minds of killers. [Online] CNN, July 25, 2012. http://www.cnn.com/2012/07/24/opinion/swanson-colorado-shooting/index.html.
84. **Wasielewski, Philip G.** Defining the War on Terror. [Online] Joint Force Quarterly, Issue 44, 1st Quarter 2007, pp. 13-18, 2007. http://www.ndu.edu/press/lib/pdf/jfq-44/JFQ-44.pdf.
85. **Murray, Rebecca.** Scarred by Sri Lanka's war with Tamil Tigers, female ex-fighters build new lives. [Online] The Christian Science Monitor, October 29, 2010. http://www.csmonitor.com/World/Asia-South-Central/2010/1029/Scarred-by-Sri-Lanka-s-war-with-Tamil-Tigers-female-ex-fighters-build-new-lives.
86. **Sky News.** Terror Police Arrest Teen At Stansted Airport. [Online] Sky News, January 16, 2015. http://news.sky.com/story/1409457/terror-police-arrest-teen-at-stansted-airport.
87. **Damon, Arwa and Tuysuz, Gul.** How she went from a schoolteacher to an ISIS member. [Online] CNN, October 6, 2014. http://www.cnn.com/2014/10/06/world/meast/isis-female-fighter/index.html.
88. **Hunter, Isabel.** Paris terror attacks: Turks piece together missing days of deli gunman's partner. [Online] The Independent, January 14, 2015. Paris terror attacks: Turks piece together missing days of deli gunman's partner.
89. **Glazer, Ilsa M.** Armies of the Young: Child Soldiers in War and Terrorism (review). [Online] Anthropological Quarterly, Volume 79, Number 2, Spring 2006, pp. 373-384, 2006. http://dx.doi.org/10.1353%2Fanq.2006.0021.
90. **Mohsin, Saima and Khan, Shaan.** Police: Kids young as 8 used as bombers in Pakistan. [Online] CNN, March 14, 2013. http://www.cnn.com/2013/03/14/world/asia/pakistan-child-bombers/index.html.
91. **Abubakr, Aminu and Almasy, Steve.** Girl, 13: Boko Haram tried to force me to become a suicide bomber. [Online] CNN, January 4, 2015. http://www.cnn.com/2014/12/26/world/africa/nigeria-teenage-girl-suicide-bombing/index.html.
92. **Brown, Pamela and Bruer, Wesley.** FBI official: ISIS is recruiting U.S. teens. [Online] CNN, February 3, 2015. http://www.cnn.com/2015/02/03/politics/fbi-isis-counterterrorism-michael-steinbach/index.html.
93. **Husain, Ed.** Israel, face new reality: Talk to Hamas. [Online] CNN, November 21, 2012. http://www.cnn.com/2012/11/20/opinion/husain-hamas-israel/index.html.
94. **Bergen, Peter.** The British connection to ISIS beheadings . [Online] CNN, September 16, 2014. http://www.cnn.com/2014/09/14/opinion/bergen-british-connection-isis-beheadings/index.html.
95. **Bergen, Peter and Sterman, David.** When Americans leave for jihad. [Online] CNN, August 29, 2014. http://www.cnn.com/2014/08/27/opinion/bergen-sterman-isis-american/index.html.
96. **Somra, Gena.** Parents despair for 'most wanted' terrorist son. [Online] CNN, June 7, 2013. http://www.cnn.com/2013/06/07/us/us-somalia-family-despair/index.html.
97. **BBC News.** Al-Qaeda's origins and links. [Online] BBC News, July 20, 2004. http://news.bbc.co.uk/2/hi/middle_east/1670089.stm.
98. **Chief of Naval Operations.** The United States Navy In "Desert Shield" / "Desert Storm". [Online] Naval History & Heritage, May 15, 1991. http://www.history.navy.mil/wars/dstorm/index.html.
99. **Public Broadcasting Service.** Bin Laden's Fatwa. [Online] PBS Newshour, August 23, 1996. http://www.pbs.org/newshour/updates/military/july-dec96/fatwa_1996.html.
100. **BBC.** In pictures: Atrocities' aftermath. [Online] BBC News, September 12, 2001. http://news.bbc.co.uk/2/hi/americas/1538664.stm.
101. **McVeigh, Timothy.** McVeigh's Apr. 26 Letter to Fox News. [Online] Fox News, April 26, 2001. http://www.foxnews.com/story/0,2933,17500,00.html.
102. **Fonda, Daren.** How Luck Ran Out For A Most Wanted Fugitive. [Online] Time Magazine, June 9, 2003. http://www.time.com/time/magazine/article/0,9171,1004966,00.html.
103. **Olsen, Ted.** Is Eric Rudolph a Christian Terrorist? [Online] Christianity Today, June 1, 2003. http://www.christianitytoday.com/ct/2003/juneweb-only/6-2-22.0.html.

104. **Rudolph, Eric.** Full text of Eric Rudolph's written statement. [Online] Army of God, April 13, 2005. http://www.armyofgod.com/EricRudolphStatement.html.
105. **Morrison, Blake.** Special report: Eric Rudolph writes home. [Online] USA Today, July 5, 2005. http://usatoday30.usatoday.com/news/nation/2005-07-05-rudolph-cover-partone_x.htm.
106. **BBC.** Norway police say 85 killed in island youth camp attack. [Online] BBC News, July 23, 2011. http://www.bbc.co.uk/news/world-europe-14259356.
107. **Boston, William.** Killer's Manifesto: The Politics Behind the Norway Slaughter. [Online] Time Magazine, July 24, 2011. http://www.time.com/time/world/article/0,8599,2084901,00.html.
108. **Breivik, Anders Behring.** Behring Breivik kopierte Una-bomberen . [Online] DOCUMENT. no, July 24, 2011. http://www.document.no/2011/07/behring-breivik-kopierte-una-bomberen/.
109. **Kane, Alex.** Breivik manifesto outlines virulent right-wing ideology that fueled Norway massacre. [Online] Mondoweiss, July 24, 2011. http://mondoweiss.net/2011/07/breivik-manifesto-outlines-virulent-right-wing-ideology-that-fueled-norway-massacre.html.
110. **Cruickshank, Paul, Lister, Tim and Robertson, Nic.** The Danish agent, the Croatian blonde and the CIA plot to get al-Awlaki. [Online] CNN, October 24, 2012. http://www.cnn.com/2012/10/15/world/al-qaeda-cia-marriage-plot/index.html.
111. **Cockburn, Patrick.** Bereaved Iraqi mother vows revenge on US. [Online] The Independent, March 13, 2008. http://www.independent.co.uk/news/world/middle-east/bereaved-iraqi-mother-vows-revenge-on-us-795018.html.
112. **boston.com.** Victims of the Marathon bombings. [Online] boston.com, April 2013. http://www.boston.com/news/local/massachusetts/specials/boston_marathon_bombing_victim_list/.
113. **Pearson, Michael.** Official: Suspect says Iraq, Afghanistan drove Boston bombings. [Online] CNN, April 23, 2013. [Cited:].
114. —. Official: Suspect says Iraq, Afghanistan drove Boston bombings. [Online] CNN, April 23, 2013. http://www.cnn.com/2013/04/23/us/boston-attack/index.html.
115. **Lister, Tim and Cruickshank, Paul.** Dead Boston bomb suspect posted video of jihadist, analysis shows. [Online] CNN, April 22, 2013. http://www.cnn.com/2013/04/20/us/brother-religious-language/index.html.
116. **Reitman, Janet.** Jahar's World. [Online] Rolling Stone, July 17, 2013. http://www.rollingstone.com/culture/news/jahars-world-20130717.
117. **McGovern, Ray.** Boston Suspect's Writing on the Wall. [Online] Consortiumnews.com, May 17, 2013. https://consortiumnews.com/2013/05/17/boston-suspects-writing-on-the-wall/.
118. **Chrisafis, Angelique.** Charlie Hebdo attackers: born, raised and radicalised in Paris. [Online] The Guardian, January 12, 2015. http://www.theguardian.com/world/2015/jan/12/-sp-charlie-hebdo-attackers-kids-france-radicalised-paris.
119. **BBC News.** French terror attacks: Victim obituaries. [Online] BBC News Europe, January 13, 2015. http://www.bbc.com/news/world-europe-30724678.
120. **Willsher, Kim.** Charlie Hebdo killings: 'Don't be afraid. I won't kill you. You're a woman'. [Online] The Guardian, January 14, 2015. http://www.theguardian.com/world/2015/jan/14/charlie-hebdo-killings-survivor-story.
121. **Bilefsky, Dan and Baume, Maïa De La.** Terrorists Strike Charlie Hebdo Newspaper in Paris, Leaving 12 Dead. [Online] The New York Times, January 7, 2015. http://www.nytimes.com/2015/01/08/world/europe/charlie-hebdo-paris-shooting.html.
122. **Callimachi, Rukmini and Yardley, Jim.** From Scared Amateur to Slaughterer. [Online] The New York Times, January 17, 2015. http://www.nytimes.com/2015/01/18/world/europe/paris-terrorism-brothers-said-cherif-kouachi-charlie-hebdo.html?_r=0.
123. **Specia, Megan.** Muslims gather after Friday prayers to protest Charlie Hebdo's latest issue. [Online] Mashable, January 16, 2015. http://mashable.com/2015/01/16/muslims-charlie-hebdo-protest/.
124. **Levs, Josh.** 10 killed, churches torched in protests over Charlie Hebdo. [Online] CNN, January 21, 2015. http://www.cnn.com/2015/01/20/world/charlie-hebdo-violence/index.html.
125. **Levs, Josh, Atay-Alam, Hande and Bilginsoy, Zeynep.** Turkey bans Charlie Hebdo cover, newspaper gets death threats. [Online] CNN, January 14, 2015. http://www.cnn.com/2015/01/14/world/turkey-charlie-hebdo/index.html.

126. **Burke, Daniel.** Muslims' mixed response to new Mohammed cover. [Online] CNN, January 13, 2015. http://www.cnn.com/2015/01/13/living/muslims-respond-hebdo/index.html.

127. **Adichie, Chimamanda.** Chimamanda Adichie: The danger of a single story. [Online] TED, October 2009. http://www.ted.com/talks/chimamanda_adichie_the_danger_of_a_single_story. html.

128. **Pillar, Paul R.** The Decline of State-Sponsored Terrorism. [Online] The Atlantic, May 22, 2012. http://www.theatlantic.com/international/archive/2012/05/the-decline-of-state-sponsored-terrorism/257515/.

129. **Chomsky, Noam.** The United States is a Leading Terrorist State. [Online] Monthly Review, November 2001. http://monthlyreview.org/2001/11/01/the-united-states-is-a-leading-terrorist-state.

130. **Risen, James.** Secrets of History: The C.I.A. in Iran. [Online] The New York Times, April 2000. http://www.nytimes.com/library/world/mideast/041600iran-cia-index.html.

131. **The Guardian.** 60 killed by Beirut car bomb. [Online] The Guardian, March 9, 1985. http://www.guardian.co.uk/theguardian/1985/mar/09/fromthearchive.

132. **Office of the President of the United States.** Executive Order 12333: United States Intelligence Activities. [Online] Central Intelligence Agency, December 4, 1981. https://www.cia.gov/about-cia/eo12333.html#2.11.

133. **Hampson, Rick.** 25years later, bombing in Beirut still resonates. [Online] USA Today, October 18, 2008. http://usatoday30.usatoday.com/news/military/2008-10-15-beirut-barracks_N.htm.

134. **107th Congress Public Law 40.** Authorization for the Use of Military Force (AUMF). [Online] U.S. Government Printing Office, September 18, 2001. http://www.gpo.gov/fdsys/pkg/PLAW-107publ40/html/PLAW-107publ40.htm.

135. **David, Leigh and Wilson, James.** Counting Iraq's victims. [Online] The Gudardian, October 10, 2001. http://www.guardian.co.uk/world/2001/oct/10/iraq.socialsciences.

136. **Albright, Madeleine Korbel and Woodward, William.** Madam Secretary: A Memoir. [Online] Miramax Books, April 6, 2005. http://books.google.com/books?id=RBuEq2f5U_QC.

137. **Burns, John F.** AFTER THE ATTACKS: IN ISLAMABAD; Pakistan Antiterror Support Avoids Vow of Military Aid. [Online] The New York Times, September 16, 2001. http://www.nytimes.com/2001/09/16/us/after-attacks-islamabad-pakistan-antiterror-support-avoids-vow-military-aid.html?src=pm&pagewanted=all.

138. **Maher, Bill.** Spacial Delivery. [Online] HBO, November 30, 2012. http://www.real-time-with-bill-maher-blog.com/real-time-with-bill-maher-blog/2012/11/30/spacial-delivery.html.

139. **Gettell, Oliver.** North Korea files U.N. complaint over Seth Rogen film "The Interview". [Online] Los Angeles Times, July 9, 2014. http://www.latimes.com/entertainment/movies/moviesnow/la-et-mn-north-korea-the-interview-un-complaint-20140709-story.html.

140. **Chilton, Martin.** How the CIA brought Animal Farm to the screen. [Online] The Telegraph, November 5, 2014. http://www.telegraph.co.uk/culture/film/11209390/How-the-CIA-brought-Animal-Farm-to-the-screen.html.

141. **Associated Press in Jerusalem.** Israeli newspaper edits out Angela Merkel from front page on Paris march . [Online] The Guardian, January 14, 2015. http://www.theguardian.com/world/2015/jan/14/israeli-newspaper-hamevaser-merkel-women-charlie-hebdo-rally.

142. **Senior, Jennifer.** In Conversation: Antonin Scalia. [Online] New York Magazine. http://nymag.com/news/features/antonin-scalia-2013-10/index1.html.

Chapter 4
Cures for Terrorism

We do not have the right to resort to violence — or the threat of violence — when we don't get our way.
— President Bill Clinton (April 18, 2010)

Our job as citizens is to ask questions.
— Thomas Blanton, National Security Archive
George Washington University (December 16, 2010)

The tools to change the world are in everybody's hands, and how we use them is...up to all of us. ... Public safety is too important to leave to the professionals.
— Marc Goodman at TEDGlobal 2012 (June 28, 2012)

It's a different type of war. Dealing with terror is going to be more like managing disease.
— Henry "Hank" Crumpton, deputy director of the CIA
Counterterrorism Center (July 27, 2012)

Peace is the only path to true security. ... No wall is high enough, and no Iron Dome is strong enough, to stop every enemy from inflicting harm.
— President Barack Obama (March 21, 2013)

The bottom line is, you don't beat an idea by beating a person. You beat an idea by beating an idea.
— Jon Lovett, former speechwriter to President Obama
(April 7, 2014)

4.1 Terrorism as a Disease

Two days after the terrorist attack on Charlie Hebdo in Paris on January 7, 2015, a senior U.S. intelligence official told CNN: "We've expected this. The boundaries between all of these affiliates are seemingly breaking down and the threat is

© Springer International Publishing Switzerland 2015   73
N. Lee, *Counterterrorism and Cybersecurity*, DOI 10.1007/978-3-319-17244-6_4

metastasizing and turning into a global network" [1]. In other words, terrorism is metastasizing like cancer in the global body of humanity.

Dracunculiasis, also known as Guinea worm disease (GWD), is caused by the parasite Dracunculus medinensis. The disease affects communities that do not have safe water to drink. There is no vaccine or drug therapy for Guinea worm disease. Through health education and innovative low-cost water treatments, the Carter Center has led the effort to eradicate the disease in collaboration with the Centers for Disease Control and Prevention (CDC), the World Health Organization (WHO), the United Nations Children's Fund (UNICEF), and the Bill & Melinda Gates Foundation. Two decades of eradication efforts have successfully reduced Guinea worm disease infection cases from 3.5 million worldwide in 1986 to a miniscule 148 cases in 2013 [2]. The Carter Center has predicted that "Guinea worm disease is poised to be the next human disease after smallpox to be eradicated" [3].

Henry "Hank" Crumpton, former deputy director of the Central Intelligence Agency (CIA) Counterterrorism Center, led an insurgent to overthrow the Taliban and to attack al-Qaeda in Afghanistan just after 9/11. Crumpton spoke at the Aspen Security Forum in July 2012 about the war on terror: "It's a different type of war. Dealing with terror is going to be more like managing disease" [4].

There are two fundamental ways to manage disease: treat the symptoms or remedy the root causes. Crumpton chose the former, the symptomatic treatment. In his 2010 interview on *60 Minutes*, Crumpton told CBS correspondent Lara Logan, "[My] orders were fairly simple: Find al-Qaeda and kill them, especially leadership. Destroy command and control. … If they kill me, I have told my family and my friends not to complain about anything, because I have killed many of them with pride" [5].

In spite of the operational successes, Crumpton admitted, "There will be an attack in the homeland. And sadly I think we face that prospect in the future. I think we'll be hit again." When Logan asked if such an attack would be on the scale of 9/11, he responded, "It's certainly possible. Or perhaps even greater" [5].

In his 2012 autobiography *American Sniper*, Navy SEAL marksman Chris Kyle expressed his only regret that "he didn't kill more" after having more than 160 confirmed kills during his four combat tours in Iraq [6]. In February 2015, CNN Pentagon Correspondent Barbara Starr revealed that the U.S. government has a secret 'hit list' of ISIS suspects: "The United States has already killed a dozen or so ISIS operatives on the list, including an ISIS chemical weapons expert, the senior official says. But others are added to the list as intelligence is gained about their role in ISIS" [7]. In other words, the list is getting longer as we know more about ISIS operations.

Trying to get rid of the symptoms (terrorists) without paying attention to the root causes (terrorist motives) does not eradicate the disease but may instead exacerbate it. American author and philosopher Henry David Thoreau wrote in his book *Walden; or, Life in the Woods* that "there are a thousand hacking at the branches of evil to one who is striking at the root" [8]. Jon Lovett, former speechwriter to President Barack Obama, talked about important issues facing America and he said, "The bottom line is, you don't beat an idea by beating a person. You beat an idea by beating an idea" [9].

At the TEDGlobal 2012 conference in Edinburgh, global security advisor and futurist Marc Goodman stated that "the tools to change the world are in everybody's hands, and how we use them is…up to all of us. … Public safety is too important to leave to the professionals" [10]. This chapter explores the cures for terrorism through education, communications, and innovative treatments:

1. "Revenge is sour" — George Orwell
2. "Govern your passions or they will be your undoing" — Mr. Spock
3. "Impossible to carry a grudge and a big dream at the same time" — Unknown
4. "Every truth has two sides" — Aesop
5. "Give everyone a voice" — Mark Zuckerberg
6. "The only security of all is in a free press" — Thomas Jefferson
7. "Free speech would not protect a man falsely shouting fire" — Oliver Wendell Holmes, Jr.
8. "198 methods of nonviolent action" — Gene Sharp
9. "We do not have the right to resort to violence when we don't get our way" — President Bill Clinton
10. "Peace is the only path to true security" — President Barack Obama

4.2 "Revenge Is Sour" — George Orwell

The ancient Code of Hammurabi dates back to 1792-50 B.C. when the Mesopotamian king Hammurabi ruled the Babylonian Empire. Out of the collection of 282 laws, the most famous Hammurabi's Code is "an eye for an eye, and a tooth for a tooth." However, the law of retaliation treated various social classes of society differently. Hammurabi wrote, "If a man has destroyed the eye of a man of the gentleman class, they shall destroy his eye. If he has destroyed the eye of a commoner, he shall pay one mina of silver. If he has destroyed the eye of a gentleman's slave, he shall pay half the slave's price" [11].

Fast forward 3,800 years to the present time, double standards continue to prevail in our modern world. Age, gender, race, ethnicity, religion, politics, sexual orientation, and ideological belief never cease to affect human judgment. A ruling party may label its opponents "terrorists" in order to delegitimize them, but an oppressive regime may commit acts of violence against its own people. From Timothy McVeigh's point of view, the U.S. government was the villain in the Waco siege. He carried out his revenge by the Oklahoma City bombing without realizing that murdering innocent people is not really a punishment for the U.S. government.

In his essay "Revenge is Sour," George Orwell argued that "the whole idea of revenge and punishment is a childish daydream. Properly speaking, there is no such thing as revenge. Revenge is an act which you want to commit when you are powerless and because you are powerless: as soon as the sense of impotence is removed, the desire evaporates also" [12].

Osama bin Laden was powerless in persuading his countryman Saudi King Fahd and Defense Minister Prince Sultan bin Abdul Aziz to deny American intervention in the Persian Gulf War [13]. Codenamed "Operation Desert Storm," the multinational coalition forces from the United States, Saudi Arabia, the United Kingdom, and Egypt liberated Kuwait from the Iraqi invasion in less than 100 h after the ground campaign started in February 1991 [14]. Although the U.S. military forces remained in Saudi Arabia after the war to maintain the stability of the region, President George H.W. Bush refused to capture Baghdad and overthrow Saddam Hussein. Meanwhile, Osama bin Laden, powerless to remove the American troops from his home country, resorted to terrorism against the United States in Saudi Arabia, Tanzania, Kenya, Yemen, and eventually New York.

In April 2003, U.S. Defense Secretary Donald Rumsfeld and Saudi Defense Minister Sultan bin Abdul Aziz announced the withdrawal of all U.S. combat forces from Saudi Arabia [15]. It was not because of the 9/11 attacks in September 2001, in which 15 of the 19 hijackers involved were Saudis. Osama bin Laden and terrorism did not win. Instead, the people of the Middle East and North Africa have influenced American foreign policy by voicing their strong disapproval of the U.S. presence in Saudi Arabia.

Reasoning is better than revenge. According to a 2008 Gallup poll, 60 % of Egyptians, 52 % of Saudis, 55 % of Tunisians, and 29 % of Turks agreed that removing all U.S. military bases from Saudi Arabia would significantly improve their opinion of the United States [16]. The relative friendliness of Turks towards Americans allowed the U.S. to deploy 400 American forces and two Patriot missile batteries in Turkey to counter the Syrian threat [17].

Revenge can backfire:

On September 11, 2012, Ansar al-Sharia attacked the U.S. Consulate in Benghazi, Libya to avenge the death of Abu Yahya al-Libi, al-Qaeda's second in command killed by a U.S. drone strike in Pakistan [18]. Their vengeance resulted in the death of U.S. ambassador J. Christopher Stevens, a well-liked figure in Libya. Demonstrations erupted in Benghazi and thousands of Libyans stormed the Ansar al-Sharia headquarters, chanting "You terrorists, you cowards. Go back to Afghanistan" [19].

On January 7, 2015, two masked gunmen, Chérif and Saïd Kouachi, forced their way into the offices of the French satirical weekly magazine *Charlie Hebdo* in Paris, killing 12 people including magazine editor Stéphane "Charb" Charbonnier, 4 cartoonists, 2 columnists, and a copy editor [20]. An amateur video of the assailants' gunfight with the police showed the men shouting, "We have avenged the Prophet Muhammad. We have killed Charlie Hebdo!" [21].

On January 11, Palestinian Authority President Mahmoud Abbas, Israeli Prime Minister Benjamin Netanyahu, Grande Mosquée de Paris Rector Dalil Boubakeur, and many world leaders joined 3.7 million people in anti-terrorism rallies in Paris and elsewhere in France as many participants carried the signs that read "Je suis Charlie" [22]. Anonymous posted a YouTube video stating that "We, Anonymous around the world, have decided to declare war on you the terrorists… we are going to shut down your accounts on all social networks" [23]. It also brought hacktivist

"Jester" out of retirement and back into hacking websites that support jihadist agenda and recruitment efforts. He defaced The Global Islamic Media Front website by planting two images of Charlie Hebdo covers showing Muhammad and a Muslim man kissing a magazine cartoonist [24].

A week after the massacre, the normal print run of 60,000 weekly *Charlie Hebdo* was extended to 5 million sold-out copies, which again angered some Muslims for the depiction of Prophet Muhammad on the magazine cover [25]. To the disappointment of Islamic extremists who tried to shut down *Charlie Hebdo* for lampooning Prophet Muhammad, the unexpected huge sale gave a much-needed shot in the arm for the financially struggling magazine.

Revenge is sour:

Finger-pointing can derail progress and incite revenge. In the wake of Charlie Hebdo attack in Paris, HBO's *Real Time with Bill Maher* host said on ABC's *Jimmy Kimmel Live*, "I know most Muslim people would not have carried out an attack like this. But here's the important point: Hundreds of millions of them support an attack like this. They applaud an attack like this. What they say is, 'We don't approve of violence,' but you know what? When you make fun of the Prophet, all bets are off" [26].

In the same line of thought as Maher, Rupert Murdoch's tweet caused a social media uproar when he wrote, "Maybe most Moslems peaceful, but until they recognize and destroy their growing jihadist cancer they must be held responsible" [27]. In a series of witty and serious responses, author J. K. Rowling tweeted: "I was born Christian. If that makes Rupert Murdoch my responsibility, I'll auto-excommunicate. ... The Spanish Inquisition was my fault, as is all Christian fundamentalist violence. Oh, and Jim Bakker. ... Eight times more Muslims have been killed by so-called Islamic terrorists than non-Muslims" [28].

Not to blame Maher and Murdoch's finger-pointing, but revenge attacks on Muslim began in France days after the Charlie Hebdo massacre. Two Muslim mosques and an affiliated restaurant were attacked by three grenades and a hail of bullets. Al Jazeera America's co-host Wajahat Ali pleaded for restraint, "France, turning on your own Muslim citizens & 'blaming Islam' for #CharlieHebdo feeds the extremists' agenda. Don't help them" [29].

Arab-American comedian Dean Obeidallah said, "I'm an American-Muslim... I'm not going to tell you, 'Islam is a religion of peace.' Nor will I tell you that Islam is a religion of violence" [30]. Salman Rushdie, author of *The Satanic Verses*, talked about the "deadly mutation" in the middle of Islam: "Governments, from the Sunni side the Saudi government, on the Shia side the Iranian government, have been putting fortunes of money into making sure that extremist mullahs are preaching in mosques around the world, and in building and developing schools in which a whole generation is being educated in extremism — and trying to prevent other forms of education" [31].

Religious experiences are, by nature, emotional and subject to interpretations by clergy and believers throughout history. It is pointless to debate which religions promote peace or violence. Even Buddhist monks who are widely regarded as pacifists have taken up arms in strife-torn regions in Thailand, Myanmar, and

Sri Lanka [32]. The love and forgiveness taught by Jesus Christ did not stop the military campaigns sanctioned by the Roman Catholic Church during the Middle Ages. Launched by Pope Urban II in November 1095, the series of Crusades against Muslims and pagans finally ended in 1291 with a death toll of 1 to 3 million people over a period of 197 years [33].

When three Israeli teenagers and a Palestinian teen were kidnapped and murdered in the summer of 2014, tensions in Jerusalem were further inflamed. "Any act of revenge of any kind whatsoever is completely inappropriate and wrong. Murder is murder," said Yishai Frankel, uncle of one of the slain Israeli teens. "One should not differentiate between bloods, be it Arab or Jew" [34].

In the series finale of the counterterrorism television show *24*, on-screen hero Jack Bauer (played by Kiefer Sutherland) was persuaded by his most trusted ally Chloe O'Brian (played by Mary Lynn Rajskub) to stop taking his own revenge. "I thought it would make me feel better. It didn't," Bauer told O'Brian in *24: Live Another Day*.

Sir Francis Bacon wrote in *Essayes and counsels, civil and moral*: "This is certain, that a man that studies revenge, keeps his own wounds green, which otherwise would heal, and do well" [35]. In other words, a man who focuses on revenge only exacerbates his own suffering. Mahatma Gandhi once said, "An eye for an eye will only make the whole world blind." And Jesus Christ told his disciples in *Matthew* 5:39 that "if anyone slaps you on the right cheek, turn to them the other cheek also."

4.3 "Govern Your Passions or They Will Be Your Undoing" — Mr. Spock

"As a matter of cosmic history, it has always been easier to destroy than to create," Mr. Spock (played by Leonard Nimoy) told Dr. McCoy (played by DeForest Kelley) in *Star Trek II: The Wrath of Khan*. When McCoy raised the moral question about the terraforming Genesis Device, Spock replied, "You must learn to govern your passions; they will be your undoing" [36].

The undoing of Osama bin Laden was due to his decision to destroy America rather than to create a bridge between two cultures. Imagine how much good bin Laden could have done with his estimated $300 million inheritance and $7 million yearly stipend [37, 38]. He could have used that money to support the Muslim Educational Cultural Center of America (MECCA), the Council on American-Islamic Relations (CAIR), or even his own grassroots campaigns to influence U.S. foreign policy.

What can a die-hard extremist do to relieve his anger and frustration? One of the answers is cage-fighting, also known as mixed-martial arts (MMA).

Since September 2010, many of the 250-some convicted pro al-Qaeda terrorists have been released from UK prison after having served their terms. Jonathan Evans, director general of the British security service MI5, issued a warning, "We know that some of these prisoners are still committed extremists who are likely to return to their terrorist activities" [39].

Abu Bakr Mansha was one of those convicted terrorists and sentenced to jail at the age of 21. After his second release from prison in March 2011, Mansha sought help from Usman Raja — one of the most renowned cage-fighting coaches in the U.K. Since 2010, Raja has successfully rehabilitated over ten released prisoners into mainstream society [40]. Employing cage-fighting sessions and the teachings of Sheikh Aleey Qadir, Raja's MMA gym has de-radicalized terror convicts with a 100 % success rate. "Take away someone's hate and they feel liberated," explained Raja. "The key is to give them a sense of purpose" [41].

Now a transformed man, Mansha tries to prevent other young Muslims following in his footsteps of terror. "I could channel my energy straight away and build something for myself," said Mansha. "My transformation came over time" [41].

Raja and his wife Khadija hope to expand their MMA gym to wean a generation away from the lure of extremism. Khadija Raja told *CNN*, "There is a real problem here and it's growing and it would be incredibly sad that we have the cure and then it's not delivered and it's not given the platform it needs to be dispersed. At the end of the day I'm still a mother and I don't want to live a world where there is this very real fear" [40].

On May 23, 2014, 22-year-old Elliot Rodger went on a killing spree in Isla Vista, California that resulted in 7 fatalities including two women outside a sorority house near the campus of UC Santa Barbara. His motivations were mainly misogyny and self-pity. Rodger frequented online discussion boards and social networks to express his frustration and hatred of women [42]. *Slate* writer Brian Levinson opined that he could have been Elliot Rodger. "Many men—including me, once upon a time—know what it's like to be young, frustrated, and full of rage toward women," Levinson wrote. "I was just as messed up as Cho, Rodger, and Klebold. I was humiliated by my weight, which bordered on morbid obesity. I fancied myself an intellectual, but would've traded an acceptance letter to Harvard for one magical kiss from a classmate I'll call Cynthia. I was a virgin until a year after I graduated college. Klebold and Rodger had friends; Cho had a family who clearly loved him. Why did these guys pick up guns, but I never did?" [43].

A man who is at peace with himself is less likely to turn into an extremist or a terrorist. American rock singer Marilyn Manson once said, "There's a difference between a man who has everything to gain and a man who has nothing to lose. If you have nothing to lose, you're dangerous in a bad way. If you have everything to gain, you're dangerous in a good way" [44].

4.4 "Impossible to Carry a Grudge and a Big Dream at the Same Time" — Unknown

Most people are victims of bullying at some point in their lives. American actress Evan Rachel Wood said, "I was bullied in school by some kids and also by a very nasty, extremely cruel teacher who made my life hell for nothing. But I try to channel any lingering anger or frustration from that time in my work. It's also my way of telling them to (very bad word) go to hell."

No one has suffered from physical bullying more than a 7-year-old Bangladeshi boy nicknamed "Okkhoy" who was severely beaten and mutilated by the beggar mafia in late 2010 because he refused to beg.

The 2008 Oscar-winning movie *Slumdog Millionaire* depicted the horror of forced begging in Bombay, India. Some of the real Mumbai street urchins told *The Telegraph* reporter that the violence in the film was, if anything, understated [45]. Indeed, what happened in real life to the 7-year-old boy was an indescribable atrocity: The attackers cracked open Okkhoy's head, sliced his throat and chest, slashed open his stomach, and chopped off his penis and his right testicle.

Against all odds, Okkhoy survived the brutal attack. *CNN* called him "Okkhoy" — the Bengali word for "unbreakable." His real name was withheld for safety reason. Okkhoy never attended school and his only goal in life was "to be a RAB [Rapid Action Battalion] member and nothing else." He said, "When I grow up, I want to bring them [the attackers] to justice." The RAB is known for unwarranted lethal force, illegal torture, and extrajudicial killings [46].

The young boy's father expressed his serious concern: "My biggest fear is that he'll start to think, 'I will find the person who did this to me and I will do the same to him.' He will live in a world of revenge. I don't want this. I don't want to be the father of a terrorist" [47].

After *CNN* first aired the Okkhoy story in 2011, businessman Aram Kovach in Columbus, Ohio, decided to help the boy with reconstructive surgery in the United States. Okkhoy finally arrived in the U.S. a year later in 2012. At Johns Hopkins Children's Center in Baltimore, Maryland, a medical team led by Dr. John Gearhart volunteered to perform a penile surgery on Okkhoy. Gearhart told *CNN*, "As far as an injury committed by one person against another, to a child, this is the most severe genital injury that I've ever seen in 23 years of doing this." Fortunately, the operation was a resounding success.

Okkhoy and his father were touched by the kindness of strangers. Before they left Baltimore, Okkhoy changed his life goal from joining RAB to becoming a doctor. He said, "I want to become a doctor, because I want to save people. And when I do, I won't take any money from them."

It has been said that "it's nearly impossible to carry a grudge and a big dream at the same time." Okkhoy will seek justice in court against the attackers, but he has a newfound priority in life so that he is not consumed by revenge and misses the opportunity to achieve a much bigger dream.

4.5 "Every Truth Has Two Sides" — Aesop

Ancient Greek fabulist Aesop said, "Every truth has two sides; it is as well to look at both, before we commit ourselves to either."

Unlike on-screen hero Jack Bauer in the counterterrorism TV drama *24*, real-life hero and FBI agent Ali Soufan did not use torture or coercion on the al-Qaeda terrorist Abu Jandal (aka Nasser al-Bahri) in order to extract information from him about the 9/11 attacks [48]. A former chief bodyguard of Osama bin Laden, Jandal considered himself a revolutionary who believed in the radical Islamist view of history that most of the world's evil came from America, a country that he knew practically nothing about.

During interrogation, Soufan brought Jandal a history book on the United States, in the Arabic language. Jandal was amazed to learn of the American Revolution and its struggle against tyranny. Soufan also showed Jandal a local Yemeni newspaper with the headline that read, "Two Hundred Yemeni Souls Perish in New York Attack." Shaken by the horrific atrocity of 9/11, Jandal drew a breath and said, "God help us. ... What kind of Muslim would do such a thing? ... The Israelis must have committed the attacks on New York and Washington. The Sheikh [Osama bin Laden] is not that crazy" [49].

After a long debate, Soufan persuaded Jandal to identify his associates in a book of mug shots. Jandal acknowledged knowing seven of the al-Qaeda members, but insisted that bin Laden would never commit the 9/11 attacks. Soufan took the seven photographs out of the book and said, "I know for sure that the people who did this were al-Qaeda guys." Jandal asked, "How do you know? Who told you?" Soufan replied, "You did. These *are* the hijackers. You just identified them." Jandal eventually told Soufan everything he knew and sadly declared, "I think the Sheikh went crazy."

There are always two sides to every story, every political system, and every religious belief. Soufan was able to open Jandal's eyes to see both sides of the story, not by torture or coercion, but by education and debate.

In a 2011 interview with Lara Logan in CBS' *60 Minutes,* Soufan said, "You need to connect with people on a human level — regardless, if they don't like you, want to kill you. ... They were trained that we are so evil and we torture and we kill and that is the reason of the rage against us. I try to deprive them from [the rage]" [50].

U.S. Navy Admiral James Stavridis, Commander of the U.S. European Command (USEUCOM) and NATO's Supreme Allied Commander Europe (SACEUR), spoke at the TEDGlobal 2012 conference in Edinburgh about teaching Afghan soldiers to read. Stavridis reasoned that "instead of building walls to create security, we need to build bridges" [51].

In December 2012, Jesuit priest and peace activist John Dear went to Kabul to meet the Afghan Peace Volunteers, a diverse community of students ages 15 to 27 who practice peace and nonviolence [52]. "I used to detest other ethnic groups," one of the youths told Dear, "but now I'm trying to overcome hate and prejudice. You international friends give me hope and strength to do this." Another youth added, "I used to put people in categories and couldn't drink tea with anyone. Now I'm learning that we are all part of one human family. Now I can drink tea with anyone" [53].

The world needs more Ali Soufan and John Dear to encourage people from diverse cultures and religions to look at both sides of truth. As Nigerian novelist Chimamanda Ngozi Adichie told the audience at TED 2009, "The single story creates stereotypes, and the problem with stereotypes is not that they are untrue, but that they are incomplete. They make one story become the only story" [54]. Our lives and our cultures are composed of many overlapping stories, and if we hear only a single story about another person or country, we risk a critical misunderstanding.

In an attempt to mend the NSA's reputation in the eyes of hackers and cyber-security professionals, NSA director Keith Alexander appeared at the 2013 Black Hat conference in Las Vegas and he said, "The whole reason I came here was to ask you to help make it better. If you disagree with what we're doing, you should help us twice as much. … I do think it's important for us to have this discussion. Because in my opinion, what you believe is what's written in the press without looking at the facts. This is the greatest technical center of gravity in the world. I ask that you all look at those facts" [55].

MIT Professor Emeritus Noam Chomsky offered this advice: "You have to try and develop a critical, open mind, and you have to be willing to evaluate and challenge conventional beliefs — accept them if they turn out to be valid but reject them if, as is so often the case, they turn out to just reflect power structures. And then you proceed with educational and organizing activities, actions as appropriate to circumstances. There is no simple formula; rather, lots of options. And gradually over time, things improve. I mean, even the hardest rock will be eroded by steady drips of water. That's what social change comes to and there are no mysterious modes of proceeding. They're hard ones, demanding ones, challenging, often costly. But that's what it takes to get a better world" [56].

4.6 "Give Everyone a Voice" — Mark Zuckerberg

Airborne leaflets have been used for military propaganda purposes before and after World War II [57]. Pamphlets were air dropped to the enemy's territories to disseminate the other side of the story. With the advent of the Internet and social networks, both sides have the chance to reach the targeted audience or mass public in real time.

As the world is increasingly becoming connected, 75 % of world leaders utilized Twitter in 2012 to engage their citizens and the global community [58]. Their Twitter use was up from 42 % a year prior in 2011. According to *Digital Daya*, the heads of state in the top 10 list are: President Barack Obama of the United States (24.6 million Twitter followers), President Hugo Chávez of Venezuela (3.8), President Abdullah Gül of Turkey (2.6), Queen Rania of Jordan (2.5), President Dmitry Medvedev of Russia (2.1), President Dilma Rousseff of Brazil (1.8), President Cristina Fernández de Kirchner of Argentina (1.5), President Juan Manuel Santos of Colombia (1.5), President Enrique Peña Nieto of Mexico (1.4), and Sheikh Mohammed bin Rashid Al Maktoum of the United Arab Emirates (UAE) and Dubai (1.3).

British Prime Minster David Cameron who once said "too many tweets make a twat" changed his mind and sent out his first tweet in October 2012. Cameron said, "In this modern world you have got to use every means to try and communicate your message and explain to people why you are doing it" [59].

An open internet is an open platform for debating opposing views. It allows both popular and unpopular voices to be heard. It is a civilized outlet for frustrated individuals to express themselves without resorting to violence or terrorism. U.S. Air Force Senior Airman Christopher R. Atkins wrote in his email to American filmmaker Michael Moore, "Every time we voice our opinion we are promoting freedom" [60].

A light-hearted example is a September 2012 YouTube video called *We Are Hungry* created by high school students in Kansas [61]. With more than a million views, the video successfully forced the Obama administration to reverse some of the new rules by allowing more meats and grains in school lunches. "Even though we're a small town in rural western Kansas, Washington did hear us," said Superintendent Dave Porter. "Our concerns were listened to" [62].

In the January 2012 Facebook IPO letter, Mark Zuckerberg wrote [63]:

At Facebook, we're inspired by technologies that have revolutionized how people spread and consume information. We often talk about inventions like the printing press and the television — by simply making communication more efficient, they led to a complete transformation of many important parts of society. They gave more people a voice. They encouraged progress. They changed the way society was organized. They brought us closer together. Today, our society has reached another tipping point. We live at a moment when the majority of people in the world have access to the internet or mobile phones — the raw tools necessary to start sharing what they're thinking, feeling and doing with whomever they want. Facebook aspires to build the services that give people the power to share and help them once again transform many of our core institutions and industries. There is a huge need and a huge opportunity to get everyone in the world connected, to give everyone a voice and to help transform society for the future. The scale of the technology and infrastructure that must be built is unprecedented, and we believe this is the most important problem we can focus on.

…

We believe building tools to help people share can bring a more honest and transparent dialogue around government that could lead to more direct empowerment of people, more accountability for officials and better solutions to some of the biggest problems of our time. By giving people the power to share, we are starting to see people make their voices heard on a different scale from what has historically been possible. These voices will increase in number and volume. They cannot be ignored. Over time, we expect governments will become more responsive to issues and concerns raised directly by all their people rather than through intermediaries controlled by a select few.

Even terrorists embrace the idea of Facebook. An online jihadist who goes by the name Rakan al-Ashja'i said, "We will benefit from the ideas in Facebook a lot. … If I could make a social networking website with the same capabilities and everything like Facebook when it first appeared — it is a very good idea" [64].

With over one billion active monthly users, Facebook is in a unique position to influence the world through the billion-strong human rally against violence. Horrified by the Sandy Hook Elementary School massacre, Beth Howard told her Facebook friends on December 14 about heading to Newtown, Connecticut to hand out free apple pies to the grieving parents and anyone in need. Within two hours, she received $2,000 in donations. After several days of intense baking with the help of more than 60 volunteers, she drove 1,100 miles to Newtown in her RV loaded with 240 apple pies. "They [the volunteers] were making pies for Newtown because of this one Facebook comment," said Howard. "That was a powerful thing" [65].

4.7 "The Only Security of All Is in a Free Press" — Thomas Jefferson

"The only security of all is in a free press," said President Thomas Jefferson, American Founding Father and principal author of the Declaration of Independence. "The force of public opinion cannot be resisted when permitted freely to be expressed" [66]. In the case of *Ginzburg v. United States* in 1965, U.S. Supreme Court justice Potter Stewart said, "Censorship reflects a society's lack of confidence in itself."

After the 9/11 attacks, Bill Maher's comments on *Politically Incorrect* and Susan Sontag's article in *The New Yorker* challenged President George W. Bush's notion of cowardice, resulting in public uproar and backlash [67]. In a statement supporting Maher, ABC said that *Politically Incorrect* is "a show that celebrates freedom of speech and encourages the animated exchange of ideas and opinions. While we remain sensitive to the current climate following last week's tragedy ... there needs to remain a forum for the expression of our nation's diverse opinions" [68].

In 1971, *The New York Times* publisher Arthur Ochs "Punch" Sulzberger decided to publish a top-secret government history of the Vietnam War known as the Pentagon Papers [69]. A federal court ordered the newspaper to halt publishing the Pentagon Papers, citing national security concerns. However, the U.S. Supreme Court ruled on First Amendment grounds that publication could resume — a landmark ruling on press freedom. Exposing the Johnson administration's systematic lies to the American public and U.S. Congress, the Pentagon Papers leaked by Daniel Ellsberg helped hasten the end of the U.S. involvement in the Vietnam War as President Richard Nixon began to withdraw American troops from Vietnam.

President Barack Obama praised Sulzberger as "a firm believer in the importance of a free and independent press, one that isn't afraid to seek the truth, hold those in power accountable and tell the stories that need to be told" [70].

However, quote approval has become a standard practice in the Obama and Romney presidential campaigns in 2012. Under the Obama administration, reporters are forbidden to identify or quote the official speakers in a "deep-background briefing" such as the one held after the U.S. Supreme Court's health care ruling in June 2012 [71].

Former CBS news anchor Dan Rather argues that newspapers and media outlets must push back on quote approval because "submitting to these new tactics puts us more in the category of lapdogs." Rather wrote in a July 2012 *CNN* article, "A free and truly independent press — fiercely independent when necessary — is the red beating heart of freedom and democracy. One of the most important roles of our journalists is to be watchdogs" [72].

Journalists should be watchdogs, not lapdogs. In the documentary film *Fahrenheit 9/11*, director Michael Moore accused the American corporate media of being President George W. Bush's "cheerleaders," instead of providing an accurate and objective analysis of the rationale for the 2003 Iraq War to topple Saddam Hussein [73]. The Walt Disney Company blocked its Miramax division from distributing the film due to the concern that it would jeopardize tax breaks Disney was receiving for its theme park business in Florida, where Bush's brother, Jeb Bush, was governor. Moore criticized Disney's decision: "At some point the question has to be asked, 'Should this be happening in a free and open society where the monied interests essentially call the shots regarding the information that the public is allowed to see?'" [74].

After U.S. Army private Bradley Manning was arrested in February 2012 for divulging three-quarters of a million secret documents to WikiLeaks, op-ed columnist Bill Keller of *The New York Times* wrote [75]:

> In the immediate aftermath of the breach, several news organizations (including this one) considered creating secure online drop-boxes for would-be leakers, imagining that new digital Deep Throats would arise. But it now seems clear that the WikiLeaks breach was one of a kind — and that even lesser leaks are harder than ever to come by. Steven Aftergood, who monitors secrecy issues for the Federation of American Scientists, said that since WikiLeaks the government has elevated the 'insider threat' as a priority, and tightened access to classified material. Nudged by an irate Congress, the intelligence agencies are at work on an electronic auditing program that would make illicit transfer of secrets much more difficult and make tracking the leaker much easier.

A strong supporter of WikiLeaks, the Pentagon Papers whistleblower Daniel Ellsberg said, "Julian Assange is not a criminal under the laws of the United States. I was the first one prosecuted for the charges that would be brought against him. I was the first person ever prosecuted for a leak in this country — although there had been a lot of leaks before me. That's because the First Amendment kept us from having an Official Secrets Act. ... The founding of this country was based on the principle that the government should not have a say as to what we hear, what we think, and what we read. ... We're not in the mess we're in, in the world, because of too many leaks. ... I say there should be some secrets. But I also say we invaded Iraq illegally because of a lack of a Bradley Manning at that time" [76].

Thomas Blanton, director of National Security Archive at George Washington University, testified before the U.S. House of Representatives in December 2010 that "our job as citizens is to ask questions," and he said [77]:

> I wish all terrorist groups would write the local U.S. ambassador a few days before they are launching anything — the way Julian Assange wrote Ambassador Louis Susman in London on November 26 — to ask for suggestions on how to make sure nobody gets hurt.

> I wish all terrorist groups would partner up with *Le Monde* and *El Pais* and *Der Spiegel* and *The Guardian*, and *The New York Times*, and take the guidance of those professional journalists on what bombs go off and when and with what regulators.

Julian Assange said in a 2011 interview, "Our No. 1 enemy is ignorance. And I believe that is the No. 1 enemy for everyone — it's not understanding what actually is going on in the world" [78]. Indeed, the two-way street of Total Information Awareness is the road that leads to a more transparent and complete picture of ourselves, our governments, and our world. Former FBI agent Ali Soufan successfully demonstrated that a history book of America in Arabic language was one of the essential tools in disarming a terrorist. The more information that countries and peoples have about each other, the better and safer the world will become.

English author Edward Bulwer-Lytton wrote in *Richelieu, or The conspiracy: in five acts* that "the pen is mightier than the sword" [79]. After the 2012 Benghazi attack, *CNN* interviewed Ahmed Abu Khattala, who was believed to be the Benghazi leader of the al Qaeda-affiliated militia group Ansar al-Sharia. U.S. Representative Jason Chaffetz was dumbfounded as he told reporters, "One of the pertinent questions today is why we have not captured or killed the terrorist who committed these attacks? News out today that CNN was able to go in and talk to one of the suspected terrorists, how come the military hasn't been able to get after them and capture or kill the people? How come the FBI isn't doing this and yet CNN is?" [80].

The pen is mightier than the sword. Through uncensored journalistic investigations and opinion pieces presenting both sides of the coin, the press can eliminate the need for terrorists to commit violent crimes in order to get their messages across.

4.8 "Free Speech Would not Protect a Man Falsely Shouting Fire" — Oliver Wendell Holmes, Jr.

While information is the oxygen of the modern age, disinformation is the carbon monoxide that can poison generations. Oliver Wendell Holmes, Jr. was an Associate Justice of the United States Supreme Court for 30 years from 1902 to 1932. In his "clear and present danger" majority opinion in the 1919 case of *Schenck v. United States*, Holmes stated that "the most stringent protection of free speech would not protect a man in falsely shouting fire in a theatre and causing a panic. ... The question in every case is whether the words used are used in such circumstances and are of such a nature as to create a clear and present danger that they will bring about the substantive evils that Congress has a right to prevent" [81].

In 1919, the United States was marred by domestic terrorism in a series of bombings and assassination attempts orchestrated by the American followers of Italian anarchist Luigi Galleani [82]. Violent activities terrorized New York City, Boston, Pittsburgh, Cleveland, Washington D.C., Philadelphia, and other U.S. cities.

In 1969, protection for speech was raised in *Brandenburg v. Ohio* from "clear and present danger" to "imminent lawless action." Clarence Brandenburg, a Ku Klux Klan (KKK) leader in rural Ohio was convicted under the Ohio criminal syndicalism statute for advocating unlawful methods of racist terrorism as a means of accomplishing political reform. However, the U.S. Supreme Court reversed Brandenburg's conviction because "the constitutional guarantees of free speech and free press do not permit a State to forbid or proscribe advocacy of the use of force or of law violation except where such advocacy is directed to inciting or producing imminent lawless action and is likely to incite or produce such action" [83].

In September 2012, a YouTube trailer of an anti-Islamic film by Nakoula Basseley Nakoula (aka Sam Bacile) wreaked chaos in the Middle East. Despite the Islamic prohibition of portraying Prophet Muhammad, Nakoula's low-budget movie *Innocence of Muslims* depicts the prophet as a womanizer, buffoon, ruthless killer, and child molester. As a result, protestors stormed US embassies in Libya and Egypt, and thousands of demonstrators took to the streets in Afghanistan, Indonesia, Pakistan, Yemen, Lebanon, and Iraq [84]. U.S. Secretary of State Hillary Clinton promptly issued a statement on September 11, 2012 that "the United States deplores any intentional effort to denigrate the religious beliefs of others" [85].

Google temporarily blocked Nakoula's YouTube video in Libya and Egypt. "We work hard to create a community everyone can enjoy and which also enables people to express different opinions," said Google in a written statement. "This can be a challenge because what's OK in one country can be offensive elsewhere. This video — which is widely available on the web — is clearly within our guidelines and so will stay on YouTube. However, given the very difficult situation in Libya and Egypt we have temporarily restricted access in both countries" [86].

Al-Qaeda terrorist Anwar al-Awlaki was dubbed the "bin Laden of the Internet." He used a blog, a Facebook page, and YouTube videos to recruit new jihadists [87]. Before his death in 2011, more than 5,000 postings featuring al-Awlaki's videos were available on YouTube. His video sermons inspired British-born Roshonara Choudhry to become the first al-Qaeda sympathizer to attempt a political assassination in the U.K. by stabbing British MP Stephen Timms in May 2010. In response to the British government's complaint, Google said, "Community Guidelines prohibit dangerous or illegal activities such as bomb-making, hate speech, or incitement to commit specific and serious acts of violence. … We have removed a significant number of videos under these policies. We're now looking into the new videos that have been raised with us and will remove all those which break our rules" [88].

The films by Nakoula and al-Awlaki were similar in that the former's is a direct insult on Islam and the latter's is a direct attack on non-Muslims. In a telephone interview with *The Wall Street Journal*, Nakoula said that his film was "a political movie, not a religious movie" [89]. Although Nakoula did not violate any U.S. law, his action was intentional and irresponsible, akin to a man falsely shouting fire in a theatre and causing a panic.

In September 2012, the American Freedom Defense Initiative (AFDI), led by Pamela Geller and Robert Spencer, ran an ad in 10 subway stations around New York that read: "In any war between the civilized man and the savage, support the civilized man. Support Israel. Defeat Jihad" [90]. New York's Metropolitan Transportation Authority (MTA) initially rejected the ad, but a district judge ruled that the ad was protected under the First Amendment. New York mayor Michael Bloomberg defended the court's decision by saying, "Americans tolerate things they might find despicable because of the First Amendment's protection of free expression" [91].

However, many New Yorkers including Rabbi Rachel Kahn-Troster found the ad "deeply misguided and disturbing." Kahn-Troster said, "The words from our mouths have power: Once released, whether intentionally or by accident, what we say shapes reality. It can bring about healing or atonement, or it can unleash violence and hatred. Geller's ads, sharply dividing the world into civilized people and savages, are only intended to hurt and tear fragile relationships apart" [92].

In *The Story of My Experiments with Truth*, Mahatma Gandhi in 1925 warned that "just as an unchained torrent of water submerges whole countrysides and devastates crops, even so an uncontrolled pen serves but to destroy" [93]. There is a fine line between free speech and hate speech. Free speech encourages debate whereas hate speech incites violence. In September 2012, the United Methodist Women funded a counter ad in New York's subway stations. It reads, "Hate speech is not civilized. Support peace in word and deed" [94].

Satire in media such as *The Interview* and *Charlie Hebdo* walk a fine line between freedom of speech and dangerous incitement:

In July 2014, North Korea complained to the United Nations about the satire comedy *The Interview* showing a hapless TV host recruited by the CIA to assassinate North Korean dictator Kim Jong-un. "To allow the production and distribution of such a film on the assassination of an incumbent head of a sovereign state should be regarded as the most undisguised sponsoring of terrorism as well as an act of war," UN ambassador Ja Song Nam told UN secretary general Ban Ki-moon in a letter [95].

In the aftermath of Charlie Hebdo shooting in Paris on January 7, 2015, the magazine's cofounder Henri Roussel opined that the publication went too far with its provocative images [96]. Pope Francis remarked, "You don't kill in God's name. ... [But] you cannot provoke, you cannot insult the faith of others. You cannot make fun of the faith of others" [97]. Turkey's Deputy Prime Minister Yalcin Akdogan wrote on Twitter, "Those who are publishing figures referring to our supreme Prophet are those who disregard the sacred. [Such a move is] open incitement and provocation" [98]. A Muslim woman also tweeted, "Still confused by how supporting publishing material 1.6 billion peaceful people find personally offensive can bring anyone joy. 1.6 billion Muslims want to express solidarity for a crime, and yet you are telling them that to do so they must accept ridicule. No" [99]. Al Jazeera correspondent Mohamed Vall Salem emailed his coworkers, "What Charlie Hebdo did was not free speech it was an abuse of free speech in my opinion, go back to the cartoons and have a look at them. It's not about what the drawing said, it was about how they said it. I condemn those heinous killings, but I'M NOT CHARLIE" [100].

4.9 "198 Methods of Nonviolent Action" — Gene Sharp

Political scientist Gene Sharp, nominated for the 2009 Nobel Peace Prize and highly regarded as the father of nonviolent struggle, compiled 198 methods of nonviolent action in his 1973 book *The Politics of Nonviolent Action*, Vol. 2:

The Methods of Nonviolent Action [101]. The list of methods is publicly and freely available online at The Albert Einstein Institution: http://www.aeinstein. org/organizations/org/198_methods.pdf [102]. The comprehensive methods are grouped into six major categories and 37 subcategories ranging from individual to group to nationwide to international activities:

I. Nonviolent Protest and Persuasion

   a) Formal Statements

       1. Public speeches
       2. Letters of opposition or support
       3. Declarations by organizations and institutions
       4. Signed public statements
       5. Declarations of indictment and intention
       6. Group or mass petitions

   b) Communications with a Wider Audience

       7. Slogans, caricatures, and symbols
       8. Banners, posters, and displayed communications
       9. Leaflets, pamphlets, and books
       10. Newspapers and journals
       11. Records, radio, and television
       12. Skywriting and earthwriting

   c) Group Representations

       13. Deputations
       14. Mock awards
       15. Group lobbying
       16. Picketing
       17. Mock elections

   d) Symbolic Public Acts

       18. Displays of flags and symbolic colors
       19. Wearing of symbols
       20. Prayer and worship
       21. Delivering symbolic objects
       22. Protest disrobings
       23. Destruction of own property
       24. Symbolic lights
       25. Displays of portraits
       26. Paint as protest
       27. New signs and names
       28. Symbolic sounds
       29. Symbolic reclamations
       30. Rude gestures

e) Pressures on Individuals

    31. "Haunting" officials
    32. Taunting officials
    33. Fraternization
    34. Vigils

f) Drama and Music

    35. Humorous skits and pranks
    36. Performances of plays and music
    37. Singing

g) Processions

    38. Marches
    39. Parades
    40. Religious processions
    41. Pilgrimages
    42. Motorcades

h) Honoring the Dead

    43. Political mourning
    44. Mock funerals
    45. Demonstrative funerals
    46. Homage at burial places

i) Public Assemblies

    47. Assemblies of protest or support
    48. Protest meetings
    49. Camouflaged meetings of protest
    50. Teach-ins

j) Withdrawal and Renunciation

    51. Walk-outs
    52. Silence
    53. Renouncing honors
    54. Turning one's back

II. Social Noncooperation

a) Ostracism of Persons

    55. Social boycott
    56. Selective social boycott
    57. Lysistratic nonaction
    58. Excommunication
    59. Interdict

b) Noncooperation with Social Events, Customs, and Institutions

    60. Suspension of social and sports activities
    61. Boycott of social affairs
    62. Student strike
    63. Social disobedience
    64. Withdrawal from social institutions

c) Withdrawal from the Social System

    65. Stay-at-home
    66. Total personal noncooperation
    67. "Flight" of workers
    68. Sanctuary
    69. Collective disappearance
    70. Protest emigration (hijra)

III. Economic Noncooperation: Boycotts

a) Actions by Consumers

    71. Consumers' boycott
    72. Nonconsumption of boycotted goods
    73. Policy of austerity
    74. Rent withholding
    75. Refusal to rent
    76. National consumers' boycott
    77. International consumers' boycott

b) Action by Workers and Producers

    78. Workmen's boycott
    79. Producers' boycott

c) Action by Middlemen

    80. Suppliers' and handlers' boycott

d) Action by Owners and Management

    81. Traders' boycott
    82. Refusal to let or sell property
    83. Lockout
    84. Refusal of industrial assistance
    85. Merchants' "general strike"

e) Action by Holders of Financial Resources

    86. Withdrawal of bank deposits
    87. Refusal to pay fees, dues, and assessments
    88. Refusal to pay debts or interest

89.   Severance of funds and credit
90.   Revenue refusal
91.   Refusal of a government's money

f) Action by Governments

92.   Domestic embargo
93.   Blacklisting of traders
94.   International sellers' embargo
95.   International buyers' embargo
96.   International trade embargo

IV. Economic Noncooperation: Strikes

a) Symbolic Strikes

97.   Protest strike
98.   Quickie walkout (lightning strike)

b) Agricultural Strikes

99.   Peasant strike
100. Farm Workers' strike

c) Strikes by Special Groups

101. Refusal of impressed labor
102. Prisoners' strike
103. Craft strike
104. Professional strike

d) Ordinary Industrial Strikes

105. Establishment strike
106. Industry strike
107. Sympathetic strike

e) Restricted Strikes

108. Detailed strike
109. Bumper strike
110. Slowdown strike
111. Working-to-rule strike
112. Reporting "sick" (sick-in)
113. Strike by resignation
114. Limited strike
115. Selective strike

f) Multi-Industry Strikes

116. Generalized strike
117. General strike

g) Combination of Strikes and Economic Closures

    118. Closing shops or suspending work (hartal)
    119. Economic shutdown

V. Political Noncooperation

  a) Rejection of Authority

    120. Withholding or withdrawal of allegiance
    121. Refusal of public support
    122. Literature and speeches advocating resistance

  b) Citizens' Noncooperation with Government

    123. Boycott of legislative bodies
    124. Boycott of elections
    125. Boycott of government employment and positions
    126. Boycott of government departments, agencies, and other bodies
    127. Withdrawal from government educational institutions
    128. Boycott of government-supported organizations
    129. Refusal of assistance to enforcement agents
    130. Removal of own signs and placemarks
    131. Refusal to accept appointed officials
    132. Refusal to dissolve existing institutions

  c) Citizens' Alternatives to Obedience

    133. Reluctant and slow compliance
    134. Nonobedience in absence of direct supervision
    135. Popular nonobedience
    136. Disguised disobedience
    137. Refusal of an assemblage or meeting to disperse
    138. Sitdown
    139. Noncooperation with conscription and deportation
    140. Hiding, escape, and false identities
    141. Civil disobedience of "illegitimate" laws

  d) Action by Government Personnel

    142. Selective refusal of assistance by government aides
    143. Blocking of lines of command and information
    144. Stalling and obstruction
    145. General administrative noncooperation
    146. Judicial noncooperation
    147. Deliberate inefficiency and selective noncooperation by enforcement agents
    148. Mutiny

  e) Domestic Governmental Action

    149. Quasi-legal evasions and delays
    150. Noncooperation by constituent governmental units

f) International Governmental Action

    151. Changes in diplomatic and other representations
    152. Delay and cancellation of diplomatic events
    153. Withholding of diplomatic recognition
    154. Severance of diplomatic relations
    155. Withdrawal from international organizations
    156. Refusal of membership in international bodies
    157. Expulsion from international organizations

VI. Nonviolent Intervention

a) Psychological Intervention

    158. Self-exposure to the elements
    159. The fast (Fast of moral pressure, Hunger strike, Satyagrahic fast)
    160. Reverse trial
    161. Nonviolent harassment

b) Physical Intervention

    162. Sit-in
    163. Stand-in
    164. Ride-in
    165. Wade-in
    166. Mill-in
    167. Pray-in
    168. Nonviolent raids
    169. Nonviolent air raids
    170. Nonviolent invasion
    171. Nonviolent interjection
    172. Nonviolent obstruction
    173. Nonviolent occupation

c) Social Intervention

    174. Establishing new social patterns
    175. Overloading of facilities
    176. Stall-in
    177. Speak-in
    178. Guerrilla theater
    179. Alternative social institutions
    180. Alternative communication system

d) Economic Intervention

    181. Reverse strike
    182. Stay-in strike
    183. Nonviolent land seizure
    184. Defiance of blockades

185. Politically motivated counterfeiting
186. Preclusive purchasing
187. Seizure of assets
188. Dumping
189. Selective patronage
190. Alternative markets
191. Alternative transportation systems
192. Alternative economic institutions

e) Political Intervention

193. Overloading of administrative systems
194. Disclosing identities of secret agents
195. Seeking imprisonment
196. Civil disobedience of "neutral" laws
197. Work-on without collaboration
198. Dual sovereignty and parallel government

Gene Sharp's 198 methods of nonviolent action compiled in 1973 remains highly relevant today in spite of the technological advances in the past four decades. In fact, social media and the Internet can help amplify and empower nonviolent action.

While some of the methods may seem unconventional, many of them have been practiced by political dissidents around the world. For example, hundreds of thousands of demonstrators employed method #47, assemblies of protest or support, and method #18, displays of flags and symbolic colors, during the 2004 Orange Revolution in Ukraine [103]. In 2011, Egyptian and Arab Spring protesters wrote anti-government songs, posted them on the Internet, and sang them in the streets — just like Gene Sharp suggested in method #36, performances of plays and music, and method #37, singing [104]. On December 7, 2013 when demonstrations in Ukraine turned violent, Markiyan Matsekh promoted peaceful protest by placing a piano in front of the riot police in Kiev and playing Chopin's Waltz in C-sharp minor [105] (See Fig. 4.1).

Method #181, reverse strike, is exceptionally constructive. In 1958, Italian social activist Danilo Dolci, known as the Sicilian Gandhi, dispatched 150 unemployed men to repair a dirt road outside Partinico, Sicily when the government had refused to do so or to grant them permission [106]. Dolci was arrested and sentenced to eight months in jail as a result.

Method #23, destruction of own property, is undeniably destructive; but the key difference between this and terrorism is that it deals with one's own property and not with other people's property or putting others in harm's way. As long as the owner does not endanger anyone or make an insurance claim, destruction of own property can be a powerful and selfless statement.

Method #57, Lysistratic nonaction, is also known as crossed legs movement or sex strikes. Professor Michael Soussan of New York University and Elizabeth Weingarten of New America wrote on CNN, "Even terrorists have fears. And the

Fig. 4.1 Markiyan Matsekh played the piano in front of the riot police in Kiev on December 7, 2013. (Courtesy of Andrew Meakovsky, Oleg Matsekh, and Marikiyan Matsekh.)

prospect of gender equality appears to rank high on their list of worst nightmares. … Educated women and girls would fundamentally challenge the power structure of organizations like ISIS. … So, any successful strategy against these groups must put women's rights at the front and center of policy planning" [107]. In countries where women are oppressed and physically abused, Lysistratic nonaction is a dangerous but an effective method that those women can employ collectively to help stop terrorism. *Slate* magazine editor L.V. Anderson reported her findings in August 2012 that "Generally, sex strikes — known in activist circles as 'Lysistratic nonaction,' a nod to Aristophanes' ancient Greek comedy — appear to be more successful when the women involved have little economic autonomy, when their demands are specific and realistic, and when they possess endurance and strength in numbers" [108].

4.10 "We Do not Have the Right to Resort to Violence When We Don't Get Our Way" — President Bill Clinton

The 2011 revolt in Syria began with peaceful protests but it turned into a civil war after the government waged a brutal crackdown on dissent, killing more than 40,000 activists [109]. Armed conflict may be inevitable in some circumstances against an oppressive authoritative regime. However, there is no excuse for violence in a democratic society that welcomes free speech and open debates. In an April 2010 article published in *The New York Times*, President Bill Clinton wrote [110]:

> We should never forget what drove the bombers, and how they justified their actions to themselves. They took to the ultimate extreme an idea advocated in the months and years before the bombing by an increasingly vocal minority: the belief that the greatest threat to American freedom is our government, and that public servants do not protect our freedoms, but abuse them. On that April 19, the second anniversary of the assault of the Branch Davidian compound near Waco, deeply alienated and disconnected Americans decided murder was a blow for liberty.
>
> Americans have more freedom and broader rights than citizens of almost any other nation in the world, including the capacity to criticize their government and their elected officials. But we do not have the right to resort to violence — or the threat of violence — when we don't get our way. Our founders constructed a system of government so that reason could prevail over fear. Oklahoma City proved once again that without the law there is no freedom.
>
> Criticism is part of the lifeblood of democracy. No one is right all the time. But we should remember that there is a big difference between criticizing a policy or a politician and demonizing the government that guarantees our freedoms and the public servants who enforce our laws. Civic virtue can include harsh criticism, protest, even civil disobedience. But not violence or its advocacy. That is the bright line that protects our freedom. It has held for a long time, since President George Washington called out 13,000 troops in response to the Whiskey Rebellion. Fifteen years ago, the line was crossed in Oklahoma City. In the current climate, with so many threats against the president, members of Congress and other public servants, we owe it to the victims of Oklahoma City, and those who survived and responded so bravely, not to cross it again.

Oklahoma City bomber Timothy McVeigh never expressed any remorse despite facing execution by lethal injection in May 2001 [111]. However, Centennial Olympic Park bomber Eric Rudolph wrote his mother from jail: "Perhaps I should have found a peaceful outlet for my opposition to the government in Washington: maybe I should have been a lawyer and fought [for] decency in the face of this rotten system; perhaps I could have taken up teaching and sought to inculcate a healthy outlook in a decidedly unhealthy society. But I didn't do any of these things, and I resorted to force to have my voice heard" [112].

Terrorism does not advance the cause but only serves to hurt the intention of its perpetrators. Terrorism ought to be replaced by nonviolent civil disobedience. Peace journalist Robert Koehler posted in his blog on March 3, 2011: "The 'street' and the 'masses' have actually found — maybe rediscovered is the right word — the power of nonviolent collective action. And the cradle of this new approach to civilization is not in the comfortably snoozing West but in the impoverished and long-brutalized Middle East, where every despot has either tumbled or is shaking in his boots — and where violence has suddenly been stripped of its righteousness and been exposed as weakness, no matter how much mayhem it produces" [113].

Adolf Hitler, the epitome of terrorism, once said, "Demoralize the enemy from within by surprise, terror, sabotage, assassination. This is the war of the future." Hitler lost the war nonetheless. When terrorists realize that there are many more negatives in conducting terrorism and many more positives in abandoning terrorism, the horrible disease will be cured. In an unauthenticated document dated

December 14, 2001 (three months after the 9/11 attacks), Osama bin Laden predicted in his will that he would be killed as a result of a "betrayal" and he instructed his children not to continue to wage the holy war against America and Israel [114].

4.11 "Peace Is the Only Path to True Security" — President Barack Obama

Five days after the Sandy Hook Elementary School massacre in Newtown, Connecticut, President Barack Obama spoke in a White House press conference on December 19, 2012: "It's encouraging that people of all different backgrounds and beliefs and political persuasions have been willing to challenge some old assumptions and change some long-standing positions. That conversation has to continue, but this time the words need to lead to action" [115]. Obama's words apply not only to finding long-term solutions to gun violence in the U.S. but also to domestic and international terrorism around the globe.

During his visit to Israel and the West Bank in March 2013, President Obama told a captive audience at the International Convention Center in Jerusalem [116]:

> Peace is necessary. Peace is the only path to true security. … There is no question that Israel has faced Palestinian factions who turned to terror, and leaders who missed historic opportunities. That is why security must be at the center of any agreement. And there is no question that the only path to peace is through negotiation. … The only way to truly protect the Israeli people over the long term is through the absence of war - because no wall is high enough, and no Iron Dome is strong enough or perfect enough, to stop every enemy from inflicting harm. … Peace must be made among peoples, not just governments. No single step can change overnight what lies in the hearts and minds of millions. No single step is going to erase years of history and propaganda. But progress with the Palestinians is a powerful way to begin, while sidelining extremists who thrive on conflict and division. It would make a difference!

Indeed, it will make a positive difference in world security and counterterrorism by setting our mind on pursuing peaceful solutions rather than escalating the war on terror. After the Charlie Hebdo shooting in Paris on January 7, 2015, a Nihaad Hosany emailed BBC News: "It's so awful. Because of three idiots, three terrorists, the Muslim community will suffer again. Islam is a religion of peace and understanding. Not this monstrosity" [117]. A week after the massacre, the magazine cover of the survivors' issue of *Charlie Hebdo* depicts the Prophet Muhammad with a tear falling from his cheek, holding a sign that reads, "Je suis Charlie (I am Charlie)." Above Muhammad are the words "Tout est pardonné (All is forgiven)."

Prof. Hussein Rashid at Hofstra University commented on the magazine cover, "My initial thought is that the cover is a near perfect response to the tragedy. They are not backing down from the depiction of Muhammad, exercising their free speech rights. At the same time, the message is conciliatory, humble, and will hopefully reduce the anger directed to the Muslim communities of France" [99]. Australian prime minister Tony Abbott concurred, "I rather like that cartoon. I'm not sure that I would have liked everything that Charlie Hebdo produced but this is a cartoon of the prophet with a tear streaming down his face saying all is forgiven. That spirit of forgiveness is what we need more and more in this rancorous modern world" [118].

Mahatma Gandhi once said, "You must not lose faith in humanity. Humanity is like an ocean; if a few drops of the ocean are dirty, the ocean does not become dirty." Besides, two wrongs do not make a right. President Obama is correct: Peace is the only path to true security; and peace requires both free speech and willingness to listen. When a heckler interrupted President Barack Obama numerous times during his speech in May 2013, Obama retorted, "This is part of free speech, is you being able to speak but also you listening and me being able to speak, alright?" [119].

Bibliography

1. **Brown, Pamela.** U.S. official on terror attacks: 'This isn't going to stop'. [Online] CNN, January 9, 2015. http://www.cnn.com/2015/01/09/us/france-attacks-u-s-/index.html.
2. **CDC.** Parasites - Dracunculiasis (also known as Guinea Worm Disease). [Online] Centers for Disease Control and Prevention, April 14, 2014. http://www.cdc.gov/parasites/guineaworm/.
3. **The Carter Center.** Guinea Worm Disease Eradication. [Online] The Carter Center, 2011. http://www.cartercenter.org/health/guinea_worm/mini_site/index.html.
4. **Dougherty, Jill.** Experts: No easy cure for the disease of terror. [Online] CNN, July 27, 2012. http://security.blogs.cnn.com/2012/07/27/experts-no-easy-cure-for-the-disease-of-terror/.
5. **CBS News.** Ex-CIA Operative Comes Out of the Shadows. [Online] CBS News, August 2, 2010. http://www.cbsnews.com/8301-18560_162-6014887.html.
6. **Jett, Dennis.** The Real 'American Sniper' Had No Remorse About the Iraqis He Killed. [Online] New Republic, January 13, 2015. http://www.newrepublic.com/article/120763/american-sniper-clint-eastwood-biopic-misrepresents-chris-kyle.
7. **Starr, Barbara.** Official: U.S. keeping ISIS kill list. [Online] CNN, February 18, 2015. http://www.cnn.com/2015/02/18/politics/us-isis-kill-list/index.html.
8. **Thoreau, Henry David.** Walden; or, Life in the Woods. [Online] Ticknor and Fields, August 9, 1854. http://.
9. **Lovett, Jon.** The Culture of Shut Up. [Online] The Atlantic, April 7, 2014. http://www.theatlantic.com/politics/archive/2014/04/the-culture-of-shut-up/360239/.
10. **TED.** The technological future of crime: Marc Goodman at TEDGlobal 2012. [Online] TED, June 28, 2012. http://blog.ted.com/2012/06/28/the-technological-future-of-crime-marc-goodman-at-tedglobal-2012/.
11. **Hammurabi.** Hammurabi's Code: An Eye for an Eye. [Online] ushistory.org, 1792-50 B.C. http://www.ushistory.org/civ/4c.asp.
12. **Orwell, George.** Revenge is Sour. [Online] The Tribune, November 9, 1945. http://www.george-orwell.org/Revenge_is_Sour/0.html.
13. **Robinson, Adam.** Bin Laden: Behind the Mask of a Terrorist. [Online] Arcade Publishing, November 22, 2002. http://books.google.com/books?id=zTGPHuW4qGIC&pg=PT118.
14. **Chief of Naval Operations.** The United States Navy In "Desert Shield" / "Desert Storm". [Online] Naval History & Heritage, May 15, 1991. http://www.history.navy.mil/wars/dstorm/index.html.
15. **Schmitt, Eric.** U.S. to Withdraw All Combat Forces From Saudi Arabia. [Online] The New York Times, April 29, 2003. http://www.nytimes.com/2003/04/29/international/worldspecial/29CND-RUMS.html.

16. **Ray, Julie.** Opinion Briefing: U.S. Image in Middle East/North Africa. [Online] Gallup, January 27, 2009. http://www.gallup.com/poll/114007/opinion-briefing-image-middle-east-north-africa.aspx.

17. **Starr, Barbara.** How the Patriot deployment to Turkey will work. [Online] CNN, December 14, 2012. http://security.blogs.cnn.com/2012/12/14/how-the-patriot-deployment-to-turkey-will-work/.

18. **Quilliam.** The Attack on the US Consulate Was A Planned Terrorist Assault Against US and Libyan Interests. [Online] Quilliam, September 12, 2012. http://www.quilliamfoundation.org/press-releases/the-attack-on-the-us-consulate-was-a-planned-terrorist-assault-against-us-and-libyan-interests/.

19. **Zway, Suliman Ali and Fahim, Kareem.** Angry Libyans Target Militias, Forcing Flight. [Online] The New York Times, September 21, 2012. http://www.nytimes.com/2012/09/22/world/africa/pro-american-libyans-besiege-militant-group-in-benghazi.html.

20. **BBC News.** French terror attacks: Victim obituaries. [Online] BBC News Europe, January 13, 2015. http://www.bbc.com/news/world-europe-30724678.

21. **Bilefsky, Dan and Baume, Maïa De La.** Terrorists Strike Charlie Hebdo Newspaper in Paris, Leaving 12 Dead. [Online] The New York Times, January 7, 2015. http://www.nytimes.com/2015/01/08/world/europe/charlie-hebdo-paris-shooting.html.

22. **Fantz, Ashley.** Array of world leaders joins 3.7 million in France to defy terrorism. [Online] CNN, January 12, 2015. http://www.cnn.com/2015/01/11/world/charlie-hebdo-paris-march/.

23. **Thompson, Mark.** Anonymous declares war over Charlie Hebdo attack. [Online] CNN, January 9, 2015. http://money.cnn.com/2015/01/09/technology/anonymous-charlie-hebdo-terrorists/index.html.

24. **Pagliery, Jose.** Meet the vigilante who hacks jihadists . [Online] CNN, January 16, 2015. http://money.cnn.com/2015/01/16/technology/security/jester-hacker-vigilante/index.html.

25. **BBC News.** Charlie Hebdo attack: Print run for new issue expanded. [Online] BBC News Europe, January 14, 2015. http://www.bbc.com/news/world-europe-30808284.

26. **Grove, Lloyd.** Bill Maher: Hundreds of Millions of Muslims Support Attack on 'Charlie Hebdo'. [Online] The Daily Beast, January 8, 2015. http://www.thedailybeast.com/articles/2015/01/08/bill-maher-hundreds-of-millions-of-muslims-support-attack-on-charlie-hebdo.html.

27. **Murdoch, Rupert.** Maybe most Moslems peaceful... [Online] Twitter, January 9, 2015. https://twitter.com/rupertmurdoch/status/553734788881076225.

28. **Kennedy, Maev.** JK Rowling attacks Murdoch for tweet blaming all Muslims for Charlie Hebdo deaths. [Online] The Guardian, January 11, 2015. http://www.theguardian.com/books/2015/jan/11/jk-rowling-condemns-murdoch-tweet-charlie-hebdo-harry-potter-news-corp-muslims-christian.

29. **Blumberg, Antonia.** Mosques Attacked In Wake Of Charlie Hebdo Shooting. [Online] The Huffington Post, January 9, 2015. http://www.huffingtonpost.com/2015/01/08/mosque-attacks-charlie-hebdo_n_6436224.html.

30. **Obeidallah, Dean.** I'm Muslim, and I hate terrorism. [Online] CNN, April 24, 2013. http://www.cnn.com/2013/04/24/opinion/obeidallah-muslims-hate-terrorism/index.html.

31. **Guardian staff.** Murdoch says Muslims must be held responsible for France terror attacks . [Online] The Guardian, January 10, 2015. http://www.theguardian.com/world/2015/jan/10/rupert-murdoch-muslims-must-be-held-responsible-for-france-terror-attacks.

32. **Shadbolt, Peter.** Conflict in Buddhism: 'Violence for the sake of peace?'. [Online] CNN, April 23, 2013. http://www.cnn.com/2013/04/22/world/asia/buddhism-violence/index.html.

33. **The Metropolitan Museum of Art.** The Crusades (1095–1291). [Online] The Metropolitan Museum of Art . [Cited: December 9, 2012.] http://www.metmuseum.org/toah/hd/crus/hd_crus.htm.

34. **Khadder, Kareem, Botelho, Greg and Levs, Josh.** Palestinian teen's abduction, killing intensifies tensions in Mideast. [Online] CNN, July 3, 2014. http://www.cnn.com/2014/07/02/world/meast/mideast-tensions/index.html.

35. **Bacon, Francis Sir.** Essayes and counsels, civil and moral. [Online] 1664. http://www.folge
r.edu/eduPrimSrcDtl.cfm?psid=123.

36. **IMDb.** Memorable quotes for Star Trek II: The Wrath of Khan. [Online] IMDb, 1982.
http://www.imdb.com/title/tt0084726/quotes.

37. **Ackman, Dan.** The Cost Of Being Osama Bin Laden . [Online] Forbes, September 14,
2001. http://www.forbes.com/2001/09/14/0914ladenmoney.html.

38. **The Economist.** Osama bin Laden. [Online] The Economist, May 5, 2011. http://www.
economist.com/node/18648254.

39. **Corera, Gordon.** MI5 head warns of serious risk of UK terrorist attack. [Online] BBC
News, September 16, 2010. http://www.bbc.co.uk/news/uk-11335412.

40. **Robertson, Nic and Cruickshank, Paul.** Cagefighter 'cures' terrorists. [Online] CNN, July
23, 2012. http://www.cnn.com/2012/07/20/world/europe/uk-caging-terror-main/index.html.

41. **—.** Convicted terrorist calmed by cagefighting. [Online] CNN, July 28, 2012. http://
edition.cnn.com/2012/07/22/world/europe/uk-caging-terror-mansha/index.html.

42. **Woolf, Nicky.** 'PUAhate' and 'ForeverAlone': inside Elliot Rodger's online life.
[Online] The Guardian, May 30, 2014. http://www.theguardian.com/world/2014/may/30/
elliot-rodger-puahate-forever-alone-reddit-forums.

43. **Levinson, Brian.** I Could Have Been Elliot Rodger. [Online] Slate, May 31, 2014.
http://www.slate.com/articles/life/dispatches/2014/05/i_could_have_been_elliot_rodger_
young_frustrated_and_full_of_rage_toward.single.html.

44. **Ryzik, Melena.** A Dark Prince Steps Into the Light. [Online] The New York Times,
January 15, 2015. http://www.nytimes.com/2015/01/18/arts/music/a-dark-prince-steps-into-
the-light-.html.

45. **Nelson, Dean.** Slumdog Millionaire: Meet the real Mumbai street urchins. [Online]
The Telegraph, January 18, 2009. http://www.telegraph.co.uk/news/worldnews/asia/
india/4280812/Slumdog-Millionaire-Meet-the-real-Mumbai-street-urchins.html.

46. **msnbc.com.** WikiLeaks: U.K. trained Bangladeshi 'death squad'. [Online] MSNBC,
December 21, 2010. http://www.msnbc.msn.com/id/40773855/ns/us_news-wikileaks_in_
security/.

47. **Ahmed, Saeed, Cohen, Lisa and Sidner, Sara.** From horror to hope: Boy's miracle recovery from brutal attack. [Online] CNN, December 7, 2012. http://www.cnn.com/2012/12/06/
world/freedom-project-operation-hope/index.html.

48. **Ghosh, Bobby.** After Waterboarding: How to Make Terrorists Talk? [Online] Time
Magazine, June 8, 2009. http://www.time.com/time/magazine/article/0,9171,1901491,00.h
tml.

49. **Wright, Lawrence.** The Agent. Did the C.I.A. stop an F.B.I. detective from preventing 9/11? [Online] The New Yorker, July 10, 2006. http://www.newyorker.com/
archive/2006/07/10/060710fa_fact_wright?currentPage=all.

50. **60 Minutes.** Ex-FBI agent who interrogated Qaeda members speaks out. [Online] CBS
Interactive Inc, September 9, 2011. http://www.cbsnews.com/8301-18560_162-20104007/
ex-fbi-agent-who-interrogated-qaeda-members-speaks-out-/.

51. **Stavridis, James.** James Stavridis: A Navy Admiral's thoughts on global security. [Online] TED,
June 2012. http://www.ted.com/talks/james_stavridis_how_nato_s_supreme_commander_thinks_
about_global_security.html.

52. **Dear, John S.J.** Afghanistan journal, part one: Learning a nonviolent lifestyle in Kabul.
[Online] National Catholic Reporter, December 11, 2012. http://ncronline.org/blogs/
road-peace/afghanistan-journal-part-one-learning-nonviolent-lifestyle-kabul.

53. **—.** Afghanistan journal, part two: bearing witness to peacemaking in a war-torn country. [Online] National Catholic Reporter, December 18, 2012. http://ncronline.org/blogs/
road-peace/afghanistan-journal-part-two-bearing-witness-peacemaking-war-torn-country.

54. **Adichie, Chimamanda.** Chimamanda Adichie: The danger of a single story. [Online]
TED, October 2009. http://www.ted.com/talks/chimamanda_adichie_the_danger_of_a_
single_story.html.

55. **Greenberg, Andy.** NSA Director Heckled At Conference As He Asks For Security Community's Understanding. [Online] Forbes, July 31, 2013. http://www.forbes.com/sites/ andygreenberg/2013/07/31/nsa-director-heckled-at-conference-as-he-asks-for-security-com- munitys-understanding/.

56. **Chomsky, Noam and Schivone, Gabriel Matthew.** United States of Insecurity: Interview with Noam Chomsky. [Online] Monthly Review, May 2008. http://monthlyreview. org/2008/05/01/united-states-of-insecurity-interview-with-noam-chomsky.

57. **Engber, Daniel.** I'm Covered in Leaflets! [Online] Slate, July 18, 2006. http://www. slate.com/articles/news_and_politics/explainer/2006/07/im_covered_in_leaflets.html.

58. **Digital Daya.** Research Note: World Leaders on Twitter. [Online] Digital Daya, December 2012. http://www.digitaldaya.com/admin/modulos/galeria/pdfs/69/156_biqz7730.pdf.

59. **Press Association.** David Cameron gets 100,000 Twitter followers days after starting account. [Online] The Guardian, October 9, 2012. http://www.guardian.co.uk/politics/2012/ oct/09/david-cameron-100000-twitter-followers.

60. **Atkins, Christopher R.** "I solemnly swear to defend the Constitution of the United States of America against all enemies, foreign and domestic.". [Online] Michael Moore, January 28, 2005. http://www.michaelmoore.com/words/soldiers-letters/i-solemnly-swear-to-defend- the-constitution-of-the-united-states-of-america-against-all-enemies-foreign-and-domestic.

61. **Kansas high school students.** We Are Hungry. [Online] YouTube, September 17, 2012. http ://www.youtube.com/watch?v=2IB7NDUSBOo.

62. **Cohen, Elizabeth.** Peanut butter, garlic bread back on school plates. [Online] CNN, December 12, 2012. http://www.cnn.com/2012/12/12/health/school-lunch-changes/index.html.

63. **Facebook.** Form S-1 Registration Statement. [Online] United States Securities and Exchange Commission, February 1, 2012. http://sec.gov/Archives/edgar/ data/1326801/000119312512034517/d287954ds1.htm.

64. **Levine, Adam.** A social network site for jihadists? [Online] CNN, April 5, 2002. http://secu rity.blogs.cnn.com/2012/04/05/faqebook-dreams-of-a-jihadi-social-network/.

65. **Drash, Wayne.** Bringing healing to Newtown, one pie at a time. [Online] CNN, December 19, 2012. http://eatocracy.cnn.com/2012/12/19/bringing-healing-to-newtown- one-pie-at-a-time/.

66. **Frontline.** News War. [Online] wgbh educational foundation. [Cited: December 11, 2012.] http://www.pbs.org/wgbh/pages/frontline/teach/newswar/hand1.html.

67. **Bohlen, Celestine.** In New War on Terrorism, Words Are Weapons, Too. [Online] The New York Times, September 29, 2001. http://www.nytimes.com/2001/09/29/arts/ think-tank-in-new-war-on-terrorism-words-are-weapons-too.html.

68. **ABC News.** Maher Apologizes for 'Cowards' Remark. [Online] ABC News, September 20, 2001. http://abcnews.go.com/Entertainment/story?id=102318&page=1.

69. **Apple, R. W. Jr.** Pentagon Papers. [Online] The New York Times, June 23, 1996. http:// topics.nytimes.com/top/reference/timestopics/subjects/p/pentagon_papers/index.html.

70. **Haberman, Clyde.** Arthur O. Sulzberger, Publisher Who Transformed The Times for New Era, Dies at 86. [Online] 2012, 29 September. http://www.nytimes.com/2012/09/30/nyreg ion/arthur-o-sulzberger-publisher-who-transformed-times-dies-at-86.html?pagewanted=all.

71. **Peters, Jeremy W.** Latest Word on the Trail? I Take It Back. [Online] The New York Times, July 15, 2012. http://www.nytimes.com/2012/07/16/us/politics/latest-word-on-the- campaign-trail-i-take-it-back.html?pagewanted=all.

72. **Rather, Dan.** Dan Rather: 'Quote approval' a media sellout. [Online] CNN, July 19, 2012. http://www.cnn.com/2012/07/19/opinion/rather-quote-approval-reporting/index.html.

73. **IMDb.** Fahrenheit 9/11. [Online] IMDb, June 25, 2004. http://www.imdb.com/title/tt0361596/.

74. **Rutenberg, Jim.** Disney Is Blocking Distribution of Film That Criticizes Bush. [Online] The New York Times, May 5, 2004. http://www.nytimes.com/2004/05/05/ us/disney-is-blocking-distribution-of-film-that-criticizes-bush.html.

75. **Keller, Bill.** WikiLeaks, a Postscript. [Online] The New York Times, February 19, 2012. http ://www.nytimes.com/2012/02/20/opinion/keller-wikileaks-a-postscript.html.

76. **Ellsberg, Daniel.** Daniel Ellsberg on Colbert Report: Julian Assange is Not a Criminal Under the Laws of the United States. [Online] Ellsberg.Net, December 10, 2010. http://www.ellsberg.net/archive/daniel-ellsberg-on-colbert-report.

77. **Blanton, Thomas.** Hearing on the Espionage Act and the Legal and Constitutional Implications of Wikileaks. [Online] Committee on the Judiciary, U.S. House of Representatives, December 16, 2010. http://www.gwu.edu/~nsarchiv/news/20101216/Blanton101216.pdf.

78. **RT.** WikiLeaks revelations only tip of iceberg – Assange. [Online] RT, May 3, 2011. http://rt.com/news/wikileaks-revelations-assange-interview/.

79. **Bulwer-Lytton, Edward.** Richelieu, or The conspiracy: in five acts. [Online] Chapman and Hall, 1856. http://books.google.com/books?id=FfktAAAAYAAJ.

80. **King, John and Carter, Chelsea J.** Lawmaker: If CNN can interview suspect in Benghazi attack, why can't FBI? [Online] CNN, August 7, 2013. http://www.cnn.com/2013/07/31/politics/benghazi-investigation-suspect/index.html.

81. **Holmes, Oliver Wendell Jr.** Schenck v. United States. [Online] Cornell University Law School, March 3, 1919. http://www.law.cornell.edu/supct/html/historics/USSC_CR_0249_0047_ZO.html.

82. **FBI.** 1919 Bombings. [Online] Federal Bureau of Investigation. [Cited: December 12, 2012.] http://www.fbi.gov/philadelphia/about-us/history/famous-cases/famous-cases-1919-bombings.

83. **Supreme Court of the United States.** Brandenburg v. Ohio. [Online] Cornell University Law School, June 9, 1969. http://www.law.cornell.edu/supct/html/historics/USSC_CR_0395_0444_ZO.html.

84. **CNN Wire Staff.** No let-up in protests over anti-Islam film. [Online] CNN, September 18, 2012. http://www.cnn.com/2012/09/17/world/film-protests/index.html.

85. U.S. Missions Stormed in Libya, Egypt. [Online] Bradley, Matt; Nissenbaum, Dion, September 12, 2012. http://online.wsj.com/article/SB100008723963904440175045776456810574982 66.html.

86. **York, Jillian C.** Should Google censor an anti-Islam video? . [Online] CNN, September 16, 2012. http://www.cnn.com/2012/09/14/opinion/york-libya-youtube/index.html.

87. **Madhani, Aamer.** Cleric al-Awlaki dubbed 'bin Laden of the Internet'. [Online] USA Today, August 24, 2010. http://usatoday30.usatoday.com/news/nation/2010-08-25-1A_Awlaki25_CV_N.htm.

88. **Gardham, Duncan, Rayner, Gordon and Bingham, John.** Why hasn't YouTube taken down terror videos? [Online] The Telegraph, November 3, 2010. http://www.telegraph.co.uk/news/politics/8106672/Why-hasnt-YouTube-taken-down-terror-videos.html.

89. **Basu, Moni.** New details emerge of anti-Islam film's mystery producer. [Online] CNN, September 14, 2012. http://www.cnn.com/2012/09/13/world/anti-islam-filmmaker/index.html.

90. **Cawthon, Erinn.** Controversial 'Defeat Jihad' ad to appear in NYC subways. [Online] CNN, September 19, 2012. http://www.cnn.com/2012/09/19/us/new-york-controversial-subway-ad/index.html.

91. **CBS Radio New York.** Bloomberg Weighs In On Provocative Subway Ad. [Online] CBS Radio New York, September 21, 2012. http://newyork.cbslocal.com/2012/09/21/bloomberg-weighs-in-on-provocative-subway-ad/.

92. **Kahn-Troster, Rachel.** Subway ads: A right to hate speech, a duty to condemn. [Online] CNN, September 25, 2012. http://www.cnn.com/2012/09/25/opinion/kahn-troster-anti-islam-hate-ads/index.html.

93. **Gandhi, Mohandas K.** An Autobiography: The Story of My Experiments with Truth. [Online] Beacon Press, 1993. http://books.google.com/books/about/An_Autobiography.html?id=VsMLYjEsyaEC.

94. **Sgueglia, Kristina.** Interfaith group protests ad that says 'Support Israel. Defeat Jihad'. [Online] CNN, September 25, 2012. http://religion.blogs.cnn.com/2012/09/25/interfaith-group-protests-ad-that-says-support-israel-defeat-jihad/.

95. North Korea complains to UN about Seth Rogen comedy The Interview. [Online] The Guardian, July 10, 2014. http://www.theguardian.com/film/2014/jul/10/north-korea-un-the-interview-seth-rogen-james-franco.
96. **Stelter, Brian.** Charlie Hebdo co-founder: Prophet cartoons went too far. [Online] CNNMoney, January 17, 2015. http://money.cnn.com/2015/01/16/media/charlie-hebdo-henri-roussel/.
97. **McKenna, Josephine.** Pope Francis on free speech: 'You cannot insult the faith of others'. [Online] Religion News Service, January 15, 2015. http://www.religionnews.com/2015/01/15/pope-francis-free-speech-cannot-insult-faith-others/.
98. **Levs, Josh, Atay-Alam, Hande and Bilginsoy, Zeynep.** Turkey bans Charlie Hebdo cover, newspaper gets death threats. [Online] CNN, January 14, 2015. http://www.cnn.com/2015/01/14/world/turkey-charlie-hebdo/index.html.
99. **Burke, Daniel.** Muslims' mixed response to new Mohammed cover. [Online] CNN, January 13, 2015. http://www.cnn.com/2015/01/13/living/muslims-respond-hebdo/index.html.
100. **Bordelon, Brendan.** 'I AM NOT CHARLIE': Leaked Newsroom E-mails Reveal Al Jazeera Fury over Global Support for Charlie Hebdo. [Online] National Review Online, January 9, 2015. http://m.nationalreview.com/corner/396131/i-am-not-charlie-leaked-newsroom-e-mails-reveal-al-jazeera-fury-over-global-support.
101. **Sharp, Gene.** 198 Methods of Nonviolent Action. [Online] Porter Sargent Publishers, 1973. http://www.aeinstein.org/organizations103a.html.
102. —. 198 Methods of Nonviolent Action. [Online] The Albert Einstein Institution. [Cited: December 27, 2012.] http://www.aeinstein.org/organizations/org/198_methods.pdf.
103. **Meek, James.** Divided they stand. [Online] The Guardian, December 9, 2004. http://www.guardian.co.uk/world/2004/dec/10/ukraine.jamesmeek.
104. **Lee, Amy.** Egypt's Revolutionary Music, And 7 Other Revolutions That Turned To Song. [Online] The Huffington Post, January 25, 2012. http://www.huffingtonpost.com/2012/01/25/egypt-revolution-january-25_n_1229332.html.
105. **Buist, Erica.** That's me in the picture: Markiyan Matsekh plays the piano for riot police in Kiev, 7 December 2013. [Online] The Guardian, December 7, 2013. http://www.theguardian.com/artanddesign/2014/dec/05/thats-me-in-picture-ukraine-protest-piano-matsekh.
106. **Tagliabue, John.** Danilo Dolci, Vivid Voice Of Sicily's Poor, Dies at 73. [Online] The New York Times, December 31, 1997. http://www.nytimes.com/1997/12/31/world/danilo-dolci-vivid-voice-of-sicily-s-poor-dies-at-73.html.
107. **Soussan, Michael and Weingarten, Elizabeth.** What really scares terrorists. [Online] CNN, December 26, 2014. http://www.cnn.com/2014/12/26/opinion/soussan-weingarten-gender-equality/index.html.
108. **Anderson, L.V.** Do Sex Strikes Ever Work? [Online] Slate, August 27, 2012. http://www.slate.com/articles/news_and_politics/explainer/2012/08/sex_strike_in_togo_do_sex_strikes_ever_work_.html.
109. **The Associated Press.** Syria shuts down Internet access as country imposes nationwide online blackout. [Online] New York Daily News, November 29, 2012. http://www.nydailynews.com/news/world/nationwide-internet-blackout-syria-article-1.1210074.
110. **Clinton, Bill.** What We Learned in Oklahoma City. [Online] The New York Times, April 18, 2010. http://www.nytimes.com/2010/04/19/opinion/19clinton.html.
111. **Borger, Julian.** McVeigh brushes aside deaths. [Online] The Guardian, March 29, 2001. http://www.guardian.co.uk/world/2001/mar/30/julianborger.
112. **Morrison, Blake.** Special report: Eric Rudolph writes home. [Online] USA Today, July 5, 2005. http://usatoday30.usatoday.com/news/nation/2005-07-05-rudolph-cover-partone_x.htm.
113. **Koehler, Robert.** You Can't Kill an Idea. [Online] The Huffington Post, March 3, 2011. http://www.huffingtonpost.com/robert-koehler/you-cant-kill-an-idea_b_830881.html.
114. **Flock, Elizabeth.** Osama bin Laden tells his children not to fight jihad in his will. [Online] The Washington Post, May 4, 2011. http://www.washingtonpost.com/blogs/blogpost/post/osama-bin-laden-tells-children-not-to-fight-jihad-in-his-will/2011/05/04/AFDP4UmF_blog.html.

115. **Obama, Barack.** Remarks by the President in a Press Conference. [Online] The White House, December 19, 2012. http://www.whitehouse.gov/the-press-office/2012/12/19/remarks-president-press-conference.
116. **Miller, Sara.** Obama: Peace is the only path to true security. [Online] The Jerusalem Post, March 21, 2013. http://www.jpost.com/Diplomacy-and-Politics/Obama-Peace-is-the-only-path-to-true-security-307323.
117. **Madi, Mohamed, et al.** As it happened: Charlie Hebdo attack. [Online] BBC News, January 7, 2015. http://www.bbc.com/news/live/world-europe-30710777.
118. **Akkoc, Raziye.** 'Insulting and provocative': how the world reacted to Charlie Hebdo's Prophet Mohammed cover. [Online] The Telegraph, January 14, 2015.
119. **Monroe, Bryan.** 5 things we learned from Obama's speech. [Online] CNN, May 24, 2013. http://www.cnn.com/2013/05/23/politics/5-things-obama-terror/index.html.

Chapter 5
War and Peace

World peace is as simple and elegant as $E = mc^2$.
—Newton Lee

Peace cannot be kept by force. It can only be achieved by understanding.
—Albert Einstein (December 14, 1930)

Too many of us think [peace] is impossible. Too many think it unreal. But that is a dangerous, defeatist belief. ... Our problems are manmade—therefore they can be solved by man.
—President John F. Kennedy (June 10, 1963)

We must either love each other, or we must die.
—President Lyndon Johnson in "Peace Little Girl (Daisy)" (September 7, 1964)

Violence never brings permanent peace. It solves no social problem: it merely creates new and more complicated ones.
—Martin Luther King Jr.
Nobel Peace Prize acceptance speech (December 11, 1964)

The belief that peace is desirable is rarely enough to achieve it. Peace requires responsibility. Peace entails sacrifice.
—President Barack Obama,
Nobel Peace Prize acceptance speech (December 10, 2009)

If there is an Internet connection, my camera is more powerful [than my AK-47].
—Syrian dissident Abu Ghassan (June 2012)

Instead of building walls to create security, we need to build bridges.
—U.S. Navy Admiral and former NATO's Supreme Allied Commander James Stavridis, TEDGlobal 2012 (June 2012)

© Springer International Publishing Switzerland 2015
N. Lee, *Counterterrorism and Cybersecurity*, DOI 10.1007/978-3-319-17244-6_5

We believe that relationships between nations aren't just about
relationships between governments or leaders—they're about
relationships between people, particularly young people.
                                    —First Lady Michelle Obama (March 25, 2014)

The meaning of our whole life and existence is love.
                                    —Russian President Vladimir Putin (November 7, 2014)

5.1 War as State-Sponsored Terrorism

The Federal Bureau of Investigation (FBI) defines terrorism as "the unlawful use
of force or violence against persons or property to intimidate or coerce a government, the civilian population, or any segment thereof in furtherance of political
or social objectives" [1]. MIT Professor Emeritus Noam Chomsky believes that
the U.S. official doctrine of low-intensity warfare is almost identical to the official
definition of terrorism [2]. Political commentator Bill Maher equates U.S. drone
attacks with terrorist acts [3].

While a terrorist act causes innocent people pain, suffering, and even death, war
is legitimized state-sponsored terrorism in a grand scale. In World War II, 15 million soldiers died in battles while 45 million civilians perished under war-related
circumstances [4]. Between five and six million Jews were killed in the Holocaust
[5]. Over 27 % of the civilian population in Hiroshima and 24 % of the residents
in Nagasaki were wiped out by atomic bombs [6].

In war-torn countries, people live in constant fear. Jesuit priest and peace activist
John Dear interviewed families at the refugee camps in Afghanistan in December 2012.
Raz Mohammad, a member of Afghan Peace Volunteers, told his somber story [7]:

> My brother-in-law was killed by a U.S. drone in 2008. He was a student, and visiting
> some friends one summer evening when they decided to walk to a garden and sit there
> and talk. They were enjoying the evening, sitting in the garden, when a drone flew by
> and dropped a bomb. Everyone was incinerated. We couldn't find any remains. My sis-
> ter was left behind with her baby boy. I think the drone attacks were first begun in my
> province. We hear them about every three nights. They have a low, buzzing sound, like a
> mosquito. They hover over us. They fly over us during the day, and fly over us during the
> night, when we can see the spotlight at the front of the drone. Occasionally, the large U.S.
> fighter bombers fly over, and they make a huge noise. All the people of the area, espe-
> cially the children, are afraid of the U.S. soldiers, the U.S. tanks, the U.S. drones, and the
> U.S. fighter bombers. They fear being killed.
>
> …
>
> No one I know wants the war to continue. Ordinary people everywhere are sick and tired
> of war, yet we're demonized as warriors and terrorists. None of us can tell who is a mem-
> ber of the Taliban and who isn't. If we can't tell who is a member of the Taliban, how can
> anyone in the U.S. claim to know who is in the Taliban? Meanwhile, our schools, hospi-
> tals and local services have all collapsed. The U.S./NATO forces are not helping anyone,
> only bringing fear and death to the people.

At a women's sewing cooperative in Afghanistan, a woman expressed her frustration and pleaded: "I thought President Obama would care for the oppressed, but he

has made things much worse for us. He is even worse than President Bush. Please ask the people of the U.S. to take to the streets again and do what they can to stop this war now" [7].

The woman's plea did not fall on deaf ears. In January 2013, White House officials said that the Obama administration was considering the withdrawal of all U.S. troops from Afghanistan after Operation Enduring Freedom in Afghanistan (OEF-A) officially ended in late 2014 [8]. However, the White House announced in May 2014 that a residual force of 9,800 U.S. troops would remain in Afghanistan until the end of 2015, and the number would be reduced at the end of 2016 to a small military presence at the U.S. Embassy [9].

In a way, President Barack Obama acknowledged his predicament when he accepted the 2009 Nobel Peace Prize amid controversy and in the midst of two wars. Obama said, "We are at war, and I'm responsible for the deployment of thousands of young Americans to battle in a distant land. Some will kill, and some will be killed. And so I come here with an acute sense of the costs of armed conflict — filled with difficult questions about the relationship between war and peace, and our effort to replace one with the other. ... To say that force may sometimes be necessary is not a call to cynicism — it is a recognition of history; the imperfections of man and the limits of reason" [10].

5.2 Complacency in War

Leo Tolstoy wrote in his 1869 novel *War and Peace*, "Война не любезность, а самое гадкое дело в жизни, и надо понимать это и не играть в войну. (War is not a courtesy but the most horrible thing in life; and we ought to understand that, and not play at war.)" [11].

At the Battle of Fredericksburg on December 13, 1862, U.S. Confederate General Robert E. Lee said to his men, "It is well that war is so terrible, otherwise we should grow too fond of it" [12]. Notwithstanding the horrors of war, desensitization to violence has perpetuated armed conflicts around the world, and the public has become inured to all but the most catastrophic or apocalyptic events.

Take the decades-long Gaza-Israel conflict as an example. Both Israelis and Palestinians have died from rocket attacks, airstrikes, and shootings. When the conflict escalated and death toll rose in November 2012, U.S. Secretary of State Hillary Clinton went to Jerusalem to meet with Israel Prime Minister Benjamin Netanyahu to push for a truce [13]. When a ceasefire was announced in Cairo, both sides claimed victory and accused the other as the oppressor. Israel killed Hamas' military leader Ahmed al-Jaabari and significantly weakened their offensive capability [14], whereas Hamas declared triumph with diplomatic support from Egypt, Turkey, and Qatar [15].

Albert Einstein said in his speech to the New History Society on December 14, 1930 that "Peace cannot be kept by force. It can only be achieved by understanding." In his Nobel Peace Prize acceptance speech on December 11, 1964, Martin

Luther King Jr. said, "Nations have frequently won their independence in battle. But in spite of temporary victories, violence never brings permanent peace. It solves no social problem: it merely creates new and more complicated ones. Violence is impractical because it is a descending spiral ending in destruction for all. It is immoral because it seeks to humiliate the opponent rather than win his understanding: it seeks to annihilate rather than convert" [16].

Win or lose, innocent civilians are caught in the middle of the deadly violence [17]. "Innocent people, including children, have been killed or injured on both sides. Families on both sides were forced to cower in fear as the violence raged around them," said U.N. Secretary-General Ban Ki-moon [18].

In an armed conflict, the International Humanitarian Law protects "those who do not take part in the fighting, such as civilians and medical and religious military personnel" [19]. A major part of the law is contained in the four Geneva Conventions of 1949 and nearly every country in the world has agreed to be bound by them. With that in mind, the United Nations fact-finding mission headed by Judge Richard Goldstone published a 574-page report in September 2009 accusing both Hamas and Israel of deliberately targeting civilians [20]. Goldstone later retracted the findings, but a senior Israel military official described in November 2012 that civilian casualties as "regrettable" but unavoidable because the "terrorist infrastructure is embedded inside the population" [21].

Politicians and the military-industrial complex have grown to be complacent about war. Violence has become the first course of action and the first choice of reaction in dealing with international conflicts, even at the expense of innocent civilians. In fact, according to American and Israeli officials, the November 2012 Gaza-Israel conflict was considered "a practice run for any future armed confrontation with Iran, featuring improved rockets that can reach Jerusalem and new antimissile systems to counter them" [22].

War has become a popular option in the political playbook, a child's play with deadly consequences. An elder in an Afghanistan refugee camp put it this way, "The powers that be have turned Afghanistan into a killing field, their personal playground of war" [7]. Or as Paul the Apostle wrote in *I Corinthians* 4:9, "For it seems to me that God has put us apostles on display at the end of the procession, like those condemned to die in the arena. We have been made a spectacle to the whole universe, to angels as well as to human beings."

5.3 The Warrior's Code

Prof. Shannon E. French at Case Western Reserve University and the U.S. Naval Academy spoke of the warrior's code: "The code is designed to prevent soldiers from becoming monsters. Butchering civilians, torturing prisoners, desecrating the enemies' bodies — are all battlefield behaviors that erode a soldier's humanity. … There is something worse than death, and one of those things is to completely lose your humanity" [23].

In the 2013 movie *Horns* starring Daniel Radcliffe and Juno Temple, the protagonist said, "Man is not born evil. The Devil himself was a fallen angel" [24].

Indeed, there are hope and kindness in everyone if we dig deep enough. There are true stories about how soldiers risk their lives for the enemy in acts of military chivalry through the ages. In the book *A Higher Call*, the authors told an incredible story of American lieutenant Charlie Brown and German lieutenant Franz Stigler for their combat and chivalry in the war-torn skies of World War II. CNN's John Blake mentioned "Union and Confederate troops risked their lives to aid one another during the Civil War" and "one renowned American general traveled back to Vietnam to meet the man who almost wiped out his battalion, and the two men hugged and prayed together" [25].

On February 11, 2013, Army Staff Sgt. Clint Romesha became the fourth living person to receive the Medal of Honor for his uncommon valor in Afghanistan [26]. Romesha told reporters after the White House ceremony, "I stand here with mixed emotions of both joy and sadness today. The joy comes from recognition from us doing our jobs as soldiers on distant battlefields. But it is countered by the constant reminder of the loss of our battle buddies — my battle buddies, my soldiers, my friends" [27]. Romesha declined First Lady Michelle Obama's invitation to be her guest at the 2013 State of the Union Address. Instead he spent the evening with families and friends from his former unit, Black Knight Troop, 3-61 CAV [28].

In the 2013 science fiction film *Ender's Game* based on the novel of the same name by Orson Scott Card, protagonist Ender Wiggin (played by Asa Butterfield) was devastated to learn that his final test of a war game simulation was real, resulting in actual deaths of his comrades and total annihilation of the enemy's home planet [29]. The novel is one of the required books and annual reading requirements for all active duty and reserve, officer and enlisted Marines. The United States Marine Corporation University (USMCU) describes the novel as follows:

"In this science fiction novel, child genius Ender Wiggin is chosen by international military forces to save the world from destruction by a deadly alien race. His skills make him a leader yet Ender suffers from isolation and rivalry from his peers, pressure from adults, and fear of the enemy. His psychological battles include loneliness and fear that he is becoming like his cruel brother. The novel's major theme is the concept of a 'game' and all of the other important ideas in the novel are interpreted through this concept. Some of the important ideas in the book include: the relationship between children and adults, compassion, ruthlessness, friends and enemies, and the question of humanity: what it means to be human" [30].

5.4 Civilians Wanting Peace

In the 2011 film *Sherlock Holmes: A Game* of Shadows, Professor James Moriarty postulates, "You see, hidden within the unconscious is an insatiable desire for conflict. So you're not fighting me so much as you are the human condition. All I want is to supply the bullets and the bandages. War on an industrial scale is inevitable" [31].

Perhaps Professor Moriarty strikes a chord with the military-industrial complex, but civilians want peace more than anything else. Afghan Peace Volunteer Raz Mohammad told Jesuit priest John Dear, "We should not accept these drone attacks if we are human beings. They are killing innocent human beings. Humanity should not allow this to happen. No one I know wants the war to continue. Ordinary people everywhere are sick and tired of war" [7].

In November 2012, doctors at the Sheba Medical Center at a Tel Aviv hospital treated a four-year-old boy from Israel and an eight-year-old girl from Gaza who both lost fingers from rocket blasts. Dr. Batia Yaffe remarked, "I come to think about what is it about this piece of land that everybody is fighting about it all the time. This is what comes to my mind: whether this is our lot for eternity from now on. Always have injuries on both sides, always fighting — what's the point?" [32].

In an April 2013 interview with *The Daily Beast*, South Korean singer Psy (best known for "Gangnam Style") was asked by Marlow Stern, "There's a pretty dicey situation going on right now with North Korean leader Kim Jong-un, who keeps threatening to destroy not just South Korea but America as well. As a South Korean, what's your take on the situation?" To which Psy replied, "Well, as an entertainer I don't want to talk about politics. As a Korean citizen, I want peace. That's all I can say. I want permanent peace" [33].

2014 Nobel Peace Prize recipient Malala Yousafzai of Pakistan said in her acceptance speeches in Oslo, Norway, "[The Nobel Prize] is not just for me. It is for those forgotten children who want education. It is for those frightened children who want peace. It is for those voiceless children who want change" [34].

May El-Khalil, founding president of the Beirut Marathon Association, advocates peace through sports and "peace run" when the Lebanese put aside their differences on Marathon day. El-Khalil wrote, "Peace is an emotionally charged word. It is something that everyone wants, and it is so elusive to so many. I come from Lebanon, a country that has seen more than its share of conflict and war, but it is also a country that embraces life and peace" [35].

During the Miss Universe pageant in January 2015, Miss Israel Doron Matalon posted an Instagram selfie with Miss Lebanon Saly Greige, Miss Slovenia Urska Bracko, and Miss Japan Keiko Tsuji (see Fig. 5.1). The photo ignited a controversy for Miss Lebanon who tried to distance herself from Miss Israel. However, an Instagram user echoed the voices of many peace-loving citizens, "As a Lebanese, I really don't see anything wrong with this picture. It's time we promote peace over violence as the latter has gotten us nowhere. And if you really wanna discuss politics, it surely shouldn't be under the picture of 4 beautiful ladies who have nothing to do with your b******t. Good luck ladies!" [36].

5.5 Peace Entailing Sacrifice

"A winning strategy sometimes necessitates sacrifice," Sherlock Holmes replies to Professor James Moriarty. "A war has been averted" [31].

Fig. 5.1 An Instagram Selfie of Miss Israel, Miss Lebanon, Miss Slovenia, and Miss Japan at the 2015 Miss Universe Pageant

In his Nobel Peace Prize acceptance speech on December 10, 2009, President Barack Obama said, "In many countries, there is a disconnect between the efforts of those who serve and the ambivalence of the broader public. I understand why war is not popular, but I also know this: The belief that peace is desirable is rarely enough to achieve it. Peace requires responsibility. Peace entails sacrifice" [10].

Being the President of the United States is itself a sacrifice. The job is highly competitive and extremely stressful, and yet it does not pay as well as most CEO's of the Fortune 500 companies. The President enjoys very little privacy in his personal life. Taking a leisure stroll outside the White House or dining in a restaurant requires Secret Service protection during the time of presidency as well as for a period of 10 years from the date the president leaves office [37].

Serving in the military is another sacrifice. Many soldiers are deployed in foreign lands under harsh and life-threatening environments, far away from their friends and families. Many veterans suffer from Post-Traumatic Stress Disorder (PTSD). In 2011, there were 165 active-duty suicides in the U.S. Army [38]. In 2012, the Army reported a record high of 325 suicides among active and non-active military personnel, exceeding the total number of U.S. military combat casualties — 313 — in Operation Enduring Freedom in Afghanistan [39]. "Suicide is the toughest enemy I have faced in my 37 years in the Army," remarked Gen. Lloyd J. Austin III, vice chief of staff of the Army, in a 2012 news release from the U.S. Department of Defense [40].

We all want peace, but wanting alone is not enough to actuate peace. After the 9/11 terrorist attacks, Disney Online General Manager Ken Goldstein said at a senior staff meeting, "Before 9/11, no one in this room knew what al-Qaeda was."

Fig. 5.2 United States Constitution (Page 1 of 4)

The room was filled with highly-educated and talented senior executives, producers, and engineers at The Walt Disney Company. Busy with work and going
about our everyday lives, we did not make time to look around the world. We rely
too much on our government to make the right decisions and law enforcement to
protect us. Had we been more vigilant, inquisitive, attentive to world affairs, and
proactive in politics, the 9/11 attacks might have been averted. As global security advisor and futurist Marc Goodman said at the TEDGlobal 2012 conference,
"Public safety is too important to leave to the professionals" [41].

   In his 1961 inaugural address, President John F. Kennedy said, "And so, my fellow Americans: ask not what your country can do for you — ask what you can do
for your country" [42]. Your country is you, your family, friends, neighbors, and
coworkers. The first three words in the Preamble to the United States Constitution
are "We the People" — not "We the Government" [43] (see Fig. 5.2).

Kennedy continued, "My fellow citizens of the world: ask not what America will do for you, but what together we can do for the freedom of man. Whether you are citizens of America or citizens of the world, ask of us the same high standards of strength and sacrifice which we ask of you" [42].

Although not every citizen can be as brave as Pakistani schoolgirl and 2014 Nobel Peace Prize recipient Malala Yousafzai who became an activist for education and women's rights at the tender age of 11, everyone should be inspired by her courage and perseverance [44]. In October 2012, the 14-year-old Yousafzai was shot in the head and neck by Taliban gunmen who boarded her school bus [45]. The abhorrent assassination attempt prompted the United Nations to launch the online petition "I Am Malala" to honor Yousufzai and call for Pakistan and countries worldwide to ensure that all children have access to education [46].

U.S. First Lady Laura Bush wrote, "Malala is the same age as another writer, a diarist, who inspired many around the world. From her hiding place in Amsterdam, Anne Frank wrote, 'How wonderful it is that nobody need wait a single moment before starting to improve the world.' Today, for Malala and the many girls like her, we need not and cannot wait. We must improve their world" [47].

5.6 Attainable Peace

World peace is as simple and elegant as $E = mc^2$. Peace is not a mystery. We know what peace is. The questions are: How do we get there? How long does it take to get there? Do we need another Einstein to help figure this out?

During 2012, Facebook asked thousands of its users in their own language: "Do you think we will achieve world peace within 50 years?" Only a tiny average of 9.76 % of U.S. respondents believe that world peace is possible [48].

Charity begins at home. We must improve U.S. education. When I was in college, I once played a question card game with my friends. I chose a card in random and the question was: What would be your first order of business if you were elected President of the United States? I gave a succinct answer: Improve the educational system. One of my American-born friends rebuked me immediately, saying that the U.S. education was perfectly fine as is. A few years later in 1987, University of Chicago Professor Allan Bloom published the bestselling book *The Closing of the American Mind* in which he described how "higher education has failed democracy and impoverished the souls of today's students" [49].

Universities are good at producing engineers, doctors, lawyers, and such, but they often gloss over difficult moral and philosophical subjects such as the meaning of life, love, prejudice, war, and peace.

In a LinkedIn article titled "Why Daydreamers Will Save the World," IDEO's CEO Tim Brown suggested, "Carry a notebook with you at all times, even by your bedside. You never know when good ideas will come. Having something nearby to jot them down ensures they won't be forgotten along the way" [50]. When I was a young teenager, I carried a small notebook with me everywhere I went so that I could jot down new ideas and discoveries of the world around me. The cover of every one of my notebooks had spaces for me to fill in my name and address. I wrote my real name of course but I put down the imaginary address of a city and country named "Peace." I was obsessed with the notion of peace after having read many books on the history of World War II.

Some politicians such as 2012 presidential candidate Mitt Romney gave up on peace at the onset and hoped for a miracle down the road. Romney said, "I look at the Palestinians not wanting to see peace anyway, for political purposes, committed to the destruction and elimination of Israel, and these thorny issues, and I say, 'There's just no way.' ... You hope for some degree of stability, but you recognize that this is going to remain an unsolved problem. We live with that in China and Taiwan. All right, we have a potentially volatile situation, but we sort of live with it, and we kick the ball down the field and hope that ultimately, somehow, something will happen and resolve it" [51].

Romney's view was in sharp contrast to President John F. Kennedy's speech at the 1963 commencement address at the American University in Washington, D.C. [52]:

> First: Let us examine our attitude toward peace itself. Too many of us think it is impossible. Too many think it unreal. But that is a dangerous, defeatist belief. It leads to the conclusion that war is inevitable — that mankind is doomed — that we are gripped by forces we cannot control. We need not accept that view. Our problems are manmade — therefore they can be solved by man. And man can be as big as be wants. No problem of human destiny is beyond human beings. Man's reason and spirit have often solved the seemingly unsolvable, and we believe they can do it again.

Martin Luther King Jr. echoed Kennedy's belief in his 1964 Nobel Peace Prize acceptance speech in which he said, "I refuse to accept the view that mankind is so tragically bound to the starless midnight of racism and war that the bright daybreak of peace and brotherhood can never become a reality" [53].

There are recent examples confirming the optimism of Kennedy and King that peace can prevail: In response to the September 2012 protests across the Arab countries against the anti-Islamic film *Innocence of Muslims*, the Grand Mufti of Egypt Ali Gomaa said, "My message to those who want [strife] between Muslims and Christians in Egypt, I tell them, 'You will not succeed, because we are one people that have been living together for more than 1,400 years'" [54].

In October 2012, Philippines President Benigno Aquino announced a historic peace deal after 15 years of negotiations between the government and the Moro Islamic Liberation Front. Aquino said, "This means that hands that once held rifles will be put to use tilling land, selling produce, manning work stations and opening doorways of opportunity for other citizens" [55].

In July 2013, the Israeli government approved freeing 104 Palestinian prisoners ahead of a peace talk between the Israelis represented by Justice Minister Tzipi Livni and Yitzhak Molcho, and the Palestinians represented by Chief Negotiator Saeb Erekat and Mohammad Shtayyeh [56].

In December 2013, Iranian President Hassan Rouhani reached out on Twitter at Christmastime: "May Jesus Christ, Prophet of love & peace, bless us all on this day. Wishing Merry #Christmas to those celebrating, esp Iranian Christians." And the Shiite Muslim country's hard-line Supreme Leader, Ayatollah Ali Khamenei, tweeted back: "No doubt that Jesus #Christ has no less value among Muslims than among the pious Christians".

World peace is certainly attainable when people and their leaders are willing to give peace a chance.

5.7 A Just and Lasting Peace

At the 1963 American University commencement address, President John F. Kennedy said that peace is "not a Pax Americana enforced on the world by American weapons of war" [52]. Instead, peace is the result of international cooperation and mutual tolerance:

Let us focus instead on a more practical, more attainable peace — based not on a sudden revolution in human nature but on a gradual evolution in human institutions — on a series of concrete actions and effective agreements which are in the interest of all concerned. There is no single, simple key to this peace–no grand or magic formula to be adopted by one or two powers. Genuine peace must be the product of many nations, the sum of many acts. It must be dynamic, not static, changing to meet the challenge of each new generation. For peace is a process — a way of solving problems.

With such a peace, there will still be quarrels and conflicting interests, as there are within families and nations. World peace, like community peace, does not require that each man love his neighbor — it requires only that they live together in mutual tolerance, submitting their disputes to a just and peaceful settlement. And history teaches us that enmities between nations, as between individuals, do not last forever.

In his 2009 Nobel Peace Prize acceptance speech, President Barack Obama outlined three ways to build a "just and lasting" peace [10]:

First, in dealing with those nations that break rules and laws, I believe that we must develop alternatives to violence that are tough enough to actually change behavior — for if we want a lasting peace, then the words of the international community must mean something. Those regimes that break the rules must be held accountable. Sanctions must exact a real price. Intransigence must be met with increased pressure — and such pressure exists only when the world stands together as one. ... The closer we stand together, the less likely we will be faced with the choice between armed intervention and complicity in oppression.

This brings me to a second point — the nature of the peace that we seek. For peace is not merely the absence of visible conflict. Only a just peace based on the inherent rights and dignity of every individual can truly be lasting. It was this insight that drove drafters of the Universal Declaration of Human Rights after the Second World War. In the wake of devastation, they recognized that if human rights are not protected, peace is a hollow promise. ... I believe that peace is unstable where citizens are denied the right to speak freely or worship as they please; choose their own leaders or assemble without fear. ... Only when Europe became free did it finally find peace.

Third, a just peace includes not only civil and political rights — it must encompass economic security and opportunity. For true peace is not just freedom from fear, but freedom from want. It is undoubtedly true that development rarely takes root without security; it is also true that security does not exist where human beings do not have access to enough food, or clean water, or the medicine and shelter they need to survive. It does not exist where children can't aspire to a decent education or a job that supports a family. The absence of hope can rot a society from within. And that's why helping farmers feed their own people — or nations educate their children and care for the sick — is not mere charity.

To recap and expound on President Obama's talking points:

1. A just and lasting peace demands apologies and forgiveness.

In October 1990, U.S. Attorney General Dick Thornburgh presented an entire nation's apology to 120,000 Japanese American internees and their descendants. In an emotional reparation ceremony in Washington, Thornburgh said, "By finally admitting a wrong, a nation does not destroy its integrity but, rather, reinforces the sincerity of its commitment to the Constitution and hence to its people" [57].

10 years later, President Barack Obama publicly acknowledged the United States' involvement in the coup during a 2009 speech in Cairo, "In the middle of the Cold War, the United States played a role in the overthrow of a democratically elected Iranian government" [58].

In December 2014, Army veteran and Prof. Eric Fair at Lehigh University admitted guilt and asked for forgiveness: "Teaching a class about war reminded me daily that I am no college professor. I was an interrogator at Abu Ghraib. I tortured. Abu Ghraib dominates every minute of every day for me. ... Abu Ghraib will fade. My transgressions will be forgotten. But only if I allow it. I've published articles in newspapers detailing our abusive treatment of Iraqi detainees. I've done interviews on TV and radio. I've spoken to groups from Amnesty International, and I've confessed everything to a lawyer from the Department of Justice and two agents from the Army's Criminal Investigation Command. I've said everything there is to say" [59].

Let bygones be bygones. President John F. Kennedy said in his 1961 inaugural address, "So let us begin anew — remembering on both sides that civility is not a sign of weakness, and sincerity is always subject to proof. ... Let both sides explore what problems unite us instead of belaboring those problems which divide us" [42].

2. A just and lasting peace embraces gender equality.

Conciliation Resources and Saferworld published a joint report titled "Gender, violence and peace: a post-2015 development agenda" in which it states that "Gender and peace are closely linked: peace is vital to promote gender equality, while gender inequality can also undermine peace and drive conflict and violence" [60]. U.S. Secretary of State John Kerry said in September 2014, "The United States believes gender equality is critical to our shared goals of prosperity, stability, and peace, and ... investing in women and girls worldwide is critical to advancing U.S. foreign policy" [61]. Ambassador Catherine M. Russell heads the Secretary's Office of Global Women's Issues (S/GWI) to promote stability, peace, and development by empowering women politically, socially, and economically around the world.

3. A just and lasting peace calls for economic reforms.

Prof. Michael Jerryson of Eckerd College said that "with economic strength and vitality comes less likelihood of civil insurrections and wars" [62]. The Marshall

Plan of 1948 (named after U.S. Secretary of State George Marshall) helped to rebuild the European economy after World War II, and the belief was that an "orderly, prosperous Europe requires the economic contributions of a stable and productive Germany" [63]. In East Asia, Army General Douglas MacArthur oversaw the reconstruction of Japan which ultimately becomes one of the world's leading economic powers. In January 2014, amid the controversy of actress Scarlett Johansson's endorsement deal with SodaStream and parting ways with Oxfam International, Johansson issued a public statement: "[I am] a supporter of economic cooperation and social interaction between a democratic Israel and Palestine. SodaStream is a company that is not only committed to the environment but to building a bridge to peace between Israel and Palestine, supporting neighbors working alongside each other, receiving equal pay, equal benefits and equal rights" [64].

President Obama concluded his speech by saying, "As the world grows smaller, you might think it would be easier for human beings to recognize how similar we are; to understand that we're all basically seeking the same things; that we all hope for the chance to live out our lives with some measure of happiness and fulfillment for ourselves and our families ... The non-violence practiced by men like Gandhi and King may not have been practical or possible in every circumstance, but the love that they preached — their fundamental faith in human progress — that must always be the North Star that guides us on our journey. For if we lose that faith — if we dismiss it as silly or naïve; if we divorce it from the decisions that we make on issues of war and peace — then we lose what's best about humanity. ... Clear-eyed, we can understand that there will be war, and still strive for peace. We can do that — for that is the story of human progress; that's the hope of all the world; and at this moment of challenge, that must be our work here on Earth" [10].

5.8 Peace and Friendships on Facebook

The world is getting smaller thanks to social media. Peace on Facebook https://peace.facebook.com/ in 2013 stated that "Facebook is proud to play a part in promoting peace by building technology that helps people better understand each other. By enabling people from diverse backgrounds to easily connect and share their ideas, we can decrease world conflict in the short and long term" [65].

According to a Pew Internet report in March 2012, 18 % of social networking site users have blocked, unfriended, or hidden someone on the site because of their differences in political views [66]. On the other hand, Friendships on Facebook publishes daily "friending" numbers between people of different regions, religions, and political affiliations.

Figure 5.3 shows that new friendships are formed everyday regardless of the prevailing political climate. New geographic connections are on the upswing from July 2012 to January 2013 across the board, with India-Pakistan leading the chart by an increase of 120 %. Religious and political connections, however, are down overall by an average of 20 %.

Geographic Friendships	Connections on July 6, 2012	Connections on January 24, 2013
India-Pakistan	168,999	371,790
Albania-Serbia	21,804	38,572
Israel-Palestine	18,321	24,044
Greece-Turkey	5,970	9,695

Religious Friendships	Connections on July 6, 2012	Connections on January 24, 2013
Muslim-Christian	120,542	107,741
Christian-Atheist	32,381	25,638
Muslim-Jewish	545	501
Sunni-Shiite	164	103

Political Friendships	Connections on July 6, 2012	Connections on January 24, 2013
U.S. Conservative/Liberal	8,034	6,064

Fig. 5.3   Facebook daily friendships statistics on July 6, 2012 and January 24, 2013

Religious Friendships	Connections on November 13	Connections on November 22
Muslim-Jewish	644	470

Fig. 5.4   Facebook daily Muslim-Jewish friendships the day before and after the Israel-Hamas conflict in November 2012

Political Friendships	Connections on November 4	Connections on November 8
U.S. Conservative/Liberal	9,148	6,716

Fig. 5.5   Facebook daily U.S. Conservative/Liberal friendships two days before and after the U.S. Presidential election in November 2012

One can observe a 27 % drop in Facebook daily Muslim-Jewish friendships in the aftermath of the Israel-Hamas conflict in November 2012 (see Fig. 5.4). Similarly, the Facebook daily U.S. Conservative/Liberal friendships also decreased by 27 % after the 2012 U.S. Presidential election (see Fig. 5.5).

In June 2012, *Time Magazine*'s congressional correspondent Jay Newton-Small asked the Syrian dissident Abu Ghassan whether his AK-47 or his video camera was the more powerful weapon. Ghassan replied, "My AK!" But he paused for a few seconds, and said, "Actually, if there is an Internet connection, my camera is more powerful" [67]. Partially aided by the Internet Freedom Grants from the U.S. State Department, Syrian rebels have been filming the protests and posting them on the Internet. Newton-Small called Ghassan a "cyber warrior" [68].

Indeed, the power of the Internet to change the world cannot be understated. The Internet has empowered individuals to start their own grassroots movements. Amid the longstanding Arab-Israeli conflict since 1920 that has taken more than 107,000 lives [69] and has cost more than $12 trillion dollars [70], Ronny Edry, an Israeli graphic designer based in Tel Aviv, reached out to the people of Iran on Facebook in March 2012. Edry and his wife uploaded posters on the Facebook page of Pushpin Mehina with the resounding message "IRANIANS. We will never bomb your country. We *Heart* You" [71].

Edry shared his Facebook experience: "My idea was simple. I was trying to reach the other side. There are all these talks about war, Iran is coming to bomb us and we bomb them back, we are sitting and waiting. I wanted to say the simple words that this war is crazy. In a few hours, I had hundreds of shares and thousands of likes. ... I think it's really amazing that someone from Iran poked me and said 'Hello, I'm from Iran, I saw your poster on Facebook.' ... I got a private message from Iran: 'We love you too. Your word reaches out there, despite the censorship. And Iranian people, aside from the regime, have no hard feelings or animosity towards anybody, particularly Israelis'" [72].

Time Magazine named Facebook co-founder and CEO Mark Zuckerberg its Person of the Year 2010 for connecting more than half a billion people and mapping the social relations among them, for creating a new system of exchanging information and for changing how we live our lives. *Time* summarized Facebook's mission: "Facebook wants to populate the wilderness, tame the howling mob and turn the lonely, antisocial world of random chance into a friendly world, a serendipitous world. You'll be working and living inside a network of people, and you'll never have to be alone again. The Internet, and the whole world, will feel more like a family, or a college dorm, or an office where your co-workers are also your best friends" [73].

With over a billion active users, Facebook is in a unique position to influence the world by enabling Facebook users to create grassroots movements for peace. U.S. Navy Admiral James Stavridis, Commander of the U.S. European Command (USEUCOM) and NATO's Supreme Allied Commander Europe (SACEUR), spoke at the TEDGlobal 2012 conference in Edinburgh about reaching out to people through social networks and providing services such as teaching Afghan soldiers to read. Stavridis said, "Instead of building walls to create security, we need to build bridges. ... The six largest nations in the world in descending order: China, India, **Facebook**, the United States, Twitter, and Indonesia. ... Moving that message [of friendship] is how we connect international, interagency, private, public, and the social net, to help create security. ... No one person, no one alliance, no one nation, no one of us is as smart as all of us thinking together" [74].

On February 11, 2015, Peace on Facebook https://peace.facebook.com/ displayed a simple webpage showing "A World of Friends: Facebook connects people

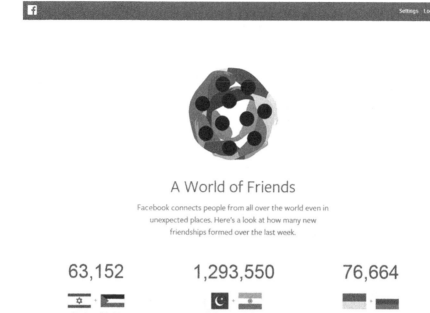

Fig. 5.6 Peace on Facebook homepage (February 11, 2015)

from all over the world even in unexpected places. Here's a look at home many new friendships formed over the last week… 63,152 Israel + Palestinian Territory, 1,293,550 Pakistan + India, 76,664 Ukraine + Russia" (see Fig. 5.6).

5.9 One Small Step for an Individual; One Giant Leap for Humankind

On July 20, 1969, Apollo 11 landed on the moon when an estimated 600 million people around the world were watching on live television. Astronaut Neil Armstrong climbed down the *Eagle*'s ladder and proclaimed: "That's one small step for a man, one giant leap for mankind" [75].

Each and everyone on Earth can make one small step, which will cumulatively result in a giant leap for mankind towards world peace. In addition to getting involved in social, political, economic, or environmental activism, here are some small steps that we can make:

1. Listen to foreign music.

"Music has a powerful effect on everybody: Children, adults, animals, plants, everything," said Quincy Jones in an *ACM Computers in Entertainment* interview in November 2003. "It's, I think, one of the abstract miracles and phenomena. … It can have a physical effect on your soul, in terms of strength and belief system" [76]. At AT&T Bell Laboratories in the late eighties, I produced the first-ever "Intergalactic Music Festival" to promote peace and friendship by showcasing international songs and cultural dances performed by fellow AT&T employees [77]. At the 2015 music symposium and pre-Grammy party hosted by *Hollywood Weekly*, I led a panel with Grammy-winning Ian Boxill, Jeff Broadbent, Isaac Schankler, and Shele Sondheim in the discussion of music as a universal language that breaks through national and cultural boundaries [78].

The beauty of music is that we do not necessarily need to know a language to enjoy a song. We hear the phonemes, emotion, melody, and instrumentals regardless of the vocal language. Who in a million years would have predicted that a Korean pop video — "Gangnam Style" / "강남스타일" by Psy — became so popular worldwide that it "broke" YouTube's view counter? "We saw this coming a couple months ago and updated our systems to prepare for it," said a Google spokesperson in December 2014. "We updated to a 64-bit counter (9,223,372,036,854,775,808 max — that's 9 quintillion)" [79].

To start listening to foreign music seriously, multilingual songs are great choices. Some of my favorites include "Moon of Dreams" / "Відлуння мрій" in Ukrainian and English by Ruslana featuring T-Pain [80], "Pure Love" / "خالص عشق" in Persian and English by Arash featuring Helena, [81] and "7 Seconds" in African Wolof, French, and English by Youssou N'Dour featuring Neneh Cherry [82].

2. Watch foreign films.

Academy Award-winning *Crouching Tiger, Hidden Dragon* (Mandarin) directed by Ang Lee is by far the most well-known foreign-language film in America. Some of my other favorite foreign movies are *Run Lola Run* (German) directed by Tom Tykwer, *Amelie* (French) directed by Jean-Pierre Jeunet, and *Hero* (Mandarin) directed by Zhang Yimou.

For better or worse, Hollywood dominates the global film industry. U.S. Department of Commerce reported that "U.S. exports of film and entertainment media often attain shares in international markets in excess of 90 percent due to high global interest in U.S. filmed entertainment.... The U.S. filmed entertainment sector enjoyed a trade surplus of $14.3 billion in 2011" [83].

In February 2015, Netflix launched its movie subscription service in Cuba. "We are delighted to finally be able to offer Netflix to the people of Cuba, connecting them with stories they will love from all over the world," said Netflix chief executive Reed Hastings. "Cuba has great filmmakers and a robust arts culture and one day we hope to be able to bring their work to our global audience of over 57 million members" [84].

I certainly hope that Netflix and other streaming services will help us discover and experience more foreign cultures in the comfort of our own home.

3. Learn a foreign language.

"I feel pretty stupid that I don't know any foreign languages," Bill Gates admitted in an online Reddit chat in January 2015. "I took Latin and Greek in High School and got A's and I guess it helps my vocabulary but I wish I knew French or Arabic or Chinese. I keep hoping to get time to study one of these — probably French because it is the easiest. I did Duolingo for awhile but didn't keep it up. Mark Zuckerberg amazingly learned Mandarin and did a Q&A with Chinese students — incredible" [85].

Never underestimate the power of languages. A Pope has to be multilingual in order to govern the Catholic Church with more than 1.2 billion members across 195 countries. The long-term unification of China was mainly due to one common Chinese language imposed by Emperor Qin Shi Huang (秦始皇) in spite of the rise and fall of subsequent dynasties. In the Book of Genesis, men came together to build a city and the Tower of Babel in order to make a name for themselves. God said in response, "Behold, the people are one and they have all one language, and this they begin to do; and now nothing will be withheld from them which they have imagined to do. Come, let Us go down, and there confound their language, that they may not understand one another's speech."

4. Befriend a foreigner on Facebook or other social networks.

First lady Michelle Obama said during her 2014 goodwill tour in China, "You don't need to get on a plane to be a citizen diplomat. If you have an Internet connection in your home, school or library, within seconds you can be transported anywhere in the world and meet people on every continent" [86].

It is much easier and more fun to communicate with someone in a foreign country if we already have some basic knowledge of their culture and language. Section 5.8 talks about an individual who uses Facebook to start his own grassroots movement for peace.

5. Collaborate with a foreigner on a project or a hobby.

Martin Barstow, president of the Royal Astronomical Society and pro-vice chancellor at the University of Leicester, wrote about the European Space Agency's groundbreaking Rosetta mission in November 2014, "Space research is one of the best examples of peaceful international cooperation and of the ability of humans to do marvelous things when they work together towards a common goal. I am not the only one to have worked on several successful space projects during my career, and to have made many friends around the world as a consequence, often in countries that would have once been regarded as political rivals (or even some that still are)" [87].

6. Travel to foreign countries, study abroad, or take a sabbatical overseas.

Notwithstanding the advances of Google Earth, Oculus Rift, and other immersive technologies popularized in science fiction films such as *Total Recall* and *Star Trek's* Holodeck, there is no substitute for a truly immersive experience by living in a foreign country, tasting the authentic local cuisines, and interacting with the native people face-to-face.

First Lady Michelle Obama lauds studying aboard as "citizen diplomacy." During her goodwill tour in China in March 2014, the first lady told an audience at Peking University, "I'm here today because I know that our future depends on connections like these among young people like you across the globe. We believe that relationships between nations aren't just about relationships between governments or leaders — they're about relationships between people, particularly young people" [86].

In a 2001 interview with NYRock, Academy Award-winning actress Angelina Jolie told Prairie Miller about her experience filming *Lara Croft:* Tomb Raider in a war-torn country, "Cambodia is the most beautiful place I've ever been to. I don't want to get into the heaviness of it here, but I discovered things about what's happening in the world. Like my eyes started to open. When I was in Cambodia, I learned so much about these people and what they had been through. I expected to meet a certain kind of people because of that. And when I met them, they were so generous, spiritual, and open and kind. So I couldn't believe that they would have such patience with us, and such openness after all of that. There were areas we shot in, that we could only be in certain places, because they hadn't been de-mined yet. And to know that there are hospitals where kids are still being affected by stepping on landmines every day, was horrifying and so sad. You never hear about that. To discover that kind of stuff was to really understand people in the rest of the world. So Cambodia was really eye opening for me" [88]. Ever since the Cambodia experience, Jolie has been active in humanitarian work and become an ambassador of the United Nations High Commissioner for Refugees (UNHCR).

7. Adopt a child from a foreign country.

There is nothing like having a United Nations at home. In 2002, Angelina Jolie adopted her first child, 7-month-old Maddox Chivan, from an orphanage in Cambodia. In 2005, she adopted a daughter, 6-month-old Zahara Marley, from an orphanage in Ethiopia. A year after giving birth to a daughter, Shiloh Nouvel, Jolie adopted a son, 3-year-old Pax Thien, from an orphanage in Vietnam in 2007. She gave birth to twins — a boy, Knox Léon, and a girl, Vivienne Marcheline — in the following year. Angelina Jolie calls herself "a citizen of the world" [89].

Families with foreign adopted children add extra special meanings to "It's a Small World" boat ride at Disneyland which represents world peace through the eyes of children.

8. Marry a foreigner.

In ancient times, countries and tribes form alliances through arranged marriages. Nowadays, it is up to eligible bachelors and bachelorettes to forge friendship, peace, and understanding among nations and races through civil unions.

In July 2014, Arab-American journalist Sulome Anderson gave her Jewish boyfriend Jeremy a kiss on Twitter to show their support for a New York-based social media campaign called "Jews and Arabs refuse to be enemies." Anderson tweeted, "He calls me neshama, I call him habibi. Love doesn't speak the language of occupation #JewsAndArabsRefuseToBeEnemies [90]".

Experiencing a melting pot of cultures within an immediate or extended family on a daily basis is nothing less than marvelous, stimulating, and conducive to personal growth. President Barack Obama was born to Ann Dunham of Kansas and Barack Obama, Sr. of Kenya who met each other as students at the University of Hawaii. The President referred to his mother as "the dominant figure in my formative years… The values she taught me continue to be my touchstone when it comes to how I go about the world of politics" [91].

Fig. 5.7 Apollo 11 plaque left on the ladder of Eagle on the moon

Astronauts Neil Armstrong, Buzz Aldrin, and Michael Collins walked on the moon and left behind a plaque that reads, "Here men from the planet Earth first set foot upon the moon. July 1969 A.D. We came in peace for all mankind" [92]. (See Fig. 5.7) What will the extraterrestrials expect to see when they come visit Earth?

5.10 A Recipe for Peace

I presented a mathematical formula for peace during my job interview with the Microelectronics and Computer Consortium (MCC) in the mid-eighties. It did not help me land a job at MCC, but fortunately my other choice — Bell Laboratories — worked out.

In 2010, Power to Change Ministries published a recipe for peace as follows [93]:

"1 cup of friendship
1/2 cup of hope
2 cups of love
5 tbsp of respect
1/2 cup of kindness

1 cup of joy
3 tsp of understanding
1 1/2 cups honesty

Mix friendship, love, and kindness in a large bowl. Add understanding a few drops at a time. Then stir in honesty and joy for good firm dough. Sprinkle half of respect over it and mix well.

Pour into a cake pan and bake at 350 °F. When it is ready pour the hope and the rest of respect on top and share with everyone you know."

Notice in the recipe that every ingredient is filling up at least a full cup or teaspoon except hope and kindness. 1/2 cup of hope and 1/2 cup of kindness seem like a litmus test by asking "Is the glass half empty or half full?" Nature shows us in every waking moment that where there is life, there is hope. Humankind too often cooks up an excuse to start war instead of making peace. Rather than developing better strategies to win wars, we should focus on better recipes to attain peace.

2014 Nobel Peace Prize recipient Kailash Satyarthi of India said in his acceptance speeches in Oslo, Norway: "I refuse to accept that the world is so poor when just one week of military expenditures can bring all children to classrooms" [34].

The enormous amount of financial resources and creative energy that nations have spent on wars and weapons could have been redirected to curing deadly diseases, feeding the hungry, eliminating poverty, promoting art and culture, investing in renewable clean energy, and solving a host of other important challenges facing humanity. In October 2014, Todd Harrison of the Center for Strategic and Budgetary Assessments estimated that the annual bill for U.S. military operations alone would range from $4 billion to $22 billion [94], which amounted to $11 million to $60 million a day on average.

Activist and commentator Sally Kohn gave us a perspective on the better use of money when she analyzed in January 2015 what $889 million could buy: "559,119,496 meals for the homeless — equal to 916 meals for every homeless person in America for one year; 19,525 new teachers for New York City public school students — or 24,170 new teachers in Birmingham, Alabama; 15,180 entry-level psychologists at Veterans Administration hospitals; or clean water for 29,633,333 people around the globe;" and so on [95]. For a $22 billion annual military spending, we multiply the list of numbers by a factor of nearly 25: "13.8 billion meals for the homeless; half a million new teachers; a third of a million psychologists; or clean water for two third of a billion people in the world."

In the midst of the Russia-Ukraine conflict, Russian President Vladimir Putin briefly digressed from politics and told his audience at the 15th Congress of the Russian Geographical Society on November 7, 2014: "The meaning of our whole life and existence is love. It is love for the family, for the children, for the motherland. This is a multifaceted phenomenon; it lies at heart of any of our behaviors" [96].

Martin Luther King, Jr. said that "love is the only force capable of transforming an enemy into a friend." And President Lyndon Johnson ominously declared in the "Peace Little Girl (Daisy)" political ad on September 7, 1964: "We must either love each other, or we must die" [97].

Bibliography

1. **FBI.** What We Investigate. [Online] Federal Bureau of Investigation. [Cited: December 28, 2012.] http://www.fbi.gov/albuquerque/about-us/what-we-investigate.
2. **Chomsky, Noam.** The United States is a Leading Terrorist State. [Online] Monthly Review, November 2001. http://monthlyreview.org/2001/11/01/the-united-states-is-a-leading-terrorist-state.
3. **Maher, Bill.** Spacial Delivery. [Online] HBO, November 30, 2012. http://www.real-time-with-bill-maher-blog.com/real-time-with-bill-maher-blog/2012/11/30/spacial-delivery.html.
4. **The National WWII Museum.** By the numbers: world-wide deaths. [Online] The National WWII Museum, 2015. http://www.nationalww2museum.org/learn/education/for-students/ww2-history/ww2-by-the-numbers/world-wide-deaths.html.
5. **The Telegraph.** The Holocaust death toll. [Online] The Telegraph, January 26, 2005. http://www.telegraph.co.uk/news/1481975/The-Holocaust-death-toll.html.
6. **Yamazaki, James N.** Hiroshima and Nagasaki Death Toll. [Online] University of California Los Angeles, October 10, 2007. http://www.aasc.ucla.edu/cab/200708230009.html.
7. **Dear, John S.J.** Afghanistan journal, part two: bearing witness to peacemaking in a war-torn country. [Online] National Catholic Reporter, December 18, 2012. http://ncronline.org/blogs/road-peace/afghanistan-journal-part-two-bearing-witness-peacemaking-war-torn-country.
8. **Mount, Mike.** U.S. may remove all troops from Afghanistan after 2014. [Online] CNN, January 8, 2013. http://security.blogs.cnn.com/2013/01/08/u-s-may-remove-all-troops-from-afghanistan-after-2014/.
9. **DeYoung, Karen.** Obama to leave 9,800 U.S. troops in Afghanistan. [Online] The Washington Post, May 27, 2014. http://www.washingtonpost.com/world/national-security/obama-to-leave-9800-us-troops-in-afghanistan-senior-official-says/2014/05/27/57f37e72-e5b2-11e3-a86b-362fd5443d19_story.html.
10. **Obama, Barack.** Remarks by the President at the Acceptance of the Nobel Peace Prize. [Online] The White House, December 10, 2009. http://www.whitehouse.gov/the-press-office/remarks-president-acceptance-nobel-peace-prize.
11. **Tolstoy, Leo.** War and Peace. [Online] 1869. books.google.com/books?id = jhZzwKsi0OsC.
12. **Davis, William C.** The Battlefields of the Civil War: The Bloody Conflict of North Against South Told Through the Stories of Its Battles. [Online] University of Oklahoma Press, 1991. books.google.com/books?isbn = 0806128828.
13. **Bronner, Ethan and Kirkpatrick, David D.** U.S. Seeks Truce on Gaza as Enemies Step Up Attacks. [Online] The New York Times, November 20, 2012. http://www.nytimes.com/2012/11/21/world/middleeast/israel-gaza-conflict.html?pagewanted=all.
14. **Newton, Paula.** Analysis: Conflict shifts balance of power in the Middle East. [Online] CNN, November 22, 2012. http://edition.cnn.com/2012/11/21/world/meast/middle-east-balance-power/index.html.
15. **Greenwood, Phoebe.** Gaza declares 'victory' as ceasefire holds, but the most evident triumph is one of survival . [Online] The Telegraph, November 22, 2012. http://www.telegraph.co.uk/news/worldnews/middleeast/palestinianauthority/9697326/Gaza-declares-victory-as-ceasefire-holds-but-the-most-evident-triumph-is-one-of-survival.html.
16. **King, Martin Luther Jr.** The Nobel Peace Prize 1964: Martin Luther King Jr.: Nobel Lecture. [Online] The Nobel Foundation, December 11, 1964. http://www.nobelprize.org/nobel_prizes/peace/laureates/1964/king-lecture.html.
17. **D'Agata, Charlie.** Civilians caught in Israel-Gaza conflict. [Online] CBS News, November 18, 2012. http://www.cbsnews.com/8301-18563_162-57551583/civilians-caught-in-israel-gaza-conflict/.
18. **CNN Wire Staff.** Cease-fire appears to be holding in Gaza. [Online] CNN, November 22, 2012. http://www.cnn.com/2012/11/21/world/meast/gaza-israel-strike/index.html.
19. **ICRC Resource Center.** What is international humanitarian law? [Online] International Committee of the Red Cross, July 31, 2004. http://www.icrc.org/eng/resources/documents/legal-fact-sheet/humanitarian-law-factsheet.htm.

20. **Urquhart, Conal.** The Goldstone report: a history. [Online] The Guardian, April 14, 2011. http://www.guardian.co.uk/world/2011/apr/14/goldstone-report-history.

21. **Rudoren, Jodi and Kershner, Isabel.** Israel Broadens Its Bombing in Gaza to Include Government Sites. [Online] The New York Times, November 17, 2012. http://www.nytimes.com/2012/11/18/world/middleeast/israel-gaza-conflict.html?pagewanted=all.

22. **Sanger, David E. and Shanker, Thom.** For Israel, Gaza Conflict Is Test for an Iran Confrontation. [Online] The New York Times, November 22, 2012. http://www.nytimes.com/2012/11/23/world/middleeast/for-israel-gaza-conflict-a-practice-run-for-a-possible-iran-confrontation.html?pagewanted=all.

23. **French, Shannon E.** The Warrior's Code. U.S. Naval Academy. [Online] International Society for Military Ethics, 2001. http://isme.tamu.edu/JSCOPE02/French02.html.

24. **IMDb.** Horns. [Online] IMDb, October 3, 2014. http://www.imdb.com/title/tt1528071/.

25. **Blake, John.** Two enemies discover a 'higher call' in battle. [Online] CNN, March 9, 2013. http://www.cnn.com/2013/03/09/living/higher-call-military-chivalry/index.html.

26. **Tapper, Jake and Carter, Chelsea J.** An American hero: The uncommon valor of Clint Romesha. [Online] CNN, February 8, 2013. http://www.cnn.com/2013/02/11/politics/medal-of-honor/index.html.

27. **Tapper, Jake and Cohen, Tom.** Medal of Honor recipient conflicted by joy, sadness. [Online] CNN, February 12, 2013. http://www.cnn.com/2013/02/11/politics/medal-of-honor/index.html.

28. **Tapper, Jake.** Medal of Honor recipient declines invitation to State of the Union. [Online] CNN, February 12, 2013. http://www.cnn.com/2013/02/12/politics/sotu-invite-declined/index.html.

29. **IMDb.** Ender's Game. [Online] IMDb, November 1, 2013. http://www.imdb.com/title/tt1731141/.

30. **Library of the Marine Corps Research Guides Portal.** Commandant's Professional Reading List - Official Site. [Online] United States Marine Corporation University, January 24, 2015. http://guides.grc.usmcu.edu/content.php?pid=408059&sid=4909977.

31. **IMDb.** Sherlock Holmes: A Game of Shadows. [Online] IMDb, December 16, 2011. http://www.imdb.com/title/tt1515091/.

32. **Sidner, Sara.** Children of the conflict: Innocence interrupted by war. [Online] CNN, November 26, 2012. http://www.cnn.com/2012/11/25/world/meast/sidner-children-israel-gaza-conflict/index.html.

33. **Stern, Marlow.** Psy on New Single 'Gentleman,' Kim Jong-un, Justin Bieber & More. [Online] The Daily Beast, April 29, 2013. http://www.thedailybeast.com/articles/2013/04/29/psy-on-new-single-gentleman-kim-jong-un-justin-bieber-more.html.

34. **Botelho, Greg.** Malala, Satyarthi accept Nobel Peace Prize, press children's rights fight. [Online] CNN, December 10, 2014. http://www.cnn.com/2014/12/10/world/asia/nobel-peace-prize-awarded/.

35. **El-Khalil, May.** How marathons can bring peace. [Online] CNN, November 3, 2013. http://www.cnn.com/2013/11/03/opinion/el-kahlil-ted-beirut-marathon/index.html.

36. **Grinberg, Emanuella.** Miss Lebanon distances herself from photo with Miss Israel. [Online] CNN, January 19, 2015. http://www.cnn.com/2015/01/18/world/feat-miss-lebanon-miss-israel-picture/index.html.

37. **U.S. Secret Service.** Protective Mission. [Online] United States Secret Service. [Cited: December 29, 2012.] http://www.secretservice.gov/protection.shtml.

38. **U.S. Department of Defense.** Army Releases November Suicide Data. [Online] U.S. Department of Defense, December 13, 2012. http://www.defense.gov/releases/release.aspx?releaseid=15741.

39. **Watkins, Tom and Schneider, Maggie.** 325 Army suicides in 2012 a record. [Online] CNN, February 2, 2013. http://www.cnn.com/2013/02/02/us/army-suicides/index.html.

40. **U.S. Department of Defense.** Army Releases July Suicide Data. [Online] U.S. Department of Defense, August 16, 2012. http://www.defense.gov/releases/release.aspx?releaseid=15517.

41. **TED.** The technological future of crime: Marc Goodman at TEDGlobal 2012. [Online] TED, June 28, 2012. http://blog.ted.com/2012/06/28/the-technological-future-of-crime-marc-goodman-at-tedglobal-2012/.

42. **Kennedy, John F.** Ask not what your country can do for you. [Online] The Guardian, January 20, 1961. http://www.guardian.co.uk/theguardian/2007/apr/22/greatspeeches.

43. **The Constitutional Convention .** Constitution of the United States. [Online] U.S. National Archives and Records Administration, September 17, 1787. http://www.wdl.org/en/item/2708/.

44. **BBC.** Malala Yousafzai: Portrait of the girl blogger . [Online] BBC News Magazine, October 10, 2012. http://www.bbc.co.uk/news/magazine-19899540.

45. **Leiby, Richard and Leiby, Michele Langevine.** Taliban says it shot Pakistani teen for advocating girls' rights. [Online] The Washington Post, October 9, 2012. http://www.washingtonpost.com/world/asia_pacific/taliban-says-it-shot-infidel-pakistani-teen-for-advocating-girls-rights/2012/10/09/29715632-1214-11e2-9a39-1f5a7f6fe945_story.html.

46. **Hays, Julie.** Pakistani teen inspires others to fight for education. [Online] CNN, October 15, 2012. http://www.cnn.com/2012/10/15/world/iyw-support-for-malala/index.html.

47. **Bush, Laura.** A girl's courage challenges us to act. [Online] The Washington Post, October 10, 2012. http://www.washingtonpost.com/opinions/laura-bush-malala-yousafzais-courage-challenges-us-to-act/2012/10/10/9cd423ea-1316-11e2-ba83-a7a396e6b2a7_story.html.

48. **Facebook.** Peace on Facebook. [Online] Facebook. [Cited: January 1, 2013.] https://peace.facebook.com/.

49. **Kimball, Roger.** The Groves of Ignorance. [Online] 1987, 5 April. http://www.nytimes.com/1987/04/05/books/the-groves-of-ignorance.html?pagewanted=all.

50. **Brown, Tim.** Why Daydreamers Will Save the World. [Online] LinkedIn, February 24, 2014. https://www.linkedin.com/pulse/20140224153333-10842349-why-daydreamers-will-save-the-world.

51. **Wilson, Scott and O'Keefe, Ed.** Mitt Romney: 'Palestinians have no interest whatsoever in establishing peace'. [Online] The Washington Post, September 18, 2012. http://articles.washingtonpost.com/2012-09-18/politics/35497236_1_peace-talks-mitt-romney-palestinians.

52. **Kennedy, John F.** Commencement Address at American University, June 10, 1963. [Online] John F. Kennedy Presidential Library and Museum, June 10, 1963. http://www.jfklibrary.org/Researchold/Ready-Reference/JFK-Speeches/Commencement-Address-at-American-University-June-10-1963.aspx.

53. **King, Martin Luther Jr.** Martin Luther King's Acceptance Speech. [Online] The Nobel Foundation, 1964. http://www.nobelprize.org/nobel_prizes/peace/laureates/1964/king-acceptance_en.html.

54. **Lee, Ian and Fahmy, Mohamed Fadel.** Sunni Islam leader calls for peace, urges Muslims to have 'patience and wisdom' . [Online] CNN, November 19, 2012. http://www.cnn.com/2012/09/22/world/world-film-protests/index.html.

55. **CNN Wire Staff.** Philippines, Muslim rebels reach peace deal . [Online] CNN, October 7, 2012. http://www.cnn.com/2012/10/07/world/asia/philippines-peace-deal/index.html.

56. **Schwartz, Michael.** Israeli-Palestinian peace talks set to resume Monday. [Online] CNN, July 29, 2013. http://www.cnn.com/2013/07/28/world/meast/israel-prisoner-release/index.html.

57. **Ostrow, Ronald J.** First 9 Japanese WWII Internees Get Reparations. [Online] Los Angeles Times, October 10, 1990. http://articles.latimes.com/1990-10-10/news/mn-1961_1_japanese-wwii-internees.

58. **Merica, Dan and Hanna, Jason.** In declassified document, CIA acknowledges role in '53 Iran coup. [Online] CNN, August 19, 2013. http://www.cnn.com/2013/08/19/politics/cia-iran-1953-coup/index.html.

59. **Fair, Eric.** I Can't Be Forgiven for Abu Ghraib. [Online] The New York Times, December 9, 2014. http://www.nytimes.com/2014/12/10/opinion/the-torture-report-reminds-us-of-what-america-was.html.

60. **Wright, Hannah.** Gender equality and peace are linked – the post-2015 agenda should reflect it . [Online] The Guardian, March 3, 2014. http://www.theguardian.com/global-development/poverty-matters/2014/mar/03/equality-peace-millennium-development-goals-2015-agenda.

61. **Office of Global Women's Issues.** U.S. Department of State Policy Guidance: Promoting Gender Equality and Advancing the Status of Women and Girls. [Online] U.S. Department of State, September 8, 2014. http://www.state.gov/s/gwi/.

62. **Shadbolt, Peter.** Conflict in Buddhism: 'Violence for the sake of peace?'. [Online] CNN, April 23, 2013. http://www.cnn.com/2013/04/22/world/asia/buddhism-violence/.

63. **TIME.** CONFERENCES: Pas de Pagaille! [Online] TIME Magazine, July 28, 1947. http://content.time.com/time/magazine/article/0,9171,887417,00.html.

64. **Kershner, Isabel.** Scarlett Johansson and Oxfam, Torn Apart by Israeli Company Deal. [Online] The New York Times, January 30, 2014. http://www.nytimes.com/2014/01/31/world/middleeast/scarlett-johansson-and-oxfam-torn-apart-by-israeli-company-deal.html?_r=1.

65. **Facebook.** Peace on Facebook. [Online] Facebook. [Cited: January 1, 2013.] https://peace.facebook.com/.

66. **Rainie, Lee and Smith, Aaron.** Social networking sites and politics . [Online] Pew Internet, March 12, 2012. http://www.pewinternet.org/Reports/2012/Social-networking-and-politics/Main-findings/Social-networking-sites-and-politics.aspx.

67. **CNN Editors.** Syria's 'cyber warriors' choose cameras over guns. [Online] CNN, June 14, 2012. http://globalpublicsquare.blogs.cnn.com/2012/06/14/syrias-cyber-warriors-choose-cameras-over-guns/.

68. **Newton-Small, Jay.** Hillary's Little Startup: How the U.S. Is Using Technology to Aid Syria's Rebels. [Online] Time Magazine, June 13, 2012. http://world.time.com/2012/06/13/hillarys-little-startup-how-the-u-s-is-using-technology-to-aid-syrias-rebels/.

69. **The American-Israeli Cooperative Enterprise.** The Arab-Israeli Conflict: Total Casualties (1920–2012). [Online] Jewish Virtual Library. [Cited: April 20, 2012.] http://www.jewishvirtuallibrary.org/jsource/History/casualtiestotal.html.

70. **Strategic Foresight Group.** Cost of Conflict in the Middle East. [Online] Strategic Foresight Group Report Excerpts, January 2009. http://www.strategicforesight.com/Cost%20of%20Conflict%20-%206%20pager.pdf.

71. **Mehina, Pushpin.** Pushpin Mehina. [Online] Facebook. [Cited: April 20, 2012.] http://www.facebook.com/pushpin.

72. **Said, Samira.** Peace-minded Israeli reaches out to everyday Iranians via Facebook. [Online] CNN, March 20, 2012. http://www.cnn.com/2012/03/19/world/meast/israel-iran-social-media/index.html.

73. **Grossman, Lev.** Person of the Year 2010. Mark Zuckerberg. [Online] Time Magazine, December 15, 2010. http://www.time.com/time/specials/packages/article/0,28804,2036683_2037183_2037185,00.html.

74. **Stavridis, James.** James Stavridis: A Navy Admiral's thoughts on global security. [Online] TED, June 2012. http://www.ted.com/talks/james_stavridis_how_nato_s_supreme_commander_thinks_about_global_security.html.

75. **Jones, Eric M.** One Small Step. [Online] NASA, January 30, 2015. http://www.hq.nasa.gov/alsj/a11/a11.step.html.

76. **Lee, Newton.** Interviews with Quincy Jones. [Online] ACM Computers in Entertainment, January 2004. http://dl.acm.org/citation.cfm?doid=973801.973815.

77. **—.** Digital Da Vinci: Computers in Music. [Online] Springer Science + Business Media, April 12, 2014. http://www.amazon.com/Digital-Da-Vinci-Computers-Music/dp/149390535X/.

78. **Wilhelm, Laura.** Being All We Can Be at the Hollywood Weekly Pre-Grammy Party. [Online] The Hollywood Times, February 10, 2015. http://thehollywoodtimes.net/2015/02/10/being-all-we-can-be-at-the-hollywood-weekly-pre-grammy-party/.

79. **Savov, Vlad.** Gangnam Style broke YouTube's view counter. [Online] The Verge, December 3, 2014. http://www.theverge.com/2014/12/3/7325819/gangnam-style-broke-youtube-view-counter.

80. **Ruslana.** Ruslana feat. T- Pain "Відлуння мрій" . [Online] YouTube, March 8, 2008. https://www.youtube.com/watch?v=t-nU6bECSoc.

81. **Arash.** ARASH feat Helena "Pure Love" (Official video). [Online] Warner Music Sweden, November 26, 2008. https://www.youtube.com/watch?v=7oHObnP1sGE.

82. **N'Dour, Youssou and Cherry, Neneh.** Youssou N'Dour - 7 Seconds ft. Neneh Cherry. [Online] YouTube, October 25, 2009. https://www.youtube.com/watch?v=wqCpjFMvz-k.

83. **U.S. Department of Commerce.** The Media & Entertainment Industry in the United States. [Online] U.S. Department of Commerce, 2014. http://selectusa.commerce.gov/industry-snapshots/media-entertainment-industry-united-states.

84. **Glenza, Jessica.** Netflix launches $7.99 service for Cuba despite average wage of $17 a month . [Online] The Guardian, February 9, 2015. http://www.theguardian.com/world/2015/feb/09/netflix-launches-streaming-service-cuba.

85. **Gates, Bill.** Hi Reddit, I'm Bill Gates and I'm back for my third AMA. Ask me anything. [Online] Reddit, January 28, 2015. http://www.reddit.com/r/IAmA/comments/2tzjp7/hi_reddit_im_bill_gates_and_im_back_for_my_third/.

86. **Watkins, Tom.** Michelle Obama lauds study abroad as 'citizen diplomacy'. [Online] CNN, March 25, 2014. http://www.cnn.com/2014/03/22/politics/michelle-obama-china/index.html.

87. **Barstow, Martin.** Opinion: The big lesson from comet landing. [Online] CNN, November 14, 2014. http://www.cnn.com/2014/11/14/opinion/barstow-comet-space-agency/index.html.

88. **Miller, Prairie.** Angelina Jolie on Filling Lara Croft's Shoes and D-size Cups. [Online] NYRock, June 2001. http://www.nyrock.com/interviews/2001/jolie_int.asp.

89. **Junod, Tom.** Angelina Jolie Dies for Our Sins. [Online] Esquire, July 20, 2010. http://www.esquire.com/women/women-we-love/angelina-jolie-interview-pics-0707.

90. **Kuruvilla, Carol.** Arab-Jewish couple kiss in Twitter picture to support peace in Gaza. [Online] Daily News (New York), July 22, 2014. http://www.nydailynews.com/news/national/couple-kiss-viral-show-jews-arabs-refuse-enemies-article-1.1876242.

91. **Gammell, Caroline.** Academic prowess of Barack Obama's mother disclosed. [Online] The Telegraph, September 16, 2009. http://www.telegraph.co.uk/news/worldnews/barackobama/6196237/Academic-prowess-of-Barack-Obamas-mother-disclosed.html.

92. **NASA.** July 20, 1969: One Giant Leap For Mankind. [Online] National Aeronautics and Space Administration, July 8, 2009. http://www.nasa.gov/mission_pages/apollo/apollo11_40th.html.

93. **Power to Change Ministries.** Recipe for Peace. [Online] Power to Change Ministries, 2010. http://powertochange.com/life/peacerecipe/.

94. **Bilmes, Linda J.** Fighting the Islamic State — how much will it cost? [Online] The Boston Globe, October 8, 2014. http://www.bostonglobe.com/opinion/2014/10/07/fighting-islamic-state-how-much-will-cost/xub6sT2eWP1k67t1HWBsFL/story.html.

95. **Kohn, Sally.** A better way for the Kochs to spend their millions. [Online] CNN, January 28, 2015. http://www.cnn.com/2015/01/28/opinion/kohn-how-kochs-should-spend-their-money/index.html.

96. **RT News.** Putin: 'Love is the meaning of life'. [Online] Russia Today, November 7, 2014. http://rt.com/news/203339-putin-love-meaning-life/.

97. **Johnson, Lyndon.** Peace Little Girl (Daisy). [Online] The Living Room Candidate, September 7, 1964. http://www.livingroomcandidate.org/commercials/1964/peace-little-girl-daisy.

Part III
Counterterrorism Technologies: Total Information Awareness

Chapter 6
The Rise and Fall of Total Information Awareness

Let us not look back in anger, nor forward in fear, but around in awareness.
—American cartoonist and writer James Grover Thurber

Vital information for the millions outweighs the privacy of the few.
—Newton Lee

Scientia potentia est. (Knowledge is power.)
—Thomas Hobbes in *De Homine* (*Man*) (1658)

Information is the oxygen of the modern age.
—President Ronald Reagan (June 14, 1989)

It would be no good to solve the security problem and give up the privacy and civil liberties that make our country great.
—Admiral John Poindexter (August 12, 2003)

6.1 President Ronald Reagan and Admiral John Poindexter

American cartoonist and writer James Grover Thurber once said, "Let us not look back in anger, nor forward in fear, but around in awareness." President Ronald Reagan had long recognized the vital importance of communications technology and information sharing as he said in June 1989 after having served two terms as the President of the United States [1]:

> Information is the oxygen of the modern age…. It seeps through the walls topped with barbed wire. It wafts across the electrified, booby-trapped borders. Breezes of electronic beams blow through the Iron Curtain as if it was lace…. The Goliath of totalitarian control will rapidly be brought down by the David of the microchip.

© Springer International Publishing Switzerland 2015
N. Lee, *Counterterrorism and Cybersecurity*, DOI 10.1007/978-3-319-17244-6_6

Back in April 1984, President Reagan signed the National Security Decision Directive (NSDD) 138: Combating Terrorism, which authorized the increase of intelligence collection directed against groups or states involved in terrorism [2].

Reagan appointed Navy Vice Admiral John Poindexter as the National Security Advisor in December 1985. With a Ph.D. in Nuclear Physics from California Institute of Technology (Caltech), Poindexter had been a strong advocate of new computer technology and distributed data management system during his tenure in the U.S. military. In November 1986, however, Poindexter was forced to resign from the White House Office and retire as Rear Admiral due to his role in the Iran-Contra Affair [3].

After a 3-month investigation by the Tower Commission headed by former Senator John Tower, President Reagan addressed the nation in March 1987 acknowledging the danger of unchecked covert operations and the need for stronger presidential oversight [4]:

> A few months ago I told the American people I did not trade arms for hostages. My heart and my best intentions still tell me that's true, but the facts and the evidence tell me it is not.... I'm taking action in three basic areas: personnel, national security policy, and the process for making sure that the system works.... I have had issued a directive prohibiting the NSC [National Security Council] staff itself from undertaking covert operations — no ifs, ands, or buts.

In March 1988, Poindexter and Lieutenant Colonel Oliver North were indicted on charges of conspiracy to defraud the United States by illegally providing the Nicaraguan rebels with profits from the sale of American weapons to Iran [5]. In April 1990, Poindexter was convicted on five counts of lying to Congress and obstructing the Congressional investigation of the Reagan Administration's covert arms sales to Iran and the diversion of some proceeds to rebels fighting the Marxist Government in Nicaragua. However, in November 1991, the District of Columbia Circuit Court overturned Poindexter's conviction by a vote of two to one [6].

A day after September 11, 2001, Poindexter lamented with his close friend Brian Sharkey that they had not prevented the terrorist attacks [7]. Sharkey was a former program manager at the Defense Advanced Research Projects Agency (DARPA). Poindexter was working for BMT Syntek Technologies, a defense contractor that was developing Project Genoa, a data-mining decision-support system for DARPA. Genoa provided analyst tools to augment human cognitive processes and aid understanding of complex arguments [8]. After the 9/11 attacks, Poindexter wanted to put Project Genoa on steroids.

6.2 Defense Advanced Research Projects Agency (DARPA)

Defense Advanced Research Projects Agency (DARPA) was created as Advanced Research Projects Agency (ARPA) in 1958 by President Dwight Eisenhower in response to the surprise Sputnik launch by the Soviet Union a year before [9].

Since its formation, DARPA has made significant contributions to science and technology in collaboration with universities and research organizations. J. C. R. Licklider, director of DARPA's Information Processing Techniques Office (IPTO) in 1962, was one of the fathers of modern computer science such as the graphical user interface and personal workstations [10]. Licklider's vision of an interactive worldwide computer network led to the creation of Advanced Research Projects Agency Network (ARPANET), the predecessor to the Internet [11]. DARPA has been involved in many research and development projects such as the first satellite positioning system, stealth technology, next-generation supercomputers, and alternative energy [12].

Former DARPA director Regina Dugan spoke at TED 2012 in Long Beach, California where she introduced a robotic hummingbird, a prosthetic arm controlled by thought, and other DARPA inventions. Dugan said, "Scientists and engineers changed the world. I'd like to tell you about a magical place called DAPRA where scientists and engineers defy the impossible and refuse to fear failure" [13].

6.3 Information Awareness Office (IAO)

In January 2002, retired Admiral John Poindexter returned to the U.S. government to serve as the director of the newly-established Information Awareness Office (IAO) at the Defense Advanced Research Projects Agency (DARPA) of the U.S. Department of Defense (DoD) [14].

Dr. Tony Tether, director of DARPA, established the IAO whose official seal featured a pyramid topped with an all-seeing eye, similar to the one on the back of a U.S. dollar bill. The seal also had a Latin inscription, SCIENTIA EST POTENTIA, which means that science has a lot of potential, or in other words, knowledge is power. (See Fig. 6.1).

Fig. 6.1 Official Seal of the Information Awareness Office (IAO)

IAO's mission was to "develop new tools to detect, anticipate, train for, and provide warnings about potential terrorist attacks" [14]. In his April 2002 statement to the U.S. Senate Committee on Armed Services, Tether explained the rationale behind the creation of the IAO to "find, identify, track, and understand terrorist networks" [15]:

> One of the great challenges in the war on terrorism is to know our enemy - who he is, where he is, and what he's doing. In order to focus our efforts, I established another new DARPA office, the Information Awareness Office (IAO). IAO is developing the information systems needed to find, identify, track, and understand terrorist networks and vastly improve what we know about our adversaries. We will use the light of information technology to take away the shadows they hide in.

> For example, IAO's Evidence Extraction and Link Discovery program is aimed at finding terrorist networks hidden in the mountains of diverse data that we collect. The Wargaming the Asymmetric Environment program is explicitly aimed at predicting the behavior of terrorist groups in some detail, an extremely difficult challenge. Usually what we do now is issue broad warnings to the public to be on guard, like the several that were announced following September 11th. Wargaming the Asymmetric Environment seeks to move from those broad warnings to more specific predictions. In short, we want to go from predicting the terrorist "climate" to predicting the terrorist "weather." Some would argue that this is an outrageous goal, one that is not possible to achieve. I agree it sounds outrageous, but what if we can do it? That is why it is a DARPA program.

> In addition, IAO's **Total Information Awareness** program is now setting up a testbed at the Army's Intelligence and Security Command to test our new technologies on real-world threat data.

6.4 Perception of Privacy Invasion

Although Dr. Tony Tether's intention for Total Information Awareness (TIA) was to prevent future terrorist attacks, the American public was startled by *The New York Times* headline on November 9, 2002: "Pentagon Plans a Computer System That Would Peek at Personal Data of Americans." In his article, senior writer John Markoff expressed his deep concerns about privacy invasion by the U.S. government [16]:

> The Pentagon is constructing a computer system that could create a vast electronic dragnet, searching for personal information as part of the hunt for terrorists around the globe — including the United States.

> As the director of the effort, Vice Adm. John M. Poindexter, has described the system in Pentagon documents and in speeches, it will provide intelligence analysts and law enforcement officials with instant access to information from Internet mail and calling records to credit card and banking transactions and travel documents, without a search warrant.

> Historically, military and intelligence agencies have not been permitted to spy on Americans without extraordinary legal authorization. But Admiral Poindexter, the former

national security adviser in the Reagan administration, has argued that the government needs broad new powers to process, store and mine billions of minute details of electronic life in the United States.

In order to deploy such a system, known as **Total Information Awareness**, new legislation would be needed.... That legislation would amend the Privacy Act of 1974, which was intended to limit what government agencies could do with private information.

Senior editor Hendrik Hertzberg of *The New Yorker* concurred with Markoff in his December 9, 2002 article comparing the Information Awareness Office with Dr. Strangelove's vision [17]:

The [Information Awareness] Office's main assignment is, basically, to turn everything in cyberspace about everybody — tax records, driver's license applications, travel records, bank records, raw FBI files, telephone records, credit card records, shopping mall security camera videotapes, medical records, every e-mail anybody ever sent — into a single, humongous, multi-googolplexibyte database that electronic robots will mine for patterns of information suggestive of terrorist activity. Dr. Strangelove's vision — "a chikentic gomplex of gumbyuders" — is at last coming into its own.

"This could be the perfect storm for civil liberties in America," said Marc Rotenberg, president and executive director of the Electronic Privacy Information Center (EPIC) in Washington. "The vehicle is the Homeland Security Act, the technology is DARPA and the agency is the FBI. The outcome is a system of national surveillance of the American public" [16].

6.5 Privacy Protection in Total Information Awareness (TIA)

Since the establishment of the Information Awareness Office in January 2002, Dr. Tony Tether was aware of potential privacy issues. In March 2002, DARPA issued a formal call for proposals (BAA02-08) on Information Awareness [18]. The purpose was to "develop information technologies to help prevent continued terrorist attacks on the citizens, institutions, and property of the United States and its allies." Information repositories and privacy protection technologies were the number 1 priority on the three stated objectives:

1. Development of revolutionary technology for ultra-large all-source information repositories and associated **privacy protection technologies**;
2. Development of collaboration, automation, and cognitive aids technologies that allow humans and machines to think together about complicated and complex problems more efficiently and effectively; and
3. Development and implementation of an end-to-end, closed-loop prototype system to aid in countering terrorism through prevention by integrating technology and components from existing DARPA programs such as: Genoa, EELD

(Evidence Extraction and Link Discovery), WAE (Wargaming the Asymmetric Environment), TIDES (Translingual Information Detection, Extraction and Summarization), HID (Human Identification at Distance), Bio-Surveillance; as well as programs resulting from the first two areas of this BAA and other programs.

Figure 6.2 is a diagram of the Total Information Awareness (TIA) system designed by the Information Awareness Office (IAO) showing the workflow from detection, classification, identification, tracking, understanding, to preemption. "Privacy and Security" is highlighted in red, filtering the data needed for repositories and counterterrorism analysis.

To accommodate "ultra-large all-source information," the database envisioned is "of an unprecedented scale, will most likely be distributed, must be capable of being continuously updated, and must support both autonomous and semi-automated analysis. The latter requirement implies that the representation used must, to the greatest extent possible, be interpretable by both algorithms and human analysts. The database must support change detection and be able to execute automated procedures implied by new information."

DARPA acknowledged that "the reduced signature and misinformation introduced by terrorists who are attempting to hide and deceive imply that uncertainty must be represented in some way." The call for proposals stressed the importance "to protect the privacy of individuals not affiliated with terrorism" by seeking "technologies for controlling automated search and exploitation algorithms and for purging data structures appropriately."

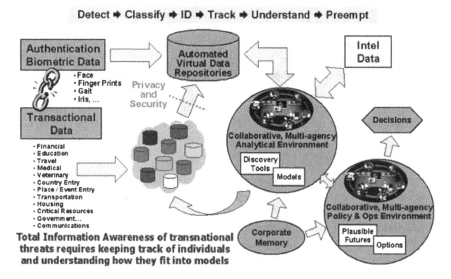

Fig. 6.2 Total Information Awareness of transnational threats requires keeping track of individuals and understanding how they fit into models (Courtesy of the Defense Advanced Research Projects Agency)

In fact, the TIA program included the Genisys Privacy Protection Program. The goal of Genisys was to make databases easy to use and simple to integrate. The Genisys privacy protection program would ensure personal privacy and protect sensitive intelligence sources [19]:

> Information systems and databases have the potential to identify terrorist signatures through the transactions they make, but Americans are rightfully concerned that data collection, integration, analysis, and mining activities implicate privacy interests. The Genisys Privacy Protection Program aims to provide security with privacy by providing certain critical data to analysts while controlling access to unauthorized information, enforcing laws and policies through software mechanisms, and ensuring that any misuse of data can be quickly detected and addressed. Research being conducted under other IAO programs may indicate that information about terrorist planning and preparation activities exists in databases that also contain information about U.S. persons. Privacy protection technologies like those being developed under the Genisys Privacy Protection Program would be essential to protect the privacy of U.S. citizens should access to this sort of information ever be contemplated.

Barbara Simons, computer scientist and past president of the Association for Computing Machinery (ACM), was nonetheless highly skeptical: "I'm just not convinced that the TIA will give us tools for catching terrorists that we don't already have or that could be developed with far less expensive and less intrusive systems" [20].

On the particular issue of database security and privacy, Simons said, "Even if one were able to construct a system which did protect privacy in some sense, we certainly have not been very successful with building humongous databases that are secure.... A lot of my colleagues are uncomfortable about this and worry about the potential uses that this technology might be put to, if not by this administration then by a future one. Once you've got it in place you can't control it" [16].

The economist warned of an Orwellian future in its 2003 special report on the Internet society: "As more human interactions are conducted and recorded electronically, as the ability to analyse databases grows and as video and other offline surveillance technologies become cheaper and more effective, it will become ever easier for authoritarian governments to set up systems of widespread surveillance. George Orwell's Big Brother of '*1984*' might yet become a reality, a few decades later than he expected" [21].

6.6 Opposing Views on TIA

In response to continuing criticism and public uproar, Dr. Tony Tether testified before the U.S. House of Representatives House Armed Services Committee in March 2003 [22]:

> The goal of our Information Awareness program is to create information technology that America's national security community can use to detect and defeat terrorist networks before they can attack us.

One of our Information Awareness programs is Total Information Awareness (TIA), around which there has been much controversy. If I knew only what I read in the press about TIA, I would be concerned too. So I'd like to briefly address some of the main concerns.

No American's privacy has changed in any way as a result of DARPA's Information Awareness programs, including the TIA. The Department of Defense *is not* developing technology so it can maintain dossiers on every American citizen. The Department of Defense *is not* assembling a giant database on Americans.

Instead, the TIA program is designed as an experimental, multi-agency prototype network that participating agencies can use to better share, analyze, understand, and make decisions based on whatever data to which they currently have *legal* access. ITA will integrate three broad categories of information technologies from DARPA and elsewhere: advanced collaboration and decision support tools, language translation, and data search and pattern recognition.

On February 7, 2003, the DoD announced the establishment of two boards to oversee TIA. These boards, an internal oversight board and an outside advisory committee, will work with DARPA as we continue our research. They will help ensure that TIA develops and disseminates its tools to track terrorists in a manner consistent with U.S. constitutional law, U.S. statutory law, and American values.

Tether and Poindexter believed that the 9/11 attacks could have been averted if intelligence information was collected and analyzed in time, especially about the known al-Qaeda terrorists Khalid al-Mihdhar and Nawaf al-Hazmi who entered the United States in January 2000.

Former *Newsweek* investigative correspondent Michael Isikoff said, "What's stunning is that, from that moment on, they [al-Mihdhar and al-Hazmi] lived entirely out in the open. They opened up bank accounts, they got a California driver's license, they opened up credit cards and they interacted with at least five other of the hijackers on 9/11" [23].

Proponents of total information awareness believe that TIA can eliminate the element of surprise so that we can prevent future 9/11 and Pearl Harbor attacks. Opponents disagree. Amy Belasco, specialist in national defense at the Foreign Affairs, Defense, and Trade Division, summarized the opposing views of TIA in her March 2003 report [14]:

To proponents, TIA R&D holds out the promise of developing a sophisticated system that would develop new technologies to find patterns from multiple sources of information in order to give decision makers new tools to use to detect, preempt and react to potential terrorist attacks.

To opponents, TIA has the potential to violate the privacy of individuals by giving the government access to vast amounts of information about individuals as well as possibly misidentifying individuals as potential terrorists.

In the 1982 science fiction movie *Star Trek II: The Wrath of Khan*, Mr. Spock says, "Logic clearly dictates that the needs of the many outweigh the needs of the few" [24]. Spock's logic makes sense in a free democratic society where vital information for the millions outweighs the privacy of the few.

6.7 Demystifying IAO and TIA

The Total Information Awareness (TIA) program involves multiple research and development (R&D) programs and the integration of these programs into a prototype TIA system. In July 2002, the Information Awareness Office (IAO) published a TIA System Description Document [24], and DARPA dedicated $137.5 million to the R&D programs and $10 million to the system integration for fiscal year 2003 [14].

In April 2003, the IAO issued an elaborate, year-long call for proposals (BAA03-23) on Information Awareness [25]. The IAO was soliciting ideas that "will imagine, develop, apply, integrate, demonstrate and transition information technologies and components for possible use in prototype closed-loop information systems to counter asymmetric threats." The proposal explained the goal of the Total Information Awareness program [25]:

> Program outputs will exploit information to significantly improve preemption capabilities, national security warning, and national security decision-making. The most serious asymmetric threat facing the United States is terrorism, a threat characterized by collections of people loosely organized in shadowy networks that are difficult to identify and define. IAO plans to develop technology that will allow understanding of the intent of these networks, their plans, and potentially define opportunities for disrupting or eliminating the threats. To effectively and efficiently carry this out, we must promote sharing, collaborating and reasoning to convert nebulous data to knowledge and actionable options. IAO will accomplish this by pursuing the development of technologies, components, and applications that may become integrated in a prototype [Total Information Awareness] system.

The IAO had no interest in developing information collection technology. Instead, its primary interest was in the following topic areas [26]:

1. Collaborative Reasoning and Decision Support Technologies

   - Detect terrorist planning and preparation activities.
   - Facilitate information sharing.
   - Conduct simulation and risk analysis.
   - Investigate structured argumentation, evidential reasoning, storytelling, change detection, and truth maintenance.

2. Language Translation Technologies

   - Detect, extract, summarize, and translate information.
   - Develop speech-to-text transcription technologies for English, Chinese, and Arabic languages.
   - Port applications to new languages within one month.

3. Pattern Recognition and Predictive Modeling Technologies

   - Extract evidence and find patterns from vast amounts of unstructured textual data (such as intelligence messages or news reports that are legally available and obtainable by the U.S. Government).
   - Discover critical information from speech and text of multiple languages.

- Develop threat-specific tools to enable analysts and decision makers to predict terrorist attacks and to simulate potential intervention strategies.
- Identify abnormal health detectors indicative of a biological attack.

4. Data Search and Privacy Protection Technologies

- Exploit distributed databases, information repositories, and sensor feeds.
- Represent uncertainty in structured data.
- Develop privacy protection technologies including immutable audit, self-reporting data, tamper-proof accounting system, anonymization and inferencing techniques, use of filtering and expunging software agents, and selective revelation concepts.

5. Biometric Technologies

- Develop automated, multimodal, biometric technologies to detect, recognize, and identify humans, alone or grouped, in disguise or not, at a distance, day or night, and in all weather conditions.
- Investigate 3D morphable modeling approaches, the feasibility of networking and fusing multiple biometric sensors, and activity recognition monitoring concepts.

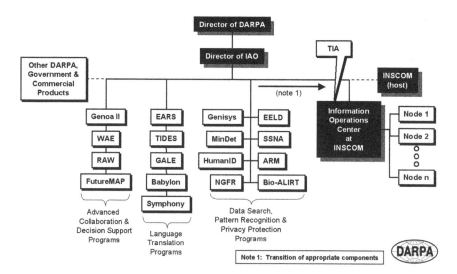

Fig. 6.3 Information Awareness Office (IAO) (Courtesy of the Defense Advanced Research Projects Agency)

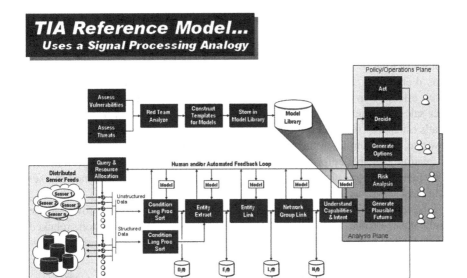

Fig. 6.4 Total Information Awareness (TIA) reference model (Courtesy of the Defense Advanced Research Projects Agency)

Figure 6.3 shows the overall organization and activities of the Information Awareness Office (IAO) dated May 2003 [19]. The IAO was responsible for transitioning appropriate technologies to the Total Information Awareness (TIA) system from R&D programs in advanced collaboration and decision support, language translation, data search, pattern recognition, and privacy protection. The TIA would be hosted by the Information Operations Center at INSCOM (U.S. Army's Intelligence and Security Command).

Figure 6.4 is a TIA reference model using a signal processing analogy to show how the software components from IAO and other government programs and from commercial sources fit together. TIA provides the analysts with the capability to discover the plans and intentions of potential terrorist activities by building and refining models of terrorist attacks based on available information.

6.8 Demise of IAO and TIA

Despite the support of the Bush administration and the last-ditched effort by IAO renaming Total Information Awareness to Terrorism Information Awareness in May 2003 [19], adverse media reaction and public distrust of Admiral John Poindexter proved to be too strong to overcome.

In July 2003, Poindexter faced harsh criticism from the media about IAO's Policy Analysis Market (PAM), part of the Futures Markets Applied to Prediction (FutureMAP) project. PAM was an online futures trading market in which anonymous speculators would bet on forecasting terrorist attacks, assassinations, and coups d'état [27]. IAO justified PAM by stating that such futures trading had proven effective in predicting other events like oil prices, elections, and movie ticket sales. Business journalist James Surowiecki of *The New Yorker* remarked, "That's especially important in the case of the intelligence community because we know that, for example, in the case of 9/11 there was lots of valuable and relevant information available before the attack took place. What was missing was a mechanism for aggregating that information in a single place. A well-designed market might have served as that mechanism" [28].

However, a sample bet on the assassination of Yasser Arafat proved to be simply unacceptable to the U.S. Congress [29]. Many U.S. Senators rebuked Poindexter for applying economic theory of efficient markets and market discovery to national security [30]:

Hillary Clinton of New York: "It's a futures market on death, and not in keeping with our values."

Tom Daschle of South Dakota: "I couldn't believe that we would actually commit $8 million to create a Web site that would encourage investors to bet on futures involving terrorist attacks and public assassinations."

Ron Wyden of Oregon: "The idea of a federal betting parlor on atrocities and terrorism is ridiculous and it's grotesque."

Byron Dorgan of North Dakota: "Can you imagine if another country set up a betting parlor so that people could go in and bet on the assassination of an American political figure?"

FutureMAP and PAM were the last straw for Poindexter, IAO, and TIA. A month later in August 2003, Poindexter resigned with an open letter to DARPA director Tony Tether, in which he vehemently defended his actions and viewpoints [31]:

[On first premise/research path — TIA:]

As you know as our research has evolved we have had basically two research paths — each in the context of a premise. The first premise is that the U.S. government has all of the data it needs to find information that would allow us to detect foreign terrorists and their plans and thus enable the prevention of attacks against U.S. interests…. On this first research path we created an experimental network called TIA and partnered with nine foreign intelligence, counter-intelligence and military commands for testing experimental tools using foreign intelligence data that is currently available to them…. The work under this premise should not be controversial in the U.S. since the tools are being applied using foreign intelligence data and as I have said is completely responsive to the problems the Congress has raised with respect to 9/11.

[On second premise/research path — FutureMAP:]

If we are wrong on the first premise and the U.S. government does not have all of the data it needs to find the terrorists and prevent their attacks, we felt it prudent to explore a second research path. This is the controversial one. In terms of the recent flap over FutureMap - did we want to bet the safety of thousands if not millions of Americans that our first premise was correct? Since we didn't want to make that bet, we devoted a relatively small portion of the funds that had been made available to us to this second research path.

[On privacy issue:]

We knew from the beginning that this second research path would be controversial and if the research proved successful, we would have to solve the privacy issue if it were ever to be deployed. We did not want to make a tradeoff between security and privacy. It would be no good to solve the security problem and give up the privacy and civil liberties that make our country great.... We needed to find a solution for all three concerns: privacy of US citizens, privacy of foreign citizens and privacy of sources and methods.

In early 2002, shortly after the new office was formed, we began a study called Security with Privacy to imagine ways technology could be developed to preserve the privacy of individuals and still search through data that is not currently available to the government looking for specific patterns of activity that are related to terrorist planning and preparation activities.

[On portrayal by major media:]

In November 2002 after our work had been badly misrepresented in the major media, it was decided that I should not speak publicly to provide a defense and explanation of our work since I was such a "lightning rod" (not my words).... I regret we have not been able to make our case clear and reassure the public that we do not intend to spy on them....

[On closing plea:]

In my opinion, the complex issues facing this nation today may not be solved using historical solutions and rhetoric that has been applied in the past, and that it may be useful to explore complex solutions that sometimes involve controversial technical concepts in order to rediscover the privacy foundations of this nation's strength and the basis for its freedoms.

Poindexter's plea fell on deaf ears. A month later in September 2003, the U.S. Congress axed the Information Awareness Office and terminated the funding for TIA: "The conferees agree with the Senate position which eliminates funding for the Terrorism Information Awareness (TIA) program within the Defense Advanced Research Projects Agency (DARPA). The conferees are concerned about the activities of the Information Awareness Office (IAO) and direct that the Office be terminated immediately. The only research projects previously under the jurisdiction of the Information Awareness Office (IAO) that may continue under DARPA are: Bio-Event Advanced Leading Indicator Recognition Technology, Rapid Analytic Wargaming, Wargaming the Asymmetric Environment, and Automated Speech and Text Exploitation in Multiple Languages (including Babylon and Symphony)" [32].

Bibliography

1. **Associated Press**. Reagan Urges 'Risk' on Gorbachev : Soviet Leader May Be Only Hope for Change, He Says. [Online] Los Angeles Times, June 13, 1989. http://articles.latimes. com/1989-06-13/news/mn-2300_1_soviets-arms-control-iron-curtain.
2. **Office of Science and Technology Policy**. Obama Administration Unveils "Big Data" Initiative: Announces $200 Million In New R&D Investments. [Online] Executive Office of the President, March 29, 2012. http://www.whitehouse.gov/sites/default/files/microsites/ostp/ big_data_press_release.pdf.

3. **The New York Times**. IRAN-CONTRA REPORT; Arms, Hostages and Contras: How a Secret Foreign Policy Unraveled. [Online] The New York Times, November 19, 1987. http://www.nytimes.com/1987/11/19/world/iran-contra-report-arms-hostages-contras-secret-foreign-policy-unraveled.html.

4. **Reagan, Ronald**. Address to the Nation on Iran-Contra. [Online] University of Virginia Miller Center, March 4, 1987. http://millercenter.org/scripps/archive/speeches/detail/3414.

5. **Shenon, Philip**. North, Poindexter and 2 Others Indicted on Iran-Contra Fraud and Theft Charges. [Online] The New York Times, March 17, 1988. http://www.nytimes.com/1988/03/17/world/north-poindexter-and-2-others-indicted-on-iran-contra-fraud-and-theft-charges.html?pagewanted=all.

6. **Greenhouse, Linda**. Supreme Court Roundup; Iran-Contra Appeal Refused by Court. [Online] The New York Times, December 8, 1992. http://www.nytimes.com/1992/12/08/us/supreme-court-roundup-iran-contra-appeal-refused-by-court.html.

7. **Harris, Shane**. Lightning Rod. [Online] Government Executive, July 15, 2004. http://www.govexec.com/magazine/features/2004/07/lightning-rod/17199/.

8. **DARPA**. Genoa. [Online] mirror of decommissioned Federal government site www.darpa.mil/iao/Genoa.htm. http://infowar.net/tia/www.darpa.mil/iao/Genoa.htm.

9. **Defense Advanced Research Projects Agency** . History. [Online] Defense Advanced Research Projects Agency . [Cited: November 29, 2012.] http://www.darpa.mil/about/history/history.aspx.

10. **Waldrop, Mitchell M**. No, This Man Invented The Internet . [Online] Forbes, November 27, 2000. http://www.forbes.com/asap/2000/1127/105.html.

11. **Van Atta, Richard**. 50 Years of Bridging the Gap. [Online] Defense Advanced Research Projects Agency. [Cited: November 29, 2012.] http://www.darpa.mil/WorkArea/DownloadAsset.aspx?id=2553.

12. **Defense Advanced Research Projects Agency**. First 50 Years. [Online] Defense Advanced Research Projects Agency. [Cited: November 29, 2012.] http://www.darpa.mil/About/History/First_50_Years.aspx.

13. **Dugan, Regina**. Regina Dugan: From mach-20 glider to humming bird drone. [Online] TED, March 2012. http://www.ted.com/talks/regina_dugan_from_mach_20_glider_to_humming_bird_drone.html.

14. **Belasco, Amy**. Total Information Awareness Programs: Funding, Composition, and Oversight Issues. [Online] The Air University, March 21, 2003. http://www.au.af.mil/au/awc/awcgate/crs/rl31786.pdf.

15. **Tether, Tony**. Statement by Dr. Tony Tether to the U.S. Senate Committee on Armed Services. [Online] U.S. Senate Committee on Armed Services, April 10, 2002. http://www.armed-services.senate.gov/statemnt/2002/April/Tether.pdf.

16. **Markoff, John**. Pentagon Plans a Computer System That Would Peek at Personal Data of Americans. [Online] The New York Times, November 9, 2002. http://www.nytimes.com/2002/11/09/politics/09COMP.html?pagewanted=all.

17. **Hertzberg, Hendrik**. Too Much Information. [Online] The New Yorker, December 9, 2002. http://www.newyorker.com/archive/2002/12/09/021209ta_talk_hertzberg.

18. **Defense Advanced Research Projects Agency**. INFORMATION AWARENESS Solicitation Number: BAA02-08. [Online] Federal Business Opportunities, March 21, 2002. https://www.fbo.gov/index?s=opportunity&mode=form&tab=core&id=82cd202bb8a0528c389f11d00dad8514&_cview=0.

19. **Electronic Privacy Information Center**. Report to Congress regarding the Terrorism Information Awarenss. [Online] May 20, 2003. http://epic.org/privacy/profiling/tia/may03_report.pdf.

20. **Manjoo, Farhad**. Total Information Awareness: Down, but not out . [Online] Salon, January 29, 2003. http://www.salon.com/2003/01/29/tia_privacy/.

21. **The Economist**. Caught in the net. [Online] The Economist, January 23, 2003. http://www.economist.com/node/1534249.

22. **Tether, Tony**. Statement by Dr. Tony Tether to the U.S. House of Representatives House Armed Services Committee. [Online] DARPA, March 27, 2003. www.darpa.mil/WorkArea/ DownloadAsset.aspx?id=1778.

23. **Koch, Kathleen**. White House downplays Newsweek report. [Online] CNN, June 3, 2002. http://articles.cnn.com/2002-06-03/politics/white.house.newsweek_1_bin-laden-operatives-qaeda-bush-officials.

24. **Gregory, Mack**. Total Information Awareness Program (TIA) System Description Document (SDD). [Online] Hicks and Associates, Inc, July 19, 2002. http://epic.org/privacy/profiling/tia /tiasystemdescription.pdf.

25. **Defense Advanced Research Projects Agency**. INFORMATION AWARENESS Solicitation Number: BAA03-23. [Online] Federal Business Opportunities, April 15, 2003. https://www.fbo.gov/index?s=opportunity&mode=form&id=2c05fdcf3acfc3ceb88b1edce3 2ecb7c&tab=core&_cview=1.

26. **Defense Advanced Research Projects Agency**.—. Information Awareness Proposer Information Pamphlet. [Online] Federal Business Opportunities, April 15, 2003. https://www. fbo.gov/utils/view?id=fb42bb6199c56c7c465308f4344826fc.

27. **Hulse, Carl**. THREATS AND RESPONSES: PLANS AND CRITICISMS; Pentagon Prepares A Futures Market On Terror Attacks. [Online] July 29, 2003. http://www.nytimes. com/2003/07/29/us/threats-responses-plans-criticisms-pentagon-prepares-futures-market-terr or.html?pagewanted=all&src=pm.

28. **Looney, Robert**. DARPA's Policy Analysis Market for Intelligence: Outside the Box or Off the Wall? . [Online] Strategic Insights, September 2003. http://www.au.af.mil/ au/awc/awcgate/nps/pam/si_pam.htm.

29. **Scheiber, Noam**. 2003: THE 3rd ANNUAL YEAR IN IDEAS; Futures Markets in Everything. [Online] The New York Times, December 14, 2003. http://www.nytime s.com/2003/12/14/magazine/2003-the-3rd-annual-year-in-ideas-futures-markets-in-everything.html.

30. **Starr, Barbara**. Pentagon folds bets on terror. [Online] CNNMoney, July 29, 2003. http:// money.cnn.com/2003/07/29/news/terror_futures/?cnn=yes.

31. **Poindexter, John**. John M. Pondexter Resignation Letter. [Online] The Washington Post, August 12, 2003. http://www.washingtonpost.com/wp-srv/nation/transcripts/poindexterletter.pdf.

32. **U.S. Congress**. Committee Reports. 108th Congress (2003-2004). House Report 108-283. [Online] The Library of Congress, September 2003. http://thomas.loc.gov/cgi-bin/cpquery/? &sid=cp108alJsu&refer=&r_n=hr283.108&db_id=108&item=&&sid=cp108alJsu&r_n= hr283.108&dbname=cp108&&sel=TOC_309917.

Chapter 7
The Afterlife of Total Information Awareness and Edward Snowden's NSA Leaks

By finally admitting a wrong, a nation does not destroy its integrity but, rather, reinforces the sincerity of its commitment to the Constitution and hence to its people.
— U.S. Attorney General Dick Thornburgh (October 10, 1990)

Congress gave me the authority to use necessary force to protect the American people, but it didn't prescribe the tactics.
— President George W. Bush (January 23, 2006)

Technology is a two-edged sword for the intelligence community. For instance, with biology, there could be a time in the not distant future when teenagers can design biological components just as they do computer viruses today.
— ODNI Director of Science and Technology Steven Nixon (2008)

It's important to recognize that you can't have 100 % security and also then have 100 % privacy and zero inconvenience.
— President Barack Obama (June 7, 2013)

We strongly encourage all governments to be much more transparent about all programs aimed at keeping the public safe. It's the only way to protect everyone's civil liberties and create the safe and free society we all want over the long term.
— Facebook CEO Mark Zuckerberg (June 7, 2013)

Being a patriot means knowing when to protect your country, when to protect your Constitution, when to protect your countrymen from the violations and encroachments of adversaries.
— Edward Snowden (May 2014)

© Springer International Publishing Switzerland 2015   151
N. Lee, *Counterterrorism and Cybersecurity*, DOI 10.1007/978-3-319-17244-6_7

7.1 NSA's Terrorist Surveillance Program

Although the U.S. Congress axed the Information Awareness Office (IAO) and dismantled Total Information Awareness (TIA) in September 2003, TIA did not really cease to exist. Five years later in March 2008, a *Wall Street Journal* article reported that the National Security Agency (NSA) has been building essentially the same system as TIA for its Terrorist Surveillance Program and other U.S. governmental agencies. Wall Street Journal intelligence correspondent Siobhan Gorman wrote [1]:

> According to current and former intelligence officials, the spy agency now monitors huge volumes of records of domestic emails and Internet searches as well as bank transfers, credit-card transactions, travel and telephone records. The NSA receives this so-called "transactional" data from other agencies or private companies, and its sophisticated software programs analyze the various transactions for suspicious patterns.

> Two current officials also said the NSA's current combination of programs now largely mirrors the former TIA project. But the NSA offers less privacy protection. TIA developers researched ways to limit the use of the system for broad searches of individuals' data, such as requiring intelligence officers to get leads from other sources first. The NSA effort lacks those controls…

The NSA uses its own high-powered version of social-network analysis to search for possible new patterns and links to terrorism. Former NSA director Gen. Michael Hayden explained, "The program … is not a driftnet over [U.S. cities such as] Dearborn or Lackawanna or Fremont, grabbing conversations that we then sort out by these alleged keyword searches or data-mining tools or other devices… This is not about intercepting conversations between people in the United States. This is hot pursuit of communications entering or leaving America involving someone we believe is associated with al-Qaeda. … This is focused. It's targeted. It's very carefully done. You shouldn't worry" [2].

In spite of Hayden's assurance, the American Civil Liberties Union (ACLU) issued a statement accusing the NSA of reviving TIA to be an Orwellian domestic spying program [3]:

> "Congress shut down TIA because it represented a massive and unjustified governmental intrusion into the personal lives of Americans," said Caroline Fredrickson, Director of the Washington Legislative Office of the ACLU. "Now we find out that the security agencies are pushing ahead with the program anyway, despite that clear congressional prohibition. The program described by current and former intelligence officials in Monday's Wall Street Journal could be modeled on Orwell's Big Brother."

> "Year after year, we have warned that our great nation is turning into a surveillance society where our every move is tracked and monitored," said Barry Steinhardt, Director of the ACLU's Technology and Liberty Project. "Now we have before us a program that appears to do that very thing. It brings together numerous programs that we and many others have fought for years, and it confirms what the ACLU has been saying the NSA is up to: mass surveillance of Americans."

The mass surveillance of Americans is a direct violation of the Fourth Amendment to the U.S. Constitution — a Bill of Rights that guards against unreasonable searches and seizures, along with requiring any warrant to be judicially sanctioned and supported by probable cause.

In September 2012, NSA whistleblower William Binney filed a sworn declaration that the agency has installed within the U.S. no fewer than 10 and possibly in excess of 20 intercept centers [4] including the AT&T center on Folsom Street in San Francisco. Binney's testimony supports the revelation from AT&T whistleblower Mark Klein in 2006 [5]. Klein installed inside AT&T's San Francisco switching office a Semantic Traffic Analyzer — an Internet monitoring tool that can reconstruct all of the e-mails sent along with attachments, see what web pages have been clicked on, capture instant messages, and record video streams and VoIP (Voice over Internet Protocol) phone calls [6]. In addition, federal law enforcement agents have been flying small Cessna airplanes across the U.S. to spy on mobile phone calls, tricking the phones into sharing their location data as well as their identities by mimicking actual cell towers [7].

"The danger here is that we fall into a totalitarian state," warned Binney. "This is something the KGB, the Stasi or the Gestapo would have loved to have had" [8].

7.2 President George W. Bush and NSA Warrantless Wiretapping

The National Security Agency (NSA) was created by President Harry Truman in 1952 after World War II to continue the code-breaking work and to prevent another surprise like the attack on Pearl Harbor [9]. The Central Security Service (CSS) was established by President Richard Nixon in 1972 to promote full partnership between NSA and the Service Cryptologic Components of the U.S. Armed Forces [10]. The mission for NSA/CSS is as follows:

> The National Security Agency/Central Security Service (NSA/CSS) leads the U.S. Government in cryptology that encompasses both Signals Intelligence (SIGINT) and Information Assurance (IA) products and services, and enables Computer Network Operations (CNO) in order to gain a decision advantage for the Nation and our allies under all circumstances. The Information Assurance mission confronts the formidable challenge of preventing foreign adversaries from gaining access to sensitive or classified national security information. The Signals Intelligence mission collects, processes, and disseminates intelligence information from foreign signals for intelligence and counterintelligence purposes and to support military operations. This Agency also enables Network Warfare operations to defeat terrorists and their organizations at home and abroad, consistent with U.S. laws and the protection of privacy and civil liberties.

Notwithstanding the constitutional rights of American citizens, the NSA has acted in accordance to a presidential order signed in 2002, shortly after the 9/11 terrorist attacks. *The New York Times* revealed in December 2005 that President George W. Bush "secretly authorized the National Security Agency to eavesdrop on Americans and others inside the United States to search for evidence of terrorist activity without the court-approved warrants ordinarily required for domestic spying" [11]:

> While many details about the program remain secret, officials familiar with it say the NSA eavesdrops without warrants on up to 500 people in the United States at any given time. The list changes as some names are added and others dropped, so the number

monitored in this country may have reached into the thousands since the program began, several officials said. Overseas, about 5,000 to 7,000 people suspected of terrorist ties are monitored at one time, according to those officials.

USA Today also reported that NSA has been secretly collecting the phone call records of tens of millions of Americans, using data provided by AT&T, Verizon, and BellSouth [2]:

The NSA program reaches into homes and businesses across the nation by amassing information about the calls of ordinary Americans — most of whom aren't suspected of any crime. This program does not involve the NSA listening to or recording conversations. But the spy agency is using the data to analyze calling patterns in an effort to detect terrorist activity. … It's the largest database ever assembled in the world. The agency's goal is "to create a database of every call ever made" within the nation's borders.

In response to the harsh criticisms from the media, the NSA disclosed some success stories of the domestic eavesdropping program, including foiling a plot by Iyman Faris, a Pakistani American truck driver in Ohio, who wanted to bring down the Brooklyn Bridge in 2002 with blowtorches [12]. The NSA Terrorist Surveillance Program also claimed to have helped thwart the fertilizer bomb attacks on British pubs and train stations in 2004 [13].

In January 2006, President George W. Bush publicly defended NSA's warrantless terrorist surveillance program (aka warrantless wiretapping) that bypassed the 1978 Foreign Intelligence Surveillance Act (FISA). Signed into law by President Jimmy Carter, FISA was introduced by Senator Ted Kennedy to provide judicial and congressional oversight of the government's covert surveillance activities of foreign entities and individuals in the United States without violating the Fourth Amendment to the U.S. Constitution [14].

"If I wanted to break the law, why was I briefing Congress?" Bush told an audience at Kansas State University. "Congress gave me the authority to use necessary force to protect the American people, but it didn't prescribe the tactics" [15]. Bush argued that bypassing the courts fell within presidential power during a time when the country is fighting terrorism. The U.S. Congress sided with Bush and the NSA warrantless wiretapping.

In July 2008, Congress passed the FISA Amendments Act of 2008, which grants retroactive immunity to complicit telecoms and allows eavesdropping in emergencies without court approval for up to seven days [16, 17]. The American Civil Liberties Union (ACLU) filed a lawsuit against FISA Amendments Act of 2008, calling it an "unconstitutional dragnet wiretapping law" [18].

Constitutional rights should be upheld under all circumstances; lest we forget the Japanese American internment during World War II. In October 1990, U.S. Attorney General Dick Thornburgh presented an entire nation's apology to 120,000 Japanese American internees and their descendants. In an emotional reparation ceremony in Washington, Thornburgh said, "By finally admitting a wrong, a nation does not destroy its integrity but, rather, reinforces the sincerity of its commitment to the Constitution and hence to its people" [19].

7.3 Poindexter's Policy Analysis Market

Although Poindexter's controversial Policy Analysis Market (PAM) program was terminated by the U.S. Congress in 2003, a San Diego based private firm that helped develop the software revived PAM in 2004 without involvement from the U.S. government. The new version of PAM would "allow traders to buy and sell contracts on political and economic events in the Middle East, but only on questions that have a positive or neutral slant, such as 'Iraqi oil exports will exceed 2 million barrels a day during the third quarter of 2004'" [20].

Economic journalist Noam Scheiber wrote in *The New York Times*, "The beauty of futures markets like PAM is that they're among the most meritocratic institutions ever devised" [21]. He cited the success of Hewlett-Packard's futures market in predicting actual sales [22], the high accuracy of the Iowa Electronic Markets (IEM) in forecasting presidential elections [23], and the impressive performance of the Hollywood Stock Exchange (HSX) in picking Oscar winners [24].

Unlike the controversial PAM that was open to public trading, restricting its access to intelligence officers with credible information can make PAM a powerful decision support tool for the U.S. intelligence community.

7.4 Project Argus: Bio-surveillance Priming System

After the 9/11 terrorist attacks, Dr. Eric Haseltine left his position as Executive Vice President of R&D at Walt Disney Imagineering to join the National Security Agency (NSA) as Director of Research in 2002. From 2005 to 2007, Haseltine was Associate Director for Science and Technology at the newly established Office of the Director of National Intelligence (ODNI).

In collaboration with Georgetown University researchers, Haseltine and his successor Steven Nixon at ODNI oversaw the development of Argus, a bio-surveillance AI program that monitors foreign news reports and other open sources looking for anything that could provide an early warning of an epidemic, nuclear accident, or environmental catastrophe.

In a statement by Dr. James Wilson of Georgetown University before the U.S. Senate Committee on Homeland Security, Wilson testified in October 2007 [25]:

> Argus is designed to detect and track early indications and warnings of foreign biological events that may represent threats to global health and national security. Argus serves a "tipping function" designed to alert its users to events that may require action. It is not in the business of determining whether, or what type of actions should be taken.
>
> Argus is based on monitoring social disruption through native language reports in electronic local sources around the globe. Argus specifically focuses on taxonomy of direct and indirect types of indications and warnings including:
>
> • Environmental conditions thought to be conducive to support outbreak triggering;
> • Reports of disease outbreaks in humans or animals; and
> • Markers of social disruption such as school closings or infrastructure overloads.

> We estimate we are accessing over a million pieces of information daily covering every country in the world which results in producing, on average, 200 reports per day. Using a disease event warning system modeled after NOAA's National Weather Service, we issue Warnings, Watches, and Advisories in accordance with guidelines agreed upon by our research partners in the federal government. On average, we have 15 Advisories, 5 Watches, and 2 Warnings active on our Watchboard at any given time, with 2,200 individual case files of socially disruptive biological events maintained and monitored daily in over 170 countries involving 130 disease entities affecting humans or animals.

> Since the program began, we have logged over 30,000 biological events in varying stages of social disruption throughout the world involving pathogens such as H5N1 avian influenza, other influenza strains, Ebola virus, cholera, and other exotic pathogens.

> We have discovered the Argus methodology can be made sensitive to events involving nuclear and radiological, chemical, terrorist, political instability, genocide and conflict, crop surveillance, and natural disasters.

Wilson cited the successful tracking of the H3N2 influenza virus spreading from China to Chile, Argentina, Australia, and several other countries. However, Argus was unable to monitor that strain of influenza within the U.S. because domestic monitoring was prohibited

Eric Haseltine said in a 2006 *U.S. News & World Report* interview, "I sleep a little easier at night knowing that Argus is out there" [26]. In 2008, ODNI Director of Science and Technology Steven Nixon told Lawrence Wright of *The New Yorker*, "Technology is a two-edged sword for the intelligence community. For instance, with biology, there could be a time in the not distant future when teenagers can design biological components just as they do computer viruses today. That's why I think intelligence is as critical now as at any time in our nation's history" [27].

7.5 President Barack Obama's Big Data R&D Initiative

At the Defense Advanced Research Projects Agency (DARPA), the dismantled Information Awareness Office (IAO) has been replaced by the Information Innovation Office (I2O). I2O has been carrying on research projects that were not shut down by the U.S. Congress since the termination of Total Information Awareness (TIA) [28]. They include Bio-Event Advanced Leading Indicator Recognition Technology, Rapid Analytic Wargaming, Wargaming the Asymmetric Environment, and Automated Speech and Text Exploitation in Multiple Languages (including Babylon and Symphony) [29].

Total Information Awareness requires efficient and effective data mining In March 2012, the Obama administration announced more than $200 million in funding for the "Big Data Research and Development Initiative" [30]. The first wave of agency commitments includes National Science Foundation (NSF), National Institutes of Health (NIH), Department of Energy (DOE), U.S. Geological Survey, and Department of Defense (including DARPA) [31].

Among the funded DARPA programs is Anomaly Detection at Multiple Scales (ADAMS), one of several key technologies that were directly applicable to Total Information Awareness [32].

"In the same way that past Federal investments in information-technology R&D led to dramatic advances in supercomputing and the creation of the Internet, the initiative we are launching today promises to transform our ability to use Big Data for scientific discovery, environmental and biomedical research, education, and national security," said Dr. John P. Holdren, Assistant to the President and Director of the White House Office of Science and Technology Policy [31].

7.6   Palantir Technologies Funded by CIA's in-Q-Tel

A palantír is a magical artifact from J. R. R. Tolkien's *The Lord of the Rings*. Shaped as a spherical stone that resembles a crystal ball, a palantír is used for both communication and as a means of seeing events in other parts of the world.

Palantir Technologies, named after the magical spying stone, was founded in 2004 by Peter Thiel (PayPal cofounder), Alex Karp, Joe Lonsdale, Stephen Cohen, and Nathan Gettings. With early investments from Thiel and the Central Intelligence Agency (CIA) venture arm In-Q-Tel, Palantir develops software applications for integrating, visualizing and analyzing big data that is structured, unstructured, relational, temporal, and geospatial [33].

Built upon PayPal's fraud detection algorithms, Palantir excels in discovering connections between seemingly unrelated incidents as well as the people involved. Palantir has helped "identify the core group of 27 men behind the killing of American journalist Daniel Pearl in 2002, uncover the GhostNet computer network that was infecting the systems of many countries' embassies, and enable several U.S. police departments to do predictive policing using data analysis to take a proactive approach to patrol deployment" [34]. Besides counterterrorism and law enforcement, Palantir applications have been deployed in banks, hospitals, law firms, insurance companies, pharmaceuticals, and other organizations [35].

The National Center for Missing and Exploited Children (NCMEC) has also used Palantir software to solve child abuse and abduction cases. Ernie Allen, CEO of NCMEC, praised Palantir for "the ability to do the kind of link-and-pattern analysis we need to build cases, identify perpetrators, and rescue children" [36].

However, Palantir's involvement in a convoluted plot to bring down WikiLeaks in 2011 raised some eyebrows [37]. Furthermore, Palantir's senior counsel Bryan Cunningham was former National Security Council legal adviser who supported President George W. Bush's authorization of the NSA warrantless wiretapping [38].

Christopher Soghoian, principal technologist at the American Civil Liberties Union (ACLU) voiced his concerns while he was a graduate fellow at Indiana University: "I don't think Palantir the firm is evil. I think their clients could be using it for evil things." In regard to Palantir's built-in privacy protection features, Soghoian said, "If you don't think the NSA can disable the piece of auditing functionality, you have to be kidding me. They can do whatever they want, so it's ridiculous to assume that this audit trail is sufficient" [36].

"Using Palantir technology," *Bloomberg Businessweek* reported in November 2011, "the FBI can now instantly compile thorough dossiers on U.S. citizens, tying together surveillance video outside a drugstore with credit-card transactions, cell-phone call records, e-mails, airplane travel records, and Web search information" [36].

Peter Thiel told *Businessweek* in defense of Palantir, "We cannot afford to have another 9/11 event in the U.S. or anything bigger than that. That day opened the doors to all sorts of crazy abuses and draconian policies. The best way to avoid such scenarios in the future would be to provide the government the most cutting-edge technology possible and build in policing systems to make sure investigators use it lawfully" [36].

7.7 Microsoft and NYPD's Domain Awareness System

In April 2009, the New York Police Department (NYPD) has developed a real-time networked Domain Awareness System (DAS) to detect, deter, and prevent potential terrorist activities in New York City [39]. As part of the counterterrorism program of the NYPD's Counterterrorism Bureau, the deployed DAS technology includes closed-circuit televisions (CCTVs), License Plate Readers (LPRs), and other domain awareness devices.

The CCTVs are operated by NYPD as well as partnering companies and government agencies that provide feeds from their proprietary CCTVs into the Lower Manhattan Security Coordination Center. License plate data are collected by fixed or mobile LPR devices. Other domain awareness devices gather environmental data and detect hazards.

In August 2012, Microsoft and NYDP jointly announced bringing the DAS technology to law enforcement agencies around the world. According to retired U.S. Army Lt. Gen. Mike McDuffie, DAS "aggregates and analyzes public safety data in real time and combines artificial intelligence analytics with video from around a jurisdiction to identify potential threats and protect critical infrastructure" [40].

7.8 NSA's Utah Data Center: A $1.5 Billion Data-Mining and Spy Center

On September 6, 2009, an email was sent by a suspected al-Qaeda member "Ahmad" from Pakistan to a previously unknown man in Denver, Colorado. It was instantly logged by the NSA computers in Fort Meade, Maryland. The Denver-based emailer, Najibullah Zazi, was later convicted of a suicide bombing plot against the New York subway [41].

"Forty years ago there were 5,000 stand-alone computers, no fax machines and not one cellular phone," said former NSA director Gen. Michael Hayden. "Today there are over 180 million computers — most of them networked. There

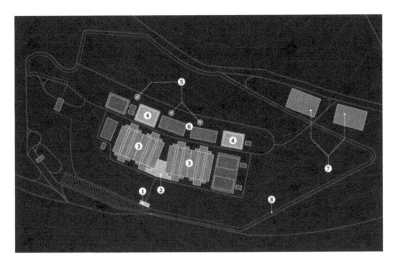

Fig. 7.1 NSA's Utah Data Center Conceptual Site Plan (Courtesy of the U.S. Army Corps of Engineers)

are roughly 14 million fax machines and 40 million cellphones, and those numbers continue to grow" [42]. With more than a dozen listening posts around the world, the NSA intercepts about two million phone calls, e-mail messages, faxes and other types of communications every hour.

The skyrocketing volume of information is to be stored and handled at the nation's largest data-mining center at Camp Williams National Guard Training Site in Bluffdale, Utah. The new construction was given the "green light" when President Barack Obama signed the 2009 Supplemental-War Funding Bill [43]. The NSA broke ground on the facility in January 2011. The United States Army Corps of Engineers (USACE) was the construction agent for the NSA. The 1-million-square-foot, $1.5 billion data center went online in December 2013 and was completed and fully operational in May 2014 [44].

Figure 7.1 shows the conceptual site plan for the NSA Utah Data Center with eight main areas:

1. Visitor control center — A $9.7 million facility for ensuring that only authorized personnel gain access.
2. Administration — A 900,000-square-foot space for technical support and administrative personnel.
3. Data halls — Four 25,000-square-foot facilities house rows of servers.
4. Backup generators and fuel tanks — They can power the center for three days in an emergency.
5. Water storage and pumping — They are able to pump 1.7 million gallons of liquid per day.
6. Chiller plant — About 60,000 tons of cooling equipment keep servers from overheating.

7. Power substation — An electrical substation to meet the center's estimated 65-megawatt demand.
8. Security — Video surveillance, intrusion detection, and other antiterrorism protection at a cost of $10 million.

James Bamford, investigative journalist and former Navy intelligence analyst, wrote in a March 2012 issue of the *Wired* magazine, "Once it's operational, the Utah Data Center will become, in effect, the NSA's cloud. The center will be fed data collected by the agency's eavesdropping satellites, overseas listening posts, and secret monitoring rooms in telecom facilities throughout the US. All that data will then be accessible to the NSA's code breakers, data-miners, China analysts, counterterrorism specialists, and others working at its Fort Meade headquarters and around the world" [45].

NSA deputy director John Chris Inglis said, "It's a state-of-the-art facility designed to support the intelligence community in its mission to, in turn, enable and protect the nation's cybersecurity." But an unnamed senior intelligence officer told Bamford, "This is more than just a data center. It is also critical for breaking codes. … Everybody's a target; everybody with communication is a target" [45].

In July 2014, Greenpeace, Electronic Frontier Foundation (EFF), and the Tenth Amendment Center (TAC) flew a 135-foot-long thermal airship over the NSA Utah Data Center in Bluffdale, carrying the message: "NSA Illegal Spying Below" [46]. Figure 7.2 is a photograph of the Utah Data Center dated July 9, 2014.

The NSA is taking advantage of Utah's cheap electric power and a cut-rate deal for millions of gallons of local water used to cool the computer servers. In

Fig. 7.2 NSA's Utah Data Center (Courtesy of Electronic Frontier Foundation)

November 2014, however, Utah lawmaker Marc Roberts proposed a bill to "refuse support to any federal agency which collects electronic data within this state" and therefore prohibit the NSA from negotiating new water deals when its current Bluffdale agreement runs out in 2021 [47].

Meant to be a parody of nsa.gov, nsa.gov1.info/utah-data-center/index.html offers details, photos, and links to news media, privacy groups, and government websites concerning the NSA Utah Data Center.

7.9   Global Surveillance and Abuse of Power

Investigative journalist Shane Harris compared Poindexter's TIA with the NSA data-mining project in his 2012 article "Giving In to the Surveillance State" published in *The New York Times* [48]:

> Today, this global surveillance system continues to grow. It now collects so much digital detritus — e-mails, calls, text messages, cellphone location data and a catalog of computer viruses — that the NSA is building a 1-million-square-foot facility in the Utah desert to store and process it.

> What's missing, however, is a reliable way of keeping track of who sees what, and who watches whom. After TIA was officially shut down in 2003, the NSA adopted many of Mr. Poindexter's ideas except for two: an application that would "anonymize" data, so that information could be linked to a person only through a court order; and a set of audit logs, which would keep track of whether innocent Americans' communications were getting caught in a digital net.

Without data anonymizer and audit logs, abuse of power is bound to happen. In the 2012 romantic comedy *This Means War* starring Reese Witherspoon, Chris Pine, and Tom Hardy, two CIA agents misappropriated company technology and assets to spy on the same woman in a love triangle story [49]. In real life, a number of male and female NSA employees have been found guilty of "Loveint" violations for misusing the agency's surveillance power to spy on their love interests [50].

Besides being morally wrong, unsanctioned or extracurricular spying activities risk the unnecessary exposure of technological apparatus and the potential compromise of the espionage network. When a covert action is not fully supported by a spying organization from top to bottom, an unintentional ripple effect can jeopardize ongoing and future legitimate operations (to the happy tune of adversaries' counterintelligence).

Illegal spying in the international arena happens more often than not, to various degrees depending on the perpetrators and the persons of interest. A close relative of mine was first cyberstalked on a social media platform, which is rather common in this Digital Age. However, it quickly escalated into eavesdropping on mobile phone calls and Skype video conferencing. She was shocked to learn that her stalker knew so much about her private conversations with her friends and business associates. The truly startling moment came when she called the same taxi service that she normally uses. Instead of a regular taxi, a nice town car showed up and confirmed the call. The well-dressed chauffeur was polite but he was way too inquisitive to be a regular cab driver. (Come to think of it, Uber and its mobile app would make spying so much easier.) Since that taxi incident, she has noticed similar oddities with a waitress at an upscale hotel restaurant, a steward on a popular cruise ship, and a system administrator at a social networking company. Given that all electronic and physical communications can be monitored and traced by both allies and adversaries, whoever is orchestrating the illegal spying is unjustifiably risking the exposure of undercover agents and surveillance protocols, thereby compromising other covert operations in the long run.

7.10 Edward Snowden's NSA Leaks and PRISM

On May 20, 2013, Edward Snowden, an employee of defense contractor Booz Allen Hamilton at the National Security Agency, arrived in Hong Kong from Hawaii with four laptop computers in his possession. The computer hard drives contained some of the U.S. government's top secret programs [51].

On June 1, *Guardian* journalists Glenn Greenwald and Ewen MacAskill and documentary maker Laura Poitras flew from New York to Hong Kong to begin a week of interviews with Snowden.

On June 5, *The Guardian* published a top-secret document by the U.S. Foreign Intelligence Surveillance Court. The April 2013 court order forced Verizon to hand over the phone records of millions of Americans, including but not limited to session identifying information (e.g., originating and terminating telephone number, International Mobile Subscriber Identity (IMSI) number, and International Mobile station Equipment Identity (IMEI) number) [52].

On June 6, a second *Guardian* story unveiled the top-secret PRISM program that allows officials to collect without court orders any data including emails, chat, videos, photos, stored data, voice-over-IP, file transfers, video conferencing, logins, and online social networking details from communication providers and social networks (See Figs. 7.3, 7.4, 7.5, 7.6, 7.7, 7.8, 7.9, 7.10, 7.11, 7.12, 7.13, 7.14, 7.15, 7.16, 7.17, 7.18 and 7.19, courtesy of *The Guardian* and nsa.gov1.info — a parody of nsa.gov). Under the PRISM program, the NSA spends about $20 million annually to access the servers of the participating companies, beginning in 2007 with Microsoft, followed by Yahoo! in 2008, Google, Facebook, and PalTalk in 2009, YouTube in 2010, Skype and AOL in 2011, and Apple in 2012.

In response to the revelation of PRISM, a senior U.S. official issued a statement: "*The Guardian* and *Washington Post* articles refer to collection of communications pursuant to Sect. 702 of the Foreign Intelligence Surveillance Act. This law does not allow the targeting of any US citizen or of any person located within the United State" [53].

On June 7, President Barack Obama defended the NSA dragnet surveillance programs: "They help us prevent terrorist attacks. … It's important to recognize

PRISM/US-984XN
Overview

OR

*The SIGAD Used **Most** in NSA Reporting*
Overview

April 2013

Derived From: NSA/CSSM 1-52
Dated: 20070108
Declassify On: 20350901
TOP SECRET//SI//ORCON//NOFORN

Fig. 7.3 PRISM Overview

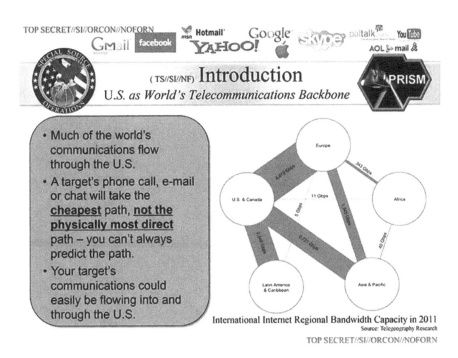

Fig. 7.4 PRISM Introduction

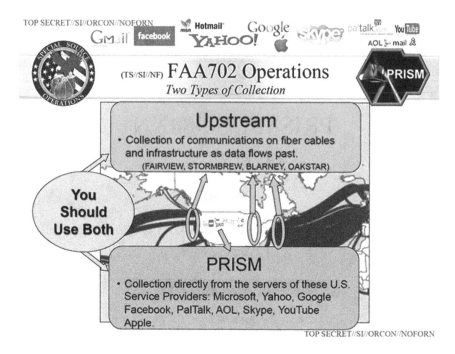

Fig. 7.5 PRISM FAA702 Operations (Two Types of Collection)

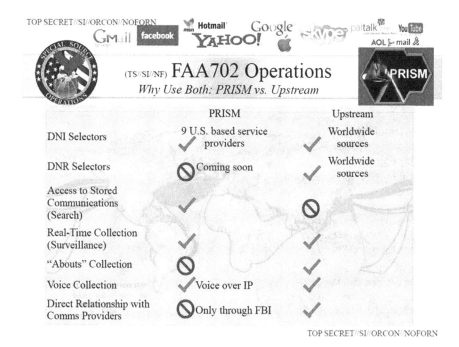

Fig. 7.6 PRISM FAA702 Operations (Why Use Both: PRISM vs. Upstream)

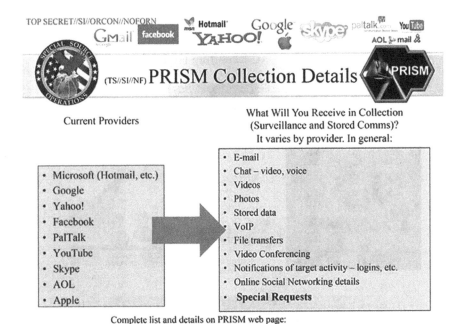

Fig. 7.7   PRISM Collection Details

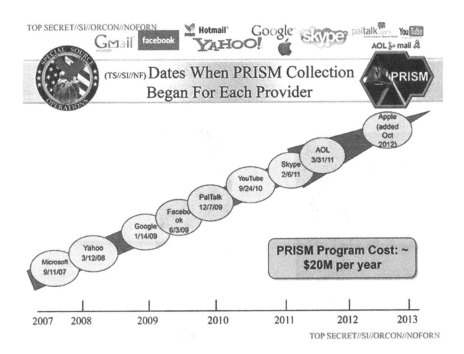

Fig. 7.8   Dates When PRISM Collection Began For Each Provider

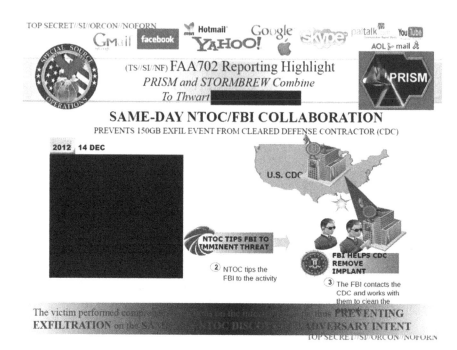

Fig. 7.9   FAA702 Reporting Highlight (PRISM and STORMBREW Combine To Thwart <Redacted>)

Fig. 7.10   Some Higher Volume Domains Collected from FAA Passive (Heavily Redacted)

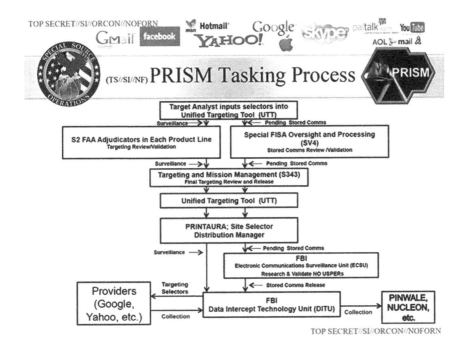

Fig. 7.11   PRISM Tasking Process

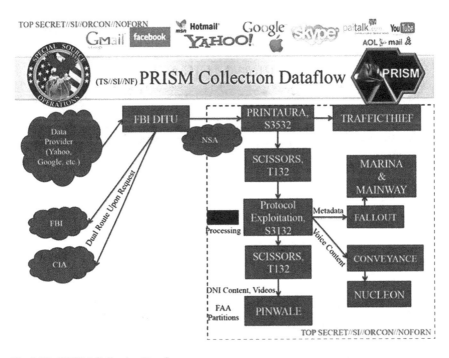

Fig. 7.12   PRISM Collection Dataflow

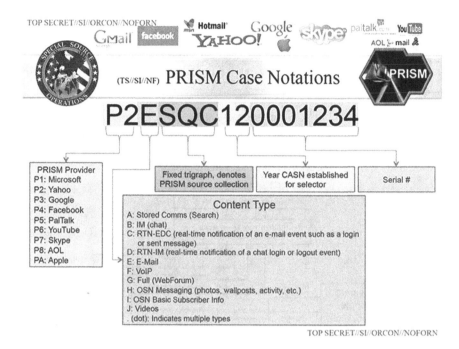

Fig. 7.13 PRISM Case Notations

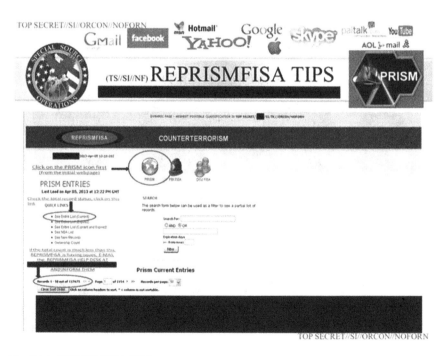

Fig. 7.14 REPRISMFISA TIPS

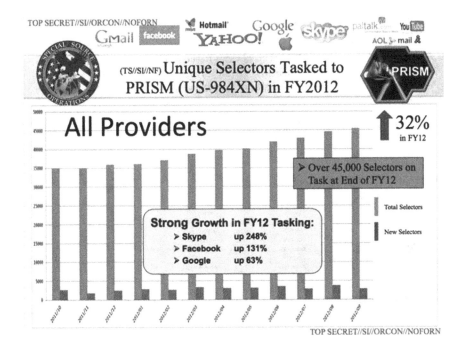

Fig. 7.15 Unique Selectors Tasked to PRISM (US-984XN) in FY2012

Fig. 7.16 A Week in the Life of PRISM Reporting (Sampling of Reporting Topics from 2-8 Feb 2013)

that you can't have 100 % security and also then have 100 % privacy and zero inconvenience" [54]. A former senior U.S. intelligence official explained then NSA director Gen. Keith Alexander's rationale behind the surveillance programs: "Rather than look for a single needle in the haystack, his approach was, 'Let's

TOP SECRET//SI//ORCON//NOFORN   Hotmall Google paltalk YouTube

Gmail facebook YAHOO! AOL mail

(TS//SI//NF) PRISM (US-984XN) Based Reporting:
June 2011 – May 2012
Sorted By # of PRISM-Based Reports Per OPI   PRISM

OPI - Top Producers Issuing	PRISM-Based Reports	% Increase in PRISM-Based Reports Compared to June 2010-May2011	% Of All OPI Reporting Which is PRISM-Based	% Points Change from June 2010 - May 2011 period	All Reports By OPI	Single-Source to PRISM	% of PRISM-Based Reports Which are Single Source
SCS (F6*, US-96*, US-97*, US-3219)	3723	Up 67%	20	+7 (up 54%)	18640	3040	82
S2I - Counterterrorism	3493	Up 5%	42	-2 (down 5%)	8242	2074	60
S2E - Middle East & Africa	2574	Up 47%	16	+2 (up 14%)	16537	1959	76
S2G - Combating Prolif	2092	Up 49%	30	+3 (up 11%)	6872	1395	67
NSAT (USJ-783*)	1690	Up 20%	30	+3 (up 11%)	5713	1319	78
S2A -	1389	Up 8%	11	-1 (down 8%)	12445	1196	86
NSAG (USJ-800*)	1255	Down 8%	11	0 (no change)	11741	883	70
ECC (ESOC) (USJ-753*, USM-44)	1147	Up 6%	52	+2 (up 4%)	2217	922	80
S2C - Intl Sec Issues	1147	Up 75%	13	+5 (up 63%)	8989	861	75
S2D - Countering Frgn Intel	862	Up 40%	12	-5 (down 29%)	7089	545	63
S2F - Intl Crime & Narc	666	Up 41%	16	+2 (up 14%)	4122	497	75
S2B -	634	Down 10%	13	-3 (down 19%)	4842	452	71
NTOC (V*)	455	Up 237%	21	+8 (up 62%)	2195	355	78
DSD	310	Down 15%	4	0 (no change)	7511	296	95
NSAH (USJ-750*)	237	Down 10%	2	+1 (up 50%)	12023	155	65
S2J - Weapons and Space	225	Up 221%	33	+11 (50%)	692	186	83
GCHQ	197	Up 137%	2	+1.9 (up 1900%)	11257	170	86
S2H -	176	Up 159%	5	+3 (up 150%)	3353	155	88
SSG	16	Up 60%	17	-19 (down 52%)	92	14	88
Utah Regional Ops Cntr (USJ-755)	12	Up 20%	6	-17 (down 74%)	207	12	100

Source: PLUS - 11-13 June 2012   TOP SECRET//SI//ORCON//NOFORN

Fig. 7.17 PRISM (US-984XN) Based Reporting: June 2011 — May 2012 (Sorted By # of PRISM-Based Reporting Per OPI)

collect the whole haystack. Collect it all, tag it, store it. … And whatever it is you want, you go searching for it" [55].

On July 31, a third *Guardian* story revealed the top-secret XKeyscore program touted as the "widest-reaching" system for developing intelligence from the Internet [56]. XKeyscore covers "nearly everything a typical user does on the Internet" including browsing history, searches, content of emails, online chats, and social media activity. It also allows an analyst to learn the IP addresses of every person who visits any website the analyst specifies. In his first video interview with Laura Poitras and Glenn Greenwald, Snowden said, "I, sitting at my desk, could wiretap anyone, from you or your accountant, to a federal judge or even the president, if I had a personal email" [57].

In August 2013, the NSA released a memorandum indicating that the agency monitored a mere 1.6 % of the world's Internet traffic. However, a shrewd commentator by the name "Pavel87" made the observation that "if you take away porn, advertisement, streaming video and music, you are left with the 1.6 % of ACTUAL information such as email, VoIP and chats. So the NSA actually filters out 98.4 % of irrelevant content and reads 100 % of your actual info" [58].

Apart from Boundless Informant, MUSCULAR, PRISM, QUANTUM, Steller Wind, Tempora, Turbulence, Xkeyscore and spying on foreign countries by the NSA and British's Government Communications Headquarters (GCHQ), there is

OPI - Top Producers Issuing	PRISM-Based Reports	% Increase in PRISM-Based Reports Compared to June 2010 - May2011	% Of All OPI Reporting Which is PRISM-Based	% Points Change from June 2010 - May 2011 period	All Reports By OPI	Single-Source to PRISM	% of PRISM-Based Reports Which are Single Source
ECC (ESOC) (USJ-753*, USM-44)	1147	Up 6%	52	+2 (up 4%)	2217	922	80
S2I - Counterterrorism	3493	Up 5%	42	-2 (down 5%)	8242	2074	60
S2J - Weapons and Space	225	Up 221%	33	+11 (50%)	692	186	83
S2G - Combating Prolif	2092	Up 49%	30	+3 (up 11%)	6872	1395	67
NSAT (USJ-783*)	1690	Up 20%	30	+3 (up 11%)	5713	1319	78
NTOC (V*)	455	Up 237%	21	+8 (up 62%)	2195	355	78
SCS (F6*, US-96*, US-97*, US-3219)	3723	Up 67%	20	+ 7 (up 54%)	18640	3040	82
SSG	16	Up 60%	17	-19 (down 52%)	92	14	88
S2E - Middle East & Africa	2574	Up 47%	16	+2 (up 14%)	16537	1959	76
S2F - Intl Crime & Narc	666	Up 41%	16	+2 (up 14%)	4122	497	75
S2C - Intl Sec Issues	1147	Up 75%	13	+5 (up 63%)	8989	861	75
S2B -	634	Down 10%	13	-3 (down 19%)	4842	452	71
S2D - Countering Frgn Intel	862	Up 40%	12	-5 (down 29%)	7089	545	63
S2A -	1389	Up 8%	11	-1 (down 8%)	12445	1196	86
NSAG (USJ-800*)	1255	Down 8%	11	0 (no change)	11741	883	70
Utah Regional Ops Cntr (USJ-755)	12	Up 20%	6	-17 (down 74%)	207	12	100
S2H -	176	Up 159%	5	+3 (up 150%)	3353	155	88
DSD	310	Down 15%	4	0 (no change)	7511	296	95
NSAH (USJ-750*)	237	Down 10%	2	+1 (up 50%)	12023	155	65
GCHQ	197	Up 137%	2	+1.9 (up 1900%)	11257	170	86

Source: PLUS - 11 -13 June 2012   TOP SECRET//SI//ORCON//NOFORN

Fig. 7.18 PRISM (US-984XN) Based Reporting: June 2011 — May 2012 (Sorted By % of PRISM-Based Reporting Per OPI)

much more top-secret information that has not been made public by the journalists in the know. Snowden even "demanded" the journalists that they should consult the government before publishing [59].

The Washington Post Editorial Board opined that "the first U.S. priority should be to prevent Mr. Snowden from leaking information that harms efforts to fight terrorism and conduct legitimate intelligence operations." It did not regret what had been disclosed to the public because "documents published so far by news organizations have shed useful light on some NSA programs and raised questions that deserve debate, such as whether a government agency should build a database of Americans' phone records." However, it warned that "Mr. Snowden is reported to have stolen many more documents, encrypted copies of which may have been given to allies such as the WikiLeaks organization" [60].

In July 2013, *The Guardian* in London destroyed the hard drives containing top-secret documents leaked by Edward Snowden after a threat of legal action by the British government. Guardian editor Alan Rusbridger said in an interview, "I explained to British authorities that there were other copies in America and Brazil so they wouldn't be achieving anything. But once it was obvious that they would be going to law I preferred to destroy our copy rather than hand it back or allow the courts to freeze our reporting. … I don't think we had Snowden's consent to hand the material back, and I didn't want to help the UK authorities to know what he had given us" [61].

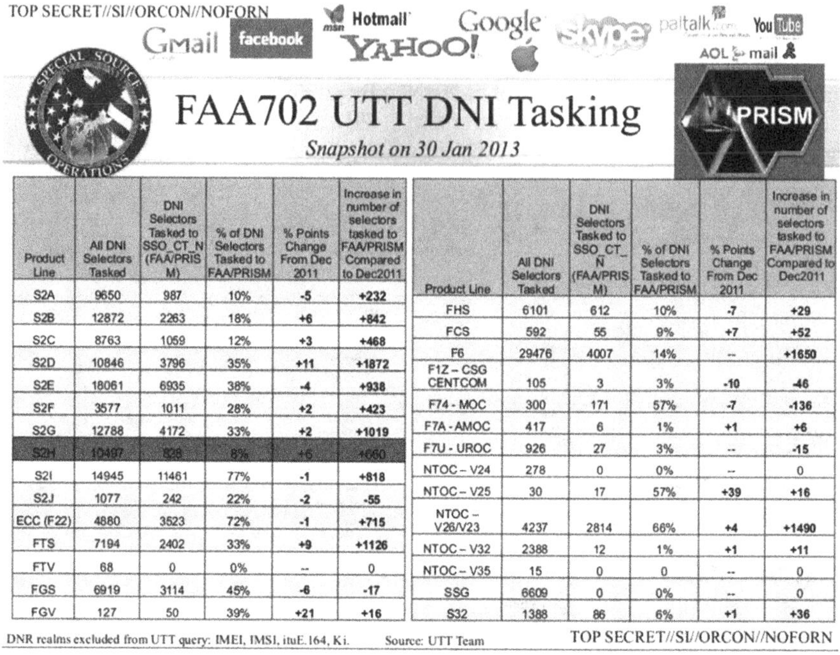

TOP SECRET//SI//ORCON//NOFORN

FAA702 UTT DNI Tasking
Snapshot on 30 Jan 2013

Product Line	All DNI Selectors Tasked	DNI Selectors Tasked to SSO_CT_N (FAA/PRISM)	% of DNI Selectors Tasked to FAA/PRISM	% Points Change From Dec 2011	Increase in number of selectors tasked to FAA/PRISM Compared to Dec2011
S2A	9650	987	10%	-5	+232
S2B	12872	2263	18%	+6	+842
S2C	8763	1059	12%	+3	+468
S2D	10846	3796	35%	+11	+1872
S2E	18061	6935	38%	-4	+938
S2F	3577	1011	28%	+2	+423
S2G	12788	4172	33%	+2	+1019
S2H	10497	838	8%	+6	+060
S2I	14945	11461	77%	-1	+818
S2J	1077	242	22%	-2	-55
ECC (F22)	4880	3523	72%	-1	+715
FTS	7194	2402	33%	+9	+1126
FTV	68	0	0%	--	0
FGS	6919	3114	45%	-6	-17
FGV	127	50	39%	+21	+16

Product Line	All DNI Selectors Tasked	DNI Selectors Tasked to SSO_CT_N (FAA/PRIS M)	% of DNI Selectors Tasked to FAA/PRISM	% Points Change From Dec 2011	Increase in number of selectors tasked to FAA/PRISM Compared to Dec2011
FHS	6101	612	10%	-7	+29
FCS	592	55	9%	+7	+52
F6	29476	4007	14%	--	+1650
F1Z – CSG CENTCOM	105	3	3%	-10	-46
F74 - MOC	300	171	57%	-7	-136
F7A - AMOC	417	6	1%	+1	+6
F7U - UROC	926	27	3%	--	-15
NTOC – V24	278	0	0%	--	0
NTOC – V25	30	17	57%	+39	+16
NTOC – V26/V23	4237	2814	66%	+4	+1490
NTOC – V32	2388	12	1%	+1	+11
NTOC – V35	15	0	0%	--	0
SSG	6609	0	0%	--	0
S32	1388	86	6%	+1	+36

DNR realms excluded from UTT query: IMEI, IMSI, ituE.164, Ki.   Source: UTT Team   TOP SECRET//SI//ORCON//NOFORN

Fig. 7.19 FAA702 UTT DNI Tasking (Snapshot on 30 Jan 2013)

In August 2013, pro-privacy email service provider Lavabit with 350,000 users including Edward Snowden was shut down by its owner Ladar Levison amid court battle. "I have been forced to make a difficult decision: to become complicit in crimes against the American people or walk away from nearly 10 years of hard work by shutting down Lavabit," Levison wrote in a statement. "After significant soul searching, I have decided to suspend operations" [62].

While Americans are equally divided on labeling Edward Snowden a hero or a traitor [63], Snowden justified his actions by saying that "I can't in good conscience allow the U.S. government to destroy privacy, Internet freedom, and basic liberties for people around the world with this massive surveillance machine they're secretly building" [64]. President Jimmy Carter weighed in, "He's obviously violated the laws of America, for which he's responsible, but I think that the invasion of human rights and American privacy has gone too far. I think that the secrecy that has been surrounding this invasion of privacy has been excessive, so I think that the bringing of it to the public notice has probably been, in the long term, beneficial" [65].

In a May 2014 interview with Brian Williams at *NBC News*, Snowden said, "Being a patriot doesn't mean prioritizing service to government above all. Being a patriot means knowing when to protect your country, when to protect your Constitution, when to protect your countrymen from the violations and encroachments of adversaries. And those adversaries don't have to be foreign countries. They can be bad policies. ... When you look at the carefulness of the programs that have been disclosed... the way these have all been filtered through the most trusted journalist institutions... the way the government has had a chance to chime in on these and to make their case... How can it be said that I did not serve my government and how can it be said that this harmed the country when all three branches of government have made reforms as a result of it?" [66].

7.11 Social Networks' Responses to NSA Leaks and PRISM

WikiLeaks founder Julian Assange said in a May 2011 interview, "Facebook in particular is the most appalling spying machine that has ever been invented. Here we have the world's most comprehensive database about people: their relationships, their names, their addresses, their locations and their communications with each other, their relatives — all sitting within the United States, all accessible to US intelligence" [67].

Edward Snowden's NSA leaks in June 2013 have reinvigorated Assange's bold accusation [68]. However, Facebook has vehemently denied giving the NSA and FBI backdoor access to their servers [69]. Nevertheless, military and civilian technologies have interwoven into every fabric of our society. NSA's research has contributed significantly to the development of the supercomputer, the cassette tape, the microchip, quantum mathematics, nanotechnology, biometrics, and semiconductor technology [70]. The algorithm that powers Roomba, a household robot vacuum cleaner, was originally developed for clearing minefields [71]. Facebook co-founder and CEO Mark Zuckerberg himself said at the 2011 annual f8 conference, "We exist at the intersection of technology and social issues" [72].

On June 7, 2013, Zuckerberg responded personally to the allegedly "outrageous" press reports about PRISM [73]:

> Facebook is not and has never been part of any program to give the US or any other government direct access to our servers. We have never received a blanket request or court order from any government agency asking for information or metadata in bulk, like the one Verizon reportedly received. And if we did, we would fight it aggressively. We hadn't even heard of PRISM before yesterday.

> When governments ask Facebook for data, we review each request carefully to make sure they always follow the correct processes and all applicable laws, and then only provide the information if is required by law. We will continue fighting aggressively to keep your information safe and secure.

> We strongly encourage all governments to be much more transparent about all programs aimed at keeping the public safe. It's the only way to protect everyone's civil liberties and create the safe and free society we all want over the long term.

Along with Facebook, Google and Apple also denied any knowledge of PRISM. Google told *The Guardian*: "Google cares deeply about the security of our users' data. We disclose user data to government in accordance with the law, and we review all such requests carefully. From time to time, people allege that we have created a government backdoor into our systems, but Google does not have a backdoor for the government to access private user data" [69]. Whereas Apple told CNBC: "We have never heard of PRISM. We do not provide any government agency with direct access to our servers, and any government agency requesting customer data must get a court order" [74].

For damage control as well as to bolster their public image, Apple, Facebook, Google, LinkedIn, Microsoft, and Yahoo! all have released Transparency Reports showing criminal, Foreign Intelligence Surveillance Act (FISA), and other requests for information from the U.S. government. For the six-month period from January to June 2013, Apple received FISA and law enforcement requests affecting between 2,000 and 3,000 accounts [75], Facebook between 5,000 and 5,999 accounts [76], Google between 9,000 and 9,999 accounts [77], LinkedIn under 250 accounts [78], Microsoft between 15,000 and 15,999 accounts [79], and Yahoo! between 30,000 and 30,999 accounts [80].

Facebook said that it would "push for even more transparency, so that our users around the world can understand how infrequently we are asked to provide user data on national-security grounds." However, Microsoft conceded, "What we are permitted to publish continues to fall short of what is needed to help the community understand and debate these issues" [81].

At the TechCrunch Disrupt Conference in September 2013, Yahoo! CEO Marissa Mayer gave an honest answer — a welcoming change from all the standard denial statements by most technology companies. Mayer said, "In terms of the NSA, we can't talk about those things because they are classified. Releasing classified information is treason, and you're incarcerated. … It makes more sense for us to work within the system. We file suit against the government … asking to be able to be more transparent with the numbers on the NSA requests" [82].

In October 2013, *The Washington Post* published the NSA documents about the MUSCULAR project that secretly taps into the communications links connecting Yahoo and Google data centers around the world. Figure 7.20 shows a NSA presentation on "Google Cloud Exploitation," courtesy of *The Washington Post*. In response, Google's chief legal officer David Drummond said, "We are outraged at the lengths to which the government seems to have gone to intercept data from our private fiber networks, and it underscores the need for urgent reform" [83].

In November 2013, German newspaper *Der Spiegel* reported on Snowden's leaks about NSA's QUANTUM program that redirects some users to fake LinkedIn and Slashdot websites to plant malware on the users' computers. LinkedIn denied any knowledge of the program and said that it "would never approve such activity" [84].

At the 2013 RSA conference, an FBI agent propositioned Wickr cofounder Nico Sell about adding a backdoor to her mobile app that provides military-grade encryption of text, picture, audio and video messages as well as anonymity and secure file shredding features. Sell told Meghan Kelly at *Venture Beat*, "I think he was trying to intimidate me. He just caught me really off-guard. I feel like he was mad. I mean, I said that I was doing a no-backdoor guarantee in the presentation. That was one of the major messages there" [85].

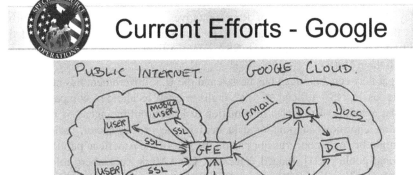

Fig. 7.20   Google Cloud Exploitation

F-Secure and some fellow companies decided to boycott the 2014 RSA conference in protest of the alleged $10 million dollar payment from the NSA to RSA
for inserting a backdoor into its encryption products. Nico Sell questioned if those
boycotters were also boycotting Google, Facebook, and Microsoft: "I bet those
guys were also paid for backdoors. It's hard to boycott everybody who is paid for
a backdoor" [85].

While privacy advocates, media skeptics, and conspiracy theorists continue to
question the real involvement of technology companies with the NSA, Google
has beefed up the security of Gmail by encrypting communications between
Google's servers as well as to and from end users [86]. Yahoo! has also encrypted
all information that moves between its data centers and to/from its consumers
[87]. Microsoft has expanded encryption across its services and reassured customers that its products do not contain backdoors [88]. And Facebook CEO Mark
Zuckerberg called President Barack Obama on March 12, 2014 to express his
frustration about the government's spying and hacking programs. Zuckerberg said
after the phone call, "When our engineers work tirelessly to improve security, we
imagine we're protecting you against criminals, not our own government. ... The
U.S. government should be the champion for the Internet, not a threat. They need
to be much more transparent about what they're doing, or otherwise people will
believe the worst. ... Unfortunately, it seems like it will take a very long time for
true full reform" [89].

7.12 Reform Government Surveillance and Reset the Net

In May 2014, the U.S. House of Representatives approved a new bill to curb NSA's domestic dragnet surveillance. Rep. Jim Sensenbrenner who shepherded the 2001 USA Patriot Act said, "The NSA might still be watching us, but now we can watch them" [90].

The spat between American businesses and the U.S. government has now come full circle, as ex-NSA official "Chris" Inglis warned technology companies that amass vast amounts of personal information to learn from the agency's mistakes. "There's an enormous amount of data held in the private sector," said Inglis. "There might be some concerns not just on the part of the American public, but the international public" [91]. Indeed, Facebook CEO Mark Zuckerberg told Farhad Manjoo of *The New York Times* in an April 2014 interview: "Understanding who you serve is always a very important problem, and it only gets harder the more people that you serve" [92].

In June 2014, a year after Snowden's NSA leaks, the Reform Government Surveillance coalition of nine technology companies published an open letter to the members of the U.S. Senate, urging them to pass a version of the USA Freedom Act that offers more transparency than the one passed by the U.S. House of Representatives [93]:

Dear Members of the Senate:

It's been a year since the first headlines alleging the extent of government surveillance on the Internet.

We understand that governments have a duty to protect their citizens. But the balance in many countries has tipped too far in favor of the state and away from the rights of the individual. This undermines the freedoms we all cherish, and it must change.

Over the last year many of our companies have taken important steps, including further strengthening the security of our services and taking action to increase transparency. But the government needs to do more.

In the next few weeks, the Senate has the opportunity to demonstrate leadership and pass a version of the USA Freedom Act that would help restore the confidence of Internet users here and around the world, while keeping citizens safe.

Unfortunately, the version that just passed the House of Representatives could permit bulk collection of Internet "metadata" (e.g. who you email and who emails you), something that the Administration and Congress said they intended to end. Moreover, while the House bill permits some transparency, it is critical to our customers that the bill allows companies to provide even greater detail about the number and type of government requests they receive for customer information.

It is in the best interest of the United States to resolve these issues. Confidence in the Internet, both in the U.S. and internationally, has been badly damaged over the last year. It is time for action. As the Senate takes up this important legislation, we urge you to ensure that U.S. surveillance efforts are clearly restricted by law, proportionate to the risks, transparent, and subject to independent oversight.

Signed,

Tim Armstrong, *AOL*
Tim Cook, *Apple*
Drew Houston, *Dropbox*
Mark Zuckerberg, *Facebook*
Larry Page, *Google*
Jeff Weiner, *LinkedIn*
Dick Costolo, *Twitter*
Satya Nadella, *Microsoft*
Marissa Mayer, *Yahoo!*

On June 5, 2014, Edward Snowden came out to support the launch of "Reset the Net" — a day of action to raise public awareness of online privacy tools [94]. It was organized by the nonprofit organization Fight for the Future led by co-founder Tiffiniy Cheng."The Snowden leaks have taught tech companies a hard lesson that not only do they have to secure their services against Chinese spies and hackers, but they also need to treat their own government as a security threat," said Kevin Bankston, policy director of the New America Foundation's Open Technology Institute. "In response to that lesson we have seen a substantial improvement around Internet security" [95].

Bibliography

1. **Gorman, Siobhan.** NSA's Domestic Spying Grows As Agency Sweeps Up Data. [Online] The Wall Street Journal, March 10, 2008. http://online.wsj.com/article/SB120511973377523 845.html.
2. **Cauley, Leslie.** NSA has massive database of Americans' phone calls . [Online] USA Today, May 11, 2006. http://www.usatoday.com/news/washington/2006-05-10-nsa_x.htm.
3. **American Civil Liberties Union .** Stunning New Report on Domestic NSA Dragnet Spying Confirms ACLU Surveillance Warnings. [Online] American Civil Liberties Union, March 12, 2008. http://www.aclu.org/technology-and-liberty/stunning-new-report-domestic-nsa-dragnet-spying-confirms-aclu-surveillance-wa.
4. **CASE NO. CV-08-04373-JSW.** Declaration Of William E. Binney In Support Of Plaintiffs' Motion For Partial Summary Judgment Rejecting The Government Defendants' State Secret Defense. [Online] United States District Court for the Northern District of California, September 28, 2012. http://info.publicintelligence.net/NSA-WilliamBinneyDeclaration.pdf.
5. **C-06-0672-VRW.** Declaration of Mark Klein in Support of Plaintiffs' Motion for Preliminary Injunction. [Online] United States District Court Northern District of California, June 8, 2006. https://www.eff.org/files/filenode/att/SER_klein_decl.pdf.
6. **Poe, Robert.** The Ultimate Net Monitoring Tool. [Online] Wired, May 17, 2006. http://www.wired.com/science/discoveries/news/2006/05/70914.
7. **Pagliery, Jose.** U.S. planes spy on American phones. [Online] CNNMoney, November 14, 2014. http://money.cnn.com/2014/11/13/technology/security/federal-planes-spy/index.html.
8. **Kelley, Michael.** NSA Whistleblower Details How The NSA Has Spied On US Citizens Since 9/11. [Online] Business Insider, August 24, 2012. http://www.businessinsider.com/nsa-whistleblower-william-binney-explains-nsa-surveillance-2012-8.
9. **National Security Agency.** Our History. [Online] National Security Agency, January 15, 2009. http://www.nsa.gov/public_info/speeches_testimonies/nsa_videos/history_of_nsa.shtml.
10. —. Central Security Service (CSS). [Online] National Security Agency, November 21, 2012. http://www.nsa.gov/about/central_security_service/index.shtml.

11. **Risen, James and Lichtblau, Eric.** Bush Lets U.S. Spy on Callers Without Courts . [Online] The New York Times, December 16, 2005. http://www.nytimes.com/2005/12/16/politics/16pr ogram.html?pagewanted=all.

12. **Department of Justice.** Iyman Faris Sentenced for Providing Material Support to Al Qaeda. [Online] U.S. Department of Justice, October 28, 2003. http://www.justice.gov/opa/pr/2003/October/03_crm_589.htm.

13. **Summers, Chris and Casciani, Dominic.** Fertiliser bomb plot: The story. [Online] BBC News, April 30, 2007. http://news.bbc.co.uk/2/hi/uk_news/6153884.stm.

14. **epic.org.** Foreign Intelligence Surveillance Act (FISA). [Online] Electronic Privacy Information Center. http://epic.org/privacy/terrorism/fisa/.

15. **Sanger, David E. and O'Neil, John.** White House Begins New Effort to Defend Surveillance Program. [Online] The New York Times, January 23, 2006. http://www.nytimes.com/2006/01/23/politics/23cnd-wiretap.html?pagewanted=all.

16. **th Congress.** Foreign Intelligence Surveillance Act Of 1978 Amendments Act Of 2008. [Online] U.S. Senate Select Committee on Intelligence, July 10, 2008. http://www.intelligence.senate.gov/laws/pl110261.pdf.

17. **Lichtblau, Eric.** Senate Approves Bill to Broaden Wiretap Powers. [Online] The New York Times, July 10, 2008. http://www.nytimes.com/2008/07/10/washington/10fisa.html.

18. **ACLU.** ACLU Sues Over Unconstitutional Dragnet Wiretapping Law . [Online] American Civil Liberties Union, July 10, 2008. http://www.aclu.org/national-security/aclu-sues-over-unconstitutional-dragnet-wiretapping-law.

19. **Ostrow, Ronald J.** First 9 Japanese WWII Internees Get Reparations. [Online] Los Angeles Times, October 10, 1990. http://articles.latimes.com/1990-10-10/news/mn-1961_1_japanese-wwii-internees.

20. **Gongloff, Mark.** Middle East futures market returns. Private firm will restart Pentagon project, but without contracts for violence, in 2004. [Online] CNNMoney, November 18, 2003. http://money.cnn.com/2003/11/17/news/terror_futures/index.htm.

21. **Scheiber, Noam.** 2003: THE 3rd ANNUAL YEAR IN IDEAS; Futures Markets in Everything. [Online] The New York Times, December 14, 2003. http://www.nytimes.com/2003/12/14/magazine/2003-the-3rd-annual-year-in-ideas-futures-markets-in-everything.html.

22. **Chen, Kay-Yut and Plott, Charles R.** Information Aggregation Mechanisms: Concept, Design and Implementation for a Sales Forecasting Problem. [Online] California Institute of Technology, March 2002. www.hpl.hp.com/personal/Kay-Yut_Chen/paper/ms020408.pdf.

23. **The University of Iowa.** Iowa Electronic Markets. [Online] University of Iowa Henry B. Tippie College of Business. http://tippie.uiowa.edu/iem/.

24. **HSX.** Hollywood Stock Exchange: The Entertainment Market. [Online] http://www.hsx.com/.

25. **Wilson, James M.** Statement by James M. Wilson V, MD. [Online] U.S. Senate Committee on Homeland Security and Governmental Affairs, October 4, 2007. http://www.hsgac.senate.gov//imo/media/doc/WilsonTestimony.pdf.

26. **U.S. News & World Report.** Q&A: DNI Chief Scientist Eric Haseltine. [Online] U.S. News & World Report, November 3, 2006. http://www.usnews.com/usnews/news/articles/061103/3qahaseltine_6.htm.

27. **Wright, Lawrence.** The Spymaster. Can Mike McConnell fix America's intelligence community? [Online] The New Yorker, January 21, 2008. http://www.newyorker.com/reporting/2008/01/21/080121fa_fact_wright?currentPage=all.

28. **DARPA.** Information Innovation Office. [Online] Defense Advanced Research Projects Agency. http://www.darpa.mil/Our_Work/I2O/.

29. **U.S. Congress.** Committee Reports. 108th Congress (2003-2004). House Report 108-283. [Online] The Library of Congress, September 2003. http://thomas.loc.gov/cgi-bin/cpquery/?&sid=cp108alJsu&refer=&r_n=hr283.108&db_id=108&item=&&sid=cp108alJsu&r_n=hr283.108&dbname=cp108&&sel=TOC_309917.

30. **Kalil, Tom.** Big Data is a Big Deal. [Online] The White House, March 29, 2012. http://www.whitehouse.gov/blog/2012/03/29/big-data-big-deal.

31. **Office of Science and Technology Policy.** Obama Administration Unveils "Big Data" Initiative: Announces $200 Million In New R&D Investments. [Online] Executive Office of the President, March 29, 2012. http://www.whitehouse.gov/sites/default/files/microsites/ostp/big_data_press_release.pdf.

32. **Executive Office of the President.** Big Data Across the Federal Government. [Online] The White House, March 29, 2012. http://www.whitehouse.gov/sites/default/files/microsites/ostp/big_data_fact_sheet_final_1.pdf.

33. **Gorman, Siobhan.** How Team of Geeks Cracked Spy Trade . [Online] The Wall Street Journal, September 4, 2009. http://online.wsj.com/article/SB125200842406984303.html.

34. **Bansal, Manju.** The Butterfly Effect: Predicting Tsunamis from Ripples. [Online] MIT Technology Review, October 6, 2014. http://www.technologyreview.com/view/531431/the-butterfly-effect-predicting-tsunamis-from-ripples/.

35. **Palantir Technologies .** Industries & Solutions. [Online] Palantir Technologies . [Cited: November 14, 2012.] http://www.palantir.com/solutions/.

36. **Vance, Ashlee and Stone, Brad.** Palantir, the War on Terror's Secret Weapon. [Online] Bloomberg Businessweek, November 22, 2011. http://www.businessweek.com/magazine/palantir-the-vanguard-of-cyberterror-security-11222011.html.

37. **Anderson, Nate.** Spy Games: Inside the Convoluted Plot to Bring Down WikiLeaks. [Online] Wired, February 14, 2011. http://www.wired.com/threatlevel/2011/02/spy/.

38. **Angle, Jim and Herridge, Catherine.** Debate Rages Over Legality of NSA Wiretap Program. [Online] Fox News, December 21, 2005. http://www.foxnews.com/story/0,2933,179323,00.html.

39. **New York City Police Department.** Public Security Privacy Guidelines. [Online] New York City Police Department, April 2, 2009. http://www.nyc.gov/html/nypd/downloads/pdf/crime_prevention/public_security_privacy_guidelines.pdf.

40. **McDuffie, Mike.** Microsoft and NYPD Announce Partnership Providing Real-Time Counterterrorism Solution Globally. [Online] Microsoft, August 8, 2012. http://www.microsoft.com/government/en-us/state/brightside/Pages/details.aspx?Microsoft-and-NYPD-Announce-Partnership-Providing-Real-Time-Counterterrorism-Solution-Globally&blogid=697.

41. **Cruickshank, Paul.** Inside the plot to devastate New York. [Online] CNN, May 2, 2012. http://security.blogs.cnn.com/2012/05/02/time-line-for-a-terror-plot/.

42. **Bamford, James.** War of Secrets; Eyes in the Sky, Ears to the Wall, and Still Wanting. [Online] The New York Times, September 8, 2002. http://www.nytimes.com/2002/09/08/weekinreview/war-of-secrets-eyes-in-the-sky-ears-to-the-wall-and-still-wanting.html?pagewanted=all.

43. **Draughn, Katisha.** Federal partners break ground on $1.5 billion center. [Online] US Army Corps of Engineers Baltimore District, February 10, 2011. http://www.nab.usace.army.mil/News/20110210_Federal%20partners%20break%20ground%20on%20$1.5%20billion%20center.htm.

44. **Office of the Under Secretary of Defense for Acquisition, Technology and Logistics.** MilCon Status Report - August, 2014 - Under Secretary of Defense for AT&L. [Online] Office of the Under Secretary of Defense for Acquisition, Technology and Logistics, September 17, 2014. http://www.acq.osd.mil/ie/fim/library/milcon/MILCON_EOM-AUG_Report_2014-09-17.xlsx.

45. **Bamford, James.** The NSA Is Building the Country's Biggest Spy Center (Watch What You Say). [Online] Wired, March 15, 2012. http://www.wired.com/threatlevel/2012/03/ff_nsadatacenter/all/.

46. **Electronic Frontier Foundation.** Diverse Groups Fly Airship Over NSA's Utah Data Center to Protest Illegal Spying. [Online] Electronic Frontier Foundation, June 27, 2014. https://www.eff.org/press/releases/diverse-groups-fly-airship-over-nsas-utah-data-center-protest-illegal-internet-spying.

47. **McMillan, Robert.** Utah Considers Cutting Off Water to the NSA's Monster Data Center. [Online] Wired Magazine, November 20, 2014. http://www.wired.com/2014/11/utah-considers-cutting-water-nsas-monster-data-center/.

48. **Harris, Shane.** Giving In to the Surveillance State. [Online] The New York Times, August 22, 2012. http://www.nytimes.com/2012/08/23/opinion/whos-watching-the-nsa-watchers.html.
49. **IMDb.** This Means War. [Online] IMDb, February 17, 2012. http://www.imdb.com/title/tt1596350/.
50. **Lewis, Paul.** NSA employee spied on nine women without detection, internal file shows . [Online] The Guardian, September 27, 2013. http://www.theguardian.com/world/2013/sep/27/nsa-employee-spied-detection-internal-memo.
51. **Gidda, Mirren.** Edward Snowden and the NSA files – timeline. [Online] The Guardian, July 25, 2013. http://www.theguardian.com/world/2013/jun/23/edward-snowden-nsa-files-timeline.
52. **theguardian.com.** Verizon forced to hand over telephone data – full court ruling. [Online] The Guardian, June 5, 2013. http://www.theguardian.com/world/interactive/2013/jun/06/verizon-telephone-data-court-order.
53. **Greenwald, Glenn and MacAskill, Ewen.** NSA Prism program taps in to user data of Apple, Google and others. [Online] The Guardian, June 6, 2013. http://www.theguardian.com/world/2013/jun/06/us-tech-giants-nsa-data.
54. **Jackson, David.** Obama defends surveillance programs. [Online] USA Today, June 7, 2013. http://www.usatoday.com/story/news/politics/2013/06/07/obama-clapper-national-security-agency-leaks/2400405/.
55. **Nakashima, Ellen and Warrick, Joby.** For NSA chief, terrorist threat drives passion to 'collect it all'. [Online] The Washington Post, July 14, 2013. http://www.washingtonpost.com/world/national-security/for-nsa-chief-terrorist-threat-drives-passion-to-collect-it-all/2013/07/14/3d26ef80-ea49-11e2-a301-ea5a8116d211_story.html.
56. **Greenwald, Glenn.** XKeyscore: NSA tool collects 'nearly everything a user does on the internet' . [Online] The Guardian, July 31, 2013. http://www.theguardian.com/world/2013/jul/31/nsa-top-secret-program-online-data.
57. **Poitras, Laura and Greenwald, Glenn.** NSA whistleblower Edward Snowden: 'I don't want to live in a society that does these sort of things' – video . [Online] The Guardian, June 9, 2013. http://www.theguardian.com/world/video/2013/jun/09/nsa-whistleblower-edward-snowden-interview-video.
58. **Perez, Evan.** Documents shed light on U.S. surveillance programs. [Online] CNN, August 9, 2013. http://www.cnn.com/2013/08/09/politics/nsa-documents-scope/index.html.
59. **Ohlheiser, Abby.** Edward Snowden Would 'Like to Go Home'. [Online] The Wire, May 28, 2014. http://www.thewire.com/national/2014/05/edward-snowden-would-like-to-go-home/371784/.
60. **Editorial Board.** Plugging the leaks in the Edward Snowden case. [Online] The Washington Post, July 1, 2013. http://www.washingtonpost.com/opinions/how-to-keep-edward-snowden-from-leaking-more-nsa-secrets/2013/07/01/4e8bbe28-e278-11e2-a11e-c2ea876a8f30_story.html.
61. **Borger, Julian.** NSA files: why the Guardian in London destroyed hard drives of leaked files. [Online] The Guardian, August 20, 2013. http://www.theguardian.com/world/2013/aug/20/nsa-snowden-files-drives-destroyed-london.
62. **Poulsen, Kevin.** Snowden's e-mail provider shuts down amid court battle. [Online] August 9, 2013. http://www.cnn.com/2013/08/09/tech/web/snowden-email-lavabit/index.html.
63. **Edwards-Levy, Ariel and Freeman, Sunny.** Americans Still Can't Decide Whether Edward Snowden Is A 'Traitor' Or A 'Hero,' Poll Finds. [Online] The Huffington Post, October 30, 2013. http://www.huffingtonpost.com/2013/10/30/edward-snowden-poll_n_4175089.html.
64. **Smith, Matt.** NSA leaker comes forward, warns of agency's 'existential threat'. [Online] CNN, June 9, 2013. http://www.cnn.com/2013/06/09/politics/nsa-leak-identity/index.html.
65. **Watkins, Tom.** Father proposes deal for Snowden's voluntary return. [Online] CNN, June 30, 2013. http://www.cnn.com/2013/06/28/us/snowden-lawyer-offer/index.html.
66. **NBC News.** Inside the Mind of Edward Snowden. [Online] NBC News, May 28, 2014. http://www.nbcnews.com/feature/edward-snowden-interview/snowden-being-patriot-means-knowing-when-protect-your-country-n117151.
67. **RT News.** WikiLeaks revelations only tip of iceberg – Assange. [Online] RT News, May 2, 2011. http://www.rt.com/news/wikileaks-revelations-assange-interview/.

68. **Taylor, Chris.** Through a PRISM, Darkly: Tech World's $20 Million Nightmare. [Online] Mashable, June 6, 2013. http://mashable.com/2013/06/06/through-a-prism-darkly-techs-20-million-nightmare-is-our-fault/.
69. **Fitzpatrick, Alex.** Facebook, Google, Apple, Yahoo Make Similar PRISM Denials. [Online] Mashable, June 6, 2013. http://mashable.com/2013/06/06/facebook-google-apple-prism/#.
70. **National Security Agency.** Our History Video Transcript. [Online] National Security Agency, January 15, 2009. http://www.nsa.gov/public_info/speeches_testimonies/nsa_videos/history_of_nsa.shtml.
71. **Glass, Nick and Ponsford, Matthew.** The secret military tech inside household robot vacuum cleaner. [Online] CNN, March 31, 2014. http://www.cnn.com/2014/03/31/tech/innovation/the-secret-military-technology-roomba-vacuum/index.html.
72. **Bosker, Bianca.** Facebook's f8 Conference (LIVE BLOG): Get The Latest Facebook News. [Online] The Huffington Post, September 22, 2011. http://www.huffingtonpost.com/2011/09/22/facebook-f8-conference-live-blog-latest-news_n_975704.html.
73. **Zuckerberg, Mark.** I want to respond personally to the outrageous press reports about PRISM. [Online] Facebook, June 7, 2013. https://www.facebook.com/zuck/posts/10100828955847631.
74. **CNBC.** Apple to @CNBC. [Online] Twitter, June 6, 2013. https://twitter.com/CNBC/status/342778613264945152.
75. **Apple.** Report on Government Information Requests. [Online] Apple, November 5, 2013. https://www.apple.com/pr/pdf/131105reportongovinforequests3.pdf.
76. **Stretch, Colin.** Facebook Releases New Data About National Security Requests. [Online] Facebook Newsroom, February 3, 2014. http://newsroom.fb.com/news/2014/02/facebook-releases-new-data-about-national-security-requests/.
77. **Salgado, Richard.** Shedding some light on Foreign Intelligence Surveillance Act (FISA) requests. [Online] Google Official Blog, February 4, 2014. http://googleblog.blogspot.com/2014/02/shedding-some-light-on-foreign.html.
78. **Rottenberg, Erika.** Updated LinkedIn Transparency Report: Including Requests Related to U.S. National Security-Related Matters. [Online] LinkedIn Official Blog, February 3, 2014. http://blog.linkedin.com/2014/02/03/updated-linkedin-transparency-report-including-requests-related-to-u-s-national-security-related-matters/.
79. **Smith, Brad.** Providing additional transparency on US government requests for customer data. [Online] Microsoft TechNet, February 3, 2014. http://blogs.technet.com/b/microsoft_on_the_issues/archive/2014/02/03/providing-additional-transparency-on-us-government-requests-for-customer-data.aspx.
80. **Bell, Ron and Altschuler, Aaron.** More Transparency For U.S. National Security Requests. [Online] Yahoo, February 3, 2014. http://yahoo.tumblr.com/post/75496314481/more-transparency-for-u-s-national-security-requests.
81. **Gustin, Sam.** Tech companies jockey to seem the most transparent. [Online] CNN, June 18, 2013. http://www.cnn.com/2013/06/18/tech/web/tech-companies-data-transparent/index.html.
82. **TechCrunch.** Marissa Mayer Comments on the NSA | Disrupt SF 2013. [Online] TechCrunch, September 11, 2013. http://www.youtube.com/watch?v=gS78slU6kq8.
83. **Gellman, Barton and Soltani, Ashkan.** NSA infiltrates links to Yahoo, Google data centers worldwide, Snowden documents say. [Online] The Washiongton Post, October 30, 2013. http://www.washingtonpost.com/world/national-security/nsa-infiltrates-links-to-yahoo-google-data-centers-worldwide-snowden-documents-say/2013/10/30/e51d661e-4166-11e3-8b74-d89d714ca4dd_story.html.
84. **Neal, Ryan W.** Edward Snowden Reveals 'Quantum Insert': NSA And GCHQ Used Fake LinkedIn And Slashdot Pages To Install Spyware. [Online] International Business Times, November 11, 2013. http://www.ibtimes.com/edward-snowden-reveals-quantum-insert-nsa-gchq-used-fake-linkedin-slashdot-pages-install-spyware.
85. **Kelly, Meghan.** Wickr cofounder to give Reddit AMA on what it's like to be bullied by the FBI. [Online] Venture Beat, January 10, 2014. http://venturebeat.com/2014/01/10/nico-sell-ama/.

86. **Pagliery, Jose.** Google tries to NSA-proof Gmail. [Online] CNNMoney, March 21, 2014. http://money.cnn.com/2014/03/20/technology/security/gmail-nsa/index.html.
87. **Mayer, Marissa.** Our Commitment to Protecting Your Information. [Online] Yahoo!, November 18, 2013. http://yahoo.tumblr.com/post/67373852814/our-commitment-to-protecting-your-information.
88. **Smith, Brad.** Protecting customer data from government snooping. [Online] Microsoft Blogs, December 4, 2013. http://blogs.technet.com/b/microsoft_blog/archive/2013/12/04/protecting-customer-data-from-government-snooping.aspx.
89. **Pagliery, Jose.** Mark Zuckerberg calls Obama to complain about NSA. [Online] CNNMoney, March 14, 2014. http://money.cnn.com/2014/03/13/technology/security/mark-zuckerberg-nsa/index.html.
90. **Mascaro, Lisa.** House overwhelmingly approves bill to curb NSA domestic spying. [Online] Los Angeles Times, May 22, 2014. http://www.latimes.com/nation/politics/la-na-nsa-reforms-20140523-story.html.
91. **Yadron, Danny.** Ex-NSA Official Inglis Warns Tech Firms: Be Transparent. [Online] The Wall Street Journal, March 5, 2014. http://online.wsj.com/news/articles/SB10001424052702 3047328045794217333369167364.
92. **Manjoo, Farhad.** Can Facebook Innovate? A Conversation With Mark Zuckerberg. [Online] The New York Times, April 16, 2014. http://bits.blogs.nytimes.com/2014/04/16/can-facebook-innovate-a-conversation-with-mark-zuckerberg/.
93. **Reform Government Surveillance.** Reform Government Surveillance. [Online] ReformGovernmentSurveillance.com, June 2014. https://www.reformgovernmentsurveillance.com/USA FreedomAct.
94. **Fight for the Future.** Privacy Pack. [Online] Reset the Net, June 5, 2014. https://pack.resetth enet.org/.
95. **Risen, Tom.** Cybersecurity Boosted After Snowden NSA Revelations. [Online] U.S. News & World Report, June 5, 2014. http://www.usnews.com/news/articles/2014/06/05/cybersecurity-boosted-after-snowden-nsa-revelations.

Chapter 8
A Two-Way Street of Total Information Awareness

> The two-way street of Total Information Awareness is the road that leads to a more transparent and complete picture of ourselves, our governments, and our world.
>
> —Newton Lee

> The fantasy worlds that Disney creates have a surprising amount in common with the ideal universe envisaged by the intelligence community, in which environments are carefully controlled and people are closely observed, and no one seems to mind.
>
> —Lawrence Wright, *The New Yorker*
> (January 21, 2008)

> Our job as citizens is to ask questions.
>
> — Thomas Blanton, National Security Archive
> George Washington University (December 16, 2010)

8.1 It's a Small World, with CCTVs

In January 2008, Pulitzer Prize-winner Lawrence Wright wrote in *The New Yorker* an in-depth article about the U.S. intelligence community focusing on the Office of the Director of National Intelligence (ODNI) and the necessity for interagency communications — something that Total Information Awareness (TIA) was meant to facilitate. Wright observed that "the fantasy worlds that Disney creates have a surprising amount in common with the ideal universe envisaged by the intelligence community, in which environments are carefully controlled and people are closely observed, and no one seems to mind" [1].

In addition to the bag checks at the entrances to Disney theme parks, plain clothes security officers and closed-circuit television (CCTV) hidden cameras have kept the parks safe without intruding on the privacy of the guests. Other than

© Springer International Publishing Switzerland 2015
N. Lee, *Counterterrorism and Cybersecurity*, DOI 10.1007/978-3-319-17244-6_8

a few rare incidents, Disneyland is "the happiest place on earth" [2]. Although our every move may be monitored and recorded, we feel complete freedom to do whatsoever we want other than causing harm to others or damages to properties.

In the year 2012, from ATMs to parking lots to shopping malls, there are approximately 30 million cameras in the world capturing 250 billion hours of raw footage annually [3]. In the United Kingdom, CCTV is so prevalent that some residents can expect to be captured by a camera at least 300 times a day [4]. With more than 1.85 million cameras operating in the U.K. [5], the security-camera cordon surrounding London has earned the nickname of "Ring of Steel" [6]. The U.K. first introduced the security measures in London's financial district in mid-1990s during an Irish Republican Army (IRA) bombing campaign. After the 9/11 terrorist attacks, the "Ring of Steel" was widened to include more businesses [7].

Since the 1970s, the proliferation of CCTV cameras in public places has led to some unease about the erosion of civil liberties and individual human rights, along with warnings of an Orwellian "Big Brother" culture. Nevertheless, nowadays we all have accepted the presence of CCTV in public places.

In the U.S., New York, Los Angeles, San Francisco, and Chicago are among the major cities that have implemented citywide CCTV monitoring systems. Disney theme parks, Six Flags, and other public attractions also use video surveillance systems that can see in the dark.

Tourists not only love to visit Disneyland but also flock to Las Vegas casinos and resorts, another fantasy world, where security cameras are in ample use. In March 2012, Mirage Resort in Las Vegas became the 50th casino to install facial recognition software as part of the surveillance suite of Visual Casino loss-reduction systems [8].

8.2 Facebook Nation: Total Information Awareness

President Barack Obama, in his 2011 State of the Union Address, called America "the nation of Edison and the Wright brothers" and "of Google and Facebook" [9]. Enormous amounts of information are being gathered on everyone living in the Facebook nation. For the 2012 presidential election, Obama's data-mining team created a massive database of voter information, consumer data, and social media contacts [10]. The analysis of big data enabled Obama's campaign to run computer simulations, fundraise a staggering $1 billion dollars, reach the swing-state voters more effectively, and ultimately win the reelection for President Obama.

In a pep talk at Wakefield High School in September 2009, Obama told the students, "Be careful what you post on Facebook. Whatever you do, it will be pulled up later in your life" [11]. In August 2012, Prof. Amitai Etzioni of George Washington University opined that "Facebook merely adds to the major inroads made by the CCTV cameras that are ubiquitous in many cities around the globe, along with surveillance satellites, tracking devices, spy malware and, most recently, drones used not for killing terrorists but for scrutinizing spaces heretofore considered private, like our backyards. Corporations keep detailed dossiers on what we purchase. No wonder privacy advocates argue that we live in a surveillance society and privacy 'ended with Facebook'" [12].

A year after TIA was officially shut down in 2003, Facebook was born in 2004 with about 650 users during its first week of debut. In August 2008, Facebook had grown to 100 million users. By July 2010, Facebook reached 500 million. And in October 2012, Facebook topped 1 billion monthly active users [13]. Facebook as a nation in 2015 is overtaking China as the largest country in the world.

In an interview with *Ad Age*'s Ann-Christine Diaz, Facebook's head of consumer marketing Rebecca Van Dyck linked Facebook with the innate human desire to connect. Dyck said, "We make the tools and services that allow people to feel human, get together, open up. Even if it's a small gesture, or a grand notion — we wanted to express that huge range of connectivity and how we interact with each other" [14].

On October 4, 2012, Facebook released a new 91-second video *The Things That Connect Us* depicting chairs, doorbells, airplanes, bridges, dance floors, basketball, a great nation, and the universe [15]:

Chairs. Chairs are made so that people can sit down and take a break. Anyone can sit on a chair, and if the chair is large enough, they can sit down together. And tell jokes. Or make up stories. Or just listen. Chairs are for people. And that is why **chairs are like Facebook**.

Doorbells. Airplanes. Bridges. These are things people use to get together so they can open up and connect about ideas, and music, and other things people share.

Dance floors. Basketball. A great nation. A great nation is something people build, so that they can have a place where they belong.

The universe is vast and dark and makes us wonder if we are alone. So maybe the reason we make all of these things is to remind ourselves that we are not.

Directed by acclaimed Mexican filmmaker Alejandro González Iñárritu, the cleverly crafted video has been described by some critics as "puzzling" and "disingenuous" [16]. Nonetheless, it is not difficult to see that the video alludes to the rise of Facebook nation with over 1 billion cybercitizens. It is truly a global phenomenon since the majority of Facebook users (81 %) live outside the U.S. and Canada [13].

"Chairs are like Facebook" — Chairs are the most basic, ubiquitous, and indispensable furniture in most parts of the world. Facebook is one of the most prevalent social networks today. However, sitting in stationary chairs puts stress on spinal disks and increases the chance of lower-back injury, resulting in $11 billion a year in workers' compensation claims [17]. Unlike stationary chairs, Facebook must be quick to adapt to changes.

"We are not [alone]" — Facebook users tell jokes, make up stories, or just listen to other Facebook friends. In my book *Facebook Nation: Total Information Awareness*, I portray the social media ecosystem as a world of increasing total information awareness, which is essentially a civilian version of Poindexter's TIA program [18]. On Facebook, people volunteer their personal information such as their gender, birthday, education, workplace, city of residence, interests, hobbies, photos, friends, families, schoolmates, coworkers, past histories, relationship status, likes, dislikes, and even current location. WikiLeaks founder Julian Assange told RT's Laura Emmett in a May 2011 interview [19]:

> Facebook in particular is the most appalling spying machine that has ever been invented. Here we have the world's most comprehensive database about people: their relationships, their names, their addresses, their locations and their communications with each other, their relatives — all sitting within the United States, all accessible to US intelligence. Facebook, Google, Yahoo! — all these major US organizations have built-in interfaces for US intelligence. It's not a matter of serving a subpoena. They have an interface that they have developed for U.S. intelligence to use.

Although Assange's bold accusation of Facebook and social media is subject to debate, it is open knowledge that law enforcement authorities in New York, Atlanta, San Diego, and Chicago have been using Facebook to gather evidence against gang members and criminals. The success of social media sleuthing has prompted the New York Police Department (NYPD) to double the size of its online investigators in October 2012.

"By capitalizing on the irresistible urge of these suspects to brag about their murderous exploits on Facebook, detectives used social media to draw a virtual map of their criminal activity over the last 3 years," said NYPD commissioner Raymond Kelly [20].

Donna Lieberman, executive director of the New York Civil Liberties Union, concurred with Kelly. Lieberman and said, "NYPD has the right, indeed the obligation, to pursue effective avenues for investigating criminal gang activity, and that includes using Facebook and other social media. But such methods must be closely monitored so they don't become a vehicle for entrapment or unauthorized surveillance" [20].

The truth of the matter is that there is hardly any private information on the Internet. The U.S. Library of Congress has been archiving Web content since 2000. Twitter and the federal library announced in April 2010 that every public tweet posted since 2006 would be archived digitally [21]. By January 2013, the Library of Congress has compiled a total of more than 170 billion Twitter messages and it is now processing about 500 million new tweets per day [22]. The federal library plans to make all the tweets available to researchers and the general public in the near future.

8.3 Surveillance Satellites, Tracking Devices, Spyware, and Drones

A year before Prof. Amitai Etzioni's 2012 *CNN* article about "surveillance satellites, tracking devices, spy malware, and drones used not for killing terrorists but for scrutinizing spaces heretofore considered private, like our backyards" [12], the Defense Advanced Research Projects Agency (DARPA) and AeroVironment had developed the Nano Hummingbird — a miniature drone in camouflage that looks like a hummingbird capable of maneuvering in urban areas [23]. If we really take a look around, we can find ourselves surrounded by:

1. Surveillance satellites for Google Earth and Maps

Satellite images provide the necessary database for Google Earth, Google Maps, and other useful applications. At the "Next Dimension" Google Maps press event in June 2012, Google announced that it has over 1 billion monthly users for all of Google Map Services [24]. In July 2012, Google published new high-resolution aerial and satellite imagery for 25 cities and 72 countries/regions in both Google Earth and Maps.

Geo Data Specialist Bernd Steinert wrote in a Google blog, "In our continuing effort to build the most comprehensive and accurate view of the world, the Google Earth and Maps Imagery team just published another extensive catalog of new imagery. This week we have exciting new updates to both our high resolution aerial and satellite imagery and our 45° imagery" [25].

Google Earth and Maps enable us to see not only the amazing world but also our neighbors' private backyards.

2. Tracking devices such as iPhone, iPad and Carrier IQ

In April 2011, O'Reilly Radar reported that iPhones and 3G iPads are regularly recording the position of the device into a hidden file called "consolidated.db" [26]. The secret database file has been storing the locations (latitude-longitude coordinates) and time stamps, effectively tracking the history of movement of the iPhone and 3G iPad users for a year since iOS 4 was released in 2010.

Not to be outdone by the iPhone location tracking software, the Carrier IQ software has been found on about 150 million cell phones including the iPhones, Android, BlackBerry, and Nokia phones [27]. On November 28, 2011, security researcher Trevor Eckhart posted a video on YouTube detailing hidden software installed on smartphones that secretly logs keypresses, SMS messages, and browser URLs [28].

Our trusted smartphones have unknowingly become the tracking devices for big businesses.

3. Spyware using cookies, tracking code, and mobile apps

In February 2012, Jonathan Mayer, a graduate student at Stanford University, demonstrated that four advertising companies, Google's DoubleClick, Vibrant Media, Media Innovation Group, and PointRoll, have been deliberately circumventing Apple Safari's privacy feature by installing temporary cookies on the user devices in order to track users' behavior [29]. Safari is the primary web browser on the iPhone, iPad, and Macintosh computers. The Stanford findings contradicted Google's own instructions to Safari users on how to avoid tracking.

According to a *Wall Street Journal* Research conducted by Ashkan Soltani, Google placed the tracking code within ads displayed on 29 of the top 100 most-visited U.S. websites [30]. Among them are household names YouTube, AOL, *People Magazine*, *The New York Times*, WebMD, Merriam-Webster Dictionary, Fandango.com, Match.com, TMZ, and Yellow Pages.

Also in February 2012, Twitter acknowledged that when a user taps the "Find friends" feature on its smartphone app, the company downloads the user's entire address book, including email addresses and phone numbers, and keeps the data on its servers for 18 months [31].

Off-the-shelf anti-virus software can protect us from malware but not the sophisticated spyware from trusted companies.

4. Drones operated by police, civilians, and Google

Police departments in Seattle, Miami, Little Rock, and other cities have been using unmanned drones for surveillance and law enforcement purposes. According to *U.S. News and World Report,* drones are used to "gain an aerial perspective consistent with the open view doctrine," which allows officers to monitor areas that are in "plain view" [32].

Congressman Hank Johnson of Georgia voiced his concerns, "As the number of drones rises, so, too, will the number of suspects. During the civil rights movement, would activists have left their homes if they knew they were being monitored from cameras 30,000 feet above" [33]?

On the flip side, low-cost drones under $300 each enabled protestors in Occupy Wall Street to monitor the police. In December 2011, activists remotely piloted a Parrot AR.Drone, dubbed "The Occucopter," on their smartphones to provide a live feed of Occupy Wall Street from above [34]. An increasing number of civilians including journalists, wildlife researchers, sports photographers, and real estate agents are using drones for their work [35].

Nonetheless, police and civilian drones pale in comparison to the ubiquitous Google Street View. Launched in May 2007, Google Street View has captured 20 petabytes of data in 39 countries and about 3000 cities [36]. Google uses cars, trikes, boats, snowmobiles, trolleys, and people outfitted with custom cameras to capture 360-degree panoramic images around the world.

In May 2010, however, Google made a stunning admission that for over 3 years, its camera-toting Street View cars have inadvertently collected snippets of private information that people send over unencrypted WiFi networks [37]. In October 2010, Google admitted to "accidentally" collecting and storing entire e-mails, URLs, and passwords from unsecured WiFi networks with its Street View cars in more than 30 countries, including the United States, Canada, Mexico, some of Europe, and parts of Asia [38].

At the "Next Dimension" Google Maps press event in June 2012, Google Street View engineering director Luc Vincent demonstrated a new prototype backpack with camera rig. Vincent explained that "the camera has 15, 5-megapixel lenses and the battery for the rig lasts all day." But he also quipped that "if you hike with a partner, they'll be in every scene" [36].

In October 2012, Google Street View launched its biggest update ever by doubling the number of special collections, adding 250,000 miles of roads around the world, and increasing Street View coverage in Macau, Singapore, Sweden, the U.S., Thailand, Taiwan, Italy, the United Kingdom, Denmark, Norway, and Canada.

Google Street View Program Manager Ulf Spitzer said, "Street View, as you know, is a useful resource when you're planning a route or looking for a destination, but it can also magically transport you to some of the world's picturesque and culturally significant landmarks" [39]. Google Street View is indeed handy for planning a trip to a new restaurant or an unfamiliar neighborhood as well as exploring national parks, university campuses, sports stadiums, and museums around the world. We are living in an information age where abundant data are readily available at our fingertips, whether we are individuals, businesses, or government agencies.

"I fear Google more than I pretty much fear the government," said Jeff Moss, founder of Black Hat and DEF CON, at the 2012 Black Hat Conference. "Google, I'm contractually agreeing to give them all my data" [40]. Such data may eventually include your DNA. In March 2014, the search giant launched Google Genomics to store human genomes on the cloud to facilitate medical research and doctor treatment [41].

8.4 A Two-Way Street of Total Information Awareness

Private businesses and the ubiquity of social networks are creating the necessary technologies and infrastructures for Total Information Awareness (TIA) [18]. However, unlike the government-proposed TIA which is a rigid one-way mirror, the industry-led TIA is an evolving two-way street. Facebook, Google, YouTube, Wikipedia, and even the controversial WikiLeaks all collect information from everywhere and make it available to everyone, whether they are individuals, businesses, or government agencies.

There are some safety concerns about exposing too much information online, especially high value terrorist targets. According to the Federal Bureau of Investigation (FBI), the individuals who planned the attempted car bombing of Times Square on May 1, 2010 used public web cameras for reconnaissance [42].

As a precaution, Google Maps has digitally modified or blurred its satellite imagery on some landmarks including the roof of the White House, NATO air force hub serving as a retreat for the Operation Iraqi Freedom forces, Mobil Oil facilities in Buffalo New York, and even the Dutch Royal Family's Huis Ten Bosch Palace in Netherlands [43].

Pete Cashmore, founder and CEO of *Mashable,* suggested that today's world is both reminiscent of George Orwell's *Nineteen Eighty-Four* and radically at odds with it. Cashmore wrote in a January 2012 *CNN* article, "The online world is indeed allowing our every move to be tracked, while at the same time providing a counterweight to the emergence of Big Brother. … Unlike in Orwell's dystopian world, people today are making a conscious choice to do so. The difference between this reality and Orwell's vision is the issue of control: While his Thought Police tracked you without permission, some consumers are now comfortable with sharing their every move online" [44].

Not only are consumers knowingly sharing their private information with big businesses, government officials are also leaking classified information to the media and private companies for either financial gains or political purposes. In 2012, seven U.S. Navy SEALs were reprimanded for divulging classified combat gear to Electronic Arts, maker of the multiplatform game "Medal of Honor: Warfighter" [45]. A string of leaks from high-ranking government officials resulted in the public airing of details about the U.S. cyber attack on Iran's nuclear centrifuge program [46], increased U.S. drone strikes against militants in Yemen [47] and Pakistan [48], and a double agent disrupting al-Qaeda's custom-fit underwear bombing plot [49].

Even President Barack Obama himself revealed on a Google+ video chat room interview in January 2012 that the U.S. conducted many drone strikes to hunt down al-Qaeda and Taliban in Pakistan [50]. Some former military and intelligence officers accused Obama of disclosing clandestine operations to the public. Ex-Navy SEAL Benjamin Smith voiced his opposition to leaks, "As a citizen, it is my civic duty to tell the president to stop leaking information to the enemy. It will get Americans killed." Another former Navy SEAL Scott Taylor said of the bin Laden raid, "If you disclose how we got there, how we took down the building, what we did, how many people were there, that it's going to hinder future operations, and certainly hurt the success of those future operations" [51].

CIA director David Petraeus issued a statement to his employees in January 2012 about the arrest of a former Agency officer by the FBI on charges that he illegally disclosed classified information to reporters. Petraeus wrote, "Unauthorized disclosures of any sort—including information concerning the identities of other Agency officers—betray the public trust, our country, and our colleagues" [52]. Ironically, Petraeus pleaded guilty in March 2015 for leaking classified information to his biographer and mistress Paula Broadwell.

On the other hand, Thomas Blanton, Director of National Security Archive at George Washington University, testified in December 2010 before the U.S. House of Representatives on the massive overclassification of the U.S. government's national security information. At the Judiciary Committee hearing on the Espionage Act and the legal and constitutional implications of WikiLeaks, Blanton said that "Actually our job as citizens is to ask questions…. Experts believe 50 to 90 % of our national security secrets could be public with little or no damage to real security" [53]. He cited his findings:

A few years back, when Rep. Christopher Shays (R-CT) asked Secretary of Defense Donald Rumsfeld's deputy for counterintelligence and security how much government information was overclassified, her answer was 50 %. After the 9/11 Commission reviewed the government's most sensitive records about Osama bin Laden and Al-Qaeda, the co-chair of that commission, former Governor of New Jersey Tom Kean, commented that "three-quarters of what I read that was classified shouldn't have been" — a 75 % judgment. President Reagan's National Security Council secretary Rodney McDaniel estimated in 1991 that only 10 % of classification was for "legitimate protection of secrets" — so 90 % unwarranted. Another data point comes from the Interagency Security Classification Appeals Panel, over the past 15 years, has overruled agency secrecy claims in whole or in part in some 65 % of its cases.

To deal with the over-classification problem, National Academy of Sciences researcher Herb Lin suggested that government budgets can be used to change behavior. "The incentives to classify information are many, and the incentives to refrain from classifying it are few," said Lin. "Classifying information doesn't incur any monetary cost for the classifier, and any economist will tell you that a free good will be overused" [54]. He proposed that every time a "top secret" stamp is used, the Pentagon and intelligence agencies should be charged against a classification budget.

WikiLeaks founder Julian Assange said in a 2011 interview, "Our No. 1 enemy is ignorance. And I believe that is the No. 1 enemy for everyone — it's not understanding what actually is going on in the world" [55]. Bill Keller, former executive editor of *The Times*, opines that "the most palpable legacy of the WikiLeaks campaign for transparency is that the U.S. government is more secretive than ever" [56]. U.S. Air Force Senior Airman Christopher R. Atkins wrote in his email to American filmmaker Michael Moore, "The single greatest danger to America and our way of life is ourselves. No foreign power can dictate your oppression. No foreign army can impose martial law upon us. No foreign dictator can remove the precious right that I am exercising at this moment. Militaries do not keep people free! Militaries keep us safe, but it is we citizens who ensure freedom!" [57].

Notwithstanding the potential risks and benefits of information sharing, the two-way street of Total Information Awareness is the road that leads to a more transparent and complete picture of ourselves, our governments, and our world. As Wikipedia's founder Jimmy Wales said, "Imagine a world in which every single person on the planet is given free access to the sum of all human knowledge" [58].

8.5 No Doomsday for the Internet

Governments around the world convened in Dubai for the World Conference on International Telecommunications (WCIT) from December 3 to 14, 2012 to decide on the future of the open Internet [59]. Hosted by United Nations' International Telecommunication Union (ITU), the conference was to update the 1988 International Telecommunication Regulations (ITRs) behind closed doors. However, documents have leaked, showing some government proposals that may threaten the open Internet [60]. Among them were government's "right to know how its traffic is routed" and restrictions on public access to telecommunications when used for "undermining the sovereignty, national security, territorial integrity and public safety of other States, or to divulge information of a sensitive nature."

"Proposals by various governments to treat internet connections like the telephone system are cause for concern regarding privacy and the unfettered, free flow of information," said Emma Llansó, policy counsel at the Center for Democracy & Technology. "[But] there's not going to be some kind of doomsday scenario that there's a treaty that makes the internet go dark. What we're seeing is governments putting forward visions of the internet and having discussions" [61].

The Egyptian government shut down Internet access at midnight January 28, 2011 when activists organized through Facebook and Twitter nationwide protests to call for an end to President Hosni Mubarak's government [62]. Google responded to the Internet blockade by working with Twitter and SayNow to unveil a web-free speak-to-tweet service, allowing anyone to send and receive tweets by calling a phone number [63]. The unprecedented totalitarian action failed to thwart the planned demonstrations, and Mubarak resigned as president.

To disrupt rebel communications, the Syrian government shut down the Internet across the country and cut cellphone services in select areas on November 29, 2012. *The Associated Press* reported that "the revolt in Syria began with peaceful protests but turned into a civil war after the government waged a brutal crackdown on dissent" [64]. As the Arab Spring uprising continues to spread throughout the Middle East region, governments should have learned that it is futile to suppress the public or enemies by shutting down the Internet and social networks. Doing so only ends up backfiring.

President Ronald Reagan had predicted that "technology will make it increasingly difficult for the state to control the information its people receive" [65]. He said in June 1989, "Information is the oxygen of the modern age. It seeps through the walls topped by barbed wire, it wafts across the electrified borders, the Goliath of totalitarianism will be brought down by the David of the microchip" [66].

Terry Kramer, U.S. Ambassador to 2012 WCIT, refused to sign the United Nations (U.N.) treaty on telecommunications and Internet. He said, "The Internet has given the world unimaginable economic and social benefit during these past 24 years, all without U.N. regulation" [67].

Nevertheless, the OpenNet Initiative (ONI) reported in April 2012 that 42 countries filter and censor content out of the 72 studied, while 21 countries have been engaging in substantial or pervasive filtering [68]. Freedom House published its findings that "even a number of democratic states have considered or implemented various restrictions in response to the potential legal, economic, and security challenges raised by new media" [69].

Vinton Cerf, a father of the Internet and Google's chief Internet evangelist, strongly opposes government intervention in the World Wide Web. He said that the Internet allows "all of us to reach a global audience at a click of a mouse," and he warned that "history is rife with examples of governments taking actions to 'protect' their citizens from harm by controlling access to information and inhibiting freedom of expression and other freedoms outlined in The Universal Declaration of Human Rights. We must make sure, collectively, that the Internet avoids a similar fate" [70].

8.6 Web 2.0 for Intelligence Community: Intellipedia, A-Space, Deepnet

In 2004, Calvin Andrus of the Central Intelligence Agency (CIA) wrote a paper entitled "The Wiki and the Blog: Toward a Complex Adaptive Intelligence Community," calling for the U.S. intelligence community to follow the lead of the social media industry. In 2006, Sean Dennehy and Don Burke oversaw the creation of Intellipedia, the U.S. intelligence community's version of Wikipedia on top secret, secret, and unclassified networks.

Burke said, "In addition to analysis, we need people who can create an ecosystem of knowledge that is not specifically about answering tomorrow's questions, but creating a world of information that is connected" [71].

Dennehy added, "There's too much emphasis on the analytical report. It's important to look at how we get to the finished intelligence. Intellipedia does this by making the process more social and creating a dialogue that's transparent" [71].

In 2008, the U.S. intelligence community launched A-Space (Analytic Space), an online collaboration environment modeled after MySpace and Facebook to improve analysts' abilities to share information, form communities, and collaborate [72].

"After an initial slow start in which analysts were skeptical about sharing information, they now see that collaboration reduces the time and effort they spend on analyzing data," said Michael Wertheimer of the Office of the Director of National Intelligence (ODNI) [73].

A 2010 Defense Science Board reported that the National Security Agency (NSA) has been developing tools to find and index data in the deep web (aka deepnet) that "contains government reports, databases, and other sources of information of high value to DoD [Department of Defense] and the intelligence community" [74].

It should not come as a surprise that social networking is becoming a tool of espionage. In 2009, CIA double agent Morten Storm was tasked to ensnare the American-born al-Qaeda cleric Anwar al-Awlaki, one of the most wanted terrorists. Dubbed the "bin Laden of the Internet," al-Awlaki had a blog, a Facebook page, and YouTube videos [75]. When al-Awlaki wanted to marry a white European Muslim convert, Storm stumbled across a Facebook group and found Aminah, a 33-year-old blonde who agreed to become al-Awlaki's third wife [76]. The CIA eventually tracked down al-Awlaki in Yemen and killed him by a drone strike in September 2011 [77].

Whether or not Julian Assange of WikiLeaks is correct in his assertion that "Facebook, Google, and Yahoo have built-in interfaces for the U.S. intelligence to use" [19], law enforcement agencies and the intelligence community have begun to harness the power of social networks in combating international and domestic terrorism.

Google executive chairman Eric Schmidt and Google Ideas director Jared Cohen wrote in their 2013 book *The New Digital Age: Transforming Nations, Businesses, and Our Lives*, "As the terrorists of the future are forced to live in both the physical and the virtual world, their model of secrecy and discretion will suffer. There will be more digital eyes watching, more recorded interactions, and, as careful as even the most sophisticated terrorists are, even they cannot completely hide online" [78].

Bibliography

1. **Wright, Lawrence.** The Spymaster. Can Mike McConnell fix America's intelligence community? [Online] The New Yorker, January 21, 2008. http://www.newyorker.com/reporting/2008/01/21/080121fa_fact_wright?currentPage=all.
2. **Niles, Robert.** What's the point of a bag check, anyway? [Online] Theme Park Insider, March 5, 2012. http://www.themeparkinsider.com/flume/201203/2940/.

3. **3VR Inc.** Use Video Analytics and Data Decision Making to Grow Your Business. [Online] Digital Signage Today. [Cited: May 28, 2012.] http://www.digitalsignagetoday.com/whitepap ers/4891/Use-Video-Analytics-and-Data-Decision-Making-to-Grow-Your-Business.

4. **Fussey, Pete.** An Interrupted Transmission? Processes of CCTV Implementation and the Impact of Human Agency, Surveillance & Society. [Online] Surveillance and Criminal Justice. 4(3): 229-256, 2007. http://www.surveillance-and-society.org.

5. **Reeve, Tom.** How many cameras in the UK? Only 1.85 million, claims ACPO lead on CCTV. [Online] Security News Desk, March 2011. http://www.securitynewsdesk.com/2011 /03/01/how-many-cctv-cameras-in-the-uk/.

6. **Hope, Christopher.** 1,000 CCTV cameras to solve just one crime, Met Police admits. [Online] The Telegraph, August 25, 2009. http://www.telegraph.co.uk/news/uknews/ crime/6082530/1000-CCTV-cameras-to-solve-just-one-crime-Met-Police-admits.html.

7. **BBC News.** 'Ring of steel' widened. [Online] BBC News, December 18, 2003. http://news. bbc.co.uk/2/hi/uk_news/england/london/3330771.stm.

8. **Viisage Technology, Inc.** Viisage Technology and Biometrica Systems Achieve 50th Facial Recognition Installation at Mirage Resort, Las Vegas. [Online] PR Newswire, March 29, 2012. http://www.prnewswire.com/news-releases/viisage-technology-and-biometrica-systems-achieve-50th-facial-recognition-installation-at-mirage-resort-las-vegas-73268177.html.

9. **Obama, Barack.** State of the Union 2011: President Obama's Full Speech. [Online] ABC News, January 25, 2011. http://abcnews.go.com/Politics/State_of_the_Union/state-of-the-union-2011-full-transcript/story?id=12759395&page=3.

10. **Scherer, Michael.** Inside the Secret World of the Data Crunchers Who Helped Obama Win. [Online] TIME, November 7, 2012. http://swampland.time.com/2012/11/07/inside-the-secret-world-of-quants-and-data-crunchers-who-helped-obama-win/.

11. **Times, The Washington.** Obama: Be careful what you put on Facebook. [Online] The Washington Times, September 8, 2009. http://www.washingtontimes.com/news/2009/ sep/08/obama-advises-caution-what-kids-put-facebook/?page=all.

12. **Etzioni, Amitai.** Despite Facebook, privacy is far from dead. [Online] CNN, May 27, 2012. http://www.cnn.com/2012/05/25/opinion/etzioni-facebook-privacy/index.html.

13. **Ortutay, Barbara.** Facebook tops 1 billion users. [Online] USA Today, October 4, 2012. htt p://www.usatoday.com/story/tech/2012/10/04/facebook-tops-1-billion-users/1612613/.

14. **Greenfield, Rebecca.** Facebook's New Ad Finds 'Real Human Emotion' in Chairs. [Online] The Atlantic Wire, October 4, 2012. http://www.theatlanticwire.com/technology/2012/10/ facebook-aims-real-human-emotion-new-ad/57596/.

15. **Facebook.** The Things That Connect Us. [Online] Facebook, October 4, 2012. http://www.yo utube.com/watch?v=c7SjvLceXgU.

16. **Haglund, David.** Facebook's Disingenuous New Ad. [Online] Slate, October 4, 2012. http://www.slate.com/blogs/browbeat/2012/10/04/facebook_chair_ad_why_chairs_ explained_video_.html.

17. **Takahashi, Corey.** New Office Chair Promotes Concept of 'Active Sitting'. [Online] The Wall Street Journal, July 21, 2007. http://online.wsj.com/article/SB869433022356380000.html.

18. **Lee, Newton.** Facebook Nation: Total Information Awareness. [Online] Springer, October 14, 2014. http://www.amazon.com/Facebook-Nation-Total-Information-Awareness/dp/1493917390/.

19. **RT.** WikiLeaks revelations only tip of iceberg – Assange. [Online] RT News, May 2, 2011. http://www.rt.com/news/wikileaks-revelations-assange-interview/.

20. **Hays, Tom.** NYPD is watching Facebook to Fight Gang Bloodshed. [Online] Associated Press, October 2, 2012. http://bigstory.ap.org/article/nypd-watching-facebook-fight-gang-bloodshed.

21. **Gross, Doug.** Library of Congress to archive your tweets. [Online] CNN, April 14, 2010. http:// www.cnn.com/2010/TECH/04/14/library.congress.twitter/index.html.

22. —. Library of Congress digs into 170 billion tweets. [Online] CNN, January 7, 2013. http:// www.cnn.com/2013/01/07/tech/social-media/library-congress-twitter/index.html.

23. **Hennigan, W. J.** It's a bird! It's a spy! It's both. [Online] Los Angeles Times, February 17, 2011. http://articles.latimes.com/2011/feb/17/business/la-fi-hummingbird-drone-20110217.

24. **CNet Live Blog.** Google's 'next dimension' of Maps event. [Online] CNet, June 6, 2012. http://live.cnet.com/Event/Googles_next_dimension_of_Maps_event?Page=6.
25. **Steinert, Bernd.** Imagery Update: Explore your favorite places in high-resolution. [Online] Google Maps, July 27, 2012. http://google-latlong.blogspot.co.uk/2012/07/imagery-update-explore-your-favorite.html.
26. **Allan, Alasdair.** Got an iPhone or 3G iPad? Apple is recording your moves. [Online] O'Reilly Radar, April 20, 2011. http://radar.oreilly.com/2011/04/apple-location-tracking.html.
27. **Kravets, David.** Researcher's Video Shows Secret Software on Millions of Phones Logging Everything. [Online] Wired, November 29, 2011. http://www.wired.com/threatlevel/2011/11/secret-software-logging-video/.
28. **Eckhart, Trevor.** Carrier IQ Part #2. [Online] YouTube, November 28, 2011. http://www.youtube.com/watch?v=T17XQI_AYNo.
29. **Mayer, Jonathan.** Web Policy. Safari Trackers. [Online] Web Policy Blog, February 17, 2012. http://webpolicy.org/2012/02/17/safari-trackers/.
30. **Angwin, Julia and Valentino-Devries, Jennifer.** Google's iPhone Tracking. Web Giant, Others Bypassed Apple Browser Settings for Guarding Privacy. [Online] The Wall Street Journal, February 17, 2012. http://online.wsj.com/article_email/SB10001424052970204880404577225380456599176-lMyQjAxMTAyMDEwNjExNDYyWj.html.
31. **Sarno, David.** Twitter stores full iPhone contact list for 18 months, after scan. [Online] Los Angeles Times, February 14, 2012. http://www.latimes.com/business/technology/la-fi-tn-twitter-contacts-20120214,0,5579919.story.
32. **Koebler, Jason.** Police to Use Drones for Spying on Citizens. [Online] US News and World Report, August 13, 2012. http://www.usnews.com/news/articles/2012/08/23/docs-law-enforcement-agencies-plan-to-use-domestic-drones-for-surveillance.
33. **Ahlers, Mike M.** Congress likes drones, but now looks at flip side of use. [Online] CNN, October 26, 2012. http://www.cnn.com/2012/10/25/us/drones-privacy/index.html.
34. Occupy Wall Street's New Drone: 'The Occucopter'. [Online] Time Magazine, December 21, 2011. http://techland.time.com/2011/12/21/occupy-wall-streets-new-drone-the-occucopter/.
35. **Hargreaves, Steve.** Drones go mainstream. [Online] CNNMoney, January 9, 2013. http://money.cnn.com/2013/01/09/technology/drones/index.html.
36. **Farber, Dan.** Google takes Street View off-road with backpack rig. [Online] CNet, June 6, 2012. http://news.cnet.com/8301-1023_3-57448293-93/google-takes-street-view-off-road-with-backpack-rig/.
37. **Stone, Brad.** Google Says It Inadvertently Collected Personal Data. [Online] The New York Times, May 14, 2010. http://bits.blogs.nytimes.com/2010/05/14/google-admits-to-snooping-on-personal-data/.
38. **Landis, Marina.** Google admits to accidentally collecting e-mails, URLs, passwords. [Online] CNN, October 22, 2010. http://articles.cnn.com/2010-10-22/tech/google.privacy.controls_1_wifi-data-alan-eustace-google-s-street-view?_s=PM:TECH.
39. **Spitzer, Ulf.** Making Google Maps more comprehensive with biggest Street View update ever. [Online] Google Maps, October 11, 2012. http://google-latlong.blogspot.co.uk/2012/10/making-google-maps-more-comprehensive.html.
40. **Kelly, Heather.** Is the government doing enough to protect us online? [Online] CNN, July 31, 2012. http://www.cnn.com/2012/07/25/tech/regulating-cybersecurity/index.html.
41. **Regalado, Antonio.** Google Wants to Store Your Genome. [Online] MIT Technology Review, November 6, 2014. http://www.technologyreview.com/news/532266/google-wants-to-store-your-genome/.
42. **Mueller, Robert S. III.** Combating Threats in the Cyber World: Outsmarting Terrorists, Hackers, and Spies. [Online] Federal Bureau of Investigation, March 1, 2012. http://www.fbi.gov/news/speeches/combating-threats-in-the-cyber-world-outsmarting-terrorists-hackers-and-spies.
43. **Jackson, Nicholas.** 15 High-Profile Sites That Google Doesn't Want You to See. [Online] The Atlantic, June 21, 2011. http://www.theatlantic.com/technology/archive/2011/06/15-high-profile-sites-that-google-doesnt-want-you-to-see/240766/.

44. **Cashmore, Pete.** Why 2012, despite privacy fears, isn't like Orwell's 1984. [Online] CNN, January 23, 2012. http://www.cnn.com/2012/01/23/tech/social-media/web-1984-orwell-cashmore/index.html.

45. **Mount, Mike.** Navy SEALs punished for revealing secrets to video game maker. [Online] CNN, November 9, 2012. http://security.blogs.cnn.com/2012/11/09/navy-seals-busted-for-giving-secrets-to-make-video-game-more-real/ .

46. **Sanger, David E.** Obama Order Sped Up Wave of Cyberattacks Against Iran. [Online] The New York Times, June 1, 2012. http://www.nytimes.com/2012/06/01/world/middleeast/obama-ordered-wave-of-cyberattacks-against-iran.html?pagewanted=all.

47. **Schmitt, Eric.** U.S. to Step Up Drone Strikes Inside Yemen. [Online] The New York Times, April 25, 2012. http://www.nytimes.com/2012/04/26/world/middleeast/us-to-step-up-drone-strikes-inside-yemen.html.

48. **Cloud, David S. and Rodriguez, Alex.** CIA gets nod to step up drone strikes in Pakistan. [Online] Los Angeles Times, June 8, 2012. http://articles.latimes.com/2012/jun/08/world/la-fg-pakistan-drone-surge-20120608.

49. **Shane, Scott and Schmitt, Eric.** Double Agent Disrupted Bombing Plot, U.S. Says. [Online] The New York Times, May 8, 2012. http://www.nytimes.com/2012/05/09/world/middleeast/suicide-mission-volunteer-was-double-agent-officials-say.html?pagewanted=all.

50. **Levine, Adam.** Obama admits to Pakistan drone strikes. [Online] CNN, January 30, 2012. http://security.blogs.cnn.com/2012/01/30/obama-admits-to-pakistan-drone-strikes/.

51. **McConnell, Dugald and Todd, Brian.** Former special forces officers slam Obama over leaks on bin Laden killing. [Online] CNN, August 17, 2012. http://www.cnn.com/2012/08/16/politics/former-seals-obama/index.html.

52. **Petraeus, David H.** Statement to Employees by Director of the Central Intelligence Agency David H. Petraeus on Safeguarding our Secrets. [Online] Central Intelligence Agency, January 23, 2012. https://www.cia.gov/news-information/press-releases-statements/2012-press-releasese-statements/safeguarding-our-secrets.html.

53. **Blanton, Thomas.** Hearing on the Espionage Act and the Legal and Constitutional Implications of Wikileaks. [Online] Committee on the Judiciary, U.S. House of Representatives, December 16, 2010. http://www.gwu.edu/~nsarchiv/news/20101216/Blanton101216.pdf.

54. **Sanger, David E.** A Washington Riddle: What Is 'Top Secret'? [Online] The New York Times, August 3, 2013. http://www.nytimes.com/2013/08/04/sunday-review/a-washington-riddle-what-is-top-secret.html?_r=1&.

55. **RT.** WikiLeaks revelations only tip of iceberg – Assange. [Online] RT, May 3, 2011. http://rt.com/news/wikileaks-revelations-assange-interview/.

56. **Keller, Bill.** WikiLeaks, a Postscript. [Online] The New York Times, February 19, 2012. http://www.nytimes.com/2012/02/20/opinion/keller-wikileaks-a-postscript.html.

57. **Atkins, Christopher R.** "I solemnly swear to defend the Constitution of the United States of America against all enemies, foreign and domestic.". [Online] Michael Moore, January 28, 2005. http://www.michaelmoore.com/words/soldiers-letters/i-solemnly-swear-to-defend-the-constitution-of-the-united-states-of-america-against-all-enemies-foreign-and-domestic.

58. **Wales, Jimmy.** An appeal from Wikipedia founder Jimmy Wales. [Online] Wikimedia Foundation, October 30, 2010. http://wikimediafoundation.org/wiki/Appeal2/en.

59. **Solomon, Brett.** The U.N. Shouldn't Make Decisions About an Open Internet Behind Closed Doors. [Online] Wired, November 30, 2012. http://www.wired.com/opinion/2012/11/you-cant-make-decisions-about-the-open-internet-behind-closed-doors/.

60. **Nxt.** Why we are making all WCIT documents public. [Online] .Nxt, November 22, 2012. http://news.dot-nxt.com/2012/11/23/why-we-are-making-all-wcit-doc.

61. **Kravets, David.** Internet hangs in balance as world leaders meet in secret. [Online] Wired, December 3, 2012. http://www.cnn.com/2012/12/03/tech/web/world-conference-international-telecommunications/index.html.

62. **Cowie, James.** Egypt Leaves the Internet. [Online] Renesys Blog, January 27, 2011. http://www.renesys.com/blog/2011/01/egypt-leaves-the-internet.shtml.

63. **AFP.** Google unveils Web-free 'tweeting' in Egypt move. [Online] Google, January 31, 2011. http://www.google.com/hostednews/afp/article/ALeqM5h8de3cQ8o_S2zg9s72t7sxNToBqA? docId=CNG.ddc0305146893ec9e9e6796d743e6af7.c81.

64. **The Associated Press.** Internet down nationwide in Syria. [Online] USA Today, November 30, 2012. http://www.usatoday.com/story/news/world/2012/11/29/internet-syria/1735721/.

65. **The Economist.** Caught in the net. [Online] The Economist, January 23, 2003. http://www. economist.com/node/1534249.

66. **Paulson, Matthew.** Politics of the Future - How the Internet is Changing and Will Change Politics Forever. [Online] Yahoo! News, November 6, 2006. http://voices.yahoo.com/ politics-future-internet-changing-105886.html.

67. **Fitzpatrick, Alex.** U.S. refuses to sign UN Internet treaty. [Online] CNN, December 14, 2012. http://www.cnn.com/2012/12/14/tech/web/un-internet-treaty/index.html.

68. **ONI Team.** Global Internet filtering in 2012 at a glance. [Online] OpenNet Initiative, April 3, 2012. http://opennet.net/blog/2012/04/global-internet-filtering-2012-glance.

69. Internet Freedom. [Online] Freedom House. [Cited: December 4, 2012.] http://www.freedomhouse.org/issues/internet-freedom.

70. **Cerf, Vinton.** 'Father of the internet': Why we must fight for its freedom. [Online] CNN, November 30, 2012. http://edition.cnn.com/2012/11/29/business/ opinion-cerf-google-internet-freedom/index.html.

71. **Central Intelligence Agency .** Intellipedia Celebrates Third Anniversary with a Successful Challenge. [Online] Central Intelligence Agency, April 29, 2009. https://www.cia.gov/ news-information/featured-story-archive/intellipedia-celebrates-third-anniversary.html.

72. **Bain, Ben.** A-Space set to launch this month. [Online] Federal Computer Week, September 3, 2008. http://fcw.com/articles/2008/09/03/aspace-set-to-launch-this-month.aspx.

73. **Yasin, Rutrell.** National security and social networking are compatible. [Online] Government Computer News, July 23, 2009. http://gcn.com/Articles/2009/07/23/ Social-networking-media-national-security.aspx?Page=3.

74. **Bamford, James.** The NSA Is Building the Country's Biggest Spy Center (Watch What You Say). [Online] Wired, March 15, 2012. http://www.wired.com/threatlevel/2012/03/ ff_nsadatacenter/all/.

75. **Madhani, Aamer.** Cleric al-Awlaki dubbed 'bin Laden of the Internet'. [Online] USA Today, August 24, 2010. http://usatoday30.usatoday.com/news/nation/2010-08-25-1A_Awlaki25_C V_N.htm.

76. **Cruickshank, Paul, Lister, Tim and Robertson, Nic.** The Danish agent, the Croatian blonde and the CIA plot to get al-Awlaki. [Online] CNN, October 16, 2012. http://www.cnn. com/2012/10/15/world/al-qaeda-cia-marriage-plot/index.html.

77. **Mazzetti, Mark, Schmitt, Eric and Worth, Robert F.** Two-Year Manhunt Led to Killing of Awlaki in Yemen. [Online] The New York Times, September 30, 2011. http://www.nytim es.com/2011/10/01/world/middleeast/anwar-al-awlaki-is-killed-in-yemen.html?pagewanted= all.

78. **Schmidt, Eric and Cohen, Jared.** The New Digital Age: Transforming Nations, Businesses, and Our Lives. [Online] Knopf Doubleday Publishing Group, April 23, 2013. http:// books.google.com/books/about/The_New_Digital_Age.html?id=Fl_LPoLdGKsC.

Part IV
Cybersecurity: History, Strategies, and Technologies

Chapter 9
Cyber Warfare: Weapon of Mass Disruption

This world — cyberspace — is a world that we depend on every single day. ... America's economic prosperity in the 21st century will depend on cybersecurity.
— President Barack Obama's remark from the White House
(May 29, 2009)

Terrorism does remain the FBI's top priority, but in the not too-distant-future we anticipate that the cyberthreat will pose the greatest threat to our country.
— Former FBI Director Robert Mueller (March 1, 2012)

The Internet is a haystack full of needles.
— Jeff Williams, cofounder and CTO of Aspect (March 2012)

The war is being fought on three fronts. The first is physical, the second is the world of social networks, and the third is cyber attacks.
— Carmela Avner, Israel's Chief Information Officer
(November 18, 2012)

On the scale of 1 to 10, this is an 11.
Bruce Schneier, CTO of Co3 Systems (April 9, 2014)

This is Hollywood's moment in the cyber-victim spotlight.
— Arizona Senator John McCain (December 20, 2014)

It was an act of cyber vandalism that was very costly. We take it very seriously and we will respond proportionally. ... So the key here is not to suggest that Sony was a bad actor. It's making a broader point that all of us have to adapt to the possibility of cyber attacks, we have to do a lot more to guard against them.
— President Barack Obama (December 21, 2014)

© Springer International Publishing Switzerland 2015
N. Lee, *Counterterrorism and Cybersecurity*, DOI 10.1007/978-3-319-17244-6_9

9.1 Weapon of Mass Disruption

Like counterterrorism, cybersecurity is in the forefront of the U.S. national security agenda. President Barack Obama remarked from the White House on May 29, 2009 about "a weapon of mass disruption" [1]:

> We meet today at a transformational moment — a moment in history when our interconnected world presents us, at once, with great promise but also great peril. … This world — cyberspace — is a world that we depend on every single day. It's our hardware and our software, our desktops and laptops and cell phones and Blackberries that have become woven into every aspect of our lives. It's the broadband networks beneath us and the wireless signals around us, the local networks in our schools and hospitals and businesses, and the massive grids that power our nation. It's the classified military and intelligence networks that keep us safe, and the World Wide Web that has made us more interconnected than at any time in human history.

> It's the great irony of our Information Age — the very technologies that empower us to create and to build also empower those who would disrupt and destroy. … Al-Qaeda and other terrorist groups have spoken of their desire to unleash a cyber attack on our country — attacks that are harder to detect and harder to defend against. Indeed, in today's world, acts of terror could come not only from a few extremists in suicide vests but from a few key strokes on the computer — a weapon of mass disruption.

Federal Bureau of Investigation (FBI) Director Robert Mueller spoke at the 2012 RSA Conference in San Francisco: "In one hacker recruiting video, a terrorist proclaims that cyber warfare will be the warfare of the future. … Terrorism remains the FBI's top priority. But in the not too distant future, we anticipate that the cyber threat will pose the number one threat to our country. We need to take lessons learned from fighting terrorism and apply them to cyber crime" [2].

A year before in March 2011, computer and network security firm RSA disclosed a massive data breach due to "an extremely sophisticated cyber attack" on its computer systems, compromising the effectiveness of its SecurID system that is being used by more than 25,000 corporations and 40 million users around the world [3]. RSA's executive chairman Art Coviello described the attack as an "advanced persistent threat" (APT) from cyber attackers who were skilled, motivated, organized, and well-funded.

RSA was not the only victim. In 2011, more than 760 organizations including almost 20 % of the Fortune 100 companies had their computer networks compromised by some of the same resources used to hit RSA [4]. There were financial firms (e.g. Charles Schwab, Freddie Mac, PriceWaterhouseCoopers, Wells Fargo Bank, World Bank), technology companies (e.g. Amazon.com, AT&T, Cisco, eBay, Facebook, Google, IBM, Intel, Motorola, Microsoft, Sprint, Verizon, Yahoo!), governments (e.g. U.S. Internal Revenue Service, Singapore Government Network), and universities (e.g. MIT, Princeton University, University of California, University of Virginia).

In 2013, the National Cybersecurity and Communications Integration Center (NCCIC) that operates 24/7 nonstop received over 220,000 reports of cybersecurity and communications incidents from both the public and private sectors [5].

In December 2014, a malware called SoakSoak infected more than 100,000 WordPress sites that use the Slider Revolution slideshow plug-in [6]. Since over 70 million websites use WordPress as a content management system, Google was only able to catch a small percentage of the infected sites and block 11,000 domains in an attempt to curb the widespread damage.

Though not as terrifying as weapons of mass destruction, cyber attacks can be powerful weapons of mass disruption.

9.2   Financial Disruption

In October 2010, the Federal Bureau of Investigation discovered a sophisticated malware lurking in the central servers of the American stock exchange Nasdaq. Using two zero-day vulnerabilities in combination, the digital time bomb is capable of wiping out the entire stock exchange when triggered. "We've seen a nation-state gain access to at least one of our stock exchanges, I'll put it that way, and it's not crystal clear what their final objective is," said House Intelligence Committee Chairman Mike Rogers. "The bad news of that equation is, I'm not sure you will really know until that final trigger is pulled. And you never want to get to that" [7].

In an August 2012 article on CNN, Brian Patrick Eha wrote that "stock markets have become increasingly vulnerable to bugs over the last decade thanks to financial firms' growing reliance on high-speed computerized trading. Because the trading is automated, there's nobody to apply the brakes if things go wrong" [8]. Overreliance on computerized automated trading plays into the hands of industrial terrorists.

Between December 2012 and February 2013, a cyber crime ring stole $45 million from banks across the world in broad daylight. The criminals first obtained prepaid MasterCard debit cards, then hacked into the banks' systems to drastically increase the credit available on the cards, and finally withdrew large sums of money from ATM machines worldwide [9].

At 12:07 pm on April 23, 2013, a fake tweet about White House explosions from the account of the Associated Press (AP) sent stocks tumbling more than 140 points in a matter of minutes [10]. Cyber intruders had hacked into Twitter and posted a tweet using @AP credential: "Breaking: Two Explosions in the White House and Barack Obama is injured." (See Fig. 9.1). Stocks fully recovered from the plunge soon after investors realized that the tweet was fake. We would not be surprised if some cybercriminals had just made off with millions, if not billions, of dollars during the huge stock fluctuations within a short period of time.

Since 2013, cybercriminals have infiltrated more than 100 banks in 30 countries through phishing and other social engineering schemes to steal up to $1 billion from the banks. They familiarized themselves with the banks' operations, then "use that knowledge to steal money without raising suspicions, programming ATMs to dispense money at specific times or setting up fake accounts and transferring money into them, according to Kaspersky Lab" [11].

Fig. 9.1 Fake White House explosion tweet

Cybercrime costs the global economy an estimated $400 billion a year, according a June 2014 report by the Center for Strategic and International Studies [12]. President Barack Obama said in his May 2009 remark on "Securing our Nation's Cyber Infrastructure" [1]:

> It's about the privacy and the economic security of American families. We rely on the Internet to pay our bills, to bank, to shop, to file our taxes. But we've had to learn a whole new vocabulary just to stay ahead of the cyber criminals who would do us harm — spyware and malware and spoofing and phishing and botnets. Millions of Americans have been victimized, their privacy violated, their identities stolen, their lives upended, and their wallets emptied. According to one survey, in the past two years alone cyber crime has cost Americans more than $8 billion.

In the aftermath of the 9/11 attacks, a large-scale cyber attack occurred seven days later on September 18, 2001 [13]. Nimda (admin spelled backwards) became the Internet's most widespread computer malware to-date and caught many businesses off guard. "A lot of CEOs got really pissed because they thought they had spent a lot of time and money doing cybersecurity for the company and — bang! — they got hammered, knocked offline, their records got destroyed, and it cost millions of dollars per company," said Richard Clarke, special adviser to the president for cyberspace security and terrorism czar to President Bill Clinton [14].

Nimda blends all three major methods to infect and disable computers: Worm, virus, and Trojan horse. A worm propagates across a network and reproduces itself without user interaction; a virus incorporates itself into other programs when the virus code is executed by a computer user; and a Trojan horse contains hidden code that performs malicious actions. Nimda spread through email attachments, browsing of compromised web sites, and exploitation of vulnerabilities in the Microsoft IIS web server. Microsoft senior .Net developer Michael Lane Thomas characterized the perpetrators behind Nimda as "industrial terrorists" [15].

Computer Economics put the financial damage caused by Nimda to be $635 million [16]. Nimda, Code Red, SirCam, and other cyber attacks in 2001 accounted for a total loss of $13.2 billion in businesses worldwide. Figure 9.2 shows the 12 costliest computer viruses and worms ever [17, 18].

Computer viruses and worms often grab media headlines because of the widespread infection of computers globally, resulting in potentially high cumulative economic damages. Nevertheless, the lesser known cyber attacks that target specific companies are just as damaging. In 1994-1995, the "Phonemasters" were accountable for $1.85 million in business losses by hacking into the telephone networks of AT&T, British Telecommunications, GTE, MCI, Southwestern Bell, and Sprint; accessing the credit-reporting databases at Equifax and TRW; and entering the systems of Dun & Bradstreet [31]. In 1995, Russian mathematician Vladimir Levin was convicted of stealing $2.8 million from Citibank in a series of electronic break-ins [32]. In 2000, 16-year-old Canadian teenager "Mafiaboy" launched a Distributed Denial of Service (DDoS) attack that took down Amazon.com, Buy.com, CNN, eBay, E*Trade, Excite, Yahoo!, and ZDNet.com, resulting in $1.7 billion in business losses [33].

Global security advisor and futurist Marc Goodman spoke at TEDGlobal 2012 on the technological future of crime. He said, "In the last couple hundred years, we've gone from one person robbing another to train robberies, where one gang could rob 200 people at a time. Now, that's scaled to the Sony PlayStation hack, which affected 100 million people. When in history was it ever possible for one person to rob 100 million?" [34].

Symantec released its "Cybercrime Report in 2011" showing that the net cost of cybercrime totaled $139.6 billion in the United States and $388 billion

Computer Viruses and Worms	Release Date	Financial Damage
1. MyDoom(19)	2004	$38.5 billion
2. SoBig(20)	2003	$37.1 billion
3. ILOVEYOU(21)	2000	$15 billion
4. Conficker(22)	2007	$9.1 billion
5. Code Red(23)	2001	$2 billion
6. Melissa(24)	1999	$1.2 billion
7. SirCam(25)	2001	$1 billion
8. SQL Slammer(26)	2003	$750 million
9. Nimda(27)	2001	$635 million
10. Sasser(28)	2004	$500 million
11. Blaster(29)	2003	$320 million
12. Morris(30)	1988	$10 million

Fig. 9.2 The 12 costliest computer viruses and worms

worldwide [35]. In December 2012, McAfee Labs published a report entitled "Analyzing Project Blitzkrieg, a Credible Threat" that warns of an impending cyber attack on 30 U.S. banks in Spring 2013 [36]. Previously in October 2012, RSA identified the malware as Gozi Prinimalka Trojan that lies dormant in the infested computers until the prescheduled D-day to launch its attack spree [37].

First appeared in September 2013, Cryptolocker is a "ransomware" that encrypts files on Windows computers and demands some 500,000 victims to pay ransoms in order to recover their files. The creators of Cryptolocker netted $3 million before they were arrested by police in 2014 [38]. Security firms Fox-IT and FireEye have developed a decryption assistance website to help those afflicted with the CryptoLocker malware [39].

In May 2014, the Federal Bureau of Investigation and local police in 19 countries arrested more than 90 individuals who used Blackshades malware for illegal activities including extortion and bank fraud. U.S. Attorney Preet Bharara remarked, "For just $40, the BlackShades RAT [Remote Access Trojan] enabled anyone anywhere in the world to instantly become a dangerous cybercriminal, able to steal your property and invade your privacy" [40]. BlackShades is also nicknamed "creepware" when it was used to hijack the webcams of Miss Teen USA Cassidy Wolf and other young women for the purpose of blackmailing and sextortion [41].

In November 2014, cybercriminals who called themselves the Guardians of Peace (GOP) stole from Sony Pictures 100 terabytes of data including four feature films [42], destroyed 75% of their corporate computer servers, crippled the company's data centers [43], and caused the large theater chains to call off the screenings of the $44 million movie *The Interview* [44]. In February 2015, Sony gave an official figure for damages from the massive hack: $15 million [45].

9.3 Infrastructure Disruption

Army General Keith Alexander, director of the National Security Agency (NSA) and Central Security Service (CSS), was also the commander of U.S. Cyber Command that directs the operation and defense of the military's information networks [46]. At the 2012 Aspen Security Forum in Colorado, Alexander said that there had been a 17-fold increase in computer attacks on American infrastructure between 2009 and 2011; and on a scale of 1 to 10, the U.S. scored a 3 for preventing a serious cyber attack on a critical part of its infrastructure [47].

Paul Stockton, assistant secretary for Homeland Defense and Americas' Security Affairs (DH&ASA), spoke candidly at the 2012 Aspen Security Forum: "Our adversaries, state and non-state, are not stupid. They are clever and adaptive. There is a risk that they will adopt a profoundly asymmetric strategy, reach around and attack us here at home, the critical infrastructure that is not owned by the Department of Defense" [48].

In November 2012, the National Research Council of the National Academies released the previously classified 2007 report "Terrorism and the Electric Power

Delivery System," warning that the U.S. electric power grid is "inherently vulnerable" to terrorist attacks [49]:

> The U.S. electric power delivery system is vulnerable to terrorist attacks that could cause much more damage to the system than natural disasters such as Hurricane Sandy, blacking out large regions of the country for weeks or months and costing many billions of dollars.
>
> The power grid is inherently vulnerable physically because it is spread across hundreds of miles, and many key facilities are unguarded. ... Many parts of the bulk high-voltage system are heavily stressed, leaving it especially at risk to multiple failures following an attack. Important pieces of equipment are decades old and lack improved technology for sensing and control that could help limit outages and their consequences — not only those caused by a terrorist attack but also in the event of natural disasters.
>
> There are also critical systems — communications, sensors, and controls — that are potentially vulnerable to cyber attacks, whether through Internet connections or by direct penetration at remote sites. Any telecommunication link that is even partially outside the control of the system operators could be an insecure pathway into operations and a threat to the grid.

The August 2003 Northeast blackout deprived 50 million people of electrical power [50]. Many lost water service, gas stations were forced to shut down, and the New York Police Department responded as if the massive power failure were the result of a terrorist attack. It turned out that a software bug in an electricity monitoring system failed to notify operators about a local power outage which then triggered a domino effect, cutting off power in eight U.S. states and parts of Ontario. The total impact on U.S. workers, consumers, and taxpayers was a loss of $6.4 billion [51].

"Power system disruptions experienced to date in the United States, be they from natural disasters or malfunctions, have had immense economic impacts," said Prof. M. Granger Morgan of Carnegie Mellon University and chair of the committee that wrote the 2007 classified report. "Considering that a systematically designed and executed terrorist attack could cause disruptions even more widespread and of longer duration, it is no stretch of the imagination to think that such attacks could produce damage costing hundreds of billions of dollars" [49]. Morgan urged the U.S. government and the power industry to invest in power system research and to develop recovery high-voltage transformers.

Companies are increasingly using the Internet to manage electric power generation, oil-pipeline flow, and water levels in dams remotely by means of "Supervisory Control and Data Acquisition" (SCADA) systems. A SCADA is consisted of a central host that monitors and controls smaller Remote Terminal Units (RTUs) throughout a plant or in the field. The U.S. National Communications System (NCS) released a technical information bulletin in October 2004 indicating that SCADA is susceptible to the various attacks [52]:

- Use a Denial of Service (DoS) attack to crash the SCADA server leading to shutdown condition (System Downtime and Loss of Operations)
- Delete system files on the SCADA server (System Downtime and Loss of Operations)
- Plant a Trojan and take complete control of system (Gain complete control of system and be able to issue any commands available to Operators)

- Log keystrokes from Operators and obtain usernames and passwords (Preparation for future take down)
- Log any company-sensitive operational data for personal or competition usage (Loss of Corporate Competitive Advantage)
- Change data points or deceive Operators into thinking control process is out of control and must be shut down (Downtime and Loss of Corporate Data)
- Modify any logged data in remote database system (Loss of Corporate Data)
- Use SCADA Server as a launching point to defame and compromise other system components within corporate network (IP spoofing or ARP spoofing allowing man-in-the-middle attack on a local area network)

In January 2003, the SQL Slammer worm (see Fig. 9.2) penetrated a private computer network protected by a firewall at the Davis-Besse nuclear power plant in Oak Harbor, Ohio [53]. The worm disabled the Safety Parameter Display System (SPDS) and crashed the Plant Process Computer (PPC) for almost five hours. SPDS monitors the nuclear power plant's most crucial safety indicators such as coolant systems, core temperature sensors, and external radiation sensors, whereas PPC monitors the less critical components of the nuclear power plant. An energy sector cybersecurity expert spoke on condition of anonymity, "If a non-intelligent worm can get in, imagine what an intruder can do" [53].

Indeed, it happened less than seven years later. In June 2010, the Stuxnet worm was discovered by cybersecurity experts at VirusBlokAda in Belarus. Stuxnet has "all the hallmarks of weaponized software" and it was written specifically to exploit vulnerabilities in SCADA systems that control critical equipment at electric power companies, manufacturing facilities, water treatment plants, nuclear power stations, and other industrial operations [54].

Stuxnet is highly infectious and difficult to detect. It propagates via USB storage devices, injects code into system processes, and hides itself [55]. Roel Schouwenberg, a senior antivirus researcher at Kaspersky Lab, called the Stuxnet worm a "groundbreaking" piece of malware that exploited not just one but four zero-day Windows bugs: Windows shortcuts bug, print spooler bug, and two elevation of privilege (EoP) bugs [56]. Moreover, the malware employed at least two signed digital certificates to make it look legitimate.

Before Microsoft could issue patches to fix the four Windows bugs, Stuxent had infected more than 14,000 computers — nearly 60 % of which were located in Iran, 18 % in Indonesia, 8 % in India, and less than 2 % in Azerbauan, the United States, and Pakistan [57]. The malware was programmed to specifically target SCADA systems manufactured by Siemens [58]. It wreaked havoc on Iran's nuclear facilities and destroyed nearly 1,000 or 20 % of their nuclear centrifuges [59].

David E. Sanger, chief Washington correspondent of *The New York Times*, revealed in June 2012 that President Barack Obama secretly ordered the cyber attacks on Iran's nuclear enrichment facilities by accelerating Operation Olympic Games that began under the Bush administration in 2006 [60]. A part of the

Olympic Games, the Stuxnet worm was a cyber weapon against Iran. However, an error in the program code allowed the worm to escape Iran's nuclear facilities, infect the rest of the world, and inadvertently expose Operation Olympic Games.

Since the world's most sophisticated malware, Stuxnet, has been meticulously dissected and analyzed by cybersecurity experts worldwide, the concern is that the worm can be reverse-engineered and used against American targets. "Previous cyber attacks had effects limited to other computers," said Michael Hayden, former NSA director and former CIA chief. "This is the first attack of a major nature in which a cyber attack was used to effect physical destruction. Somebody crossed the Rubicon" [60].

Between 2009 and 2011, the number of cyber attacks on infrastructure reported to the U.S. Department of Homeland Security's (DHS) Industrial Control Systems Cyber Emergency Response Team (ICS-CERT) grew by a whopping 2,200 %, from nine to 198 [61]. Attacks against the energy sector represented 41 % of the total number of incidents whereas the water sector had the second highest number of attacks, representing 15 % of all incidents. Figure 9.3 shows that buffer

Fig. 9.3 Vulnerability types affecting Industrial Control Systems in FY 2012 (Courtesy of the Department of Homeland Security ICS-CERT)

Buffer Overflow	44
Input Validation	13
Resource Exhaustion	8
Authentication	8
Cross-site Scripting	8
Path Traversal	8
Resource Management	8
Access Control	7
Hard-coded Password	7
DLL Hijacking	6
SQL Injection	4
Credentials Management	3
Cryptographic Issues	3
Insufficient Entropy	3
Use After Free	3
Use of Hard-coded Credentials	2
Cross-Site Request Forgery	2
Privilege Management	2
Write-what-where Condition	2
Integer Overflow or Wraparound	2
Inadequate Encryption Strength	2
Missing Encryption of Sensitive Data	1
Code Injection	1
Forced Browsing	1
Miscellaneous	15
Total	**171**

overflow remains the most common vulnerability type among 171 unique vulnerabilities affecting Industrial Control Systems (ICS) products in fiscal year 2012.

In March 2012, ICS-CERT identified an active series of cyber intrusions targeting natural gas pipeline sector companies [62]. There are approximately 7,200 Internet facing control system devices across the United States (see Fig. 9.4). "For a while, it was a dirty little secret that just people in the industry knew [how easy it is to take over an industrial process once you're in the system]," said Dale Peterson, chief executive of control system security firm Digital Bond. "But we can't wait another 20 years, or whatever it is that people thought we could get out of these systems. We have to upgrade them now" [63].

In May 2014, ICS-CERT revealed that "a public utility was recently compromised when a sophisticated threat actor gained unauthorized access to its control system network. … In this instance, while unauthorized access was identified, ICS-CERT was able to work with the affected entity to put in place mitigation strategies and ensure the security of their control systems before there was any impact to operations" [64].

In August 2014, a team led by computer scientist J. Alex Halderman was able to hack into traffic lights in an unnamed Michigan town due to three major weaknesses: unencrypted wireless connections; the use of default usernames and passwords; and vulnerable debugging ports. "As long as the wireless card in the hacker's computer can communicate at the same frequency that the traffic lights use, it can break into the wireless network that powers the entire system," said Halderman [65].

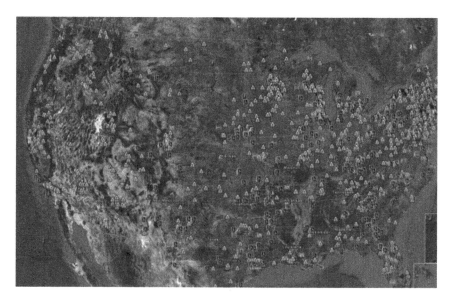

Fig. 9.4 7,200 Internet facing control system devices in the U.S. in FY 2012 (Courtesy of the Department of Homeland Security ICS-CERT)

In September and October 2014, four websites including the National Ice Center hosted by the National Oceanic and Atmospheric Administration (NOAA) were hacked, forcing the government to shut down some of its services and causing a disruption in satellite feeds [66]. The outages skewed the accuracy of National Weather Service's long-range forecasts.

In November 2014, U.S. Navy Admiral Michael Rogers who serves as director of National Security Agency (NSA), chief of Central Security Service (CSS), and commander of U.S. Cyber Command testified in Congress about the nation's power grid and other critical infrastructure: "[The detected malware] enables you to shut down very segmented, very tailored parts of our infrastructure that forestall the ability to provide that service to us as citizens. ... We see them attempting to steal information on how our systems are configured, the very schematics of most of our control systems, down to engineering level of detail so they can look at where are the vulnerabilities, how are they constructed, how could I get in and defeat them. ... It is only a matter of the when, not the if, that we are going to see something traumatic" [67].

It is more urgent now than ever before to secure the U.S. infrastructure against cyber attacks. U.S. Department of Homeland Security advocates a "defense-in-depth" approach to securing operations in vital sectors such as electricity, oil and gas, water, transportation, and chemical [68]. ICS-CERT recommends the following industry best practices:

- Minimize network exposure for all control system devices. Control system devices should not directly face the Internet.
- Locate control system networks and devices behind firewalls, and isolate them from the business network.
- If remote access is required, employ secure methods, such as Virtual Private Networks (VPNs), recognizing that VPN is only as secure as the connected devices.
- Remove, disable, or rename any default system accounts wherever possible.
- Implement account lockout policies to reduce the risk from brute forcing attempts.
- Implement policies requiring the use of strong passwords.
- Monitor the creation of administrator level accounts by third-party vendors.
- Adopt a regular patch life cycle to ensure that the most recent security updates are installed.

9.4 Government and Military Disruption

The Obama administration reported that there were almost 61,000 cyber attacks and security breaches across the entire federal government in 2013. "This is a global problem," said Tony Cole, vice president of the cyber security firm FireEye.

"We don't have a malware problem. We have an adversary problem. There are people being paid to try to get inside our systems 24/7" [69].

Cyber intruders were trying to tamper with Obama's 2008 presidential campaign, as President Barack Obama explained in his May 2009 remark on "Securing our Nation's Cyber Infrastructure" [1]:

> I know how it feels to have privacy violated because it has happened to me and the people around me. It's no secret that my presidential campaign harnessed the Internet and technology to transform our politics. What isn't widely known is that during the general election hackers managed to penetrate our computer systems. … Between August and October, hackers gained access to emails and a range of campaign files, from policy position papers to travel plans. And we worked closely with the CIA — with the FBI and the Secret Service and hired security consultants to restore the security of our systems.

In 2008, the U.S. Department of Defense (DoD) suffered a significant compromise of its classified military computer networks. It took the Pentagon nearly 14 months to clean out the computer worm agent.btz [70]. U.S. Deputy Secretary of Defense William Lynn wrote, "It began when an infected flash drive was inserted into a U.S. military laptop at a base in the Middle East. The flash drive's malicious computer code, placed there by a foreign intelligence agency, uploaded itself onto a network run by the U.S. Central Command. That code spread undetected on both classified and unclassified systems, establishing what amounted to a digital beachhead, from which data could be transferred to servers under foreign control. It was a network administrator's worst fear: a rogue program operating silently, poised to deliver operational plans into the hands of an unknown adversary" [71].

In September 2011, a keylogger virus penetrated the ground control station (GCS) at the Creech Air Force Base in Nevada and infected the cockpits of the U.S. Predator and Reaper drones [72]. The virus logged pilots' every keystroke as they remotely control the drones in their missions over Afghanistan and other countries. The virus was so resilient that it could not be wiped out by malware removal tools without completely reformatting the GCS' internal hard drives.

In December 2012, the Miami-Dade County Grand Jury reported that "someone created a computer program that automatically, systematically and rapidly submitted to the County's Department of Elections numerous bogus on-line requests for absentee ballots … from a grouping of several different Internet Protocol (IP) addresses … tracked to anonymizers overseas" [73].

Between 2011 and 2012, three companies that helped build Israel's "Iron Dome" missile defense system were broken into. Cyber intruders gained access to "huge amounts of sensitive information about Iron Dome, including schematics on Iron Dome's Arrow III missiles, drones and ballistic rockets" [74]. The Arrow III missiles are used in American defense systems as well, according to Joseph Drissel, CEO of Cyber Engineering Services Inc, who previously worked at the U.S. Department of Defense computer forensics lab.

In February 2013, personal information of more than 4,000 U.S. bank executives was stolen from the Federal Reserve System by exploiting a temporary vulnerability in a website vendor product [75]. Fortunately, the data breach incident did not affect critical operations of the U.S. central banking system.

In October 2013, Syrian Electronic Army (SEA) briefly took over Organizing for Action's (OFA) custom URL shortener, redirecting any links sent out from President Obama's Twitter and Facebook to the SEA YouTube videos [76].

In March 2014, cyber attackers accessed the federal employee database at the U.S. Office of Personnel Management (OPM) and retrieved information about employee hiring, wages, pensions, and security clearances [77]. The sensitive data can be used to target U.S. government workers who have top-secret access to classified materials.

In December 2014, cyber intruders broke into an unclassified computer network used by President Obama's top advisers [78]. The White House cybersecurity team temporarily shut down the network when they spotted the digital break-in.

John Gilligan, Chief Information Officer of the U.S. Air Force, stated that 80% of successful penetrations of federal government computers could be attributed to software bugs, trapdoors, and "Easter eggs" often found in commercial off-the-shelf (COTS) products [14].

In early 1980's, the DoD funded the development of a new programming language — Ada — that excels in mission-critical safety and security features [79]. The language was named after mathematician Ada Lovelace (1815-1852) who is often referred to as the world's first programmer. She worked with Charles Babbage, inventor of the first mechanical computer [80]. Ada was meant to be the de facto standard programming language for the U.S. government; and I was involved in drafting the Military Standard Common APSE (Ada Programming Support Environment) Interface Set (CAIS) in 1984 [81].

However, the DoD rescinded its Ada Mandate in 1997 and opted for COTS technology instead [82]. The government's shift from building expensive proprietary software to using cheaper off-the-shelf components shortens development time and hastens deployment; but at the same time the government needs to spend more in testing COTS software programs to look for vulnerabilities. In March 2012, Aspect Security reported that 26% of software libraries downloaded for use in applications had known vulnerabilities. "The problem is there are so many components that the software stack depends on," said Jeff Williams, cofounder and CTO of Aspect. "The Internet is a haystack full of needles".

Indeed, a seemingly harmless and routine Microsoft software update is the masquerade for the Flame virus that had evaded detection for years [83]. A part of Operation Olympic Games, Flame is a data-mining virus that "copies what you enter on your keyboard, monitors what you see on your computer screen, records sounds if the computer is connected to a microphone" [84]. Cybersecurity firm Kaspersky Lab announced the discovery of the Flame virus in May 2012 [85]:

> Kaspersky Lab's experts, in coordination with ITU [International Telecommunication Union], came across a new type of malware, now known as Flame. Preliminary findings indicate that this malware has been "in the wild" for more than two years — since March 2010. Due to its extreme complexity, plus the targeted nature of the attacks, no security software detected it.

The primary purpose of Flame appears to be cyber espionage, by stealing information from infected machines. Such information is then sent to a network of command-and-control servers located in many different parts of the world. The diverse nature of the stolen information, which can include documents, screenshots, audio recordings and interception of network traffic, makes it one of the most advanced and complete attack-toolkits ever discovered.

The Flame virus targeted at Iran successfully infected the computers of high-ranking Iranian officials [84]. A similar state-sponsored virus, Gauss, was discovered by Kaspersky Lab three months later in August 2012 [86]. Gauss was designed to collect computer information and steal credentials for specific banking, social network, email, and instant messaging accounts. The geographic distribution of Gauss was mostly limited to Lebanon, Israel, and the Palestinian territories, with relatively few infections in the United States and elsewhere in the world.

In 2014, three major security flaws were discovered in open-source code and COTS that affected millions of Internet browsers, computer servers, mobile devices, hardware appliances, and embedded systems:

1. Heartbleed — For more than 16 years since 1998, a programming error in a popular SSL/TLS library called OpenSSL exposes information from the private memory space such as passwords, TLS (Transport Layers Security) session keys, and long-term server private keys for decrypting past and future SLL (Secure Sockets Layer) traffic [87]. "On the scale of 1 to 10, this is an 11," said Chief Technology Officer Bruce Schneier of Co3 Systems [88]. Despite the severity of Heartbleed, it is estimated that 300,000 websites [89] and many more home automation systems and networking equipment are still not patched to fix the Heartbleed bug [90].
2. Shellshock — A flaw in the Unix Bash shell allows attackers to run malicious code on Linux servers, OS X machines, routers, and older Internet of Things devices, and more [91].
3. POODLE (Padding Oracle On Downgraded Legacy Encryption) — By forcing a connection downgrade from TLS to SSL, the vulnerability allows attackers to decrypt messages between a user's browser and an HTTPS (HTTP Secure) website [92].

To make matters worse, open-source code and COTS are often released in public beta with both known and unforeseen bugs. In February 2015, Microsoft finally fixed a critical 15-year-old security flaw known as "Jasbug" that enabled hackers to access files in a network on Windows Vista, 7, and 8 [93]. Apple announced in 2014 that the next version of its OS X operating system would release in public beta. In other words, end users have become beta testers. Adrian Covert of CNN rightfully called the current era "the end of polished and perfect software" [94].

On the bright side, the National Security Agency (NSA) has had success with open-source code notwithstanding the risk of foreign surveillance. For example, the NSA big-data system called Accumulo was built on top of open-source code because "you don't want to have to replicate everything yourself," explained Adam

Fuchs who was in charge of the project [95]. In 2011, the NSA released 200,000 lines of code to the Apache Foundation, benefiting anyone who works on big data analysis.

Last but not least, Ada is still alive today and being used in air traffic control, banking, and other high-risk industries. In 2013, the International Organization for Standardization (ISO) and the International Electrotechnical Commission (IEC) revised the Ada programming language for greater reliability and security, including a new feature known as "contract-based programming" or "design by contract" (DbC) that strengthens programs against bugs and malicious use [96]. DbC prescribes that software developers should define formal, precise, and verifiable interface specifications ("contracts") for software components, which extend the ordinary definition of abstract data types with preconditions, postconditions, and invariants.

9.5 Shodan and the Internet of Things

Developed by John Matherly in 2009, Shodan is the world's first search engine for Internet-connected devices including "the Internet of Things, Webcams, Buildings, the Web, Refrigerators, and Power Plants" according to its website [97]. Shodan collects its majority of data on web servers (port 80) as well as some data from FTP (21), SSH (22) and Telnet (23) services.

In May 2013, Security penetration tester Dan Tentler was able to use Shodan to locate a traffic light controller, red light camera, swimming pool acid pump, hydroelectric plant, hotel wine cooler, hospital heart rate monitor, home automation application, ski resort gondola ride, and car wash [98]. The traffic light controller was particularly undiscerning. It required no login credentials whatsoever; it simply displayed a warning message in uppercase letters: "DANGER! DO NOT USE WHILE CONTROLLER IS BEING USED FOR TRAFFIC CONTROL OR SERIOUS DAMAGE, INJURY OR DEATH MAY OCCUR!!!" Tentler had since notified the U.S. Department of Homeland Security about this serious lack of security in one of the American cities.

Network World magazine reported in 2013 that "Google map + TRENDnet security camera vulnerability = Hours of voyeurism and Peeping Tom paradise" [99]. In November 2014, a website called "Online IP netsurveillance cameras of the world" streamed real-time camera footages around the globe including 10,588 webcams and other video cameras in the United States, 6,228 in Korea, and 4,715 in China [100]. Its homepage stated, "Sometimes administrator (possible you too) forgets to change default password like 'admin:admin' or 'admin:12345' on security surveillance system, online camera or DVR. Such online cameras are available for all internet users. Here you can see thousands of such cameras located in cafes, shops, malls, industrial objects and bedrooms of all countries of the world. To browse cameras just select the country or camera type. This site has been designed in order to show the importance of the security settings. To remove your public camera from this site and make it private the only thing you need to do is to change your camera password" [101].

A 2014 study by HP (Hewlett-Packard) revealed that an average Internet of Things device has 25 security flaws [102]. The study involved 10 most popular IoT devices — TVs, webcams, home thermostats, remote power outlets, sprinkler controllers, hubs for controlling multiple devices, door locks, home alarms, scales, and garage door openers — that are paired with their smartphone applications. HP researchers found that 90 % of the devices collected personal information, 70 % transmitted that data on an unencrypted network, and 60 % had insecure user interfaces.

"Obviously, there are lots of benefits of connected devices in the home, but there can also be complications," said Marc Rotenberg, executive director of the Electronic Privacy Information Center. "When you worry about computer viruses, you can unplug your computer. When your house gets a virus, where do you go?" [103].

In January 2015, *MIT Technology Review* published an article titled "An Internet of Treacherous Things" to sound the alarm that millions of home and office routers need firmware upgrades in order to protect them from cyber attacks [104]. Unfortunately, many Internet-connected devices tend not to be upgraded for three reasons: manufacturers discontinue support; manufacturers go under or exit the business; and customers may be ill-equipped or lack of diligence in upgrading device firmware.

As the market research company IDC (International Data Corporation) has estimated that the universe of connected devices will total 30 billion by 2020 [105], the cybersecurity issue is rapidly growing out of hand unless every manufacturer provides an automatic firmware upgrade solution.

Although a few people may meticulously avoid connecting everyday devices to the Internet, they may not pay enough close attention to the cars that they are driving every day. Modern automobiles are pervasively computerized and are becoming a part of the Internet of Things [106].

In 2013, half of the cars broken into in London were hacked (not forced open) by spoofing the radio frequencies sent out by key fobs [107].

In August 2014, security researchers Charlie Miller and Chris Valasek reported the inherent security flaws in some new vehicles: "Both the 2014 Jeep Cherokee and the 2015 [Cadillac] Escalade ... The cars' apps, Bluetooth and telematics — which connects the car to a cellular network like OnStar — are on the same network as the engine controls, steering, brakes and tire pressure monitor system. ... In the 2014 Prius, the AM/FM/XM radio and Bluetooth are on the same network as the steering, brakes and tire pressure monitor. ... If critical functions like steering are on the same network as features that connect the car to the Internet, that can put the vehicle at risk" [108]. To protect cars from cyber attacks, Miller and Valasek built a $150 device to "listen to traffic in a car's network to understand how things are supposed to work. When an attack occurs, the device identifies traffic anomalies and blocks rogue activity" [109].

In January 2015, Elon Musk announced that over-the-air (OTA) software upgrades would soon come to Tesla Model S P85D sedans. "We have a software and firmware team that packages updates," said a Tesla spokesperson. "The packages are matched to a VIN [vehicle identification number] to ensure the car has the required hardware to receive all relevant updates" [110]. OTA software upgrades will not only affect the entertainment system but also power train and vehicle safety systems. It calls for strong cybersecurity protection.

9.6 Backdoors and Counterfeit Parts

In the 1983 film *WarGames*, a young hacker named David Lightman (played by Matthew Broderick) gained access to WOPR (War Operation Plan Response) by correctly guessing the backdoor password. Thinking that it was merely a computer game, Lightman launched "Global Thermonuclear War" that almost started World War III [111].

In real life, about two billion mobile phones, tablets, and other devices running either Android or iOS have a backdoor that allows carriers remotely install software updates without a user noticing. The poorly secured backdoor is an open invitation for intruders to silently plant malware and gain full access to a mobile device. "Even if you 'factory reset,' you still can't get rid of it," said Mathew Solnik, a research scientist at Accuvant [112].

In 2008, the National Security Agency developed a software implant code-named DROPOUTJEEP for the Apple iPhone. The iOS backdoor "utilizes modular mission applications to provide specific SIGINT (Signals Intelligence) functionality. This functionality includes the ability to remotely push/pull files from the device, SMS retrieval, contact list retrieval, voicemail, geolocation, hot mic, camera capture, cell tower location, etc. Command, control, and data exfiltration can occur over SMS messaging or a GPRS (General Packet Radio Service) data connection. All communications with the implant will be covert and encrypted" [113].

In July 2012 at the Black Hat Conference, Jonathan Brossard demonstrated a backdoor tool named Rakshasa that can be installed into the BIOS chip on a PC's motherboard to allow secret remote access over the Internet [114]. Not only is compromised hardware difficult to detect, changing hard drive or reinstalling the operating system will not get rid of the spyware. As a countermeasure, Intel has prototyped a security chip to check the BIOS for signs of tampering [115].

In October 2013, hacker /dev/ttyS0 (aka Craig) found a backdoor in a widely-deployed version of D-Link router firmware that allows the device to automatically reconfigure its settings (e.g. dynamic DNS) [116]. Thanks to this backdoor, however, arbitrary code execution is also possible [117].

In December 2013, Edward Snowden's NSA leaks revealed that the National Security Agency arranged a secret $10 million contract with security firm RSA to develop and promulgate a flawed random number generation algorithm "BSafe" in order to create a backdoor in encryption products [118].

In a January 2014 article on *Ars Technica*, security engineer Nick Sullivan describes the longstanding issue of backdoors [119]:

A backdoor is an intentional flaw in a cryptographic algorithm or implementation that allows an individual to bypass the security mechanisms the system was designed to enforce. Backdoors can be inserted by lazy programmers who want to bypass their own security systems for debugging reasons, or they can be created to intentionally weaken a system used by others.

Some recent examples:

- A flaw in a random number generator allowed people to hijack Hacker News accounts.
- A broken random number generator in Android allowed attackers to hijack thousands of dollars worth of bitcoins.
- The version of OpenSSL on the Debian distribution of the Linux operating system had a random number generator problem that could allow attackers to guess private keys created on these systems.

Backdoors can be introduced at the software, hardware, or even algorithm level. Algorithms backed by standards are not necessarily safe or free of backdoors. Some lessons to take away from this exercise are:

- Even secure cryptographic functions can be weakened if there isn't a good source of randomness.
- Randomness in deterministic systems like computers is very hard to do correctly.
- Adding unpredictable sources of entropy can help increase randomness and, in turn, secure algorithms from these types of attacks.

Apart from malware and cyber attacks, counterfeit electronic parts have infiltrated the U.S. government. Lieut. Gen. Patrick J. O'Reilly, director of the Missile Defense Agency (MDA), testified before the U.S. Senate Armed Services Committee in November 2011 [120]:

MDA has encountered incidents of counterfeit parts dating back to 2006. Total counterfeit parts found to date number about 1,300. All of them were procured from Unauthorized Distributors. We estimate the total cost to MDA for the seven instances is about $4 million. Our largest case cost the Agency $3 million to remove counterfeit parts discovered in the mission computer of our production THAAD [Theater High-Altitude Area Defense] interceptor.

Because counterfeiting continually evolves in sophistication, it is possible that electronic parts may have embedded functionality created by an enemy seeking to disable a system or obtain critical information. Detecting hidden functionality would be a difficult undertaking.

The predominant threat of counterfeit parts in missile defense systems is reduced reliability of a major DoD weapon system. We do not want to be in a position where the reliability of a $12 million THAAD interceptor is destroyed by a $2 part.

O'Reilly said that the MDA had no indication of any mission-critical hardware in the fielded BMDS (Ballistic Missile Defense System) containing counterfeit parts, but he acknowledged that detecting hidden functionality in counterfeit electronic parts poses a significant challenge to the U.S. military. The counterfeit parts can provide backdoors for computer viruses and worms to enter the system during a cyber attack. Backdoors allow cybercriminals to bypass normal authentication processes, firewalls, and other security controls.

Although the U.S. General Services Administration (GSA) has a database of about 90,000 risky suppliers that government agencies are required to check against when ordering electronic parts [121], more than 1,800 instances of counterfeit electronic parts have been found in the defense supply chain to date, and there is no telling of how many more that have not yet been discovered [122].

In 2008, FBI's Operation Cisco Raider resulted in more than 400 seizures of counterfeit Cisco network hardware and labels with an estimated retail value of over $76 million [123]. A number of government agencies bought the routers from an authorized Cisco vendor, but that legitimate vendor purchased the routers from a high-risk Chinese supplier [121]. The Cisco routers manufactured in China could have provided hackers a backdoor into secured U.S. government networks. U.S. Senator John McCain warned, "Counterfeit parts pose an increasing risk to our national security, to the reliability of our weapons systems and to the safety of our men and women in uniform" [122].

Ironically however, according to Edward Snowden's NSA leaks, the agency's Tailored Access Operations (TAO) unit and other NSA employees have been intercepting servers, routers, and other network gear being shipped to organizations targeted for surveillance, and installing covert implant firmware onto them before they are delivered to their final destinations [124]. Cisco CEO John Chambers complained in his letter to President Barack Obama in May 2014, "We simply cannot operate this way; our customers trust us to be able to deliver to their doorsteps products that meet the highest standards of integrity and security. We understand the real and significant threats that exist in this world, but we must also respect the industry's relationship of trust with our customers" [125].

9.7 Proliferation of Cyber Weapons and Reverse Engineering

Microsoft's principal security architect Roger Grimes wrote in a January 2011 *InfoWorld* article: "With the announcement of the purported success of Stuxnet, the next-generation arms race is on. Ironically, while Stuxnet has possibly slowed down the international proliferation of nuclear arms, it's also officially launched the next big weapons battle" [126].

"It take very little effort to upgrade an espionage tool to a cyber bomb," said Eugene Kaspersky, CEO and co-founder of Kaspersky Lab. He commented on the discovery of the Flame virus in May 2012: "The risk of cyber warfare has been one of the most serious topics in the field of information security for several years now. Stuxnet and Duqu belonged to a single chain of attacks, which raised cyberwar-related concerns worldwide. The Flame malware looks to be another phase in this war, and it's important to understand that such cyber weapons can easily be used against any country. Unlike with conventional warfare, the more developed countries are actually the most vulnerable in this case" [85].

On May 2, 2011, U.S. Navy SEALs destroyed an inoperable top-secret stealth Blackhawk Helicopter with explosives at the end of the bin Laden raid in Abbottabad, Pakistan. However, the tail section of the craft survived the explosion and the wreckage was hauled away by the Pakistani military. Former White House counterterrorism advisor Richard Clarke commented, "There are probably people in the Pentagon tonight who are very concerned that pieces of the helicopter may be, even now, on their way to China, because we know that China is trying to make stealth aircraft" [127]. It is very likely that other countries have been busy reverse-engineering the stealth technology from the Blackhawk Helicopter wreckage.

One should never downplay the potential of reverse engineering. In 1987 at AT&T Bell Laboratories, I worked with a team of computer scientists (Pamela Culbreth, John Goettelmann, Ron Hiller, Irvan Krantzler, Chris Macey, and Mark Tuomenoksa) in reverse engineering the Mac OS System 4.1 running on the Apple Macintosh SE. After months of development, we were able to run the Mac OS, HyperCard, and almost all other Macintosh applications on a Unix-based RISC (Reduced Instruction Set Computer) machine. Not only that, but the Mac software ran faster on Unix emulating the Mac OS than on the native Macintosh.

The R&D work at Bell Laboratories was summarized in the abstract of the United States Patent #5724590: "Application programs compiled for a first, 'source', computer are translated, from their object form, for execution on a second, 'target', computer. The translated application programs are linked or otherwise bound with a translation of the source computer system software. The translated system software operates on the image of the source computer address space in the target computer exactly as it did in the source computer. The semantics of the source computer system software are thus preserved identically. In addition, a virtual hardware environment is provided in the target computer to manage events and to deal with differences in the address space layouts between the source and target computers" [128].

With reverse engineering and clever implementations, one can conceivably modify a piece of malware to perform malicious actions against a new target or to gather intelligence from the original cyber attacker — all without any knowledge of the malware's source code.

Unless a computer virus or worm is programmed to auto self-destruct when its cover is blown without intervention from the command-and-control (CnC) servers, it is dangerous to unleash highly sophisticated and effective cyber weapons like Stuxnet, Flame, and Gauss. For that reason, some malwares are configured to self-destruct after a certain number of days. The Duqu keylogger worm, for instance, was found to automatically remove itself from an infected system after 36 days of collecting keystrokes and sending them to the perpetrators [129].

9.8 Cyber Espionage and Escalation of Cyber Warfare

Duqu was discovered in 2011 by researchers at the Laboratory of Cryptography and System Security in Budapest, Hungary [130]. The malware bears a strong resemblance to Stuxnet, except that Duqu is used for espionage instead of destruction. It transmits stolen data back to its control center by first encrypting the information and then embedding it in an innocent-looking JPEG image — a practice known as steganography. U.S. officials said that various terrorist groups had started using steganography as early as 1996 to plan their attacks [131].

In retaliation for the Stuxnet, Flame, and Duqu attacks against Iran's government and nuclear facilities, Iran launched a major cyber attack on American banks in September 2012 using botnets that involved thousands of high-powered application servers. Bank of America, JPMorgan Chase, PNC Bank, U.S. Bank, and Wells Fargo Bank were among the financial institutions that suffered slowdowns and sporadic outages. "The volume of traffic sent to these sites is frankly unprecedented," said cybersecurity expert Dmitri Alperovitch who conducted the investigation. "It's 10 to 20 times the volume that we normally see, and twice the previous record for a denial of service attack" [132]. Iran was also believed to be behind the cyber attacks on Saudi Oil company Aramco and on Exxon Mobil's subsidiary RasGas in Qatar [133].

In Syria, supporters of dictator Bashar al-Assad have impersonated opposition activists in order to pass out Remote Access Trojan (RAT) viruses on Skype calls and via emails [134]. Discovered in February 2012, the cyber espionage malware spies on opposition activists and sends information back to the government-controlled Syrian Telecommunications Establishment (STE).

In November 2012, Israeli Defense Force (IDF) live tweeted its military campaign in the Gaza strip during the weeklong Operation Pillar of Defense [135]. In response, Hamas tweeted its own account of the war along with photographs of casualties [136]. Both sides hoped to use social media to win world sympathy and shift political opinion to their sides [137].

In protest of the Israeli military action in Gaza, hacktivist group Anonymous launched a cyber attack on Israel dubbed "OpIsrael." Anonymous deleted the online databases of the Israel Ministry of Foreign Affairs and Bank of Jerusalem, took down over 663 Israel websites with DDoS attacks, and posted on the Internet over 3,000 emails, phone numbers, and addresses of "Israeli supporters" [138–140]. Anonymous announced, "November 2012 will be a month to remember for the Israeli defense forces and internet security forces. Israeli Gov. this is/will turn into a cyberwar" [141].

Israel's Chief Information Officer Carmela Avner admitted, "The war is being fought on three fronts. The first is physical, the second is the world of social networks, and the third is cyber attacks" [141]. Indeed, after the ceasefire between Israel and Hamas on November 21, 2012, supporters of both sides intensified their DDoS attacks against each other [142].

In January 2013, the battle for public opinion extended to Hollywood: Iran announced its plan to fund a high-budget movie — *The General Staff* — as a counter-story to the 1979 Iran hostage crisis depicted in the Golden Globe and Academy Award winning film *Argo* [143]. In March 2013, Iran planned to sue Hollywood filmmakers, citing that "the Iranophobic American movie attempts to describe Iranians as overemotional, irrational, insane and diabolical while at the same, the CIA agents are represented as heroically patriotic" [144].

In March 2013, a massive cyber attack against South Korean banks and broadcasters damaged 32,000 computers at their media and financial companies [145]. North Korea, the suspected culprit, has accused South Korea and the United States for daily "intensive and persistent virus attacks" on the country's Internet servers [146].

In August 2013, the Syrian Electronic Army (SEA) — a group of computer hackers who support the government of Syrian President Bashar al-Assad — caused the outage of *The New York Times* website for several hours [147]. In September 2014, SEA attacked CNBC, the Boston Globe, the Canadian Broadcasting Corp, the Independent, and the Telegraph by redirecting traffic to SEA's own website with the message "Happy thanksgiving, hope you didn't miss us! The press: Please don't pretend #ISIS are civilians. #SEA" [148].

In April 2014, Cuba accused the Obama administration for waging a cyber war by secretly financing a social network in Cuba to stir political unrest and undermine the country's communist government [149]. Dubbed "ZunZuneo," a slang for a Cuban hummingbird's tweet, the mobile app drew in some 40,000 subscribers in Cuba. The U.S. Agency for International Development (USAID) was the mastermind behind the creation of the "Cuban Twitter," hoping that the text-messaging tool would be used to organize political demonstrations.

In December 2014, University of Toronto's Citizen Lab confirmed that the ISIS opposition group "Raqqa is being Slaughtered Silently" (RSS) was hit by a bare bone virus that flew under the radar of most anti-virus scanners and that registered only a 10% detection rate [150]. The low-profile malware helped the militants to determine the computer users' physical locations by their IP addresses. In the same year, the "SandWorm Team" (based on the science fiction classic novel *Dune*) in Russia sent phishing emails to Ukrainian government officials. Embedded in the emails was a malware-laced PowerPoint attachment [151].

A more widespread and sophisticated cyber espionage malware dubbed Regin has been lurking in computers around the world since 2008. "Its capabilities and the level of resources behind Regin indicate that it is one of the main cyber espionage tools used by a nation state," said Symantec analyst Vikram Thakur. "They were trying to gain intelligence, not intellectual property. ... When a target was selected it searched airline computers to find out where the target was traveling. It scoured hotel computers to find his room number. And it tapped telecommunication computers to see who he was talking to" [152].

Security technologist Bruce Schneier argued that antivirus companies should be more open about their government malware discoveries: "Antivirus companies had tracked the sophisticated — and likely U.S.-backed — Regin malware for years. But they kept what they learned to themselves. ... Right now, antivirus companies are probably sitting on incomplete stories about a dozen more varieties of government-grade malware. But they shouldn't. We want, and need, our antivirus companies to tell us everything they can about these threats as soon as they know them, and not wait until the release of a political story makes it impossible for them to remain silent" [153].

James A. Lewis, Senior Fellow at the Center for Strategic and International Studies (CSIS), estimated that at least 12 of the world's 15 largest militaries have been building cyber warfare programs [154].

Since the 9/11 attacks, the NSA oversees "Tailored Access Operations" and the U.S. Navy operates "Computer Network Exploitation" [155]. In selecting a new service motto for the U.S. Air Force in October 2010, Education and Training Command Commander Gen. Stephen Lorenz said, "Airmen consistently told us they see themselves, and they see the heritage of the Air Force, as those entrusted by the nation to defend the modern, complex security domains — first air, then space and now cyberspace" [156].

There is no doubt that cyber war is intensifying and that nations have been stepping up their preparation for state-sponsored cyber attacks. In fact, researchers have found that malware traffic spikes can forewarn of international conflicts [157].

9.9 Cyber Cold War

"We are at war," said Russian sleeper agent Elizabeth Jennings played by Keri Russell in the TV series *The Americans* — a period drama created and produced by former CIA officer Joe Weisberg about the "Illegals Program" during the Cold War between the United States and the Soviet Union.

In real life, Operation Ghost Stories made headlines in June 2010 when 11 deep-cover Russian spies including New York real estate agent Anna Chapman were arrested by the FBI for conspiracy, money laundering, and failure to regis-ter as agents of a foreign government [158]. On the other hand, the U.S. govern-ment is much more lenient with ally spies. Hollywood producer Arnon Milchan (*12 Years a Slave, Fight Club, Pretty Woman, L.A. Confidential, JFK, Mr. and Mrs. Smith*) openly admitted, without fear of U.S. reprimand, that he had been an Israeli spy for 20 years, helping Israel develop its nuclear weapons program [159].

The Cold War officially ended in 1991, and it has moved into the cyberspace. In 2012, security company Crowdstrike uncovered the Russian-backed "Energetic Bear" operation aimed at hacking into U.S. and European energy sector, defense contractors, health care providers, manufacturers, construction companies, and uni-versity research centers. "The Russians are engaged in aggressive economic and political espionage," said Crowdstrike co-founder Dmitri Alperovitch [160].

In addition to Russia, a "cyber Cold War" has been raging under the public radar for over a decade between China and the United States, said former NSA Chief Technology Officer Prescott Winter. "It's no secret that government agencies are under attack from China. It's a significant problem, and the government has been aware of it for the past 10 to 15 years" [161].

On April 8, 2010, Chinese hackers "hijacked" the Internet for 18 minutes by redirecting 15 % of the world's online traffic to route through Chinese servers. The U.S.-China Economic and Security Review Commission released a report in November 2010: "This incident affected traffic to and from U.S. government (".gov") and military (".mil") sites, including those for the Senate, the army, the navy, the marine corps, the air force, the office of secretary of Defense, the National Aeronautics and Space Administration, the Department of Commerce, the National Oceanic and Atmospheric Administration, and many others. Certain commercial websites were also affected, such as those for Dell, Yahoo!, Microsoft, and IBM" [162].

In May 2011, People's Liberation Army (PLA) in China formally announced the deployment of a cyber security squad known as "online blue army" to protect the country from cyber attacks [163]. The Chinese blue army has built the "Great Firewall" (named after the Great Wall of China) to fence off foreign influence by censoring websites and online searches.

In January 2013, *The New York Times* and *The Wall Street Journal* accused Chinese hackers of breaking into their computer systems in order to monitor the news media coverage of China and to look for the names of informants [164]. The Chinese Foreign Ministry vehemently denied the accusations [165]. Unit 61398 headquartered on the edges of Shanghai has become the symbol of Chinese cyber-power, according to *The New York Times* [166].

In June 2013, Kaspersky Lab identified a China-based cyber espionage group called "NetTraveler" that was stealing data on space exploration, nanotechnology, energy production, nuclear power, lasers, and radio wave weapons from as many as 40 countries worldwide [167].

In January 2014, *The New York Times* revealed that the National Security Agency has implanted transceivers via USB cables into Russian military networks and nearly 100,000 computers worldwide in order to conduct surveillance over the airwave and to facilitate future cyber attacks [168]. Data can be transmitted over a covert radio frequency between a target computer and a NSA field station up to 8 miles away.

In May 2014, Attorney General Eric Holder announced a grand jury indictment against five People's Liberation Army officers for "unauthorized access to victim computers to steal information from these entities that would be useful to the victims' competitors in China" [169]. The victims included U.S. Steel Corp., Westinghouse, Alcoa, Allegheny Technologies, the United Steel Workers Union, and SolarWorld.

9.10 Psychological Cyber Warfare

On January 12, 2015, the Twitter and YouTube accounts for U.S. Central Command (CENTCOM) were hacked by ISIS sympathizers [170]. The first published tweet read: "AMERICAN SOLDIERS, WE ARE COMING, WATCH YOUR BACK. ISIS." The warning was followed by "ISIS is already here, we are in your PCs, in each military base" along with images, links, and Pentagon documents revealing contact information for some members of the military. Twitter suspended the account after 40 minutes of cyber havoc.

U.S. Central Command issued a statement: "These sites reside on commercial, non-Defense Department servers and both sites have been temporarily taken offline while we look into the incident further. CENTCOM's operational military networks were not compromised and there was no operational impact to U.S. Central Command. CENTCOM will restore service to its Twitter and YouTube accounts as quickly as possible. We are viewing this purely as a case of cybervandalism."

However, this was more than cyber vandalism. CENTCOM continued in its statement, "Additionally, we are notifying appropriate DoD and law enforcement authorities about the potential release of personally identifiable information and will take appropriate steps to ensure any individuals potentially affected are notified as quickly as possible."

It was psychological cyber warfare in addition to cyber vandalism. Two days after the CENTCOM Twitter hack, *CNN* reported that "one military wife recalls staying up all night and deleting every Facebook picture of her children, every post that mentioned them or where they went to school. She Googled herself, trying to figure out how easy it would be to find where the family lived. In the morning, she went to her car and scraped the military decal off the front window." She told *CNN*, "It's hard because I am so proud of what my husband does, but lately so many spouses that I know are actually scared that they could be targets of ISIS or someone who sympathizes with ISIS" [171].

It becomes a greater concern when psychological warfare instigates physical violence. In November 2009, U.S. Army psychiatrist Maj. Nidal Malik Hasan opened hire at Fort Hood, Texas, killing 12 people and wounding 31 others [172]. In May 2013, 25-year-old British soldier Lee Rigby was slain by two Islamic extremists, Michael Adebolajo and Michael Adebowale, in Woolwich, southeast London [173]. CNN terrorism analyst Paul Cruickshank remarked that "there's a track record of Islamist extremists inspired by al Qaeda ideology targeting soldiers in the West" [174].

9.11 Cyber Terrorism and Digital Pearl Harbor

Barry Collin coined the term "cyber terrorism" around 1987 to describe terrorism at the convergence of the physical and virtual worlds [175]. In 2002, James A. Lewis at CSIS defined cyber terrorism as "the use of computer network tools to shut down critical national infrastructures (such as energy, transportation, government operations) or to coerce or intimidate a government or civilian population" [176]. In 2011, retired FBI agent William L. Tafoya refined the definition of cyber terrorism to the criminal acts that target civilians [177]:

> Cyber terrorism is a component of information warfare, but information warfare is not cyber terrorism. ... The skills, tools, and techniques are the same, but information warfare is conducted between military combatants; cyber terrorism targets civilians. Cyber terrorists indiscriminately will attack the nation's critical infrastructure and civilians — the innocent.

> Thus, the context and targets, not the technological tools or frequency of attacks, are the more appropriate delimiters that distinguish cyber terror from information warfare. ... Attacking the largely civilian critical infrastructure is not warfare, but terrorism — cyber terror.

Tafoya pointed out that SCADA systems were vulnerable to attacks not only from computer viruses and worms but also from electromagnetic pulse (EMP) bombs and high-energy radio frequency (HERF) weapons. EMP and HERF devices use electromagnetic radiation to deliver heat, mechanical, or electrical energy to an electronic device such as a computer, a cell phone, or even an artificial cardiac pacemaker and defibrillator designed to save lives.

In April 2008, the EMP Commission issued a comprehensive 208-page report on the effects of an EMP attack on U.S. electric power, telecommunications, banking and finance, petroleum and natural gas, transportation, food infrastructure, water works, emergency services, and satellites [178]. The commission cited an incident in November 1999 when San Diego County Water Authority and San Diego Gas and Electric experienced severe electromagnetic interference to their SCADA wireless networks due to a radar operating on a ship 25 miles off the coast of San Diego. Their recommendations are three-pronged: prevention, protection, and recovery.

In October 2012, U.S. Secretary of Defense Leon E. Panetta warned that the United States was facing the possibility of a digital Pearl Harbor: "An aggressor nation or extremist group could use these kinds of cyber tools to gain control of critical switches. They could derail passenger trains, or even more dangerous, derail passenger trains loaded with lethal chemicals. They could contaminate the water supply in major cities, or shut down the power grid across large parts of the country" [179]. Cybersecurity expert Chiranjeev Bordoloi concurred, "These types of attacks could grow more sophisticated, and the slippery slope could lead to the loss of human life" [180].

The FBI has a division dedicated to combating cyber crime and cyber terrorism. Keith Lourdeau, deputy assistant director of the FBI Cyber Division, testified before the Senate Judiciary Subcommittee on Terrorism, Technology, and Homeland Security in February 2004 [181]:

> The number of individuals and groups with the ability to use computers for illegal, harmful, and possibly devastating purposes is on the rise. We are particularly concerned about terrorists and state actors wishing to exploit vulnerabilities in U.S. systems and networks. ... Counterterrorism efforts must incorporate elements from — and contribute toward — counter-intelligence, cyber, and criminal programs.

In December 2012, Dr. Mathew Burrows of the National Intelligence Council (NIC) published a 166-page report, "Global Trends 2030: Alternative Worlds," in which he expressed his outlook on the future of terrorism: "Terrorists for the moment appear focused on causing mass casualties, but this could change as they understand the scope of the disruptions that can be caused by cyber warfare" [182].

In July 2013, security researcher Barnaby Jack was inspired by an episode of the television series *Homeland*, in which a terrorist remotely hacked the pacemaker of the United States vice president. Jack passed away in San Francisco only days before his scheduled appearance at Black Hat conference to show how an ordinary pacemaker could be compromised and deliver a high-voltage shock from 50 feet away [183]. "Over the past year, we've become increasingly aware of cyber security vulnerabilities in incidents that have been reported to us," admitted William Maisel, deputy director for science at the FDA's Center for Devices and Radiological Health. "Hundreds of medical devices have been affected, involving dozens of manufacturers" [184].

In April 2014, major insurer AIG (American International Group) with 88 million customers in 130 countries announced that it would expand its cyber insurance offering to cover property damage and bodily injury [185].

In July 2014, DARPA demonstrated Extreme Accuracy Tasked Ordnance (EXACTO) that employs specially designed ammunition and real-time optical guidance system to assist military snipers to acquire moving targets in unfavorable conditions, such as high winds and dusty terrain commonly found in Afghanistan [186]. Deadly weapons as big as a missile-equipped drone and as small as a .50-caliber bullet are now under software control, which could be disastrous if they were hacked. In the television show *24: Live Another Day* starring Kiefer Sutherland as Jack Bauer, terrorists assisted by cybercriminals gain control of six U.S. military drones and use them to attack London [187].

9.12 Sony-pocalypse: from Cyber Attacks to Cyber Terrorism

In December 1941, Japan's surprise attack on Pearl Harbor caught the United States off-guard and sank four of the eight U.S. Navy battleships. 73 years later in November 2014, Sony was unprepared for the massive cyber attack that stole 100 terabytes of data [42], destroyed 75% of corporate computer servers, and crippled the company's data centers [43]. Among the stolen data were five feature films, executive emails, business contracts, company budgets, employee personal data, salary information, medical records, and celebrity secrets [188]. *The New York Times* reported that "administrators hauled out old machines that allowed them to cut physical payroll checks in lieu of electronic direct deposit" [43].

Enterprise security analyst Adrian Sanabria coined the term "Sony-pocalypse" to describe the most destructive cyber attack reported to date against a company on U.S. soil [189]. The following is a condensed timeline based on information from *Deadline Hollywood* [190], *The New York Times* [43], *The Verge* [191], and other news media outlets:

Phase 1: Cyber Attacks
November 24, 2014: Sony Pictures Entertainment's corporate computers in New York, Los Angeles, and around the world were infiltrated by cybercriminals who called themselves the Guardians of Peace (GOP). An image of a skull along with a warning message appeared on some 7,000 employees' computer screens (see Fig. 9.5). As Sony shut down all its computer systems, employees worked without

Fig. 9.5 Hacked By #GOP

computers, email, and voice mail. Executives set up a phone tree so that one person could relay a message to the next person on the tree.

November 25, 2014: A person identifying herself as one of the cyber attackers wrote in an email to *The Verge*, "We Want Equality. Sony doesn't. It's an upward battle."

Phase 2: Cyber Leaks
November 27, 2014: The GOP released Brad Pitt's *Fury* and four yet-to-be-released films — *Annie, Mr. Turner, Still Alice,* and *To Write Love On Her Arms* — for illegal downloads on the Internet.

December 1, 2014: The salaries of more than 6,000 current and former Sony employees and 17 top executives were leaked to the media.

December 2, 2014: Sony's corporate computer systems were back online. Sony acknowledged a massive data breach and informed its employees that "we unfortunately have to ask you to assume that information about you in the possession of the company might be in their possession."

December 3, 2014: The GOP revealed stolen social security numbers, salaries, home addresses, usernames and passwords of 47,000 celebrities, freelancers, and current and former Sony employees. Some of the celebrities included Angelina Jolie, Jonah Hill, Sylvester Stallone, Judd Apatow, and Rebel Wilson.

Phase 3: Cyber Extortion
December 5, 2014: The GOP threatened Sony employees and their families via an email written in broken English: "Removing Sony Pictures on earth is a very tiny work for our group which is a worldwide organization. … Please sign your name

to object the false of the company at the email address below if you don't want to suffer damage. If you don't, not only you but your family will be in danger."

December 8, 2014: The GOP condemned the controversial satire comedy *The Interview* about a hapless TV host recruited by the CIA to assassinate North Korean dictator Kim Jong-un. The group demanded Sony to "stop immediately showing the movie of terrorism which can break regional peace and cause the War!"

Phase 4: Cyber Warfare

December 10, 2014: Sony fought back by launching denial of service (DoS) attacks on websites that were leaking the stolen data, as well as spreading bogus files to the torrent network to fool downloaders.

December 11, 2014: Sony Pictures Entertainment chairman Amy Pascal and producer Scott Rudin (*The Social Networks* and *The Addams Family*) publicly apologized for their racially insensitive cracks about what black movies President Barack Obama might like in a series of email exchanges disseminated to media outlets by the GOP.

December 12, 2014: The GOP leaked highly confidential medical records of 34 Sony employees, their spouses, and their children who were undergoing medical treatments.

December 14, 2014: The GOP released an early version of the script for *Spectre*, the 2015 James Bond film. Sony sent legal warnings to about 40 news organizations to discard any stolen data given to them by the GOP.

December 15, 2014: Screenwriter Aaron Sorkin accused American journalists of abetting criminals in disseminating stolen information. Former Sony employees filed a class-action lawsuit against the studio for not taking adequate safeguards to protect their personal data. Sony Pictures CEO Michael Lynton asked employees not to read the anticipated next waves of leaked emails, saying that "I'm concerned, very concerned, that if people continue to read these emails, relationships will be damaged and hurt here at the studio."

Phase 5: Cyber Terrorism

December 16, 2014: Cyber warfare escalated to cyber terrorism. The GOP threatened 9/11-style attacks on movie theaters that planned to show *The Interview* on Christmas Day. The GOP wrote in an email to reporters, "The world will be full of fear. Remember the 11th of September 2001. We recommend you to keep yourself distant from the places at that time. (If your house is nearby, you'd better leave.)" The U.S. Department of Homeland Security issued a statement saying that "at this time there is no credible intelligence to indicate an active plot against movie theaters within the United States." Three largest theater chains — AMC, Regal, and Cinemark — along with the majority of 3,000 cinemas called off the screenings of *The Interview* in fear of physical violence and collateral damages to other movies shown in the same theaters.

December 17, 2014: Sony canceled the Christmas Day theatrical release for *The Interview*. Press screenings for the movie were also canceled. Sony pulled all TV advertising for the film. Reddit shut down the subreddit forum SonyGOP that housed links to stolen Sony Pictures documents.

December 18, 2014: Several independent theaters planned to show *Team America: World Police*, a 2004 film from the creators of *South Park*. However, Paramount Pictures refused to offer the film for redistribution in theaters. The GOP unveiled the stolen documents on "Project Goliath" — a secret campaign led by the Motion Picture Association of America (MPAA) to revive the failed SOPA legislation. Oscar-winning actor George Clooney and his agent Bryan Lourd from CAA circulated a letter among top Hollywood executives in support of Sony Pictures Entertainment, but not a single executive signed the letter.

December 19, 2014: President Barack Obama strongly disapproved of Sony's decision to cancel the theatrical release for *The Interview*. The FBI pinned blame on North Korea for the Sony cyber attack.

December 20, 2014: North Korea denied any involvement.

December 21, 2014: CNN's *State of the Union* aired an interview with President Barack Obama. He told news anchor Candy Crowley, "We've got very clear criteria as to what it means for a state to sponsor terrorism. And we don't make those judgments just based on the news of the day. We look systematically at what's been done and based on those facts, we'll make those determinations in the future. ... It was an act of cyber vandalism that was very costly. We take it very seriously and we will respond proportionally. If we set a precedent in which a dictator in another country can disrupt through cyber, a company's distribution chain or its products, and as a consequence we start censoring ourselves, that's a problem. And it's a problem not just for the entertainment industry; it's a problem for the news industry. ... So the key here is not to suggest that Sony was a bad actor. It's making a broader point that all of us have to adapt to the possibility of cyber attacks, we have to do a lot more to guard against them. ... It is a matter of setting a tone, being clear that we are not going to be intimidated by some cyber hackers."

December 23, 2014: With the support of about 300 independently-owned theaters across the country, Sony reversed its decision to cancel the theatrical release.

Phase 6: Aftermath

December 25, 2014: *The Interview* premiered in 331 small theaters across the country and draw many sell-out audiences. The film was simultaneously released online via streaming on YouTube, Google Play, Microsoft Xbox, and the dedicated website SeeTheInterview.com.

December 28, 2014: For the opening weekend, the film barely earned $3 million in ticket sales from 331 screens but it pulled in over $15 million from online purchases and rentals, making it the Sony Pictures #1 online film of all time.

January 2, 2015: President Barack Obama slapped sanctions on 10 North Korean officials and three state organizations responsible for "cyber attacks, weapons proliferation, and other illicit activities."

January 6, 2015: *The Interview* made $31 million in online and video-on-demand revenues and $5 million in limited theatrical release. Although the financial damage to Sony was estimated to be at least $100 million, Sony CEO Kazuo Hirari downplayed the financial impact of the cyber attacks on the company at a news conference during the 2015 Consumer Electronics Show (CES) in Las Vegas, Nevada.

February 4, 2015: Although the financial blow to Sony was estimated to be around $100 million, Sony gave an official figure for damages from the massive hack: $15 million.

February 5, 2015: Amy Pascal resigned from her co-chair job at Sony Pictures Entertainment.

Conclusions:

The Sony-pocalypse serves as a wakeup call for everyone to take cybersecurity seriously. "Media, pharmaceutical, energy companies. Everybody's got enemies," said Craig Carpenter, president of Resolution1 Security. "This Sony hack shows you can be brought to your knees if you're not capable of shutting something like this down before it gets out of hand. … Computer alarms go off maybe 5,000 times a day at a large company like Sony. … The workload outpaces the number of workers assigned to keep companies safe" [192].

President Obama considered the Sony saga "cyber vandalism" [193] whereas one senior Sony employee called it "a terrorist attack" [194]. I believe that "cyber terrorism" is a more appropriate term for the Sony saga. Cyber attacks and terrorist threats are a lethal combination that can only be resolved by aligning conscientious counterterrorism policies with cybersecurity technologies.

9.13 Good, Bad — Internal/External — People Put an End to Business as Usual (A Commentary by Andy Marken)

The attraction of the Internet of Things (IOT), a cashless society and cloud-based everything is a great dream. But we are beginning to see that the dream can quickly turn into one helluva' nightmare. With 2014 assaults directed at companies and institutions — private and public, it has gotten so attacks on infrastructure, financial institutions and energy organizations are daily occurrences.

It's interesting that the theft of millions of personal records and the loss of hundreds of millions of dollars has done nothing to slow companies' rush to put everything in the cloud and online. And the outing of people's emails and Facebook/LinkedIn posts haven't slowed individuals from looking like total idiots when their "personal/private" communications and documents have been publicly aired.

If nothing good comes from the Sony Entertainment fiasco, it reinforces the global recognition that:

- There is no such thing as privacy on the Internet.
- Complete, online security doesn't make a credible movie script.

- Cloud computing/storage security went out the window the minute you went online.
- There is no such thing as an offhand, off-the-cuff or private comment. Everything is public.

Security and law enforcement experts estimate that as many as 80 percent of large companies have suffered a security breach in the last year. The objective was simple — steal intellectual property (IP), money or content (PIN numbers, email addresses, personal data) that had value to someone else or simply test/prove their dark side software skills. Cybercriminals are very rarely caught.

Sony's hack attack - compared to most last year - was relatively small in terms of fiscal damage (O.K., $100M in lost revenues isn't small but…). Target, Home Depot, JPMorgan, Adobe, eBay, the Korea Credit Bureau and hundreds of others were larger because millions of credit cards and lots of corporate/personal data were stolen (see Fig. 9.6). Sony was different because it was all of that plus the humiliation/embarrassment of individuals by releasing emails — stored ideas/thoughts to the world. The volume of cyber attacks/warfare — espionage and IP — will increase this year and will be more visible … and there will be a rise in "collateral damage."

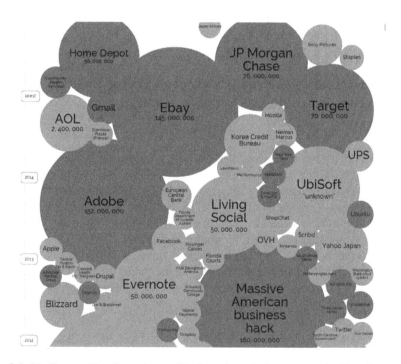

Fig. 9.6 Big Race — The size and cost of hacks and attacks have grown significantly over the past year and show little sign of shrinking. While Sony was far from being large in comparison to others, it exhibited a unique trend of being both aggressive and personal by exposing personal and corporate data (courtesy of informationisbeautiful.net)

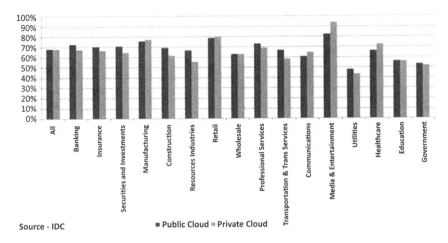

Fig. 9.7 Cloud Stuffing — Organizations in almost every sector hear the siren call of cloud computing and cloud storage and are implementing or planning their move. Low cost, easy to no recurring expenses and unlimited capacity are pretty hard to resist (courtesy of IDC)

The Internet, Web and the Cloud promised us everything; and for the most part, they deliver. Companies, organizations are busy putting themselves up in the cloud because it's cheap, it's easy and … it's cheap. It's almost irresistible, which is why cloud services hold over a quarter of the world's business data — personal medical, social media content; financial/corporate documents and business/government data (see Fig. 9.7).

The numbers are mind-boggling! According to IDC, public cloud spending will hit $127B by 2018, growing six times faster than conventional tech over the period. Gigaom Research reports that 53 percent of large enterprises are either already leveraging public cloud resources for enterprise big data or are planning to. Only 13 percent of the Gigaom respondents said they would only use private data centers. The numbers for cloud services and big data are staggering. Cloud people love talking about all of the super fantastic, productivity-leading savings:

- $300 billion a year for the US healthcare industry.
- $250 billion for the European public sector.
- 60 percent potential increase in retailers' operating margins.
- $600 billion in economic surplus for services enabled by personal-location data.

There are savings under every rock, behind every tree. Half of the world's 7.7B people will be online looking for places to store their stuff–more than 8.6ZB worth. That's equal to streaming all the movies (about 500,000) and TV shows (3M) ever made over UHD (ultra-high def) 250,000 times. By 2020, the volume of stuff will almost double (See Fig. 9.8).

To meet the demand, IDC Senior Vice President Frank Gens suggests the number of new cloud-based solutions will increase 10X, even though all of those offering the solutions will also be running hell-bent to be the best/cheapest one around.

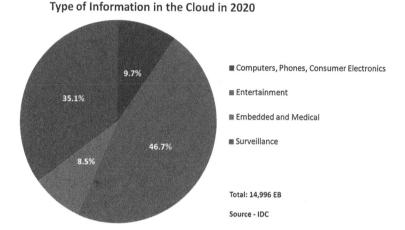

Type of Information in the Cloud in 2020

Fig. 9.8   More Data Storage — Everyone is talking the benefits of cloud storage and it's irresistible–especially when it is free to cheap. It is almost prophetic that the largest percentage of storage will be used by entertainment (courtesy of IDC)

Folks who can best afford to play the cheaper card are Amazon, Google, Apple,
Microsoft, Facebook and Alibaba.

It will be a challenging time for IT personnel because they understand they
have already been hacked and malware is already running on any number of
organization systems or it will be shortly.

The Center for Strategic and International Studies reported global cybercrime
will rack up $575B this year. Because cybercrime is so easy and bad guys/gals
are seldom prosecuted, it will far exceed that number unless management comes
to grip with how quickly their business can disappear. And the dollar value of the
losses doesn't even begin to touch the loss of corporate image/corporate equity.

By design and definition, networks and the Internet are not secure but they can
be significantly improved by IT. ISC-CERT (Industrial Control Systems Cyber
Emergency Response Team) notes that there are a number of NTP (Network Time
Protocol) security holes that can be used for distributed denial of service (DDoS)
attacks (basically shutting your organization, site down) even by individuals with
limited skills. The fix has been around for years but only now are network administrators and equipment manufacturers beginning to implement them.

The bigger issue is that IT has to balance end-user productivity with security.
Management doesn't worry about security as long as the work gets done quickly,
properly. Employees don't care as long as things are readily available. Accounting
doesn't care as long as everything comes in on budget. Lawyers care about how
much it costs in time and resources to achieve it. Until now, IT has focused on
meeting these requirements, pushing security implications into the background.
The problem with security is that everything is fine until it isn't fine. Penetrating
corporate online systems and cloud storage could well become the way companies
are crippled or even put out of business.

The Ponemon Institute found that cybercrime, which was up 10.4 percent this year, included:

- Stealing intellectual property.
- Confiscating online bank accounts.
- Creating and distributing viruses on other computers.
- Posting confidential/private personal business information on the internet.
- Disrupting a country's critical national infrastructure.

The problem is the iNet was never built for security; it was built for information exchange. As a result, attackers have been able to move freely through the organization once the network/system and its walls have been breached.

It has been made even easier as organizations have rushed to embrace BYOD (Bring Your Own Device) so people can access their content anytime, anywhere. Management initially liked the idea of BYOD with employees using their own mobile devices for work. It enabled people to be more productive and more available but the devices have shown an ugly side as well. Apps are added with malware and they are often lost/stolen with vital corporate information on them. Personal mobile devices, phones, tablets and laptops have created hundreds/thousands of network entry points that put personal information and confidential data at risk. And few of us "normal" folks understand or practice safe computing. At best, information is protected with a username and simplistic password.

The increased attention may enable IT to implement and enforce security to make hacking and cyber attacks less severe and more easily/quickly identified. IT practices such as multi-factor authentication, special privilege mandates to access content, eliminating end user admin permission, mandatory watermarking of documents/content for viewing printing, copying, forwarding as well as active logging of system activity won't eliminate network/system penetration completely; but it will make it more difficult to attack.

This year, we'll see a rush of VC investments in new automated security solution firms that will promise nearly penetration-proof solutions for organizations. The problem is the attack surfaces of most organizations are so vast that motivated hackers will always find a way in, usually courtesy of an innocent insider who fell for a spear-phishing campaign and opened the door (see Fig. 9.9). As a result, organizations have to educate employees to make them part of the solution (not part of the problem) so they can work with security in mind and:

- Not let individuals or sites persuade them to give up private/personal information such as confirming the last four digits of their social security number or corporate passwords.
- Doubt that even legitimate companies they trust will send an email requesting financial information. Don't worry, if it's real, they'll get back to you.
- Educate everyone that if they receive announcements or visit a site that has poor spelling and grammar (in your native tongue) forget it. Legitimate businesses are more professional.

"95% of all attacks on enterprise networks are the result of successful spear phishing"

Source: Allan Paller, Director of Research - SANS Institute

Fig. 9.9 Spear Phishing represents 95 % of all attacks on enterprise networks (courtesy of SANS Institute)

- Know that if they want to be paid by wire transfer or the person needs to get money out of the country or is dying of an incurable disease and wants to unload guilt money, forget it.
- Be aware that if Dr Oz wants to give hints on how to lose weight effortlessly or someone wants to help you empty your intestines, they only want to empty your wallet.
- Change their "rugged" password more often than on New Year's Day.
- Constantly check for OS/app updates (patches) and keep security/malware software up-to-date–and use it.

Once the attackers establish their beachhead, it usually becomes a trivial matter to move throughout the network and gather information. Hacking is a race between enterprise security and hackers; and the hackers are pulling ahead. Until organizations do a better job of training staff — especially individuals with sensitive data access, executives and their key staff — the attacks and losses will only grow.

The FBI stated that 90 percent of companies could not have defended against the Sony hack; which means organizations have a lot of work ahead of them. But it isn't impossible if businesses and institutions take a proactive position:

- Make security everyone's responsibility — education, awareness and personal responsibility are powerful deterrents.
- Don't rely on technology — tools are nice but reducing the amount of sensitive, private data stored in the insecure Cloud, eliminating insecure suppliers and restricting information access is a lot more effective and economic.
- Manage mobil devices — while devices can be hacked, the major issue is their being lost or stolen. While initially more costly, many organizations are returning to a COPE (corporate owned, personally enabled) program where devices have secure areas for work-related materials/information and personal areas.

Fig. 9.10 Realtime Strikes — The map may look like a pinball game with lights glittering here and there, but it's really the tracking of IP addresses and the attacks against them. The volume of activity is steadily increasing (courtesy of Norse Corp.)

- Develop and maintain a risk-aware environment — information security needs to become an integral part of every aspect of the business so it is viewed as a normal business process, not a business blocker to be circumvented.
- Top down support — If organizational data, information and content are everyone's responsibility (with the CEO leading by example) it is more readily made a part of everyone's daily activities.

Hacking tools have been around for some time and are becoming increasingly sophisticated and malicious (see Fig. 9.10). With their growing complexity and functionality, they can inflict damage on national and multi-national companies and organizations that can extend beyond cybercrime to include cyber-bullying and cyber-terrorism.

It is all empowered by our addiction to, and reliance on, technology. It can only be avoided by getting completely off the grid–using cash, dispensing with credit cards, never browsing the Internet, using pre-paid cell phones and giving up government-related document/services. It can only be put to an end by a complete revision of our technological infrastructure.

Since neither of these is likely, organizations and individuals have to improve their cyber security vigilance; knowing individuals are developing and releasing malware that can wipe data from systems beyond recovery.

Ultimately, organizations and individuals will learn to live with the constant threat, knowing that some days the good guys will be ahead and some days the bad guys will be ahead.

Being aware, prepared and practicing sound computing/security processes and procedures can level the playing field, improving your odds. Online privacy and security will have to be every employee's job because today's organizations rely so heavily on their data systems and simply adding more guards and more security personnel won't be enough.

Sony was under a serious siege, they are just the first organization wounded without a shot being fired. They won't be the last.

Bibliography

1. **Obama, Barack.** Remarks By The President On Securing Our Nation's Cyber Infrastructure. [Online] The White House, May 29, 2009. http://www.whitehouse.gov/the-press-office/remarks-president-securing-our-nations-cyber-infrastructure.
2. **Mueller, Robert S. III.** Remarks by Robert S. Mueller, III. [Online] Federal Bureau of Investigation, March 1, 2012. http://www.fbi.gov/news/speeches/combating-threats-in-the-cyber-world-outsmarting-terrorists-hackers-and-spies.
3. **Hickins, Michael and Clark, Don.** Questions Over Break-In at Security Firm RSA. [Online] The Wall Street Journal, March 18, 2011. http://online.wsj.com/article/SB10001424052748703512404576208983743029392.html.
4. **Krebs, Brian.** Who Else Was Hit by the RSA Attackers? [Online] Kerbs on Security, October 11, 2011. http://krebsonsecurity.com/2011/10/who-else-was-hit-by-the-rsa-attackers/.
5. **Zelvin, Larry and Edwards, Marty.** ICS-CERT Year in Review - 2013. [Online] National Cybersecurity and Communications Integration Center, 2013. https://ics-cert.us-cert.gov/sites/default/files/documents/Year_In_Review_FY2013_Final.pdf.
6. **Knibbs, Kate.** Report: Mysterious Russian Malware Is Infecting 100,000+ Wordpress Sites. [Online] GIZMODO, December 15, 2014. http://gizmodo.com/mysterious-russian-malware-is-infecting-over-100-000-wo-1671419522.
7. **Riley, Michael.** How Russian Hackers Stole the Nasdaq. [Online] Bloomberg, July 17, 2014. http://www.businessweek.com/printer/articles/213544-how-russian-hackers-stole-the-nasdaq.
8. **Eha, Brian Patrick.** Is Knight's $440 million glitch the costliest computer bug ever? [Online] CNNMoney, August 9, 2012. http://money.cnn.com/2012/08/09/technology/knight-expensive-computer-bug/index.html.
9. **Isidore, Chris.** 8 charged in $45 million cybertheft bank heist. [Online] CNNMoney, May 10, 2013. http://money.cnn.com/2013/05/09/technology/security/cyber-bank-heist/index.html?hpt=hp_c2.
10. **Yousuf, Hibah.** False White House explosion tweet rattles market. [Online] CNNMoney, April 23, 2013. http://buzz.money.cnn.com/2013/04/23/ap-tweet-fake-white-house/.
11. Hackers Steal Up to $1 Billion From Banks, Security Co. Says. [Online] The New York Times, February 15, 2015. http://www.nytimes.com/aponline/2015/02/15/us/ap-us-bank-hack-report.html.
12. **Center for Strategic and International Studies.** Net Losses: Estimating the Global Cost of Cybercrime. [Online] Center for Strategic and International Studies, June 2014. http://www.mcafee.com/us/resources/reports/rp-economic-impact-cybercrime2.pdf.
13. **CERT Coordination Center.** CERT® Advisory CA-2001-26 Nimda Worm. [Online] Carneige Mellon Software Engineering Institute, September 18, 2001. http://www.cert.org/advisories/CA-2001-26.html.

14. **Green, Joshua.** The Myth of Cyberterrorism. [Online] Washington Monthly, November 2002. http://www.washingtonmonthly.com/features/2001/0211.green.html.

15. **Leyden, John.** Virus writers are industrial terrorists – MS. [Online] The Register, October 23, 2001. http://www.theregister.co.uk/2001/10/23/virus_writers_are_industrial_terrorists/.

16. Malicious Code Attacks Had $13.2 Billion Economic Impact in 2001. [Online] Computer Economics, September 2002. http://www.computereconomics.com/article.cfm?id=133.

17. **Marquit, Miranda.** The 12 costliest computer viruses ever. [Online] The Fine Print, August 3, 2010. http://blog.insure.com/2010/08/03/the-12-costliest-computer-viruses-ever/.

18. **McAfee, Inc.** McAfee Looks Back on a Decade of Cybercrime. [Online] McAfee, January 25, 2011. http://www.mcafee.com/us/about/news/2011/q1/20110125-01.aspx.

19. **Help Net Security.** Mydoom.A: Timeline of an Epidemic. [Online] Help Net Security, March 2, 2004. http://www.net-security.org/malware_news.php?id=359.

20. **Roberts, Paul.** Sobig: Spam, virus or both? [Online] Computerworld, June 5, 2003. http://www.computerworld.com/s/article/81825/Sobig_Spam_virus_or_both_.

21. **CNN.** Destructive ILOVEYOU computer virus strikes worldwide. [Online] CNN, May 4, 2000. http://articles.cnn.com/2000-05-04/tech/iloveyou.01_1_melissa-virus-antivirus-companies-iloveyou-virus.

22. **Danchev, Dancho.** Conficker's estimated economic cost? $9.1 billion. [Online] ZDNet, April 23, 2009. http://www.zdnet.com/blog/security/confickers-estimated-economic-cost-9-1-billion/3207.

23. **Reuters.** 'Code Red II' spreading quickly, causing damage. [Online] USA Today, August 8, 2001. http://usatoday30.usatoday.com/life/cyber/tech/2001-08-08-code-red-2.htm.

24. **Northcutt, Stephen.** Intrusion Detection FAQ: What was the Melissa virus and what can we learn from it? [Online] SANS, March 28, 1999. http://www.sans.org/security-resources/idfaq/what_melissa_teaches_us.php.

25. **Thorsberg, Frank.** Sircam Worm: Crawling Fast but Easily Crushed. [Online] PCWorld, July 26, 2001. http://www.pcworld.com/article/56284/article.html.

26. **Lemos, Robert.** Counting the cost of Slammer. [Online] CNet, January 31, 2003. http://news.cnet.com/Counting-the-cost-of-Slammer/2100-1001_3-982955.html.

27. ─. 'Nimda' worm strikes Net, e-mail. [Online] CNet, September 18, 2001. http://news.cnet.com/2100-1001-273128.html.

28. ─. Sasser worm begins to spread. [Online] CNet, May 1, 2004. http://news.cnet.com/Sasser-worm-begins-to-spread/2100-7349_3-5203764.html.

29. **Messmer, Ellen.** Blaster Worm Racks Up Victims. [Online] PCWorld, August 15, 2003. http://www.pcworld.com/article/112047/article.html.

30. **Markoff, John.** Student, After Delay, Is Charged In Crippling of Computer Network. [Online] The New York Times, July 27, 1989. http://www.nytimes.com/1989/07/27/us/student-after-delay-is-charged-in-crippling-of-computer-network.html.

31. **Simons, John.** How an FBI Cybersleuth Busted a Hacker Ring. [Online] The Wall Street Journal, October 3, 1999. http://massis.lcs.mit.edu/archives/security-fraud/phonemasters-fraud.

32. **Johnston, David Cay.** Russian Accused of Citibank Computer Fraud. [Online] The New York Times, August 18, 1995. http://www.nytimes.com/1995/08/18/business/russian-accused-of-citibank-computer-fraud.html.

33. **Evans, James.** Mafiaboy's Story Points to Net Weaknesses. [Online] PCWorld, January 24, 2001. http://www.pcworld.com/article/39142/article.html.

34. **TED.** The technological future of crime: Marc Goodman at TEDGlobal 2012. [Online] TED, June 28, 2012. http://blog.ted.com/2012/06/28/the-technological-future-of-crime-marc-goodman-at-tedglobal-2012/.

35. **Symantec.** Cybercrime Report 2011. [Online] Symantec Corporation, 2011. http://www.symantec.com/content/en/us/home_homeoffice/html/cybercrimereport/assets/downloads/en-us/NCR-DataSheet.pdf.

36. **Sherstobitoff, Ryan.** Analyzing Project Blitzkrieg, a Credible Threat. [Online] McAfee Labs, December 13, 2012. http://www.mcafee.com/us/resources/white-papers/wp-analyzing-project-blitzkrieg.pdf.

37. **RSA FraudAction Research Labs.** Cyber Gang Seeks Botmasters to Wage Massive Wave of Trojan Attacks Against U.S. Banks. [Online] RSA, October 4, 2012. https://blogs. rsa.com/cyber-gang-seeks-botmasters-to-wage-massive-wave-of-trojan-attacks-against-u-s-banks/.

38. **Ward, Mark.** Cryptolocker victims to get files back for free. [Online] BBC News, August 6, 2014. http://www.bbc.com/news/technology-28661463.

39. **Wilhoit, Kyle and Dawda, Uttang.** Your Locker of Information for CryptoLocker Decryption. [Online] FireEye, August 6, 2014. https://www.fireeye.com/blog/executive-perspective/2014/08/your-locker-of-information-for-cryptolocker-decryption.html.

40. **Perez, Evan, Prokupecz, Shimon and Cohen, Tom.** More than 90 people nabbed in global hacker crackdown. [Online] CNN, May 19, 2014. http://www.cnn.com/2014/05/19/justice/us-global-hacker-crackdown/index.html.

41. **Botelho, Greg.** Arrest made in Miss Teen USA Cassidy Wolf 'sextortion' case. [Online] CNN, September 27, 2013. http://www.cnn.com/2013/09/26/justice/miss-teen-usa-sextortion/.

42. **Estes, Adam Clark.** The Sony Pictures Hack Was Even Worse Than Everyone Thought. [Online] GIZMODO, December 3, 2014. http://gizmodo.com/the-sony-pictures-hack-exposed-budgets-layoffs-and-3-1665739357/1666122168/+ace.

43. **Cieply, Michael and Barnes, Brooks.** Sony Cyberattack, First a Nuisance, Swiftly Grew Into a Firestorm. [Online] The New York Times, December 30, 2014. http://www.nytimes.com/2014/12/31/business/media/sony-attack-first-a-nuisance-swiftly-grew-into-a-firestorm-.html.

44. **Kastrenakes, Jacob.** The Sony hackers won: The Interview just disappeared from America's biggest theater chains. [Online] The Verge, December 17, 2014. http://www.theverge.com/2014/12/17/7411155/us-biggest-movie-theaters-wont-show-the-interview.

45. **The Associated Press.** Sony says massive hack cost the company $15 million. [Online] Mashable, February 4, 2015. http://mashable.com/2015/02/04/sony-hack-cost-15-million/.

46. What are the differences between NSA/CSS' and U.S. Cyber Command's roles? [Online] National Security Agency / Central Security Service, January 13, 2011. http://www.nsa.gov/about/faqs/about_nsa.shtml#about10.

47. **Sanger, David E. and Schmitt, Eric.** Rise Is Seen in Cyberattacks Targeting U.S. Infrastructure. [Online] The New York Times, July 26, 2012. http://www.nytimes.com/2012/07/27/us/cyberattacks-are-up-national-security-chief-says.html?_r=0.

48. **Merica, Dan.** DoD official: Vulnerability of U.S. electrical grid is a dire concern. [Online] CNN, July 27, 2012. http://security.blogs.cnn.com/2012/07/27/dod-official-vulnerability-of-u-s-electrical-grid-is-a-dire-concern/.

49. **National Research Council of the National Academies.** Electric Power Grid 'Inherently Vulnerable' to Terrorist Attacks. [Online] The National Academies Press, November 14, 2012. http://www8.nationalacademies.org/onpinews/newsitem.aspx?RecordID=12050.

50. **Barron, James.** The Blackout Of 2003: The Overview; Power Surge Blacks Out Northeast, Hitting Cities In 8 States And Canada; Midday Shutdowns Disrupt Millions. [Online] The New York Times, August 15, 2003. http://www.nytimes.com/2003/08/15/nyregion/blackout-2003-overview-power-surge-blacks-northeast-hitting-cities-8-states.html?pagewanted=all.

51. **Anderson, Patrick L. and Geckil, Ilhan K.** Northeast Blackout Likely to Reduce US Earnings by $6.4 Billion. [Online] Anderson Economic Group (AEG), August 19, 2003. http://www.andersoneconomicgroup.com/Portals/0/upload/Doc544.pdf.

52. **Communication Technologies, Inc.** Supervisory Control and Data Acquisition (SCADA) Systems. [Online] National Communications System, October 2004. http://www.ncs.gov/library/tech_bulletins/2004/tib_04-1.pdf.

53. **Poulsen, Kevin.** Slammer worm crashed Ohio nuke plant network. [Online] Security Focus, August 19, 2003. http://www.securityfocus.com/news/6767.

54. **McMillan, Robert.** New virus targets industrial secrets. [Online] Computerworld, July 17, 2010. http://www.computerworld.com/s/article/9179298/New_virus_targets_industrial_secrets.

55. **Ulasen, Sergey.** Rootkit.TmpHider. [Online] Wilders Security Forums, July 12, 2010. http://www.wilderssecurity.com/showthread.php?t=276994.

56. **Keizer, Gregg.** Is Stuxnet the 'best' malware ever? [Online] Computerworld, September 16, 2010. http://www.computerworld.com/s/article/9185919/Is_Stuxnet_the_best_malware_ever_.

57. **Thakur, Vikram.** W32.Stuxnet — Network Information. [Online] Symantec, July 23, 2010. http://www.symantec.com/connect/blogs/w32stuxnet-network-information.

58. **McMillan, Robert.** Siemens: Stuxnet worm hit industrial systems. [Online] Computerworld, September 14, 2010. http://www.computerworld.com/s/article/9185419/Siemens_Stuxnet_worm_hit_industrial_systems.

59. **Broad, William J., Markoff, John and Sanger, David E.** Israeli Test on Worm Called Crucial in Iran Nuclear Delay. [Online] The New York Times, January 15, 2011. http://www.nytimes.com/2011/01/16/world/middleeast/16stuxnet.html?pagewanted=all.

60. **Sanger, David E.** Obama Order Sped Up Wave of Cyberattacks Against Iran. [Online] The New York Times, June 1, 2012. http://www.nytimes.com/2012/06/01/world/middleeast/obama-ordered-wave-of-cyberattacks-against-iran.html?pagewanted=all.

61. **DHS ICS-CERT.** ICS-CERT Monitor. [Online] U.S. Department of Homeland Security, October/November/December 2012. http://www.us-cert.gov/control_systems/pdf/ICS-CERT_Monthly_Monitor_Oct-Dec2012.pdf.

62. **ICS-CERT.** ICS-CERT Monthly Monitor. [Online] Industrial Control Systems Cyber Emergency Response Team, April 2012. http://www.us-cert.gov/control_systems/pdf/ICS-CERT_Monthly_Monitor_Apr2012.pdf.

63. **Hargreaves, Steve.** Hackers take aim at key U.S. infrastructure. [Online] CNNMoney, February 20, 2013. http://money.cnn.com/2013/02/20/news/economy/hacking-infrastructure/.

64. **Seaby, Greg.** DHS: Hackers infiltrate public utility. [Online] CNN, May 21, 2014. http://www.cnn.com/2014/05/21/us/hackers-public-utility/index.html.

65. **Estes, Adam Clark.** Hacking Into Traffic Lights With a Plain Old Laptop Is Scary Simple. [Online] GIZMODO, August 19, 2014. http://gizmodo.com/hacking-into-traffic-lights-with-a-plain-old-laptop-is-1624102517Hacking Into Traffic Lights With a Plain Old Laptop Is Scary Simple.

66. **Flaherty, Mary Pat, Samenow, Jason and Rein, Lisa.** Chinese hack U.S. weather systems, satellite network. [Online] The Washington Post, November 12, 2014. http://www.washingtonpost.com/local/chinese-hack-us-weather-systems-satellite-network/2014/11/12/bef1206a-68e9-11e4-b053-65cea7903f2e_story.html.

67. **Crawford, Jamie.** The U.S. government thinks China could take down the power grid. [Online] CNN, November 21, 2014. http://www.cnn.com/2014/11/20/politics/nsa-china-power-grid/index.html.

68. **Control Systems Security Program, National Cyber Security Division.** Recommended Practice: Improving Industrial Control Systems Cybersecurity with Defense-In-Depth Strategies. [Online] U.S. Department of Homeland Security, October 2009. http://www.us-cert.gov/control_systems/practices/documents/Defense_in_Depth_Oct09.pdf.

69. **Frates, Chris and Devine, Curt.** Government hacks and security breaches skyrocket. [Online] CNN, December 19, 2014. http://www.cnn.com/2014/12/19/politics/government-hacks-and-security-breaches-skyrocket/.

70. **Shachtman, Noah.** Insiders Doubt 2008 Pentagon Hack Was Foreign Spy Attack (Updated). [Online] Wired, August 25, 2010. http://www.wired.com/dangerroom/2010/08/insiders-doubt-2008-pentagon-hack-was-foreign-spy-attack/.

71. **Lynn, William J. III.** Defending a New Domain. [Online] Foreign Affairs, September/October 2010. http://www.foreignaffairs.com/articles/66552/william-j-lynn-iii/defending-a-new-domain.

72. **Shachtman, Noah.** Exclusive: Computer Virus Hits U.S. Drone Fleet. [Online] Wired, October 7, 2011. http://www.wired.com/dangerroom/2011/10/virus-hits-drone-fleet/.

73. **Rundle, Katherine Fernandez, Horn, Donl and Dechovitz, Susan Leah.** Final Report Of The Miami-Dade County Grand Jury. [Online] The Circuit Court Of The Eleventh Judicial Circuit Of Florida In And For The County Of Miami-Dade, December 19, 2012. http://www.miamisao.com/publications/grand_jury/2000s/gj2012s.pdf.

74. **Segall, Laurie.** Iron Dome makers got hacked, says cybersecurity firm. [Online] CNN-Money, July 29, 2014. http://money.cnn.com/2014/07/29/technology/security/iron-dome-hack/index.html.

75. **Bull, Alister and Finkle, Jim.** Fed says internal site breached by hackers, no critical functions affected. [Online] Reuters, February 6, 2013. http://www.reuters.com/article/2013/02/06/net-us-usa-fed-hackers-idUSBRE91501920130206.

76. **Isaac, Mike.** Syrian Electronic Army Targets President Obama in Latest Hack. [Online] AllThingsD, October 28, 2013. http://allthingsd.com/20131028/syrian-electronic-army-targets-president-obama-in-latest-hack/.

77. **Pagliery, Jose.** Chinese hackers broke into U.S. federal employee network. [Online] CNN-Money, July 10, 2014. http://money.cnn.com/2014/07/10/technology/security/china-hacks-us/index.html.

78. —. White House hacked. [Online] CNNMoney, December 29, 2014. http://money.cnn.com/2014/10/29/technology/security/white-house-hack/.

79. **Ada Resource Association.** Ada Overview. [Online] Ada Information Clearinghouse. [Cited: January 8, 2013.] http://www.adaic.org/advantages/ada-overview/.

80. **Computer History Museum.** Ada Lovelace. [Online] Computer History Museum. [Cited: January 8, 2013.] http://www.computerhistory.org/babbage/adalovelace/.

81. **Ada Joint Program Office.** Military Standard Common APSE (Ada Programming Support Environment) Interface Set (CAIS). [Online] Defense Technical Information Center, 1985. http://books.google.com/books/about/Military_Standard_Common_APSE_Ada_Progra.html?id=EjEYOAAACAAJ.

82. **Keller, John.** DOD officials eye scrapping mandate to use Ada programming. [Online] Military and Aerospace Electronics, May 1, 1997. http://www.militaryaerospace.com/articles/print/volume-8/issue-5/departments/trends/dod-officials-eye-scrapping-mandate-to-use-ada-programming.html.

83. **Nakashima, Ellen, Miller, Greg and Tate, Julie.** U.S., Israel developed Flame computer virus to slow Iranian nuclear efforts, officials say. [Online] The Washington Post, June 19, 2012. http://articles.washingtonpost.com/2012-06-19/world/35460741_1_stuxnet-computer-virus-malware.

84. **Erdbrink, Thomas.** Iran Confirms Attack by Virus That Collects Information. [Online] The New York Times, May 29, 2012. http://www.nytimes.com/2012/05/30/world/middleeast/iran-confirms-cyber-attack-by-new-virus-called-flame.html.

85. **Kaspersky Lab security researchers.** Kaspersky Lab and ITU Research Reveals New Advanced Cyber Threat. [Online] Kaspersky Lab, May 28, 2012. http://www.kaspersky.com/about/news/virus/2012/Kaspersky_Lab_and_ITU_Research_Reveals_New_Advanced_Cyber_Threat.

86. **Global Research & Analysis Team (GReAT), Kaspersky Lab.** Gauss: Abnormal Distribution. [Online] Kaspersky Lab, August 9, 2012. http://www.securelist.com/en/analysis/204792238/Gauss_Abnormal_Distribution.

87. **Constantin, Lucian.** Website operators will have a hard time dealing with the Heartbleed vulnerability. [Online] PC World, April 10, 2014. http://www.pcworld.com/article/2142540/website-operators-will-have-a-hard-time-dealing-with-the-heartbleed-vulnerability.html.

88. **Foster, Rusty.** Heartbleed. [Online] Schneier on Security, April 9, 2014. https://www.schneier.com/blog/archives/2014/04/heartbleed.html.

89. **Newman, Lily Hay.** 300,000 Websites Still Haven't Patched Against Heartbleed. [Online] Slate, June 23, 2014. http://www.slate.com/blogs/future_tense/2014/06/23/_300_000_websites_are_still_vulnerable_to_heartbleed_attacks.html.

90. **Simonite, Tom.** Many Devices Will Never Be Patched to Fix Heartbleed Bug. [Online] MIT Technology Review, April 9, 2014. http://www.technologyreview.com/news/526451/many-devices-will-never-be-patched-to-fix-heartbleed-bug/.

91. **Chacos, Brad.** 'Bigger than Heartbleed' Shellshock flaw leaves OS X, Linux, more open to attack. [Online] Network World, September 25, 2014. http://www.networkworld.com/article/2687953/security0/bigger-than-heartbleed-shellshock-flaw-leaves-os-x-linux-more-open-to-attack.html.

92. **Constantin, Lucian.** The POODLE flaw returns, this time hitting TLS security protocol. [Online] Computer World, December 8, 2014. http://www.computerworld.com/article/2857113/the-poodle-flaw-returns-this-time-hitting-tls-security-protocol.html.

93. **Pagliery, Jose.** Microsoft fixes a serious 15-year-old bug. [Online] CNNMoney, February 12, 2015. http://money.cnn.com/2015/02/12/technology/security/microsoft-jasbug/index.html.

94. **Covert, Adrian.** The end of polished and perfect software. [Online] CNNMoney, May 13, 2014. http://money.cnn.com/2014/05/13/technology/innovation/beta-testing/index.html.

95. **Regalado, Antonio.** Spinoffs from Spyland. [Online] MIT Technology Review, March 18, 2014. http://www.technologyreview.com/news/525541/spinoffs-from-spyland/.

96. **Lazarte, Maria.** ISO/IEC revise Ada programming language for greater reliability and security. [Online] International Organization for Standardization, February 21, 2013. http://www.iso.org/iso/home/news_index/news_archive/news.htm?refid=Ref1707.

97. www.shodan.io. [Online] Shodan. [Cited: January 8, 2015.] https://www.shodan.io/.

98. **Goldman, David.** The Internet's most dangerous sites. [Online] CNNMoney, May 2, 2013. http://money.cnn.com/gallery/technology/security/2013/05/01/shodan-most-dangerous-internet-searches/index.html.

99. **Smith, Ms.** Unpatched TRENDnet IP cameras still provide a real-time Peeping Tom paradise. [Online] Network World, January 7, 2013. http://www.networkworld.com/article/2223785/microsoft-subnet/unpatched-trendnet-ip-cameras-still-provide-a-real-time-peeping-tom-paradise.html.

100. insecam.com. Online IP netsurveillance cameras of the world. [Online] insecam.com. [Cited: November 1, 2014.] http://insecam.com/.

101. **Cox, Joseph.** This Website Streams Camera Footage from Users Who Didn't Change Their Password. [Online] Vice Media LLC, October 31, 2014. http://motherboard.vice.com/read/this-website-streams-camera-footage-from-users-who-didnt-change-their-password.

102. **Sparkes, Matthew.** Average Internet of Things device has 25 security flaws. [Online] The Telegraph, July 30, 2014. http://www.telegraph.co.uk/technology/internet-security/11000013/Average-Internet-of-Things-device-has-25-security-flaws.html.

103. **Bilton, Nick.** Intruders for the Plugged-In Home, Coming In Through the Internet. [Online] The New York Times, June 1, 2014. http://bits.blogs.nytimes.com/2014/06/01/dark-side-to-internet-of-things-hacked-homes-and-invasive-ads/.

104. **Fleishman, Glenn.** An Internet of Treacherous Things. [Online] MIT Technology Review, January 13, 2015. http://www.technologyreview.com/news/534196/an-internet-of-treacherous-things/.

105. **Metz, Rachel.** CES 2015: The Internet of Just About Everything. [Online] MIT Technology Review, January 6, 2015. http://www.technologyreview.com/news/533941/ces-2015-the-internet-of-just-about-everything/.

106. **Checkoway, Stephen, et al.** Comprehensive Experimental Analyses of Automotive Attack Surfaces. [Online] Center for Automotive Embedded Systems Security, August 8-12, 2011. http://www.autosec.org/pubs/cars-usenixsec2011.pdf.

107. **The Huffington Post UK.** 'Half Of All Stolen Cars Are Hacked' Says Met Police. [Online] Huffington Post UK, August 5, 2014. http://www.huffingtonpost.co.uk/2014/05/08/hacked-cars_n_5286590.html.

108. **Fink, Erica.** How hackers could slam on your car's brakes. [Online] CNNMoney, August 1, 2014. http://money.cnn.com/2014/08/01/technology/security/most-hackable-cars/index.html.

109. **Finkle, Jim.** Hacking experts build device to protect cars from cyber attacks. [Online] Reuters, July 23, 2014. http://www.reuters.com/article/2014/07/23/us-cybersecurity-autos-idUSKBN0FR2FR20140723.

110. **Mearian, Lucas.** Over-the-air software coming soon to your next car. [Online] Computer World, February 5, 2015. http://www.computerworld.com/article/2880150/over-the-air-software-coming-soon-to-your-next-car.html.

111. **IMDb.** WarGames. [Online] IMDb, June 3, 1983. http://www.imdb.com/title/tt0086567/.

112. **Simonite, Tom.** Black Hat: Most Smartphones Come with a Poorly Secured Back Door. [Online] MIT Technology Review, August 7, 2014. http://www.technologyreview.com/news/529676/black-hat-most-smartphones-come-with-a-poorly-secured-back-door/.

113. **Hesseldahl, Arik.** Apple Denies Working with NSA on iPhone Backdoor. [Online] All Things D, December 31, 2013. http://allthingsd.com/20131231/apple-says-it-is-unaware-of-nsas-iphone-backdoor/.

114. **Simonite, Tom.** A Computer Infection that Can Never Be Cured. [Online] MIT Technology Review, August 1, 2012. http://www.technologyreview.com/news/428652/a-computer-infection-that-can-never-be-cured/.

115. **—.** Intel Designs a Safe Meeting Place for Private Data. [Online] MIT Technology Review, March 3, 2014. http://www.technologyreview.com/news/525131/intel-designs-a-safe-meeting-place-for-private-data/.

116. **Craig.** Reverse Engineering a D-Link Backdoor. [Online] /dev/ttyS0, October 12, 2013. http ://www.devttys0.com/2013/10/reverse-engineering-a-d-link-backdoor/.

117. **A Guest.** D-Link Backdoor Stack Overflow PoC DIR-100 v1.13. [Online] Pastebin, October 14, 2013. http://pastebin.com/vbiG42VD.

118. **Menn, Joseph.** Exclusive: Secret contract tied NSA and security industry pioneer. [Online] Reuters, December 20, 2013. http://www.reuters.com/article/2013/12/20/us-usa-security-rsa-idUSBRE9BJ1C220131220.

119. **Sullivan, Nick.** How the NSA (may have) put a backdoor in RSA's cryptography: A technical primer. [Online] Ars Technica, January 5, 2014. http://arstechnica.com/security/2014/01/how-the-nsa-may-have-put-a-backdoor-in-rsas-cryptography-a-technical-primer/.

120. **O'Reilly, Patrick J.** Lieutenant General Patrick J. O'Reilly, USA, Director, Missile Defense Agency Before the Senate Armed Services Committee. [Online] U.S. Senate Armed Services Committee, November 8, 2011. http://www.armed-services.senate.gov/statemnt/2011/11%20November/OReilly%2011-08-11.pdf.

121. **Goldman, David.** Fake tech gear has infiltrated the U.S. government. [Online] CNNMoney, November 8, 2012. http://money.cnn.com/2012/11/08/technology/security/counterfeit-tech/index.html.

122. **Carl Levin - U.S. Senate Newsroom.** Senate Approves Amendment to Strengthen Protections Against Counterfeit Electronic Parts in Defense Supply System. [Online] U.S. Senate, November 29, 2011. http://www.levin.senate.gov/newsroom/press/release/senate-approves-amendment-to-strengthen-protections-against-counterfeit-electronic-parts-in-defense-supply-system.

123. **U.S. Department of Justice.** Departments of Justice and Homeland Security Announce International Initiative Against Traffickers in Counterfeit Network Hardware. [Online] U.S. Department of Justice, February 28, 2008. http://www.justice.gov/opa/pr/2008/February/08_crm_150.html.

124. **Gallagher, Sean.** Photos of an NSA "upgrade" factory show Cisco router getting implant. [Online] Ars Technica, May 14, 2014. Photos of an NSA "upgrade" factory show Cisco router getting implant.

125. **Hesseldahl, Arik.** In Letter to Obama, Cisco CEO Complains About NSA Allegations. [Online] recode, May 18, 2014. http://recode.net/2014/05/18/in-letter-to-obama-cisco-ceo-complains-about-nsa-allegations/.

126. **Grimes, Roger A.** Stuxnet marks the start of the next security arms race. [Online] InfoWorld, January 25, 2011. http://www.infoworld.com/d/security-central/stuxnet-marks-the-start-the-next-security-arms-race-282.

127. **Ross, Brian, et al.** Top Secret Stealth Helicopter Program Revealed in Osama Bin Laden Raid: Experts. [Online] ABC Good Morning America, May 4, 2011. http://abcnews.go.com/Blotter/top-secret-stealth-helicopter-program-revealed-osama-bin/story?id=13530693&page=2.

128. **Goettelmann, John Charles, et al.** Technique for executing translated software. [Online] United States Patent and Trademark Office, March 3, 1998. http://www.google.com/patents/US5724590.

129. **Zetter, Kim.** Son of Stuxnet Found in the Wild on Systems in Europe. [Online] Wired, October 18, 2011. http://www.wired.com/threatlevel/2011/10/son-of-stuxnet-in-the-wild/.

130. **arXiv.** The Growing Threat Of Network-Based Steganography. [Online] MIT Technology Review, July 18, 2014. http://www.technologyreview.com/view/529071/the-growing-threat-of-network-based-steganography/.

131. **Kelley, Jack.** Terror groups hide behind Web encryption. [Online] USA Today, February 5, 2001. http://usatoday30.usatoday.com/life/cyber/tech/2001-02-05-binladen.htm.

132. **Goodman, David.** Major banks hit with biggest cyberattacks in history. [Online] CNNMoney, September 28, 2012. http://money.cnn.com/2012/09/27/technology/bank-cyberattacks/index.html.

133. **Goldman, David.** The real Iranian threat: Cyberattacks. [Online] CNNMoney, November 5, 2012. http://money.cnn.com/2012/11/05/technology/security/iran-cyberattack/index.html.

134. **Brumfield, Ben.** Computer spyware is newest weapon in Syrian conflict. [Online] CNN, February 17, 2012. http://www.cnn.com/2012/02/17/tech/web/computer-virus-syria/index.html.

135. **Fung, Brian.** Military Strikes Go Viral: Israel Is Live-Tweeting Its Own Offensive Into Gaza. [Online] The Atlantic, November 14, 2012. http://www.theatlantic.com/international/archive/2012/11/military-strikes-go-viral-israel-is-live-tweeting-its-own-offensive-into-gaza/265227/.

136. **Hachman, Mark.** IDF vs. Hamas War Extends to Social Media. [Online] PC Magazine, November 16, 2012. http://www.pcmag.com/slideshow/story/305065/idf-vs-hamas-war-extends-to-social-media.

137. **Sutter, John D.** Will Twitter war become the new norm? [Online] CNN, November 19, 2012. http://www.cnn.com/2012/11/15/tech/social-media/twitter-war-gaza-israel/index.html.

138. **Chan, Casey.** Anonymous Targets Israel by Taking Down Hundreds of Websites and Leaking Emails and Passwords. [Online] Gizmodo, November 16, 2012. http://gizmodo.com/5961399/anonymous-destroys-israel-by-taking-down-hundreds-of-websites-and-leaking-emails-and-passwords.

139. **Greenberg, Andy.** Anonymous Hackers Ramp Up Israeli Web Attacks And Data Breaches As Gaza Conflict Rages. [Online] Forbes, November 19, 2012. http://www.forbes.com/sites/andygreenberg/2012/11/19/anonymous-hackers-ramp-up-israeli-web-attacks-and-data-breaches-as-gaza-conflict-rages-2/.

140. **Osborne, Charlie.** Anonymous takes on Israeli websites, wipes Jerusalem bank. [Online] ZDNet, November 16, 2012. http://www.zdnet.com/anonymous-takes-on-israeli-websites-wipes-jerusalem-bank-7000007537/.

141. **Sutter, John D.** Anonymous declares 'cyberwar' on Israel. [Online] CNN, November 20, 2012. http://www.cnn.com/2012/11/19/tech/web/cyber-attack-israel-anonymous/index.html.

142. **McMillian, Robert and Ackerman, Spencer.** Despite Ceasefire, Israel-Gaza War Continues Online. [Online] Wired, November 28, 2012. http://www.wired.com/dangerroom/2012/11/israel-gaza-ddos/.

143. **Mullen, Jethro.** Coming soon: Iran's response to 'Argo'. [Online] CNN, January 14, 2013. http://www.cnn.com/2013/01/14/world/meast/iran-argo-response/index.html.

144. **Mullen, Jethro and Brumfield, Ben.** Iran to add lawsuit over 'Argo' to cinematic response. [Online] CNN, March 14, 2013. http://www.cnn.com/2013/03/13/world/meast/iran-argo-response/index.html.

145. **Kwon, K.J., Mullen, Jethro and Pearson, Michael.** Hacking attack on South Korea traced to Chinese address, officials say. [Online] CNN, March 21, 2013. http://www.cnn.com/2013/03/21/world/asia/south-korea-computer-outage/index.html.

146. **Botelho, Greg.** North Korea says it's the victim of 'intensive' cyberattacks. [Online] CNN, March 15, 2013. http://www.cnn.com/2013/03/14/world/asia/north-korea-cyberattacks/index.html.

147. **Dave, Paresh.** New York Times outage traced to phishing email to Melbourne IT partner. [Online] Los Angeles Times, August 27, 2013. http://www.latimes.com/business/technology/la-fi-tn-melbourne-it-discovers-breach-that-took-down-nytimescom-20130827-story.html.

148. **Shankland, Stephen.** Syrian Electronic Army attack spreads pop-ups across news sites. [Online] CNet, November 27, 2014. http://www.cnet.com/news/syrian-electronic-army-attack-spreads-pop-ups-across-news-sites/.

149. **Lewis, Paul and Roberts, Dan.** White House denies 'Cuban Twitter' ZunZuneo programme was covert. [Online] The Guardian, April 3, 2014. http://www.theguardian.com/world/2014/apr/03/white-house-cuban-twitter-zunzuneo-covert.
150. **Brumfield, Ben.** Study: Hack attack aimed at ISIS' opposition. [Online] CNN, December 19, 2014. http://www.cnn.com/2014/12/19/world/meast/isis-opponents-malware-attack/index.html.
151. **Pagliery, Jose.** Russian hackers exploit Windows to spy on West. [Online] CNNMoney, December 29, 2014. http://money.cnn.com/2014/10/14/technology/security/russia-hackers/index.html.
152. **Wallace, Gregory.** 'Regin' malware described as 'groundbreaking and almost peerless. [Online] CNNMoney, November 24, 2014. http://money.cnn.com/2014/11/23/technology/security/regin-malware-symantec/index.html.
153. **Schneier, Bruce.** Antivirus Companies Should Be More Open About Their Government Malware Discoveries. [Online] MIT Technology Review, December 5, 2014. http://www.technologyreview.com/view/533136/antivirus-companies-should-be-more-open-about-their-government-malware-discoveries/.
154. **Shane, Scott.** Cyberwarfare Emerges From Shadows for Public Discussion by U.S. Officials. [Online] 2012, 26 September. http://www.nytimes.com/2012/09/27/us/us-officials-opening-up-on-cyberwarfare.html?pagewanted=all.
155. **Kingsbury, Alex.** The Secret History of the National Security Agency. [Online] US News and World Report, June 19, 2009. http://www.usnews.com/opinion/articles/2009/06/19/the-secret-history-of-the-national-security-agency.
156. **Air Force News Service (AFNS).** 'Aim High … Fly-Fight-Win' to be Air Force motto. [Online] U.S. Air Force, October 7, 2010. http://www.af.mil/news/story.asp?id=123225546.
157. **Simonite, Tom.** Malware Traffic Spikes Preceded Russian and Israeli Conflicts. [Online] MIT Technology Review, August 8, 2014. http://www.technologyreview.com/news/529936/malware-traffic-spikes-preceded-russian-and-israeli-conflicts/.
158. **Associated Press.** FBI releases papers on Russian spy ring that involved Anna Chapman. [Online] The Guardian, October 31, 2011. http://www.theguardian.com/world/2011/oct/31/fbi-russian-spy-ring-anna-chapman.
159. **Duke, Alan.** 'Fight Club' producer Arnon Milchan: I helped Israeli spy agency. [Online] CNN, November 26, 2013. http://www.cnn.com/2013/11/26/showbiz/hollywood-producer-israel-spy-agency/index.html.
160. **Pagliery, Jose.** Russia attacks U.S. oil and gas companies in massive hack. [Online] CNNMoney, July 2, 2014. http://money.cnn.com/2014/07/02/technology/security/russian-hackers/index.html.
161. **Goldman, David.** China vs. U.S.: The cyber Cold War is raging. [Online] CNNMoney, July 28, 2011. http://money.cnn.com/2011/07/28/technology/government_hackers/index.htm.
162. **Ryan, Jason.** US Government and Military Websites Redirected to Chinese Servers. [Online] ABC News, November 17, 2010. http://abcnews.go.com/Technology/american-government-websites-hijacked-chinese-hackers-massive-april/story?id=12165826.
163. **Beech, Hannah.** Meet China's Newest Soldiers: An Online Blue Army. [Online] Time Magazine, May 27, 2011. http://world.time.com/2011/05/27/meet-chinas-newest-soldiers-an-online-blue-army/.
164. **Mullen, Jethro.** New York Times, Wall Street Journal say Chinese hackers broke into computers. [Online] CNN, January 31, 2013. http://www.cnn.com/2013/01/31/tech/china-nyt-hacking/index.html.
165. **Riley, Charles.** China's military denies hacking allegations. [Online] CNNMoney, February 20, 2013. http://money.cnn.com/2013/02/20/technology/china-cyber-hacking-denial/index.html.
166. **Sanger, David E. and Perlroth, Nicole.** Hackers From China Resume Attacks on U.S. Targets. [Online] The New York Times, May 19, 2013. http://www.nytimes.com/2013/05/20/world/asia/chinese-hackers-resume-attacks-on-us-targets.html?_r=0.
167. **Liberto, Jennifer.** New Chinese hacker group targets governments and nuclear facilities. [Online] CNN, June 4, 2013. http://money.cnn.com/2013/06/04/technology/security/cyber-hacker-group/index.html.

168. **Sanger, David E. and Shanker, Thom.** N.S.A. Devises Radio Pathway Into Computers. [Online] The New York Times, January 14, 2014. http://www.nytimes.com/2014/01/15/us/ nsa-effort-pries-open-computers-not-connected-to-internet.html?_r=0.

169. **Fantz, Ashley.** Chinese hackers infiltrated U.S. companies, attorney general says. [Online] CNN, May 19, 2014. http://www.cnn.com/2014/05/19/justice/china-hacking-charges/ index.html.

170. **CNN Staff.** CENTCOM Twitter account hacked, suspended. [Online] CNN, January 12, 2015. http://www.cnn.com/2015/01/12/politics/centcom-twitter-hacked-suspended/index.html.

171. **Fantz, Ashley.** After ISIS Twitter threat, military families rethink online lives. [Online] CNN, January 14, 2015. http://www.cnn.com/2015/01/14/us/social-media-military-isis/index.html.

172. **NBC News and msnbc.com.** Gunman kills 12, wounds 31 at Fort Hood. [Online] NBCNews.com, November 5, 2009. http://www.nbcnews.com/id/33678801/ns/us_news-crime_and_courts/t/gunman-kills-wounds-fort-hood/#.VNWe-i5Gwl8.

173. **BBC News.** Woolwich attack: Lee Rigby named as victim. [Online] BBC News, May 23, 2013. http://www.bbc.com/news/uk-22644857.

174. **Hunt, Nick.** London attack: Terrorists targeting soldiers at home again? [Online] CNN, May 23, 2013. http://www.cnn.com/2013/05/22/world/europe/britain-london-attack-soldiers/ index.html.

175. **Collin, Barry.** The Future of Cyberterrorism. [Online] Crime & Justice International, March 1997. http://www.cjimagazine.com/archives/cji4c18.html?id=415.

176. Assessing the Risks of Cyber Terrorism, Cyber War and Other Cyber Threats. [Online] Center for Strategic and International Studies (CSIS), December 2002. http://csis.org/files/ media/csis/pubs/021101_risks_of_cyberterror.pdf.

177. **Tafoya, William L.** Cyber Terror. [Online] Federal Bureau of Investigation, November 2011. http://www.fbi.gov/stats-services/publications/law-enforcement-bulletin/november-2011/ cyber-terror.

178. **Foster, John S., et al.** Report of the Commission to Assess the Threat to the United States from Electromagnetic Pulse (EMP) Attack. [Online] The EMP Commission, April 2008. http://www.empcommission.org/docs/A2473-EMP_Commission-7MB.pdf.

179. **Bumiller, Elisabeth and Shanker, Thom.** Panetta Warns of Dire Threat of Cyberattack on U.S. [Online] The New York Times, October 11, 2012. http://www.nytimes. com/2012/10/12/world/panetta-warns-of-dire-threat-of-cyberattack.html?pagewanted=all.

180. **Goldman, David.** Nations prepare for cyber war. [Online] CNNMoney, January 7, 2013. http://money.cnn.com/2013/01/07/technology/security/cyber-war/index.html.

181. **Lourdeau, Keith.** Keith Lourdeau, Cyber Division, FBI, Before the Senate Judiciary Subcommittee on Terrorism, Technology, and Homeland Security. [Online] 2004, 24 February. http://www.fbi.gov/news/testimony/hearing-on-cyber-terrorism.

182. **National Intelligence Council.** Global Trends 2030: Alternative Worlds. [Online] U.S. National Intelligence Council, December 2012. http://www.dni.gov/files/documents/Global Trends_2030.pdf.

183. **RT.** Hacker dies days before he was to reveal how to remotely kill pacemaker patients. [Online] RT, July 26, 2013. http://rt.com/usa/hacker-pacemaker-barnaby-jack-639/.

184. —. US warns of cyberattacks targeting medical devices. [Online] RT, June 14, 2013. http:// rt.com/usa/us-warns-medical-cyberattack-721/.

185. **Pagliery, Jose.** AIG cyber insurance covers bodily harm. [Online] CNNMoney, April 23, 2014. http://money.cnn.com/2014/04/23/technology/security/aig-cybersecurity-insurance/ index.html.

186. **Dunn, Jerome.** EXtreme ACcuracy Tasked Ordnance (EXACTO). [Online] DARPA, July 10, 2014. http://www.darpa.mil/Our_Work/TTO/Programs/Extreme_Accuracy_Tasked_Ord nance_%28EXACTO%29.aspx.

187. **IMDb.** 24: Live Another Day. [Online] IMDb, May 5, 2014. http://www.imdb.com/ title/tt1598754/.

188. **Zetter, Kim.** Sony Hackers Threaten to Release a Huge 'Christmas Gift' of Secrets. [Online] Wired, December 15, 2014. http://www.wired.com/2014/12/sony-hack-part-deux/.

189. **Pagliery, Jose.** 'Sony-pocalypse': Why the Sony hack is one of the worst hacks ever. [Online] CNNMoney, December 5, 2014. http://money.cnn.com/2014/12/04/technology/security/sony-hack/.

190. **Robb, David.** Sony Hack: A Timeline. [Online] Deadline, December 22, 2014. http://deadline.com/2014/12/sony-hack-timeline-any-pascal-the-interview-north-korea-1201325501/.

191. **Kastrenakes, Jacob and Brandom, Russell.** Sony Pictures hackers say they want 'equality,' worked with staff to break in. [Online] The Verge, December 25, 2014. http://www.theverge.com/2014/11/25/7281097/sony-pictures-hackers-say-they-want-equality-worked-with-staff-to-break-in.

192. **Pagliery, Jose.** Why North Korea's attack should leave every company scared stiff. [Online] CNNMoney, December 19, 2014. http://money.cnn.com/2014/12/19/technology/security/hacking-companies-north-korea/index.html.

193. **Bradner, Eric.** Obama: North Korea's hack not war, but 'cybervandalism'. [Online] CNN, December 21, 2014. http://www.cnn.com/2014/12/21/politics/obama-north-koreas-hack-not-war-but-cyber-vandalism/index.html.

194. **Stelter, Brian and Pallotta, Frank.** Sony zeroing in on North Korea as it investigates cyberattack. [Online] CNNMoney, December 3, 2014. http://money.cnn.com/2014/12/03/media/sony-north-korea-cyberattack/index.html.

Chapter 10
Cyber Attacks, Prevention, and Countermeasures

There is no such thing as 100 percent security, on- or offline, but we must strive to strengthen our defenses against those who are constantly working to do us harm.... The alternative could be a digital Pearl Harbor — and another day of infamy.
— U.S. Senators Joe Lieberman, Susan Collins
and Tom Carper (July 7, 2011)

There are only two types of companies: those that have been hacked, and those that will be. Even that is merging into one category: those that have been hacked and will be again.
— FBI Director Robert Mueller
RSA conference (March 1, 2012)

The attack surfaces for adversaries to get on the Internet now include all those mobile devices. The mobile security situation lags. It's far behind.
— Army Gen. Keith Alexander, Director of National Security
Agency and Commander of U.S. Cyber Command
DEF CON 20 (July 27, 2012)

The only two products not covered by product liability are religion and software, and software should not escape for much longer.
— Dan Geer, Chief Information Security Officer of In-Q-Tel
(August 6, 2014)

Overall, network security solutions haven't evolved for the past 20-plus years.
— Mike Kail, Chief Information Officer at Yahoo!
(October 22, 2014)

We are making sure our government integrates intelligence to combat cyber threats, just as we have done to combat terrorism.
— President Barack Obama in 2015 State of the Union address
(January 20, 2015)

© Springer International Publishing Switzerland 2015
N. Lee, *Counterterrorism and Cybersecurity*, DOI 10.1007/978-3-319-17244-6_10

10.1 Cybersecurity Acts

President Barack Obama said in the 2015 State of the Union address, "No foreign nation, no hacker, should be able to shut down our networks, steal our trade secrets, or invade the privacy of American families, especially our kids. We are making sure our government integrates intelligence to combat cyber threats, just as we have done to combat terrorism. And tonight, I urge this Congress to finally pass the legislation we need to better meet the evolving threat of cyber attacks, combat identity theft, and protect our children's information. If we don't act, we'll leave our nation and our economy vulnerable. If we do, we can continue to protect the technologies that have unleashed untold opportunities for people around the globe" [1].

In response to the ever-increasing number of cyber attacks on both private companies and the United States government, U.S. Congress has introduced the Cyber-Security Enhancement Act of 2007 [2], the National Commission on American Cybersecurity Act of 2008 [3], the Cybersecurity Act of 2009 [4], the Cyber Security and American Cyber Competitiveness Act of 2011 [5], the Cybersecurity Act of 2012 [6], and the Cybersecurity Information Sharing Act of 2014 [7].

In July 2011, U.S. senators Joe Lieberman, Susan Collins and Tom Carper wrote in *The Washington Post* in support of their cybersecurity bill: "There is no such thing as 100 percent security, on- or offline, but we must strive to strengthen our defenses against those who are constantly working to do us harm.... The alternative could be a digital Pearl Harbor — and another day of infamy" [8].

U.S. Senate Commerce Committee Chairman Jay Rockefeller said at a Senate Intelligence Committee hearing in January 2012, "The threat posed by cyber attacks is greater than ever, and it's a threat not just to companies like Sony or Google but also to the nation's infrastructure and the government itself. Today's cybercriminals have the ability to interrupt life-sustaining services, cause catastrophic economic damage, or severely degrade the networks our defense and intelligence agencies rely on. Congress needs to act on comprehensive cybersecurity legislation immediately" [9].

In July 2014, U.S. Senate Intelligence Committee approved the Cybersecurity Information Sharing Act authored by Intelligence Committee chairwoman Dianne Feinstein and vice chairman Saxby Chambliss. The bill encourages private companies and the federal government to share information about cyber threats with each other, and it gives the companies liability protections for sharing information about and responding to cyber threats. "We had to make compromises between what the business sector wanted and what the privacy folks wanted," Saxby Chambliss remarked. "The committee did a good job of achieving compromises on significant issues. The cyber threats to our nation are all too real" [7].

However, senators Ron Wyden and Mark Udall opposed the bill. Opponents of cybersecurity acts view the proposed legislations as digital versions of the Patriot Act of 2001, an unnecessary government intrusion into private businesses reminiscent of the Sarbanes-Oxley Act of 2002, or a justification for an overreaching "cyber-industrial complex" akin to the expansive military-industrial complex [10–12]. Privacy advocates fear that the proposed bills would give the U.S. government too much authority to examine the content of emails, file transfers, and Web searches. Former National Security Agency (NSA) consultant Ed Giorgio said, "Google has records that could help in a cyber-investigation. We have a saying in this business: 'Privacy and security are a zero-sum game" [13].

10.2 Cybersecurity Initiatives: CNCI, NICE, Presidential Executive Order

While the cybersecurity acts are embattled in heated debates, President Barack Obama signed an executive order in February 2013 to reduce cyber risk to critical infrastructure and to improve cybersecurity information sharing between the private and public sectors [14]. No one could disagree with the 2009 U.S. Cyberspace Policy Review that identified enhanced information sharing as a key component of effective cybersecurity [15]. Established by President George W. Bush in January 2008 and reconfirmed by President Barack Obama in May 2009, the Comprehensive National Cybersecurity Initiative (CNCI) has launched the following initiatives to assure a trusted and resilient information and communications infrastructure across government agencies [16]:

1. Manage the Federal Enterprise Network as a single network enterprise with Trusted Internet.
2. Deploy an intrusion detection system of sensors across the Federal enterprise.
3. Pursue deployment of intrusion prevention systems across the Federal enterprise.
4. Coordinate and redirect research and development (R&D) efforts.
5. Connect current cyber ops centers to enhance situational awareness.
6. Develop and implement a government-wide cyber counterintelligence (CI) plan.
7. Increase the security of our classified networks.
8. Expand cyber education.
9. Define and develop enduring "leap-ahead" technology, strategies, and programs.
10. Define and develop enduring deterrence strategies and programs.
11. Develop a multi-pronged approach for global supply chain risk management.
12. Define the Federal role for extending cybersecurity into critical infrastructure domains.

Education is one of the key CNCI initiatives, and its scope has expanded from a federal focus to a larger national focus. Starting in April 2010, the National Initiative for Cybersecurity Education (NICE) represents the continual evolution of CNCI, and the National Institute of Standards and Technology (NIST) has assumed the overall coordination role [17].

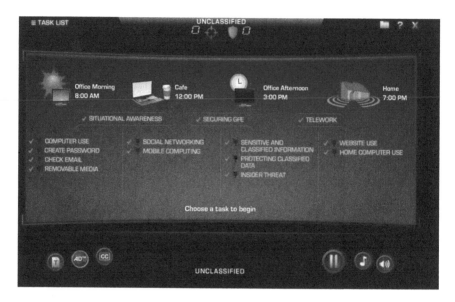

Fig. 10.1 Cyber Awareness Challenge (Courtesy of U.S. Department of Defense)

With the support of CNCI and NICE, the U.S. Department of Defense (DoD) Information Assurance Support Environment (IASE) has launched Cyber Awareness Challenge education programs for DoD and Federal employees [18]. The training programs are unclassified and accessible by the general public. The exercises cover a wide range of computer usage and serve as an excellent basic tool for everyone to learn about cybersecurity (see Fig. 10.1).

10.3 National Cyber Security Awareness Month (NCSAM)

President Barack Obama designated October as the National Cyber Security Awareness Month (NCSAM). NCSAM is designed to "engage and educate public and private sector partners through events and initiatives with the goal of raising awareness about cybersecurity and increasing the resiliency of the nation in the event of a cyber incident" [19].

October 2012 marked the ninth annual National Cyber Security Awareness Month sponsored by the U.S. Department of Homeland Security (DHS) in cooperation with the National Cyber Security Alliance (NCSA) and the Multi-State Information Sharing and Analysis Center (MS-ISAC).

The DHS encourages the entire American community to ACT — Achieve Cybersecurity Together. It reflects the interconnectedness of the modern world and the responsibility that everyone has in securing cyberspace. DHS oversees Cyber Storm, a biennial simulated cyber attacks exercise to strengthen cyber preparedness in the public and private sectors [20]. Cyber Storm I through IV were held in February 2006, March 2008, September 2010, and from Fall 2011 to 2012.

The NCSA publishes the website www.StaySafeOnline.org to educate computer users how to stay safe online, teach online safety to children and adults, keep business safe online, and get involved in National Cyber Security Awareness Month (NCSAM), Data Privacy Day (DPD), and National Cyber Security Education Council (NCEC) [21].

The MS-ISAC is a 24 × 7 cybersecurity operations center that provides real-time network monitoring, early cyber threat warnings and advisories, vulnerability identification and mitigation, and incident response for state, local, tribal, and territorial (SLTT) governments [22].

10.4 Mitigation from Denial of Service (DoS, DDoS, DRDoS) Attacks

A denial of service (DoS) attack overwhelms a website's servers by flooding them with bogus requests, making the websites unreachable or slowing the servers to a crawl. A distributed denial of service (DDoS) attack employs botnets — networks of computers infected with malware either with or without the knowledge of the computer owners. A distributed reflected denial of service (DRDoS) attack sends forged requests from a spoofed Internet Protocol (IP) address to a large number of computers that will reply to the requests [23].

In September 2012, Iran launched a major DDoS attack on American banks using botnets that involved thousands of compromised, high-powered application servers. The attackers used a new tool known as "itsoknoproblembro" (pronounced "it's OK, no problem, bro") to create peak floods exceeding 60 gigabits per second [24]. Bank of America, JPMorgan Chase, PNC Bank, U.S. Bank, and Wells Fargo Bank were among the financial institutions that suffered slowdowns and sporadic outages.

According to cybersecurity firm Prolexic Technologies, the total number of DDoS attacks increased 53 % in 2012 compared to a year earlier in 2011. Within a 3-month period from Q3 2012 to Q4 2012, the total number of infrastructure (Layer 3 and 4) attacks increased 17 %, the total number of application (Layer 7) attacks surged 72 %, the average attack duration rose 67 %, and the average bandwidth was up 20 %. Prolexic Technologies highlighted in its report that "Q4 2012 was defined by the increasing scale and diversity of DDoS attacks as well as the enduring nature of botnets" [25].

The most common network infrastructure (Layer 3 and 4) attack types are SYN floods and UDP floods, whereas the majority of flood traffic in the application (Layer 7) attack came in the form of GET floods and POST floods. Figure 10.2 shows the common DDoS attack vectors [25].

Various countermeasures against different types of DDoS attack exist in the Internet security community. Some common defenses are firewalls, routers, proxies, filtering, blackholing, and bandwidth over-provisioning [26–28]. However, as

Infrastructure (Layer 3 and 4) Attacks	Application (Layer 7) Attacks
ACK Floods	HTTP GET Floods
DNS (Domain Name System) Attacks	HTTP POST Floods
FINPUSH Floods	NTP (Network Time Protocol) Flood
ICMP (Internet Control Message Protocol) Floods	PUSH Floods
IGMP (Internet Group Management Protocol) Floods	SSL GET Floods
RIP Flood	SSL POST Floods
SYN Floods	
SYN PUSH Flood	
TCP Fragment Floods	
TCP Reset Floods	
UDP Floods	
UDP Fragment Floods	

Fig. 10.2 Common DDoS attack vectors

DDoS attacks are becoming more massive and sophisticated over the years, companies such as Prolexic Technologies, Tata Communications, and VeriSign have been offering their 24/7 monitoring and mitigation services. They provide a cloud-based DDoS mitigation service by "stopping a DDoS attack in the cloud before it reaches a customer's network" [29–31]. When a DDoS attack is detected, all traffic is routed through a "scrubbing center" which removes the bad traffic using filtering techniques and anti-DoS hardware devices, and establishes an Internet clean pipe to the customer's servers.

However, DDoS mitigation comes at a cost. In March 2013, a prolonged DDoS assault targeting The Spamhaus Project, a European spam-fighting group, resulted in a global slowdown of Internet traffic for millions of users. "It's the biggest attack we've seen," said Matthew Prince of the security firm CloudFlare. "One way to defend against DDoS attacks is to deflect some of the traffic targeted at a single server onto a bunch of other servers at different locations. That's what happened in this case, and why Web users experienced some slowdowns on other sites" [32].

Perhaps peer-to-peer content distribution is a better solution. In December 2014, BitTorrent has been testing its Maelstrom browser that can view Torrent-hosted websites across distributed peer-to-peer networks [33]. Since anyone visiting a website would automatically help share the site's content, denial of service attacks would be rendered ineffective unless it shuts down the entire Internet. BitTorrent CEO Eric Klinker said, "There'd be fewer centralized servers to get in the way of you and your data or you and the content you're interested in." Project lead Rob Velasquez added, "We've long looked at BitTorrent as a client that downloads files. We took that idea and said 'what if they weren't files? What if we put an HTML file inside of a Torrent and the browser just renders it?'"

Nonetheless, no single countermeasure or mitigation service is 100 % efficacious. To stop denial of service attacks requires the entire international community to ACT — Achieve Cybersecurity Together. As the world is increasingly interconnected, everyone shares the responsibility of securing cyberspace.

One effective solution to counter DDoS attacks is to drastically reduce the number of botnets and to eliminate their command-and-control (CnC) servers:

1. A powerful DDoS attack requires a large-scale botnet, preferably high-powered application servers or smartphones that are always on. By taking proactive measures in protecting our computer devices from malware and unauthorized access, we automatically reduce the scale of possible DDoS attacks. Proactive measures include applying system security patches promptly, installing firewalls and antivirus software, using strong passwords, and regularly checking system logs for suspicious activities.

2. A command-and-control (CnC) server is the brain of a botnet. In April 2011, the Federal Bureau of Investigation (FBI) dismantled the Coreflood botnet that had infected an estimated two million computers with malware. The FBI seized domain names, rerouted the botnet to FBI-controlled servers, and directed the zombie computers to stop the Coreflood malware [34]. In March 2012, Microsoft collaborated with the financial services industry and U.S. Marshals to seize CnC servers in Pennsylvania and Illinois, reducing the number of botnets controlling the Zeus malware that had infected more than 13 million computers worldwide [35]. Another success story — in the domain of spam attacks — is the takedown of the notorious Grum botnet in July 2012 [36]. Grum was the world's most prolific spam machine at its peak, generating 18 billion junk emails a day. Internet Service Providers (ISPs) in the Netherlands, Panama, Russia, and Ukraine joined forces to knock down all of Grum's CnC servers in three days, and 50 % of worldwide spam vanished immediately [37].

Unless everyone — individual, business, and government — heeds the proactive countermeasures, denial of service attacks are only going to get worse for everybody. In 2011, a Dirt Jumper v3 DDoS toolkit went on sale for as little as $150 on underground retail websites [38]. In 2012, a free-to-download High Orbit Ion Cannon (HOIC) program enables anyone online to launch a DDoS attack against a victim website [39]. In December 2014, Lizard Squad demonstrated its "Lizard Stresser" tool by taking down the Xbox and PlayStation networks on Christmas. The group offers paying customers denial of service attacks against any IP addresses for as low as $2.99 for 100 s of downtime [40].

10.5 Data Breach Prevention

At the 2012 RSA conference in San Francisco, Federal Bureau of Investigation (FBI) director Robert Mueller stated, "There are only two types of companies: those that have been hacked, and those that will be. Even that is merging into one category: those that have been hacked and will be again" [41]. Computer security executive Dmitri Alperovitch added, "I divide the entire set of Fortune Global 2000 firms into two categories: those that know they've been compromised and those that don't yet know" [42].

In 2014, U.S. Department of Homeland Security and the Secret Service estimated that "more than 1,000 businesses in the United States had been infected with malware that is programmed to siphon payment card details from cash registers in stores" [43]. These businesses did not realize that they had been sharing customers' credit card information with criminals due to security weakness in point-of-sale terminals [44]. In total, about 110 million Americans or 47 % of U.S. adults had their personal information exposed by data breach in a 12-month period [45].

Cybercriminals have stolen intellectual property (IP) from high-profile companies such as Google and Adobe [46]. Retired FBI agent Shawn Henry spoke at the 2012 Black Hat Conference in Las Vegas: "We worked with one company that lost $1 billion worth of IP in the course of a couple of days — a decade of research. That is not an isolated event.... Your data is being held hostage, and the life of your organization is at risk" [47].

Some of the massive data breach incidents occurring between 2007 and 2015 include:

1. In January 2007, up to 94 million Visa and MasterCard account numbers were stolen from TJX Companies dating back to 2005 and earlier, causing more than $68 million in fraud-related losses involving Visa cards alone [48].
2. In December 2009, online games service company RockYou suffered a data breach that resulted in the exposure of over 32 million usernames and passwords. Adding insult to injury, RockYou stored passwords in plain text format without any encryption [49] .
3. In January 2011, cybercriminals breached the database of online dating site PlentyofFish.com, exposing the personal and password information on nearly 30 million users [50].
4. In March 2011, computer and network security firm RSA suffered a massive data breach, jeopardizing the effectiveness of its SecurID system that is being used by more than 25,000 corporations and 40 million users around the world [51].
5. In April 2011, a cybercriminal stole the names, birth dates, and credit card numbers of 77 million customers on the Sony PlayStation Network [52].
6. In May 2011, cybercriminals illegally accessed 360,083 Citigroup customer accounts and withdrew $2.7 million from about 3,400 credit cards [53].
7. In December 2011, hackers affiliated with the Anonymous broke into the private intelligence analysis firm Strategic Forecasting (Stratfor) and obtained private information of about 860,000 people including former U.S. Vice President Dan Quayle, former Secretary of State Henry Kissinger, and former CIA Director Jim Woolsey. The group went on to publish the stolen emails and thousands of credit card numbers on the Internet [54].
8. In January 2012, Zappos.com was hacked, exposing 24 million customers' information including names, email addresses, billing and shipping addresses, phone numbers, the last four digits of credit card numbers, and encrypted passwords [55].
9. From January to February 2012, cybercriminals cracked into the Global Payments administrative account and stole more than 10 million credit and debit card transaction records [56].

10. In June 2012, cybercriminals stole 6.5 million LinkedIn passwords and posted them on an online forum. LinkedIn passwords were encoded using SHA-1 — a cryptographic hash function with weak collision resistance. About half of the encrypted passwords have been decrypted and posted online [57].

11. In June 2012, online dating site eHarmony notified its customers that a "small fraction" of its user base had been compromised [58].

12. In July 2012, Yahoo! Voices was hacked, resulting in the theft of 450,000 customer usernames and passwords [59].

13. In October 2012, Barnes and Noble disclosed that a PIN pad device used by customers to swipe credit and debit cards had been compromised at 63 of its national stores located in California, Connecticut, Florida, Illinois, Massachusetts, New Jersey, New York, Pennsylvania, and Rhode Island [60].

14. In January 2013, cybercriminals accessed Twitter user data and stole the usernames, email addresses, session tokens, and encrypted/salted versions of passwords for approximately 250,000 users [61].

15. In February 2013, personal information of more than 4,000 U.S. bank executives was stolen from the Federal Reserve System by exploiting a temporary vulnerability in a website vendor product [62].

16. Between March 2012 and February 2013, Experian — one of the three major consumer credit bureaus in the U.S. — unknowingly sold consumer data to an identity theft service, thereby exposing the social security numbers, drivers license numbers, bank accounts, and credit card data of 200 million Americans [63].

17. In March 2013, cybercriminals gained access to cloud-storage service provider Evernote's user information, including usernames, email addresses, and encrypted passwords. As a result, Evernote required all of its 50 million users to reset their passwords [64].

18. In April 2013, LivingSocial, the daily deals site owned in part by Amazon.com, suffered a massive cyber attack on its computer systems that contained more than 50 million customer names, emails, birthdates, and encrypted passwords [65].

19. In October 2013, Adobe's chief security officer Brad Arkin disclosed that "the attackers removed from our systems certain information relating to 2.9 million Adobe customers, including customer names, encrypted credit or debit card numbers, expiration dates, and other information relating to customer orders" [66].

20. From June to October 2013, about 1.1 million customers were impacted by a security breach of luxury retailer Neiman Marcus' payment systems due to a "sophisticated, self-concealing malware, capable of fraudulently obtaining payment card information" [67].

21. In November 2013, cybersecurity firm Trustwave discovered a massive data breach facilitated by keylogging malware installed on an untold number of computers around the world. Cyber intruders stole about 2 million account credentials from Facebook, Gmail, Google+, YouTube, Yahoo, Twitter, LinkedIn, and 93,000 other websites [68].

22. From November to December 2013, cybercriminals stole 40 million credit and debit card information as well as 70 million customers' information such as their name, address, phone number, and email address from discount retailer Target [69] .

23. From May 2013 to January 2014, retailer Michaels and subsidiary Aaron Brothers were hacked. Some of their 3 million customers' credit and debit card numbers and expiration dates were compromised [70].

24. In January 2014, cyber attackers posted the account information of 4.6 million Snapchat users, making usernames and partial phone numbers available for download from the website SnapchatDB.info [71].

25. In January 2014, Starbucks fixed its mobile app that is used by 10 million customers. Prior to the fix, Starbucks made the same mistake as RockYou in 2009 by storing passwords in plain text format without any encryption [72].

26. In February 2014, cyber intruders broke into Kickstarter's database and stole an unknown number of usernames, email addresses, mailing addresses, phone numbers, and encrypted passwords among its 5.7 million members. Kickstarter passed $1 billion in pledges on March 3, 2014 [73].

27. Between February and March 2014, cybercriminals gained access to eBay's corporate network and stole a database containing eBay customers' name, encrypted password, email address, physical address, phone number, and date of birth [74]. The data breach potentially affected all of eBay's 148 million customers.

28. Between March and August 2014, United Parcel Service (UPS) suffered a computer breach at 51 of its stores in 24 states [75]. Customer names, postal addresses, email addresses, and payment card information were compromised.

29. In April 2014, AOL verified that there was unauthorized access to information regarding a "significant number" of its 120 million user accounts. According to AOL, the compromised information included "AOL users' email addresses, postal addresses, address book contact information, encrypted passwords and encrypted answers to security questions that we ask when a user resets his or her password, as well as certain employee information." In addition, AOL believed that "spammers have used this contact information to send spoofed emails that appeared to come from roughly 2 % of our email accounts" [76]. (Two percent of 120 million is 2.4 million email accounts.)

30. Between April and June 2014, cybercriminals stole confidential data from 4.5 million patients who had received treatment from Community Health Systems (CHS), a company that runs more than 200 hospitals [77].

31. In June 2014, restaurant chain P.F. Chang acknowledged that at least thousands of credit and debit card numbers were stolen in early 2014 from some of the 200+ locations in the U.S. [78].

32. In August 2014, Hold Security revealed that a Russian crime ring has stolen a massive 1.2 billion username/password combinations and more than 500 million email addresses from 420,000 websites. "Hackers did not just target U.S. companies, they targeted any website they could get, ranging from Fortune 500 companies to very small websites," said Alex Holden, founder and chief information security officer of Hold Security. "And most of these sites are still vulnerable" [79].

33. In August 2014, Apple's iCloud online data backup service might have been compromised by cybercriminals who stole and shared publicly nude selfies of Jennifer Lawrence, Kirsten Dunst, Kate Upton, Mary Elizabeth Winstead, and other celebrities. Apple claimed that the celebrity accounts "were compromised by a very targeted attack on usernames, passwords and security questions" [80].
34. Between April and September 2014, cybercriminals broke into Home Depot's in-store payments systems, siphoning payment card details from 56 million customers in the United States and Canada [43].
35. In October 2014, cyber attackers broke into Staples' computers and stole 1.16 million shoppers' credit cards and debit cards information including card holder names, card numbers, expiration dates, and verification codes [81].
36. In October 2014, JPMorgan revealed that cyber intruder got hold of contact information for 76 million households and 7 million small businesses, including names, addresses, phone numbers and email addresses, as well as "internal JPMorgan Chase information relating to such users" [82].
37. In October 2014, cyber intruders breached the computer systems of USIS, a major contractor that provides background checks for the U.S. government, to access private information on U.S. government employees [83].
38. In November 2014, cybercriminals who called themselves the Guardians of Peace (GOP) stole 100 terabytes of data [84], destroyed 75 % of corporate computer servers, and crippled the company's data centers [85]. Among the stolen data were five feature films, executive emails, business contracts, company budgets, employee personal data, salary information, medical records, and celebrity secrets [86].
39. Between November and December 2014, cyber attackers got into the internal system of the Internet Corporation for Assigned Names and Numbers (ICANN) that manages global IP addresses and Domain Name System (DNS). The intruders were able to see all user names, addresses, emails, phone numbers, and encrypted passwords [87].
40. In December 2014, a Twitter account claiming affiliation with the Anonymous released a list of usernames, passwords, credit card numbers (with security codes and expiration dates) for 13,000 accounts on Amazon, PlayStation, Xbox Live, Hulu Plus, Walmart and other retail and entertainment services including pornography sites [88].
41. In December 2014, security firm FireEye confirmed that a group of cyber spies stole confidential corporate information and trade secrets from more than 100 publicly traded companies of which two-thirds are in the healthcare and pharmaceutical industries [89]. All but three of the victims are listed on the New York Stock Exchange or Nasdaq.
42. In February 2015, Anthem Inc., America's second largest health insurer, revealed that cybercriminals broke into a database containing names, birthdays, addresses, and social security numbers of 80 million customers and employees [90].

43. In March 2015, Amazon's live-stream gaming network Twitch with over 100 million members had expired all user passwords and stream keys, and disconnected all user accounts from Twitter and YouTube, due to possible "unauthorized access" by cyber intruders to "Twitch username and associated email address, your password, the last IP address you logged in from, first and last name, phone number, address, and date of birth."

According to a 2013 M-Trends report by cybersecurity firm Mandiant, it takes a company 243 days on average to discover a breach [91]. In 2012, the Verizon RISK (Response, Intelligence, Solutions, Knowledge) Team released a Data Breach Investigations Report in cooperation with the Australian Federal Police (AFP), Dutch National High Tech Crime Unit (NHTCU), Irish Reporting and Information Security Service (IRISS), U.K. Police Central e-Crime Unit (PCeU), and U.S. Secret Service (USSS) [92]. The most significant findings were:

1. The number of compromised data in 2011 skyrocketed to 174 million records, compared to 4 million a year before in 2010.
2. More than half (58 %) of the data breaches were tied to hacktivism — the use of technology and hacking skills to achieve social or political ends.
3. 81 % of data breaches were due to hacking and 69 % of that involved malware.
4. Both hacking and malware incidents were up considerably by 31 % and 20 % respectively.
5. An overwhelming 96 % of cyber attacks were unsophisticated and 97 % of breaches were avoidable through simple security measures such as firewalls and strong password protection.
6. 94 % of all data compromised involved servers.

In fact, the majority of data breach is preventable. In addition to using strong passwords, firewalls, and antivirus programs, there are three basic strategies that all businesses should follow:

1. One of the most common methods for breaching network security is SQL injection attack (SQLIA) as reported in the Yahoo! Voices data breach in July 2012 [93]. Structured Query Language (SQL) is the programming language used to manage data in a relational database management system (RDBMS). Cybercriminals inject unexpected or malformed SQL commands into the database in order to change its content or to dump its information to the attackers.

    The Open Web Application Security Project (OWASP) provides a clear, simple, and actionable guidance for preventing SQL Injection security flaws in application databases [94]. The primary defenses are using prepared statements (parameterized queries), using stored procedures, and escaping all user supplied input. Additional defenses are enforcing least privilege and performing white list input validation. Last but not least, strong encryption should be used to protect the sensitive information stored in the databases.
2. In the 2007 TJX Companies data breach incident, cybercriminals had intercepted wireless transfers of customer information at two Miami-area Marshalls stores since 2005 [95]. Global payment processing company First Data Corporation advices merchants to employ end-to-end encryption (E2EE) and

tokenization: "All merchants — whether they are brick-and-mortar, brick-and-click, or completely web-based — have both an obligation and an industry mandate to protect consumers' payment card data. The Payment Card Industry (PCI) Data Security Standards (DSS) provide guidelines on what merchants need to do to secure the sensitive data used in payment transactions. End-to-end encryption (E2EE) and tokenization solve for many of the vulnerabilities that exist in the payments processing chain. Encryption mitigates security weaknesses that exist when cardholder data has been captured but not yet authorized, and tokenization addresses security vulnerabilities after a transaction has been authorized. When combined, these two technologies provide an effective method for securing sensitive data wherever it exists throughout its lifecycle" [96].

In February 2015, Visa Inc. unveiled plans to expand tokenization in order to obfuscate customers' information during transactions. "Removing card account numbers from the processing and storage of payments represents one of the most innovative and promising technologies we've seen in decades," said Visa chief executive Charlie Scharf [97].

3. Data Loss Prevention (DLP) depends not only on cybersecurity technology but also on data management policies and human vigilance in implementing the policies. Shane MacDougall, champion of the social engineering "capture the flag" contest in DEF CON 20, made a phone call to a Wal-Mart store manager and successfully convinced him to divulge details of the store operations as well as computer security information. Wal-Mart spokesman Dan Fogleman told *CNNMoney*, "We emphasize techniques to avoid social engineering attacks in our training programs. We will be looking carefully at what took place and learn all we can from it in order to better protect our business" [98]. Perhaps one of the most heart-wrenching stories is about the security flaws in Apple and Amazon's customer service that led to a devastating hack into *Wired* technology journalist Mat Honan in August 2012 [99]:

In the space of one hour, my entire digital life was destroyed. First my Google account was taken over, then deleted. Next my Twitter account was compromised, and used as a platform to broadcast racist and homophobic messages. And worst of all, my AppleID account was broken into, and my hackers used it to remotely erase all of the data on my iPhone, iPad, and MacBook.

Getting into Amazon let my hackers get into my Apple ID account, which helped them get into Gmail, which gave them access to Twitter…. But what happened to me exposes vital security flaws in several customer service systems, most notably Apple's and Amazon's. Apple tech support gave the hackers access to my iCloud account. Amazon tech support gave them the ability to see a piece of information — a partial credit card number — that Apple used to release information. In short, the very four digits that Amazon considers unimportant enough to display in the clear on the web are precisely the same ones that Apple considers secure enough to perform identity verification. The disconnect exposes flaws in data management policies endemic to the entire technology industry, and points to a looming nightmare as we enter the era of cloud computing and connected devices.

High-profile data breaches over the years ought to remind businesses that they are responsible for the safety of their customer's information. In November 2012, researchers Luigi Auriemma and Donato Ferrante found a serious vulnerability in the game "Call of Duty: Modern Warfare 3" as well as in the CryEngine 3 graphics platform. "Once you get access to the server, which is basically the interface with the company, you can get access to all of the information on the players through the server," Ferrante said. "In general, game companies don't seem to be very focused on security but rather on performance of the game itself" [100]. This attitude has to change. The Sony PlayStation Network data breach incident in April 2011 should have been a wake-up call for all game companies.

Facebook, with over 1 billion active monthly users, is an attractive target for cybercriminals. In January 2013, a sophisticated cyber attack targeted Facebook when its employees visited a mobile developer website that was compromised [101]. The Facebook Security team flagged a suspicious domain in their corporate DNS logs and discovered the presence of malware on several company laptops. However, they found no evidence that any Facebook user data was compromised.

James Andrew Lewis, a senior fellow and program director at the Center for Strategic and International Studies, identified four measures that stop more than 80 % of all known attacks: taking inventory of hardware; taking inventory of software; limiting administrative permissions; and automating network monitoring [102].

To help reduce the incidence and extent of data breach, Illumio launched a new software solution in October 2014 to monitor data flow and connections inside a data center to look for suspicious activities [103]. Security researcher Claudio Guarnieri created a free tool called Detekt that scans Windows computers for traces of commercial surveillance spyware FinFisher and Hacking Team RCS. Detekt was released in November 2014 in partnership with Amnesty International, Digitale Gesellschaft, Electronic Frontier Foundation, and Privacy International [104].

In January 2014, Prof. Ari Juels of Cornell University and Prof. Thomas Ristenpart of the University of Wisconsin at Madison developed "honey encryption" to bamboozle cyber attackers with fake data. "Decoys and deception are really underexploited tools in fundamental computer security," said Juels. "Each decryption is going to look plausible. The attacker has no way to distinguish a priori which is correct" [105]. Previously, Juels and Ronald L. Rivest at RSA worked on the Honeywords Project that improves that security of hashed passwords by adding honeywords (aka false passwords) to the password database [106].

10.6   Fighting Back Against Phishing and Spoofing

Phishing is an example of social engineering designed to manipulate people into divulging confidential information. The most common form of phishing is an email that appears to have come from a legitimate organization or known individual. Similar to Caller ID spoofing, the sender's email address is often forged in order to hide the real identity and origin of the sender.

The recipient of a phishing email is encouraged to open an attachment that would install malware on the victim's computer, or to click on a link that directs the user to a fake website whose look and feel are almost identical to the legitimate site.

In February 2011, *Contagio* published the details of a targeted phishing attack (aka spear phishing) against personal Gmail accounts of U.S. military and government employees and associates [107]. The spoofed emails appeared to have come from .gov and .mil domains, and they contained a link to a fake Gmail timed-out re-login page for the attackers to harvest the victim's credentials.

The Anti-Phishing Working Group (APWG) released a phishing activity trends report in September 2012 [108]. The report reveals some staggering numbers:

1. The number of unique phishing sites reached an all-time monthly high of 63,253 in April 2012, with an average of 58,409 in Q2 2012.
2. The total number of URLs used to host phishing attacks increased 7 % to 175,229 in Q2 2012, up from 164,023 in Q1 2012.
3. The U.S. remained the top hosting country of phishing-based Trojan horse virus in Q2 2012: United States 46 % vs. 2nd place Russia 12 % in April, United States 78 % vs. 2nd place United Kingdom 4 % in May, and United States 55 % vs. 2nd place France 11 % in June.

In the 2014 paper titled "Handcrafted Fraud and Extortion: Manual Account Hijacking in the Wild," Google researcher Elie Bursztein and others from Google and University of California San Diego reported their new findings that included the following [109]:

1. 35 % of phishing sites target victims' email.
2. 21 % of phishing sites target banking credentials.
3. A growing number of phishing sites are targeting App Stores and social networking credentials.
4. 20 % of compromised Google accounts were logged into within 30 min.
5. The top countries of origin for hijackers were China, Ivory Coast, Malaysia, Nigeria, and South Africa.

A more sophisticated hack involves DNS spoofing, DNS cache poisoning, or DNS hijacking whereby the Domain Name System (DNS) used to translate domain names into IP addresses is compromised. In 2009, Brazil's largest bank had its domain name redirected to a criminal's computer server for four hours, fooling customers who tried to log into their accounts [110]. Between 2008 and 2012, about 4 million computers worldwide were infected with the DNSChanger malware that redirected the victims to spoofed websites [111]. In November 2012, China intentionally DNS poisoned www.google.com and all other Google subdomains for political reasons [112].

To fight back against phishing and spoofing, businesses, governments, and individuals must take all possible proactive measures:

1. Sender Policy Framework (SPF) and DomainKeys Identified Mail (DKIM) have helped reduce the ease of spoofing. SPF allows the domain owners to specify which computers are authorized to send emails with sender addresses in that domain, whereas DKIM enables an email to carry a digital signature specifying its genuine domain name. Email service providers such as Gmail

and Yahoo! have implemented DKIM since 2008. In collaboration with eBay and PayPal, Gmail automatically discards all incoming emails that claim to be coming from ebay.com or paypal.com if they cannot be verified successfully with DKIM [113]. However, mathematician Zachary Harris discovered in 2012 that weak cryptographic keys used in DKIM exposed a massive security hole in Apple, Amazon.com, eBay, Gmail, HSBC, LinkedIn, Microsoft, PayPal, Twitter, US Bank, Yahoo!, and some other large organizations [114]. U.S. Department of Homeland Security's United States Computer Emergency Readiness Team (US-CERT) issued a vulnerability note in October 2012 warning that "DomainKeys Identified Mail (DKIM) Verifiers may inappropriately convey message trust when messages are signed using keys that are too weak (<1024 bits) or that are marked as test keys" [115]. Businesses should implement stronger cybersecurity with 1024-bit or 2048-bit cryptographic keys.

2. US-CERT has been collecting phishing email messages and website locations so that the information can help people avoid becoming victims of phishing scams [116]. Recipients of phishing emails are encouraged to forward them to phishing-report@us-cert.gov. In addition, social network users should forward all suspected phishing messages to phish@fb.com for Facebook, spoof@ebay.com for eBay, and @spam for Twitter. We should also report on fraudulent web pages that are designed to look like the legitimate websites in attempt to steal users' personal information. Reports can be submitted to the Google Safe Browsing Team [117]:
http://www.google.com/safebrowsing/report_phish/

3. Commercial websites can discourage phishing by making it easier for users to distinguish between genuine and fake sites by authenticating the site to the user. The method is known as two-way authentication, mutual authentication, or bidirectional authentications. For example, Capital One 360 (formerly ING Direct) employs two-factor, two-way authentication [118]. First, a customer signs in with a username and password. Second, ING Direct displays a secret phrase and picture that only the customer knows about when setting up the account. Third, the customer enters a six-digit passcode to gain access to the account. If a commercial website registers the IP address of the customer's computer, technically it can authenticate itself to the user even before asking for their username and password. Two-way mutual authentication is an effective guard against DNS spoofing.

10.7 Password Protection and Security Questions

In September 2008, 20-year-old college student David Kernell hacked into Republican vice presidential candidate Sarah Palin's Yahoo! email account to look for information that would derail her campaign [119]. Kernell managed to reset Palin's account password by entering her birth date and correctly answering the security question "Where did you meet your spouse?" It only took Kernell 45 min on Wikipedia and Google search to find the correct answer.

In July 2012, CNet analyzed the most frequently used passwords that surfaced in the Yahoo! Voices data breach. Among more than 450,000 stolen login credentials, the most common passwords in descending order of popularity were 123456, password, 111111, welcome, ninja, freedom, f*ck, baseball, superman, 000000, America, winner, starwars, batman, spiderman, lakers, maverick, ncc1701, startrek, and ncc1701a [120].

These and many other cybercriminal stories highlight the vulnerability of weak password protection and inadequacy of security questions. The best practices for protecting our login credentials are:

1. Turn on two-factor authentication on Google, Facebook, and other websites. In October 2014, Google launched support for Security Key, an open standard that allows you to log into an account with a physical device (e.g. a USB) together with an online password [121].

2. When coming up with a password, avoid dictionary words, acronyms, and abbreviations. A password should be difficult to guess but easy to remember so that it does not need to be written down anywhere. For example, if Jack and Jill are a married couple who visit Disneyland with their three children twice every year, a strong and easy-to-remember password would be something like "J&J3di2ya" which includes mixed case letters, numbers, and punctuations. Changing your password regularly is also made simple by slightly rearranging the password phrase. For instance, "J&J3di2ya" could become "JJ&3di2ya".

3. Do not use the same password across multiple websites. Instead, create a unique password for every site. It may sound like a daunting task if you have an account on many different websites, but there is a simple two-step solution. Step one: Create a strong password stem. Step two: Append a site-specific phrase to the stem. The sample password above, "J&J3di2ya", is an example of a strong password stem. Now, if Jack and Jill share a Gmail account that Jill created on the day her mother-in-law was visiting from England, a strong and easy-to-remember password for their Gmail account would be something like "J&J3di2yaMotherEng". Similarly, one of Jack and Jill's bank accounts might have the password "J&J3di2yaBettyBoop" because they ate at a local "Betty Boop" restaurant on the day they opened that particular bank account. The basic idea is to combine a strong password stem with phrases from associative memory or good imagination.

4. Do not answer the online security questions straightforwardly. Instead, treat the answers as passwords, give a long answer, or simply be creative. The following are some excellent examples:

   a. Where did you meet your spouse? Answer: J&Jvt1980lunch
      (Jack and Jill met at Virginia Tech in 1980 during lunch.)
   b. What was the name of your first school? Answer: Ismellcheese
      (Your first school reminds you of the smell of cheese, for whatever reason.)

c. What is your pet's name? Answer: Squarerootofminus178
(The square root of a negative number is an imaginary number, meaning I don't have a pet or I have an imaginary pet.)

You can test the strength of your chosen passwords by using a Microsoft Research tool called Telepathwords that "prevents weak passwords by reading your mind" [122]. The software tries to predict the next character of your passwords by using knowledge of:

1. Common passwords, such as those made public as a result of security breaches.
2. Common phrases, such as those that appear frequently on web pages or in common search queries.
3. Common password-selection behaviors, such as the use of sequences of adjacent keys.

In addition to strong passwords, companies must beef up their password security in order to safeguard customers' passwords on their computer servers. In a 2012 interview with reporter Brian Krebs, Matasano Security researcher Thomas H. Ptacek clarified some big misconceptions and offered some practical advices [123]. In recommending bcrypt password hash developed by Coda Hale [124], Ptacek said:

> The basic mechanism by which SHA-1 passwords are cracked, or MD5 or SHA-512 hasn't changed since the early 1990s. As soon as code to implement SHA-1 came out, it was also available to John the Ripper and other password cracking tools. It's a really common misconception — including among security people — that the problem here is using SHA-1. It would not have mattered at all if they had used SHA-512, they would be no better off at all.

> UNIX passwords, and they've been salted forever, since the 70s, and they have been cracked forever. The idea of a salt in your password is a 70s solution. Back in the 90s, when people broke into UNIX servers, they would steal the shadow password file and would crack that.

> A cryptographic hash wants to do the minimum amount of work possible in order to arrive at a secure result. But a password hash wants to deliberately be designed to do the maximum amount of work. There are modern, secure password hashes that would take hundreds or thousands of years to test passwords on. If we lost 10 million bcrypted passwords, instead of 3 million of them being published and compromised, you might be looking at tens or hundreds of user passwords being compromised.

> People should use better storage algorithms if they're going to inflict passwords on their users. They should do a better job of it. But the real answer is things like two-factor authentication with smart phones. Two-factor authentication seems like the answer to me.

In December 2014, Google released FIDO (Fast Identification Online) — an open standard that provides a cryptographic backing for any service or authenticator device you want to plug in [125]. Google is hoping that FIDO-friendly fingerprint readers and secured services will one day eliminate the need for passwords altogether.

10.8 Software Upgrades and Security Patches

Upgrading the computer operating system to the latest version is highly recommended for security reasons. A 64-bit computer running Windows 7 and Internet Explorer 9, for example, is inherently more secure than an older version of Windows operating system because of new security technologies such as Address Space Layout Randomization (ASLR), Data Execution Prevention (DEP), and SmartScreen Filter [126].

Installing application software patches is as equally important as performing regular operating system updates. A vulnerability in earlier versions of Adobe Reader and Acrobat allowed cybercriminals to inject malicious code in PDF documents, and the hack became so prevalent that it represented 80 % of all exploits in 2009 [127].

In April 2012, cybercriminals exploited a Java vulnerability and infected more than half a million Apple computers by downloading malware without prompting [128]. With 3 billion devices running Java, the hack is alarming. In most cases, installing security patches is sufficient. In January 2013, US-CERT urged users to disable Java on their Web browsers even after Oracle fixed a serious security flaw in its Java software [129].

Upgrading software and applying security patches should be done at home or in the office, and never at hotels using public Wi-Fi. A virus such as "Flashback" may look like a normal Adobe Flash browser plug-in, and a malware program may be named inconspicuously like "Android System Update 4.1.2.apk." In May 2012, the FBI issued the following warning regarding hotel Wi-Fi use [130]:

> Recently, there have been instances of travelers' laptops being infected with malicious software while using hotel Internet connections. In these instances, the traveler was attempting to setup the hotel room Internet connection and was presented with a pop-up window notifying the user to update a widely-used software product. If the user clicked to accept and install the update, malicious software was installed on the laptop. The pop-up window appeared to be offering a routine update to a legitimate software product for which updates are frequently available.

> The FBI recommends that all government, private industry, and academic personnel who travel abroad take extra caution before updating software products on their hotel Internet connection. Checking the author or digital certificate of any prompted update to see if it corresponds to the software vendor may reveal an attempted attack. The FBI also recommends that travelers perform software updates on laptops immediately before traveling, and that they download software updates directly from the software vendor's Web site if updates are necessary while abroad.

The sophisticated Stuxnet worm has demonstrated that signed digital certificates can be stolen to make a malware app look legitimate [131]. The safest software updates come directly from the software vendor's website, unless the site is hacked or hijacked by DNS spoofing. Stolen digital certificates and DNS poisoning make a lethal cocktail.

Software companies must step up their efforts in securing their software products or they could be held liable for damages due to security flaws. In his keynote address at the 2014 Black Hat Conference on August 6, In-Q-Tel's Chief Information Security Officer Dan Geer said, "The only two products not covered by product liability are religion and software, and software should not escape for much longer" [132].

In December 2014, Apple pushed out its first-ever automatic security upgrade for Macintosh computers [133]. The upgrade fixed a serious flaw in the operating system that allowed a remote attacker to execute arbitrary code. Earlier in June 2013, a team of researchers from Georgia Tech demonstrated how to inject arbitrary software into an iPhone or iPad using a malicious charger [134].

10.9 Fake Software and Free Downloads

Oftentimes there a catch to free software. In December 2013, the creator of "Brightest Flashlight Free" app agreed to settle Federal Trade Commission charges that the free app, which allows a device to be used as a flashlight, deceived consumers about how their geolocation information would be shared with advertising networks and other third parties. With tens of millions of downloads, the Android app transmitted users' precise location and unique device identifier to third parties, including advertising networks, even if the user had opted out of information sharing [135].

Back in September 2011, Microsoft issued a severe security alert of a fake security protection software program that displays alerts for non-existent threats on a computer in order to entice the user to download the malware [136] (see Fig. 10.3).

Shortly after the popular game "Angry Birds Space" was released in March 2012, fake apps containing the Trojan horse virus began to surface in Android markets. One counterfeit app appears to be a fully functional Angry Birds game, but it installs a virus on the user's smartphone or tablet [137]. Another fake app disguises itself as Angry Birds and turns the infected smartphone into a SpamSoldier bot for its spam campaign [138]. Over 10,000 Facebook users have been duped into downloading fake apps that claim to be able to change the color of their profiles [139].

Soon after "Flappy Bird" creator Dong Nguyen removed the popular mobile game from Apple's App Store and Google Play in February 2014, fake versions of the game popped up almost instantly. Some of the fake apps sent messages to premium numbers, racking up unwanted charges on victim's phone bills [140]. In response, Apple and Google began rejecting all new games with "Flappy" in their titles [141].

In August 2012, *Network World* reported a recent study claiming that over 90 % of the top 100 paid apps in the Apple App Store and Google Play Android Market have been pirated, hacked, malware-laden, and then given away for free [142]. The old sayings "You get what you pay for" and "If it's too good to be true, it probably is" are certainly appropriate here. Extra precaution is warranted before downloading any "free" games, music, movies, or software.

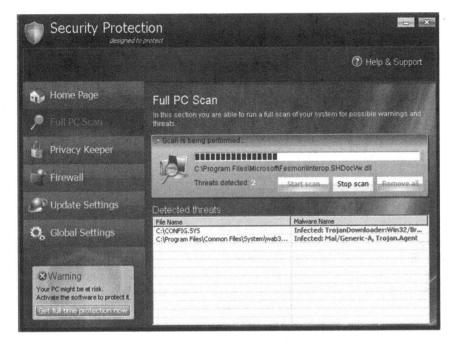

Fig. 10.3   A fake scanner interface of the "Security Protection" malware

The same vigilance applies to businesses. To attract customers, some retailers preinstall on new computers free software and games from questionable sources and distribution channels. In January 2013, Microsoft Security Intelligence Report revealed a disturbing finding that "malware has been discovered preinstalled on computers sold at retail" [143]. In October 2013, the Federal Trade Commission ordered Aaron's some 2,000 rental stores to stop renting out computers installed with spyware known as Detective Mode that allows the company to secretly collect the renters' private data through keystrokes, screen shots, and webcam images [144].

In September 2012, Microsoft disrupted the spread of the Nitol botnet malware embedded in counterfeit Windows operating system sold with some new computers [145]. "Consumers should exercise their right to demand that resellers provide them with non-counterfeit products free of malware," said Richard Domingues Boscovich, Assistant General Counsel at the Microsoft Digital Crimes Unit.

Microsoft offers a free security program — Microsoft Safety Scanner — that removes viruses, spyware, and other malicious software [146]. The safety scanner is constantly updated with the latest anti-malware definitions, therefore it expires 10 days after each download, and it has to be re-downloaded and re-run periodically.

Apple is not immune to hacking either. "The Masque bug in iOS and the corresponding WireLurker malware targeting iOS devices via Apple and Windows port-machines, had a lot of experts saying that the age of Apple malware is finally upon us," said security expert Eugene Kaspersky [147]. In November 2014, *PCWorld* reported that "some 467 Mac OS X applications offered on a Chinese third-party application store called Maiyadi were found to have been seeded with WireLurker, including 'The Sims 3,' 'International Snooker 2012' and 'Pro Evolution Soccer 2014.' Over the last six months, those applications and others have been downloaded 356,104 times and may have impacted hundreds of thousands of iPhone users" [148].

10.10   Smartphone Security Protection

In 1957, a blind seven-year old boy named Joe Engressia (aka Joybubbles) with an IQ of 172 used his unusual auditory gifts to hack the analog POTS (Plain Old Telephone Service) and became a nerve center of the "phone phreaks" subculture in 1970's [149]. Joybubbles could dial by using the hookswitch like a telegraph key. He could place free long distance phone calls by whistling the proper tones at 2600 hertz into any telephone. The world of communication has since changed to digital, and phreaks were precursors of today's computer hackers.

In February 2012, technology and market research company Forrester Research estimated that one billion people will own smartphones by 2016 [150]. These mobile phones are powerful little computers that are always on, 24/7. Consequently, they are the perfect targets for cybercriminals to steal information and to turn the phones into a botnet for launching a DDoS attack or spam campaign.

Cyber attacks on mobile phones rose by a whopping 500 % in 2012, according to McAfee [151]; and 10 % of all adults surveyed have experienced cybercrime on their mobile devices, according to Symantec [152]. "While the raw amount of Android malware continues to rise significantly, it is the increased commoditization of those malware that is the more worrying trend," said F-Secure researchers in a 2013 report. "The Android malware ecosystem is beginning to resemble that which surrounds Windows, where highly specialized suppliers provide commoditized malware services" [153].

The following are some of the popular and serious exploits that have been discovered so far:

1. March 2005: The Commwarrior.A virus replicates itself by sending multimedia messages (MMS) to people on the phone's contacts list [154].
2. April 2012: A counterfeit game based on "Angry Birds Space" installs a Trojan horse virus on victims' smartphones [137].
3. July 2012: A security flaw in the Android framework in Version 4.0.4 and below could be exploited by a rootkit, and no existing mobile security software was able to detect it [155].
4. July 2012: Vulnerabilities in the "near field communications (NFC)" features on some smartphones allow a tag with an embedded NFC chip to push a webpage to victims' phones, exploit a browser bug, and obtain unauthorized access [156].

5. December 2012: An exploit targeted at smartphones that use certain Exynos processors can let Android malware apps steal and delete all the data on victims' phones [157].

6. December 2012: Two counterfeit Android games, based on "Angry Birds Star Wars" and "The Need for Speed Most Wanted," were infected with malware that can turn victims' smartphones into a botnet for launching a mobile SMS spam campaign [158].

7. March 2013: Perkele is a toolkit designed to create malware for Android phones in order to defeat two-factor authentication used by banks, Gmail, and other services [159]. When a bank sends an SMS with a one-time code, Perkele intercepts that code and sends it to the attacker's control server to complete the unauthorized transaction.

8. April 2013: An Android Trojan called Stels harvests a victim's contact list, sends and intercepts text messages, makes phone calls to premium numbers, and installs additional malware packages [160].

9. July 2013: Cryptographer Karsten Nohl of Security Research Labs in Germany discovered a bug that could have allowed criminals to hack into hundreds of millions of cell phones. He accessed the SIM cards by exploiting flaws in the encryption keys and sending a hidden SMS text message. The phone companies fixed the bug by hacking into their own cards and rewrite parts of their operating systems [161].

10. August 2013: Jean-Pierre Seifert of the Technical University of Berlin demonstrated that a hacked GSM phone could block service to all subscribers served by base stations, preventing incoming calls and text messages from reaching other phones nearby [162].

11. October 2013: Thijs Alkemade at Utrecht University in the Netherlands discovered a major design flaw in WhatsApp's cryptographic implementation that allows cyber attackers decrypt intercepted messages [163]. WhatsApp uses the popular RC4 (Rivest Cipher 4) stream cipher and the same key for both sent and received messages.

12. February 2014: Apple found and fixed a security bug on iPhone, iPad, and iPod Touch that had allowed a man-in-the-middle attack by eavesdropping on all user communications via emails, instant messages, and social media posts [164].

13. July 2014: FireEye mobile security researchers discovered "Masque Attack" that allows hackers to replace a genuine iOS app downloaded from the App Store with a malware app through wireless networks and USB [165].

14. November 2014: Mobile security firm Lookout uncovered a sophisticated and evasive malware NotCompatible.C evolved from the original Android virus NotCompatible.A in 2012. The new incarnation of malware is "ultimately a botnet-for-rent; though the server architecture, peer-to-peer communications, and encryption make it a much more formidable threat. NotCompatible.C's use of encryption and peer-to-peer communication mirror advanced PC threats such as later Conficker" according to a Lookout report [166].

15. December 2014: German researchers revealed security flaws on Signaling System 7 (SS7) used by all major telephone companies worldwide. Functions built into SS7 enable cyber intruders to listen to phone calls, intercept text messages, collect phone numbers, and locate callers even if the cellular networks are using the most advanced encryption available [167].

"Your cell phone is communicating completely digital; it's part of the Internet," said Army Gen. Keith Alexander, director of the NSA and commander of the U.S. Cyber Command. "The attack surfaces for adversaries to get on the Internet now include all those mobile devices…. The mobile security situation lags. It's far behind" [168].

The current design of mobile devices makes differentiating legitimate sites from malicious ones a tricky task. "No matter how tantalizing a link might look on a desktop, there are cues that you shouldn't go there, such as an address that just doesn't look safe," said Hugh Thompson, Blue Coat's senior vice president and chief security strategist. "When you click a link on a mobile phone, it's harder to know what form of Russian roulette they're playing" [169].

Moreover, Google has no way to push operating software updates and security patches to millions of devices running outdated versions of Android. Dirk Sigurdson, director of engineering at IT security firm Rapid7, went so far as to say that "devices bought from companies other than Google can't be considered secure. The best bet for now is to buy Google Nexus or Google Play edition devices, which are much more quickly updated with the latest Android releases" [170].

In June 2012, the Defense Advanced Research Projects Agency (DARPA) assigned cybersecurity firm Invincea a $21-million research grant to fortify Android-based phones and tablets for use by the military personnel [171]. New research ideas include creating a virtual run-time environment separating military applications from other commercial software such as Facebook, Twitter, Skype, and games running on the smartphones. In May 2014, U.S. Navy unveiled "NeRD" (Navy eReader Device) for its sailors [172]. Preloaded with a library of 300 titles, NeRD is a secure device with no cellular connectivity or Wi-Fi.

For civilians and military personnel alike, the Federal Communications Commission (FCC) has released a "Smartphone Security Checker" to help consumers secure their mobile devices [173]. The general security checklist is as follows:

1. Set PINs and passwords to prevent unauthorized access to your phone. Configure your phone to automatically lock after five minutes or less when your phone is idle, as well as use the SIM password capability available on most smartphones.
2. Do not modify your smartphone's security settings. Tampering with your phone's factory settings, jailbreaking, or rooting your phone undermines the built-in security features offered by your wireless service and smartphone, while making it more susceptible to an attack.
3. Backup and secure all of the data stored on your phone.
4. Only install mobile apps from trusted sources.

5. Understand app permissions before accepting them. Be cautious about granting applications access to personal information on your phone or otherwise letting the application have access to perform functions on your phone.

6. Install anti-theft security protection apps that enable remote location and wiping. Some carriers offer a free "remote wipe" service that allows users to delete all of the data from a lost or stolen device to prevent data or identity theft. You cannot rely solely on passcode to protect your smartphone's content. It was discovered in January 2013 that a security flaw in Apple's iOS 6.1 allows anyone to bypass your iPhone password lock [174].

7. Keep your phone's software up-to-date by enabling automatic updates or accepting updates when prompted from your service provider, operating system provider, device manufacturer, or application provider.

8. Be smart on open Wi-Fi networks. When you access a public Wi-Fi network, your phone can be an easy target of cybercriminals. Always be aware when clicking on web links and be particularly cautious if you are asked to enter account or login information.

9. Wipe data on your old phone before you donate, resell, or recycle it. To protect your privacy, completely erase data off of your phone and reset the phone to its initial factory settings.

10. Report a stolen smartphone. The major wireless service providers, in coordination with the FCC, have established a stolen phone database. This will provide notice to all the major wireless service providers that the phone has been stolen and will allow for remote "bricking" of the phone so that it cannot be activated on any wireless network without your permission.

As if taking a page from a Hollywood script for *Mission Impossible* or a James Bond movie, GPS (Global Positioning System) spoofing and IMSI (International Mobile Subscriber Identity) catchers are security concerns for the military and law enforcement:

1. GPS spoofing broadcasts a set of normal GPS signals but different from the GPS satellites in order to deceive a GPS receiver, be it a smartphone or a wireless drone. In December 2011, Iranian electronic warfare engineers claimed to have intentionally broadcasted misguided GPS signals to hijack a CIA drone (US RQ-170 Sentinel) and guide it to an intact landing inside Iran [175]. In July 2013, a team led by Prof. Todd Humphreys of the University of Texas at Austin successfully spoofed an $80 million private yacht using the world's first openly acknowledged GPS spoofing device to knock the ship off course [176]. "Civilian GPS is not encrypted and not authenticated, so that means it's entirely predictable," said Humphreys. "Predictability is the enemy of security" [177].

2. An IMSI catcher (aka Stingray) is a virtual base transceiver station (VBTS) that can identify the IMSI of nearby GSM mobile phones and intercept their calls. Since the GSM (Global System for Mobile Communications) specification requires the handset to authenticate to the network but not vice versa, an IMSI catcher can easily trick mobile phones into thinking that they are connected to a legitimate service provider's cellular network [178].

Similar to tackling DNS spoofing, an effective countermeasure to GPS spoofing and IMSI catchers is two-way mutual authentication between the smartphones and the cell towers or base transceiver stations (BTS). In December 2014, German researcher Karsten Nohl at Security Research Labs unveiled a new tool called SnoopSnitch that can detect SMS and SS7 attacks as well as IMSI catchers [179].

Smartphone security protection and hacking will likely continue to be a cat-and-mouse game. "It's possible for something to go wrong on the scale of a big wireless network because of a coding mistake in an operating system or an application, and it's very hard to diagnose and fix," said David Fritz, senior technical staff at Sandia National Laboratories. "You can't possibly read through 15 million lines of code and understand every possible interaction between all these devices and the network" [180].

With the rise of Apple Pay and other mobile financial transactions, smartphone security is of utmost importance.

10.11 Cybersecurity Awareness: Everyone's Responsibility

In hindsight, Sandra Bullock's 1995 film *The Net* is not far-fetched by today's standard. Actress Patricia Arquette spoke of the 2015 police drama television series *CSI: Cyber*, "This cyber-stuff has really changed the face of our world. What puzzles me is the amount of brilliant people that are spending their time doing terrible things" [181].

Wired technology journalist Mat Honan wrote that besides Amazon and Apple's faults, he should have used two-factor authentication for his Google account and regularly backed up the data on his MacBook [99]. His story reminds both businesses and consumers that cybersecurity is everyone's responsibility.

We cannot rely solely on antivirus software. During a four-month long cyber attack on *The New York Times*, cybercriminals installed 45 pieces of custom malware on the network between October 2012 and January 2013. Symantec antivirus products installed at the Times were able to identify and quarantine only one piece of malware, missing all 44 others [182]. Symantec issued a follow-up statement, "Turning on only the signature-based anti-virus components of endpoint solutions alone are not enough in a world that is changing daily from attacks and threats. We encourage customers to be very aggressive in deploying solutions that offer a combined approach to security. Anti-virus software alone is not enough" [183].

"Even the most modern version of antivirus software doesn't give consumers or enterprises what they need to compete in the hacker world," said Dave Aitel, CEO of software security firm Immunity. "Deep down, nothing is as good as having a proper awareness about what's going on in your network" [184]. In February 2013, NBC.com and related sites were hacked and infected with malware that redirected visitors to malicious websites. "This morning, NBC.com was hacked and embedded with malicious iframe code that spread the Citadel Trojan. It was detected as Backdoor.Agent.RS.… The NBC web site was compromised for about 15 min and the actual iframe with the malicious redirect was embedded in a javascript file located on the NBC.com web server," said an NBC spokesperson [185]. For several days in early January 2014, Yahoo! was unknowingly serving malicious advertisements that redirected visitors to download malware on their computers [186].

More than 400 million people trust Google with their emails, and 50 million people store files in the cloud using the Dropbox service [187]. In February 2013, the U.S. Secret Service investigated the hacking and publication of private photographs and emails between members of the Bush family, including former Presidents George H.W. Bush and George W. Bush [188]. A research published by the RAND Corporation in the summer of 2014 called the World Wide Web (WWW) the "Wild Wild Web" and concluded that "for now, cybercrime has the upper hand in its duel with the law" [189].

Cyber espionage and cyber criminal activities have forced businesses to take extraordinary measures to safeguard customers' information and prevent data breach. About 50 large banks and the U.S. Treasury staged mega cyber attack drills such as Quantum Dawn in 2011 and Quantum Dawn 2 in 2013 [190]. Google hired DARPA director Regina Dugan in 2012 to fill a senior executive position. "Regina is a technical pioneer who brought the future of technology to the military during her time at DARPA," said a Google spokeswoman in an email to *Computerworld*. "She will be a real asset to Google, and we are thrilled she is joining the team" [191].

Dugan spoke at the TED 2012 conference in Long Beach, California about the spirit of scientists and engineers at DARPA: "When you remove the fear of failure, impossible things suddenly become possible" [192]. Among many of the DARPA inventions, she also talked about lightning that occurs in nature: "There are 44 lightning strikes per second around the globe. Each lightning bolt heats the air to 44,000 degrees Fahrenheit, hotter than the surface of the sun. What if we could use these electromagnetic pulses as beacons — beacons in a moving network of powerful transmitters. Experiments suggest that lightning could be the next GPS" [192].

Perhaps in the future, lightning will be used as a countermeasure to GPS spoofing. But in the meantime, companies like Facebook and Google have been enlisting hackers to find security holes in their products. In August 2011, Facebook paid out more than $40,000 within a month under its new "bug bounty" security initiative [193]. The company offers a minimum reward of $500 [194] and publicly thanks the "white-hat hackers" on the Facebook page [195]. In October 2012, Google awarded the top $60,000 prize to teenage hacker "Pinkie Pie" for uncovering a vulnerability in the Chrome browser [196]. It was the teen's second win. Ten hours after the bug was exposed, Google issued the Chrome security fixes and announced on its official blog: "Congratulations to Pinkie Pie, returning to the fray with another beautiful piece of work!" [197].

In February 2014, researchers at the University of California Berkeley and University of Maryland released a prototype browser extension called ShadowCrypt that makes it easy to send and receive encrypted text on Twitter, Facebook, or any other website [198]. Nevertheless, we must address the biggest culprit of cyber *insecurity*: human error. Regardless of how secure a communication line is and how unbreakable a cryptographic algorithm seems to be, the weakest links are often the endpoints — the sender (before encryption) and the recipient (after decryption) — unless they are code talkers using a language more obscure than Navajo.

"It doesn't matter how much money a company spends on infrastructure or technology. Until you close the human gap in the equation, you are always vulnerable to attacks." said Richard Henderson, a security strategist at FortiGuard Labs [199]. "There is no patch for a stupid user," added Denise Zheng, deputy director at the Center for Strategic and International Studies, who once oversaw a government cyber warfare program [200]. For a case in point, some 300 Instagram users posted selfies with their paychecks, enabling cybercriminals to obtain bank account and routing numbers by simply searching for Instagram photos with the hashtag #myfirstpaycheck. In October 2014, federal prosecutors charged 28 criminals who collected more than $2 million from counterfeit checks [201].

In the spirit of President John F. Kennedy, one may proclaim: "Ask not what cybersecurity can do for you, ask what you can do for cybersecurity."

Bibliography

1. **Office of the Press Secretary.** Excerpts of the President's State of the Union Address. [Online] The White House, January 20, 2015. http://www.whitehouse.gov/the-press-office/2015/01/20/excerpts-president-s-state-union-address.
2. **Schiff, et al.** H.R.2290 -- Cyber-Security Enhancement Act of 2007 (Introduced in House - IH). [Online] The Library of Congress, May 14, 2007. http://thomas.loc.gov/cgi-bin/query/z?c110:H.R.2290.
3. **Ackerman, et al.** H.R.7007 -- National Commission on American Cybersecurity Act of 2008. [Online] The Library of Congress, September 23, 2008. http://thomas.loc.gov/cgi-bin/query/z?c110:H.R.7007:.
4. **Rockefeller, et al.** S.773 -- Cybersecurity Act of 2009. [Online] The Library of Congress, April 1, 2009. http://thomas.loc.gov/cgi-bin/query/z?c111:S.773:.
5. **Reid, et al.** S.21 -- Cyber Security and American Cyber Competitiveness Act of 2011. [Online] The Library of Congress, January 25, 2011. http://thomas.loc.gov/cgi-bin/query/z?c112:S.21:.
6. **Lieberman, et al.** S.2105 -- Cybersecurity Act of 2012. [Online] The Library of Congress, February 14, 2012. http://thomas.loc.gov/cgi-bin/query/z?c112:S.2105:.
7. **Tummarello, Kate.** Intel panel approves cybersecurity bill. [Online] The Hill, July 8, 2014. http://thehill.com/policy/technology/211616-intel-panel-approves-cybersecurity-bill.
8. **Lieberman, Joe, Collins, Susan and Carper, Tom.** A gold standard in cyber-defense. [Online] The Washington Post, July 7, 2011. http://www.washingtonpost.com/opinions/a-gold-standard-in-cyber-defense/2011/07/01/gIQAjsZk2H_story.html.
9. **Nagesh, Gautham.** Sen. Rockefeller presses Congress to pass cybersecurity legislation. [Online] The Hill, January 31, 2012. http://thehill.com/blogs/hillicon-valley/technology/207729-rockefeller-presses-congress-to-pass-cybersecurity-legislation.
10. **Kain, Erik.** Does The Cybersecurity Act Of 2012 Mark The Beginning Of The War On Cyber-terrorism? [Online] Forbes, February 22, 2012. http://www.forbes.com/sites/erikkain/2012/02/22/does-the-cybersecurity-act-of-2012-mark-the-beginning-of-the-war-on-cyber-terrorism/.
11. **Stiennon, Richard.** Rockefeller's Cybersecurity Act of 2010: A Very Bad Bill. [Online] Forbes, May 4, 2010. http://www.forbes.com/sites/firewall/2010/05/04/rockefellers-cybersecurity-act-of-2010-a-very-bad-bill/.

12. **Brito, Jerry and Watkins, Tate.** Wired Opinion: Cyberwar Is the New Yellowcake. [Online] Wired, February 14, 2012. http://www.wired.com/threatlevel/2012/02/yellowc ake-and-cyberwar/.
13. **Singel, Ryan.** NSA Must Examine All Internet Traffic to Prevent Cyber Nine-Eleven, Top Spy Says. [Online] Wired, January 15, 2008. http://www.wired.com/threatlevel/2008/01/ feds-must-exami/.
14. **Obama, Barack.** Executive Order -- Improving Critical Infrastructure Cybersecurity. [Online] The White House, February 12, 2013. http://www.whitehouse.gov/ the-press-office/2013/02/12/executive-order-improving-critical-infrastructure-cybersecurity.
15. Cyberspace Policy Review: Assuring a Trusted and Resilient Information and Communications Infrastructure. [Online] The White House, May 8, 2009. http://www.whitehouse.gov/assets/documents/Cyberspace_Policy_Review_final.pdf.
16. **National Security Council.** The Comprehensive National Cybersecurity Initiative. [Online] The White House. [Cited: January 18, 2013.] http://www.whitehouse.gov/cybersecurity/ comprehensive-national-cybersecurity-initiative.
17. **The White House.** National Initiative for Cybersecurity Education (NICE) Relationship to President's Education Agenda. [Online] The White House, April 19, 2010. http://www.whitehouse.gov/sites/default/files/rss_viewer/cybersecurity_niceeducation.pdf.
18. **Information Assurance Support Environment (IASE).** Cyber Awareness Challenge. [Online] U.S. Department of Defense. [Cited: January 21, 2013.] http://iase.disa.mil/eta/cyb erchallenge/launchPage.htm.
19. **Homeland Security.** National Cyber Security Awareness Month. [Online] U.S. Department of Homeland Security. [Cited: January 18, 2013.] http://www.dhs.gov/ national-cyber-security-awareness-month.
20. —. Cyber Storm: Securing Cyber Space. [Online] U.S. Department of Homeland Security. [Cited: January 18, 2013.] http://www.dhs.gov/cyber-storm-securing-cyber-space.
21. **National Cybersecurity Alliance (NCSA).** StaySafeOnline.org. [Online] National Cybersecurity Alliance (NCSA). [Cited: January 18, 2013.] http://www.staysafeonline.org/.
22. **Center for Internet Security.** Multi-State Information Sharing and Analysis Center (MS-ISAC). [Online] Center for Internet Security. [Cited: January 18, 2013.] http://msisac.cisecurity.org/.
23. **Patrikakis, Charalampos, Masikos, Michalis and Zouraraki, Olga.** Distributed Denial of Service Attacks. [Online] The Internet Protocol Journal, December 2004. http://www.cisco.com/web/about/ac123/ac147/archived_issues/ipj_7-4/dos_attacks.html.
24. **Goodin, Dan.** DDoS attacks on major US banks are no Stuxnet—here's why. [Online] ArsTechnica, October 3, 2012. http://arstechnica.com/security/2012/10/ ddos-attacks-against-major-us-banks-no-stuxnet/.
25. **Prolexic.** Prolexic Quarterly Global DDoS Attack Report. [Online] Prolexic, Q4 2012. http://www.prolexic.com/knowledge-center-ddos-attack-report-2012-q4/pr.html.
26. **Eddy, W.** TCP SYN Flooding Attacks and Common Mitigations. [Online] The Internet Engineering Task Force (IETF), August 2007. http://tools.ietf.org/html/rfc4987.
27. **VeriSign.** DDoS Mitigation - Best Practices for a Rapidly Changing Threat Landscape Whitepaper. [Online] VeriSign, 2012. http://www.verisigninc.com/en_US/products-and-services/network-intelligence-availability/nia-information-center/ddos-best-practice-confirmation/index.xhtml.
28. **Cisco.** Defeating DDOS Attacks. [Online] Cisco. [Cited: January 20, 2013.] http://www.cisco.com/en/US/prod/collateral/vpndevc/ps5879/ps6264/ps5888/prod_white_paper0900 aecd8011e927.html.
29. **VeriSign.** VeriSign Internet Defense Network Enhanced With New DDoS Monitoring Service. [Online] Reuters, September 10, 2009. http://www.reuters.com/article/2009/09/10/ idUS126052+10-Sep-2009+MW20090910.
30. **Prolexic.** Prolexic Issues Mitigation, Detection Rules for Critical DDoS Threat Used in Banking Attacks. [Online] PresseBox, January 3, 2013. http://www.pressebox.com/ inactive/prolexic-technologies/Prolexic-Issues-Mitigation-Detection-Rules-for-Critical-DDoS-Threat-Used-in-Banking-Attacks/boxid/564817.

31. **Tata Communications.** Cloud-based security services. [Online] Tata Communications. [Cited: January 20, 2013.] http://security.tatacommunications.com/cloud.asp.
32. **Gross, Doug.** Massive cyberattack hits Internet users. [Online] CNN, March 29, 2013. http://www.cnn.com/2013/03/27/tech/massive-internet-attack.
33. **Ingraham, Nathan.** BitTorrent wants to change the way the web is built. [Online] The Verge, December 10, 2014. http://www.theverge.com/2014/12/10/7361603/bittorrenet-wants-to-change-the-way-the-web-is-built.
34. **Mueller, Robert S. III.** Robert S. Mueller, III Speech at RSA Cyber Security Conference. [Online] Federal Bureau of Investigation, March 1, 2012. http://www.fbi.gov/news/speeches/combating-threats-in-the-cyber-world-outsmarting-terrorists-hackers-and-spies.
35. **Microsoft.** Microsoft Joins Financial Services Industry to Disrupt Massive Zeus Cybercrime Operation That Fuels Worldwide Fraud and Identity Theft. [Online] Microsoft, March 25, 2012. http://www.microsoft.com/en-us/news/press/2012/mar12/03-25CybercrimePR.aspx.
36. **Mushtaq, Atif.** Grum, World's Third-Largest Botnet, Knocked Down. [Online] FireEye, July 18, 2012. http://blog.fireeye.com/research/2012/07/grum-botnet-no-longer-safe-havens.html.
37. **Cowley, Stacy.** Grum takedown: '50 % of worldwide spam is gone'. [Online] CNNMoney, July 19, 2012. http://money.cnn.com/2012/07/19/technology/grum-spam-botnet/index.htm.
38. **Prolexic.** Prolexic Issues Dirt Jumper Threat Advisory and Releases Free Security Scanner . [Online] PRWeb, December 29, 2011. http://www.prweb.com/releases/2011/12/prweb9067808.htm.
39. **Breeden, John II.** Hackers' new super weapon adds firepower to DDOS. [Online] GCN, October 24, 2012. http://gcn.com/Articles/2012/10/24/Hackers-new-super-weapon-adds-firepower-to-DDOS.aspx.
40. **Goldman, David.** Take down any website for $3 . [Online] CNNMoney, December 31, 2014. http://money.cnn.com/2014/12/31/technology/lizard-squad-attack/index.html.
41. **Cowley, Stacy.** FBI Director: Cybercrime will eclipse terrorism. [Online] CNNMoney, March 2, 2012. http://money.cnn.com/2012/03/02/technology/fbi_cybersecurity/index.htm.
42. **Perlroth, Nicole.** Some Victims of Online Hacking Edge Into the Light. [Online] The New York Times, February 20, 2013. http://www.nytimes.com/2013/02/21/technology/hacking-victims-edge-into-light.html.
43. —. Home Depot Data Breach Could Be the Largest Yet. [Online] The New York Times, September 8, 2014. http://bits.blogs.nytimes.com/2014/09/08/home-depot-confirms-that-it-was-hacked/.
44. **Pagliery, Jose.** 'Smart credit card' terminals can be hacked too. [Online] CNNMoney, August 8, 2014. http://money.cnn.com/2014/08/08/technology/security/hack-credit-card-terminal/index.html.
45. —. Half of American adults hacked this year. [Online] CNNMoney, May 28, 2014. http://money.cnn.com/2014/05/28/technology/security/hack-data-breach/.
46. **Zetter, Kim.** Google Hack Attack Was Ultra Sophisticated, New Details Show. [Online] Wired, January 14, 2010. http://www.wired.com/threatlevel/2010/01/operation-aurora/.
47. **Cowley, Stacy.** Former FBI cyber cop worries about a digital 9/11. [Online] CNN, July 25, 2012. http://money.cnn.com/2012/07/25/technology/blackhat-shawn-henry/index.htm.
48. **Jewell, Mark.** TJX breach could top 94 million accounts. [Online] NBC News, October 24, 2007. http://www.msnbc.msn.com/id/21454847/ns/technology_and_science-security/t/tjx-breach-could-top-million-accounts/.
49. **Cubrilovic, Nik.** RockYou Hack: From Bad To Worse. [Online] TechCrunch, December 14, 2009. http://techcrunch.com/2009/12/14/rockyou-hack-security-myspace-facebook-passwords/.
50. **Krebs, Brian.** PlentyofFish.com Hacked, Blames Messenger. [Online] KrebsOnSecurity.com, January 31, 2011. http://krebsonsecurity.com/2011/01/plentyoffish-com-hacked-blames-messenger/.
51. **Hickins, Michael and Clark, Don.** Questions Over Break-In at Security Firm RSA. [Online] The Wall Street Journal, March 18, 2011. http://online.wsj.com/article/SB10001424052748703512404576208983743029392.html.

52. **Wingfield, Nick, Sherr, Ian and Worthen, Ben.** Hacker Raids Sony Videogame Network. [Online] The Wall Street Journal, April 27, 2011. http://online.wsj.com/article/SB10001424 052748703778104576287362503776534.html.

53. **Smith, Aaron.** Citi: Millions stolen in May hack attack. [Online] CNNMoney, June 27, 2011. http://money.cnn.com/2011/06/27/technology/citi_credit_card/index.htm.

54. **Zakaria, Tabassum and Hosenball, Mark.** Stratfor Hack: Anonymous-Affiliated Hackers Publish Thousands Of Credit Card Numbers. [Online] Huffington Post, December 30, 2011. http://www.huffingtonpost.com/2011/12/30/stratfor-hack-anonymous_n_1176726.html.

55. **Bradley, Tony.** Zappos Hacked: What You Need to Know. [Online] PC World, January 16, 2012. http://www.pcworld.com/article/248244/zappos_hacked_what_you_need_to_know.html.

56. **Acohido, Byron.** Credit card processor hit by hackers. [Online] USA Today, March 30, 2012. http://www.usatoday.com/money/industries/banking/story/2012-03-30/mastercard-security-breach/53887854/1.

57. **Goldman, David.** More than 6 million LinkedIn passwords stolen. [Online] CNNMoney, June 7, 2012. http://money.cnn.com/2012/06/06/technology/linkedin-password-hack/index.htm.

58. **eHarmony.com.** Update on Compromised Passwords. [Online] eHarmony Blog, June 6, 2012. http://www.eharmony.com/blog/2012/06/06/update-on-compromised-passwords/#. U5ynqssU914.

59. **Gross, Doug.** Yahoo hacked, 450,000 passwords posted online. [Online] CNN, July 13, 2012. http://www.cnn.com/2012/07/12/tech/web/yahoo-users-hacked/index.html.

60. **Riley, Charles.** Barnes & Noble customer data stolen. [Online] CNNMoney, October 24, 2012. http://money.cnn.com/2012/10/24/technology/barnes--noble-hack/index.html.

61. **Lord, Bob.** Keeping our users secure. [Online] Twitter Blog, February 1, 2013. http://blog. twitter.com/2013/02/keeping-our-users-secure.html.

62. **Bull, Alister and Finkle, Jim.** Fed says internal site breached by hackers, no critical functions affected. [Online] Reuters, February 6, 2013. http://www.reuters.com/article/2013/02/06/net-us-usa-fed-hackers-idUSBRE91501920130206.

63. **Krebs on Security.** Experian Sold Consumer Data to ID Theft Service. [Online] Krebs on Security, October 20, 2013. http://krebsonsecurity.com/2013/10/experian-sold-consumer-data-to-id-theft-service/.

64. **Engberg, Dave.** Security Notice: Service-wide Password Reset. [Online] The Evernote Blog, March 2, 2013. http://blog.evernote.com/blog/2013/03/02/security-notice-service-wide-password-reset/.

65. **Swisher, Kara.** LivingSocial Hacked — More Than 50 Million Customer Names, Emails, Birthdates and Encrypted Passwords Accessed (Internal Memo). [Online] All Things D, April 26, 2013. http://allthingsd.com/20130426/livingsocial-hacked-more-than-50-million-customer-names-emails-birthdates-and-encrypted-passwords-accessed/.

66. **Arkin, Brad.** Important Customer Security Announcement. [Online] Adobe Featured Blogs, October 3, 2013. http://blogs.adobe.com/conversations/2013/10/important-customer-security-announcement.html.

67. **Wallace, Gregory.** Neiman Marcus hack hit 1.1 million customers. [Online] CNNMoney, January 23, 2014. http://money.cnn.com/2014/01/23/news/companies/neiman-marcus-hack/.

68. **Pagliery, Jose.** 2 million Facebook, Gmail and Twitter passwords stolen in massive hack. [Online] CNNMoney, December 4, 2013. http://money.cnn.com/2013/12/04/technology/security/passwords-stolen/index.html.

69. **Isidore, Chris.** Target: Hacking hit up to 110 million customers. [Online] CNNMoney, January 11, 2014. http://money.cnn.com/2014/01/10/news/companies/target-hacking/index.html.

70. **Lobosco, Katie.** Michaels hack hit 3 million. [Online] CNN, April 18, 2014. http://money.cnn.com/2014/04/17/news/companies/michaels-security-breach/.

71. **Gross, Doug.** Millions of accounts compromised in Snapchat hack. [Online] CNN, January 2, 2014. http://www.cnn.com/2014/01/01/tech/social-media/snapchat-hack/.

72. **Wallace, Gregory.** Starbucks: We fixed app that left passwords vulnerable. [Online] CNNMoney, January 17, 2014. http://money.cnn.com/2014/01/17/technology/security/starbucks-app-passwords/index.html.

73. **Kickstarter.** OMG. [Online] Kickstarter, March 3, 2014. https://www.kickstarter.com/1billion?ref=promo&ref=PromoNewsletterMar0314.

74. **eBay.** eBay Inc. To Ask eBay Users To Change Passwords. [Online] ebay inc., May 21, 2014. http://investor.ebayinc.com/releasedetail.cfm?ReleaseID=849396.

75. **Riley, Charles.** Data breach at UPS Stores in 24 states. [Online] CNNMoney, August 21, 2014. http://money.cnn.com/2014/08/21/technology/security/ups-store-data-hack/index.html.

76. **AOL Mail Team.** AOL Security Update. [Online] Aol Blog, April 28, 2014. http://blog.aol.com/2014/04/28/aol-security-update/.

77. **Orcutt, Mike.** Hackers Are Homing In on Hospitals. [Online] MIT Technology Review, September 2, 2014. http://www.technologyreview.com/news/530411/hackers-are-homing-in-on-hospitals/.

78. **Smith, Aaron.** P.F. Chang's confirms credit data was stolen. [Online] CNNMoney, June 13, 2014. http://money.cnn.com/2014/06/13/technology/security/pf-changs-security/index.html.

79. **Perlroth, Nicole and Gelles, David.** Russian Hackers Amass Over a Billion Internet Passwords. [Online] The New York Times, August 5, 2014. http://www.nytimes.com/2014/08/06/technology/russian-gang-said-to-amass-more-than-a-billion-stolen-internet-credentials.html.

80. **Cooper, Charles.** Celebs, beware: Those nude selfies will be hacked and shared. [Online] CNet, September 2, 2014. http://www.cnet.com/news/the-new-price-of-celebrity-careful-before-taking-that-nudie-selfie/.

81. **Pagliery, Jose.** Staples hack exposes 1.2 million credit cards. [Online] CNNMoney, December 20, 2014. http://money.cnn.com/2014/12/19/technology/security/staples-hack/index.html.

82. **O'Toole, James.** JPMorgan: 76 million customers hacked. [Online] CNNMoney, October 3, 2014. http://money.cnn.com/2014/10/02/technology/security/jpmorgan-hack/index.html?hpt=hp_t2.

83. **Perez, Evan.** Hackers put data of U.S. government workers at risk. [Online] CNN, October 12, 2014. http://www.cnn.com/2014/08/06/tech/hackers-security-contractor-usis/index.html.

84. **Estes, Adam Clark.** The Sony Pictures Hack Was Even Worse Than Everyone Thought. [Online] GIZMODO, December 3, 2014. http://gizmodo.com/the-sony-pictures-hack-exposed-budgets-layoffs-and-3-1665739357/1666122168/+ace.

85. **Cieply, Michael and Barnes, Brooks.** Sony Cyberattack, First a Nuisance, Swiftly Grew Into a Firestorm. [Online] The New York Times, December 30, 2014. http://www.nytimes.com/2014/12/31/business/media/sony-attack-first-a-nuisance-swiftly-grew-into-a-firestorm-.html.

86. **Zetter, Kim.** Sony Hackers Threaten to Release a Huge 'Christmas Gift' of Secrets. [Online] Wired, December 15, 2014. http://www.wired.com/2014/12/sony-hack-part-deux/.

87. **Petroff, Alanna.** The heart of the Internet has been hacked. [Online] CNNMoney, December 19, 2014. http://money.cnn.com/2014/12/19/technology/security/icann-hack-internet/index.html.

88. **Zakrzewski, Cat.** Anonymous Leaked A Massive List Of Passwords And Credit Card Numbers. [Online] TechCrunch, December 27, 2014. 1. http://techcrunch.com/2014/12/27/anonymous-leaked-a-massive-list-of-passwords-and-credit-card-numbers/.

89. **Goldman, David.** Stock market hackers steal drug company secrets. [Online] CNN, December 29, 2014. http://money.cnn.com/2014/12/01/technology/security/stock-market-hack/index.html.

90. **Mathews, Anna Wilde and Yadron, Danny.** Health Insurer Anthem Hit by Hackers. [Online] The Wall Street Journal, February 4, 2015. http://www.wsj.com/articles/health-insurer-anthem-hit-by-hackers-1423103720.

91. **Pagliery, Jose.** Why North Korea's attack should leave every company scared stiff. [Online] CNNMoney, December 19, 2014. http://money.cnn.com/2014/12/19/technology/security/hacking-companies-north-korea/index.html.

92. **Verizon RISK Team.** 2012 Data Breach Investigations Report. [Online] Verizon, 2012. http://www.verizonbusiness.com/resources/reports/rp_data-breach-investigations-report-2012_en_xg.pdf.

93. **Menegaz, Gery.** SQL Injection Attack: What is it, and how to prevent it. [Online] ZDNet, July 13, 2012. http://www.zdnet.com/sql-injection-attack-what-is-it-and-how-to-prevent-it-7000000881/.

94. **OWASP.** SQL Injection Prevention Cheat Sheet. [Online] The Open Web Application Security Project, December 6, 2012. https://www.owasp.org/index.php/SQL_Injection_Prevention_Cheat_Sheet.

95. . **Jewell, Mark.** TJX breach could top 94 million accounts. [Online] NBC News, October 24, 2007. http://www.msnbc.msn.com/id/21454847/ns/technology_and_science-security/t/tjx-breach-could-top-million-accounts/.

96. **First Data Corporation.** What Data Thieves Don't Want You to Know: The Facts About Encryption. [Online] First Data Corporation, 2012. http://www.firstdata.com/downloads/thought-leadership/TokenizationEncryptionWP.pdf.

97. **Dulaney, Chelsey.** Visa, MasterCard to Roll Out New Cybersecurity Features. [Online] The Wall Street Journal, February 13, 2015. http://www.wsj.com/articles/visa-mastercard-to-roll-out-new-cybersecurity-features-1423834542.

98. **Cowley, Stacy.** How a lying 'social engineer' hacked Wal-Mart. [Online] CNNMoney, August 8, 2012. http://money.cnn.com/2012/08/07/technology/walmart-hack-defcon/index.htm.

99. **Honan, Mat.** How Apple and Amazon Security Flaws Led to My Epic Hacking. [Online] Wired, August 6, 2012. http://www.wired.com/gadgetlab/2012/08/apple-amazon-mat-honan-hacking/all/.

100. **Kirk, Jeremy.** Researchers find vulnerability in Call of Duty: Modern Warfare 3. [Online] CSO, November 9, 2012. http://www.csoonline.com/article/721133/researchers-find-vulnerability-in-call-of-duty-modern-warfare-3.

101. **Facebook Security.** Protecting People On Facebook. [Online] Facebook, February 15, 2013. https://www.facebook.com/notes/facebook-security/protecting-people-on-facebook/10151249208250766.

102. **Lute, Jane Holl.** Is the Sony hack corporate America's cybersecurity wakeup call? [Online] Fortune Magazine, December 29, 2014. http://fortune.com/2014/12/29/is-the-sony-hack-corporate-americas-cybersecurity-wakeup-call/.

103. **Simonite, Tom.** The "Soft and Chewy Centers" That Put Your Data at Risk. [Online] MIT Technology Review, October 22, 2014. http://www.technologyreview.com/news/531931/the-soft-and-chewy-centers-that-put-your-data-at-risk/.

104. **resistsurveillance.org.** DETEKT. [Online] RESIST SURVEILLANCE. [Cited: January 26, 2015.] https://resistsurveillance.org/index.html.

105. **Simonite, Tom.** "Honey Encryption" Will Bamboozle Attackers with Fake Secrets. [Online] MIT Technology Review, January 29, 2014. http://www.technologyreview.com/news/523746/honey-encryption-will-bamboozle-attackers-with-fake-secrets/.

106. **Juels, Ari and Rivest, Ronald L.** Honeywords Project. [Online] MIT, May 2, 2013. http://people.csail.mit.edu/rivest/honeywords/.

107. **Mila.** Targeted attacks against personal accounts of military, government employees and associates . [Online] Contagio, February 17, 2011. http://contagiodump.blogspot.com/2011/02/targeted-attacks-against-personal.html.

108. **APWG.** Phishing Activity Trends Report (2nd Quarter 2012). *Anti-Phishing Working Group (APWG).* [Online] September 2012. http://docs.apwg.org/reports/apwg_trends_report_q2_2012.pdf.

109. **Bursztein, Elie, et al.** Handcrafted Fraud and Extortion: Manual Account Hijacking in the Wild. [Online] Association for Computing Machinery IMC' 14, November 5-7, 2014. http://conferences2.sigcomm.org/imc/2014/papers/p347.pdf.

110. **Kelly, Suzanne and Benson, Pam.** U.S. gears up for cyberwar amid conflicting ideas on how to fight it. [Online] CNN, February 24, 2012. http://security.blogs.cnn.com/2012/02/24/u-s-gears-up-for-cyberwar-amid-conflicting-ideas-on-how-to-fight-it/.

111. **Kim, Erin.** Internet blackout for thousands begins Monday. [Online] CNNMoney, July 9, 2012. http://money.cnn.com/2012/07/06/technology/dnschanger/index.htm.
112. **Whittaker, Zack.** Google services 'disrupted' in China; traffic declines rapidly. [Online] ZDNet, November 9, 2012. http://www.zdnet.com/google-services-disrupted-in-china-traffic-declines-rapidly-7000007195/.
113. **Taylor, Brad.** Fighting phishing with eBay and PayPal. [Online] Official Gmail Blog, July 8, 2008. http://gmailblog.blogspot.com/2008/07/fighting-phishing-with-ebay-and-paypal.html#!/2008/07/fighting-phishing-with-ebay-and-paypal.html.
114. **Zetter, Kim.** How a Google Headhunter's E-Mail Unraveled a Massive Net Security Hole. [Online] Wired, October 24, 2012. http://www.wired.com/threatlevel/2012/10/dkim-vulnerability-widespread/all/.
115. **Orlando, Michael.** Vulnerability Note VU#268267: DomainKeys Identified Mail (DKIM) Verifiers may inappropriately convey message trust. [Online] U.S. Department of Homeland Security's United States Computer Emergency Readiness Team (US-CERT), October 24, 2012. http://www.kb.cert.org/vuls/id/268267.
116. **US-CERT.** Report Phishing Sites. [Online] U.S. Department of Homeland Security's United States Computer Emergency Readiness Team (US-CERT). [Cited: January 22, 2013.] http://www.us-cert.gov/nav/report_phishing.html.
117. **The Google Safe Browsing Team.** Report Phishing Page. [Online] Google. [Cited: January 22, 2013.] http://www.google.com/safebrowsing/report_phish/.
118. **Saran, Cliff.** ING Direct implements two-factor authentication. [Online] Computer Weekly, August 17, 2006. http://www.computerweekly.com/news/2240078159/ING-Direct-implements-two-factor-authentication.
119. **Danchev, Dancho.** Attacker: Hacking Sarah Palin's email was easy. [Online] ZDNet, September 18, 2008. http://www.zdnet.com/blog/security/attacker-hacking-sarah-palins-email-was-easy/1939.
120. **Cheng, Roger and McCullagh, Declan.** Yahoo breach: Swiped passwords by the numbers. [Online] CNet, July 12, 2012. http://news.cnet.com/8301-1009_3-57470878-83/yahoo-breach-swiped-passwords-by-the-numbers/.
121. **Brandom, Russell.** Google launches support for Security Key, a simpler kind of two-factor authentication . [Online] The Verge, October 21, 2014. http://www.theverge.com/2014/10/21/7027267/google-launches-support-for-security-key-a-simpler-kind-of-two-factor.
122. **Komanduri, Saranga, et al.** Telepathwords. [Online] Microsoft Research, 2013. https://telepathwords.research.microsoft.com/.
123. **Krebs, Brian.** How Companies Can Beef Up Password Security. [Online] Krebs on Security, June 12, 2012. http://krebsonsecurity.com/2012/06/how-companies-can-beef-up-password-security/.
124. **Hale, Coda.** How To Safely Store A Password. [Online] Code Hale, January 31, 2010. http://codahale.com/how-to-safely-store-a-password/.
125. **Brandom, Russell.** Google-backed password-killer crosses major milestone . [Online] The Verge, December 9, 2014. http://www.theverge.com/2014/12/9/7359535/google-backed-password-killer-crosses-major-milestone.
126. **Microsoft.** Microsoft Security Intelligence Report. [Online] Microsoft, January-June 2012. http://download.microsoft.com/download/C/1/F/C1F6A2B2-F45F-45F7-B788-32D2CCA48D29/Microsoft_Security_Intelligence_Report_Volume_13_English.pdf.
127. **Danchev, Dancho.** Report: Malicious PDF files comprised 80 percent of all exploits for 2009. [Online] ZDNet, February 16, 2010. http://www.zdnet.com/blog/security/report-malicious-pdf-files-comprised-80-percent-of-all-exploits-for-2009/5473.
128. **Perlroth, Nicole.** Department of Homeland Security: Disable Java 'Unless It Is Absolutely Necessary'. [Online] The New York Times, January 14, 2013. http://bits.blogs.nytimes.com/2013/01/14/department-of-homeland-security-disable-java-unless-it-is-absolutely-necessary/.

129. **Dormann, Will.** Vulnerability Note VU#625617: Java 7 fails to restrict access to privileged code. [Online] U.S. Department of Homeland Security's United States Computer Emergency Readiness Team (US-CERT), January 10, 2013. http://www.kb.cert.org/vuls/id/625617.

130. **IC3.** Malware Installed on Travelers' Laptops Through Software Updates on Hotel Internet Connections . [Online] Internet Crime Complaint Center (IC3), May 8, 2012. http://www.ic3.gov/media/2012/120508.aspx.

131. **Keizer, Gregg.** Is Stuxnet the 'best' malware ever? [Online] Computerworld, September 16, 2010. http://www.computerworld.com/s/article/9185919/Is_Stuxnet_the_best_malware_ever_.

132. **Simonite, Tom.** Black Hat: More Internet-Scale Bugs Are Likely Lurking. [Online] MIT Technology Review, August 11, 2014. http://www.technologyreview.com/news/529981/black-hat-more-internet-scale-bugs-are-likely-lurking/.

133. **Wallace, Gregory.** Apple pushes out first-ever automatic security upgrade for Mac. [Online] CNNMoney, December 23, 2014. http://money.cnn.com/2014/12/23/technology/security/apple-automatic-security-upgrade/index.html.

134. **Gross, Doug.** Researchers: We can hack an iPhone through the charger. [Online] CNN, June 4, 2013. http://www.cnn.com/2013/06/03/tech/mobile/hack-iphone-charger/index.html.

135. **Federal Trade Commission.** Android Flashlight App Developer Settles FTC Charges It Deceived Consumers. [Online] Federal Trade Commission, December 5, 2013. http://www.ftc.gov/news-events/press-releases/2013/12/android-flashlight-app-developer-settles-ftc-charges-it-deceived.

136. **Fouda, Amir.** Security Protection. [Online] Microsoft Malware Protection Center, September 7, 2011. http://www.microsoft.com/security/portal/threat/encyclopedia/entry.aspx?Name=Security+Protection.

137. **Gross, Doug.** Virus found in fake Android version of 'Angry Birds: Space'. [Online] CNN, April 12, 2012. http://www.cnn.com/2012/04/12/tech/gaming-gadgets/angry-birds-virus-android/index.html.

138. **Halliday, Derek.** Security Alert: SpamSoldier. [Online] Lookout Mobile Security, December 17, 2012. https://blog.lookout.com/blog/2012/12/17/security-alert-spamsoldier/.

139. **Goldman, David.** You can't change the color of Facebook - it's a virus. [Online] CNNMoney, August 11, 2014. http://money.cnn.com/2014/08/11/technology/social/facebook-color-change/index.html.

140. **Bell, Karissa.** Fake 'Flappy Bird' Apps Are Infecting Androids With Malware. [Online] Mashable, February 12, 2014. http://mashable.com/2014/02/12/flappy-bird-malware/.

141. **Perez, Sarah.** Apple & Google Begin Rejecting Games With "Flappy" In The Title. [Online] TechCrunch, February 15, 2014. http://techcrunch.com/2014/02/15/apple-google-begin-rejecting-games-with-flappy-in-the-title/.

142. **Messmer, Ellen.** Pirated mobile Android and Apple apps getting hacked, cracked and smacked. [Online] Network World, August 20, 2012. http://www.networkworld.com/news/2012/082012-pirated-app-malware-261702.html.

143. **Microsoft Security Intelligence Report.** Deceptive Downloads: Software, Music, and Movies. [Online] Microsoft. [Cited: January 24, 2013.] http://www.microsoft.com/security/sir/story/default.aspx#!deceptive_downloads.

144. **Smith, Aaron.** Aaron's rental stores in anti-spying accord. [Online] CNNMoney, October 13, 2013. http://money.cnn.com/2013/10/23/technology/aarons-ftc-computer/index.html.

145. **Boscovich, Richard Domingues.** Microsoft Disrupts the Emerging Nitol Botnet Being Spread through an Unsecure Supply Chain. [Online] The Office Microsoft Blog, September 13, 2012. http://blogs.technet.com/b/microsoft_blog/archive/2012/09/13/microsoft-disrupts-the-emerging-nitol-botnet-being-spread-through-an-unsecure-supply-chain.aspx.

146. **Microsoft.** Microsoft Safety Scanner. [Online] Microsoft. [Cited: January 24, 2013.] http://www.microsoft.com/security/scanner/en-us/default.aspx.

147. **Dredge, Stuart.** How you could become a victim of cybercrime in 2015. [Online] The Guardian, December 24, 2014. http://www.theguardian.com/technology/2014/dec/24/cybercrime-2015-cybersecurity-ransomware-cyberwar.

148. **Kirk, Jeremy.** Chinese iOS devices fall prey to invasive WireLurker malware. [Online] PCWorld, November 6, 2014. http://www.pcworld.com/article/2844292/ apple-mobile-devices-in-china-targeted-by-wirelurker-malware.html.

149. **Martin, Douglas.** Joybubbles, 58, Peter Pan of Phone Hackers, Dies. [Online] The New York Times, August 20, 2007. http://www.nytimes.com/2007/08/20/us/20engressia.html.

150. **Chen, Brian X.** Get Ready for 1 Billion Smartphones by 2016, Forrester Says. [Online] The New York Times, February 13, 2012. http://bits.blogs.nytimes.com/2012/02/13/ get-ready-for-1-billion-smartphones-by-2016-forrester-says/.

151. **Goldman, David.** Your smartphone will (eventually) be hacked. [Online] CNNMoney, September 12, 2012. http://money.cnn.com/2012/09/17/technology/smartphone-cyberattack/ index.html.

152. **Norton.** Cybercrime Report 2011. [Online] Symantec Corporation, 2012. http://now-static.norton.com/now/en/pu/images/Promotions/2012/cybercrime/assets/downloads/ en-us/NCR-DataSheet.pdf.

153. **Constantin, Lucian.** Android threats growing in number and complexity, report says. [Online] Computer World, May 14, 2013. http://www.computerworld.com/ article/2497483/malware-vulnerabilities/android-threats-growing-in-number-and-complexity--report-says.html.

154. **Bell, Ian.** Commwarrior.A Virus Targets Cell Phones. [Online] Digital Trends, March 9, 2005. http://www.digitaltrends.com/mobile/commwarriora-virus-targets-cell-phones/.

155. **Gold, Jon.** Researchers reveal new rootkit threat to Android security. [Online] Network World, July 2, 2012. http://www.networkworld.com/news/2012/070212-android-malware-260627.html.

156. **Cowley, Stacy.** NFC exploit: Be very, very careful what your smartphone gets near. [Online] CNNMoney, July 26, 2012. http://money.cnn.com/2012/07/26/technology/nfc-hack/ index.htm.

157. **Limer, Eric.** Crazy New Exploit Can Brick Samsung Phones or Steal All Their Data. [Online] Gizmodo, December 16, 2012. http://gizmodo.com/5968879/crazy-new-exploit-can-brick-samsung-phones-or-steal-all-their-data.

158. **Kirk, Jeremy.** Android Botnet Abuses People's Phones for SMS Spam. [Online] CIO, December 17, 2012. http://www.cio.com/article/724237/Android_Botnet_Abuses_People_s_ Phones_for_SMS_Spam.

159. **Krebs, Brian.** A Closer Look: Perkele Android Malware Kit. [Online] Krebs on Security, August 19, 2013. http://krebsonsecurity.com/tag/perkele/.

160. **Ashford, Warwick.** Researchers discover new Android Trojan. [Online] Computer Weekly, April 4, 2013. http://www.computerweekly.com/news/2240180810/Researchers-discover-new-Android-Trojan.

161. **Kelly, Heather.** SIM card hack inspires quick fix by carriers. [Online] CNN, August 1, 2013. http://www.cnn.com/2013/08/01/tech/mobile/sim-card-hack/index.html.

162. **Talbot, David.** Hacked Feature Phone Can Block Other People's Calls. [Online] MIT Technology Review, August 26, 2013. http://www.technologyreview.com/news/518646/ hacked-feature-phone-can-block-other-peoples-calls/.

163. **Constantin, Lucian.** D'oh! Basic flaw in WhatsApp could allow attackers to decrypt messages. [Online] PC World, October 9, 2013. http://www.pcworld.com/article/2053480/doh-basic-flaw-in-whatsapp-could-allow-attackers-to-decrypt-messages.html.

164. **Wallace, Gregory.** Apple issues fix for security risk. [Online] CNNMoney, February 23, 2014. http://money.cnn.com/2014/02/23/technology/mobile/apple-iphone-security-hole/index.html.

165. **Xue, Hui, Wei, Tao and Zhang, Yulong.** Masque Attack: All Your iOS Apps Belong to Us. [Online] FireEye, November 10, 2014. https://www.fireeye.com/blog/threat-research/2014/11/ masque-attack-all-your-ios-apps-belong-to-us.html.

166. **Strazzere, Tim.** The new NotCompatible: Sophisticated and evasive threat harbors the potential to compromise enterprise networks. [Online] Lookout, November 19, 2014. https://blog.lookout.com/blog/2014/11/19/notcompatible/.

167. **Timberg, Craig.** German researchers discover a flaw that could let anyone listen to your cell calls. [Online] The Washington Post, December 18, 2014. http://www.washingtonpos t.com/blogs/the-switch/wp/2014/12/18/german-researchers-discover-a-flaw-that-could-let-anyone-listen-to-your-cell-calls-and-read-your-texts/.
168. **Merica, Dan.** Five things you need to know about U.S. national security. [Online] CNN, July 29, 2012. http://security.blogs.cnn.com/2012/07/29/five-things-you-need-to-know-about-u-s-national-security/.
169. **Goldman, David.** Watching porn is bad for your smartphone. [Online] CNNMoney, February 11, 2013. http://money.cnn.com/2013/02/11/technology/security/smartphone-porn/ index.html.
170. **Simonite, Tom.** Browser Exploit for Android Highlights Google's Update Problem. [Online] MIT Technology Review, February 14, 2014. http://www.technologyreview.com/ news/524631/browser-exploit-for-android-highlights-googles-update-problem/.
171. **Sengupta, Somini.** U.S. Military Hunts for Safe Smartphones for Soldiers. [Online] The New York Times, June 22, 2012. http://bits.blogs.nytimes.com/2012/06/22/u-s-military-hunts-for-safe-smartphones-for-soldiers/.
172. **Griggs, Brandon.** Meet the 'NeRD,' the Navy's new e-reader. [Online] CNN, May 8, 2014. http://www.cnn.com/2014/05/08/tech/gaming-gadgets/navy-nerd-e-reader/index.html.
173. FCC Smartphone Security Checker. [Online] FCC. [Cited: January 25, 2013.] http://www. fcc.gov/smartphone-security.
174. **Souppourison, Aaron.** iPhone lockscreen can be bypassed with new iOS 6.1 trick. [Online] The Verge, February 14, 2013. http://www.theverge.com/2013/2/14/3987830/ ios-6-1-security-flaw-lets-anyone-make-calls-from-your-iphone.
175. **Peterson, Scott.** Exclusive: Iran hijacked US drone, says Iranian engineer (Video) . [Online] The Christian Science Monitor, December 15, 2011. http://www.csmonitor.com/World/ Middle-East/2011/1215/Exclusive-Iran-hijacked-US-drone-says-Iranian-engineer-Video.
176. **Zaragoza, Sandra.** Spoofing a Superyacht at Sea. [Online] The University of Texas at Austin, July 30, 2013. http://www.utexas.edu/know/2013/07/30/spoofing-a-superyacht-at-sea/.
177. **Rutkin, Aviva Hope.** "Spoofers" Use Fake GPS Signals to Knock a Yacht Off Course. [Online] MIT Technology Review, August 14, 2013. http://www.technologyreview.com/ news/517686/spoofers-use-fake-gps-signals-to-knock-a-yacht-off-course/.
178. **Gallagher, Ryan.** FBI Accused of Dragging Feet on Release of Info About "Stingray" Surveillance Technology. [Online] Slate, October 19, 2012. http://www.slate.com/blogs/ future_tense/2012/10/19/stingray_imsi_fbi_accused_by_epic_of_dragging_feet_on_releasing_ documents.html.
179. **Nohl, Karsten.** Mobile self-defen. [Online] Security Research Labs, December 27, 2014. http://events.ccc.de/congress/2014/Fahrplan/system/attachments/2493/original/ Mobile_Self_Defense-Karsten_Nohl-31C3-v1.pdf.
180. **Sandi National Laboratories.** Sandia builds self-contained, Android-based network to study cyber disruptions and help secure hand-held devices . [Online] Sandi National Laboratories, October 2, 2012. https://share.sandia.gov/news/resources/news_releases/ sandia-builds-self-contained-android-based-network-to-study-cyber-disruptions-and-help-secure-hand-held-devices/.
181. **IMDb.** CSI: Cyber. [Online] IMDb, 2015. http://www.imdb.com/title/tt3560060/.
182. **Perlroth, Nicole.** Hackers in China Attacked The Times for Last 4 Months. [Online] The New York Times, January 30, 2013. http://www.nytimes.com/2013/01/31/technology/ chinese-hackers-infiltrate-new-york-times-computers.html.
183. **Symantec.** Symantec Statement Regarding New York Times Cyber Attack . [Online] Symantec, January 31, 2013. http://www.marketwire.com/press-release/symantec-statement-regarding-new-york-times-cyber-attack-nasdaq-symc-1751586.htm.
184. **Goldman, David.** Your antivirus software probably won't prevent a cyberattack. [Online] CNNMoney, January 31, 2013. http://money.cnn.com/2013/01/31/technology/security/ antivirus/index.html.

185. **Poeter, Damon.** NBC.com Hacked, Infected With Citadel Trojan. [Online] PC Magazine, February 21, 2013. http://www.pcmag.com/article2/0,2817,2415735,00.asp.

186. **Peterson, Andrea.** Everything you need to know about Yahoo's security breach. [Online] The Washington Post, January 6, 2014. http://www.washingtonpost.com/blogs/the-switch/wp/2014/01/06/everything-you-need-to-know-about-yahoos-security-breach/.

187. **Kelly, Heather.** Is the government doing enough to protect us online? [Online] CNN, July 31, 2012. http://www.cnn.com/2012/07/25/tech/regulating-cybersecurity/index.html.

188. **CNN Political Unit.** Investigation opened into hacked Bush family e-mails. [Online] CNN, February 8, 2013. http://politicalticker.blogs.cnn.com/2013/02/08/investigation-opened-into-hacked-bush-family-emails/.

189. **Ablon, Lillian and Libicki, Martin C.** Wild Wild Web: For Now, Cybercrime Has the Upper Hand in Its Duel with the Law. [Online] The RAND Corporation, Summer 2014. http://www.rand.org/pubs/periodicals/rand-review/issues/2014/summer/wildweb.html.

190. **Nieva, Richard.** Big banks stage mega-cyberattack drill. [Online] CNN, July 18, 2013. http://money.cnn.com/2013/07/18/technology/security/bank-cyberattack/index.html.

191. **Gaudin, Sharon.** DARPA chief leaves Pentagon for Google job. [Online] Computerworld, March 13, 2012. http://www.computerworld.com/s/article/9225156/DARPA_chief_leaves_Pentagon_for_Google_job.

192. **Dugan, Regina.** Regina Dugan: From mach-20 glider to humming bird drone. [Online] TED, March 2012. http://www.ted.com/talks/regina_dugan_from_mach_20_glider_to_humming_bird_drone.html.

193. **Segall, Laurie.** Facebook pays $40,000 to bug spotters . [Online] CNNMoney, August 30, 2011. http://money.cnn.com/2011/08/30/technology/facebook_bug_bounty/index.htm.

194. **Facebook.** Bounty. [Online] Facebook. [Cited: January 25, 2013.] http://www.facebook.com/whitehat/bounty/.

195. **—.** White Hats. [Online] Facebook. [Cited: January 25, 2013.] http://www.facebook.com/whitehat/.

196. **Pepitone, Julianne.** Google awards $60,000 prize for Chrome hack. [Online] CNNMoney, October 10, 2012. http://money.cnn.com/2012/10/10/technology/security/google-chrome-hacker-prize/index.html.

197. **Kersey, Jason.** Chrome Releases. [Online] Google, October 10, 2012. http://googlechromereleases.blogspot.com/2012/10/stable-channel-update_6105.html.

198. **Simonite, Tom.** How to Exchange Encrypted Messages on Any Website. [Online] MIT Technology Review, November 5, 2014. http://www.technologyreview.com/news/532186/how-to-exchange-encrypted-messages-on-any-website/.

199. **Robertson, Jordan.** Why Sony's Plan to Foil PlayStation-Type Attacks Faltered. [Online] Bloomberg, December 5, 2014. http://www.bloomberg.com/news/2014-12-05/why-sony-s-plan-to-foil-playstation-type-attacks-faltered.html.

200. **Frates, Chris and Devine, Curt.** Government hacks and security breaches sky-rocket. [Online] CNN, December 19, 2014. http://www.cnn.com/2014/12/19/politics/government-hacks-and-security-breaches-skyrocket/.

201. **O'Toole, James.** #StealMyIdentity: Fraudsters use paycheck selfies to steal bank details. [Online] CNNMoney, October 29, 2014. http://money.cnn.com/2014/10/29/technology/social/instagram-identity-theft/index.html.

Chapter 11
Cybersecurity Training in Medical Centers: Leveraging Every Opportunity to Convey the Message

Ray Balut and Jean C. Stanford

11.1 Introduction

Cyber security is a critical component of health information technology. As electronic health records (EHRs) become more widely adopted and as new payment models for health care require more data sharing and clinical care coordination with multiple external providers, it becomes evident that new challenges are arising. At the same time, cybercriminals are finding multiple uses for clinical data, from claims fraud to identity theft. Gangs of organized criminals are now harvesting medical data for fraud and identity theft purposes [1]. Government regulators are scrutinizing every reported health data breach with organizations potentially facing civil and criminal penalties when data is not protected properly. In such an environment, one would expect a significant cyber training budget to be available — but this is often not the case. Many health care institutions spend relatively small proportions of their annual budgets on information technology in general [2], and often only a small proportion of that budget is devoted to cybersecurity training. This disparity becomes even more pronounced when the average cost of a data breach in a health care institution in 2013 was $1,973,895 [3].

Until recently, most health records were stored on paper. Tom Sullivan, Executive Editor of HIMSS Media, wrote in January 2013, "It's harder to steal millions of paper records than electronic ones. But as more EHRs create a digitized health system where health information exchanges (HIEs) and health insurance exchanges are the norm, electronic health data is widely shared and an increasing amount of it stored in clouds and other central repositories, from where it can be accessed by a variety of mobile devices, well, that is already changing. Add to it the rocket-like proliferation of mobile devices, easily-lost and frequently unencrypted" [1].

R. Balut (✉)
Healthcare CISO, Maryland, USA

J.C. Stanford
Georgetown University, Washington, D.C., USA

© Springer International Publishing Switzerland 2015
N. Lee, *Counterterrorism and Cybersecurity*, DOI 10.1007/978-3-319-17244-6_11

11.2 Healthcare Cyber Attacks

Hospitals are focusing on creating a patient care culture. The Chief Information Security Officer (CISO) is trying, within the patient care culture, to create a broad cyber security culture. This is challenging because while healthcare has a long tradition of being a privacy culture, cyber security is a relatively new concept and on the surface seems to counter the desire to provide the users with information anywhere and anytime it might be needed. For example physicians may be sent hundreds of files in a week with critical patient information such as laboratory results or radiology interpretations. Telling a clinician not to click on an email attachment labeled "Laboratory Results" simply may not be practical as the clinician has patients to care for and until systems become truly interoperable, this may be the only way they can receive these files. Unfortunately this also leaves the caregiver vulnerable to one of the most common social engineering attacks: phishing.

In a recent health information security survey, the highest overall risk for security breaches was perceived to be errors or malicious actions by staff or employees [2]. Intrusive invasions of malware or external threat actors were the third highest concern and these attacks are often initiated when an unwary staff person clicks on an email or link that downloads malicious content. (This attack vector is known as "phishing".) Therefore, health care CISOs must use creativity to convey the most essential cyber security messages to the disparate health care user populations to ensure that users "think before they click."

11.3 Value of Medical Data to Cybercriminals

Cybercriminals have are a number of uses for medical data. At the most basic level, medical records offer all of the basics to commit various kinds of identity and insurance fraud: Names, addresses, Social Security Numbers and health insurance account information. This information can then either be used directly to commit identity theft and insurance fraud or simply packaged and resold many times over for use by others. Some of the obvious uses are to obtain payment for confidential data regarding high profile individuals such as politicians.

Adding to the value of data stolen from medical records is the potential "shelf-life" of the information. Unlike payment card data it can be very difficult to detect medical identity theft quickly. Criminals may be able to take advantage of the data for years whereas the compromise of a credit card number can be quickly detected by simply monitoring statements and the abuse stopped by canceling the number [4]. "Criminals are starting to recognize the high financial value of protected health information [and] are being more surgical about the kinds of information they're going after" [5].

Because of the long term and highly profit potential of identity theft and insurance fraud, stolen credentials from a medical record can go for many times the cost of a credit card number on underground exchanges where such information is traded [6].

Medical identity theft is a significant burden for patients who are victimized [7]. The Ponemon Institute's analysis suggests that there were at least 1,836,312 victims of medical identify theft in the United States in 2013. While some of these were victimized by close friends or family members, others were victimized due to insider breaches or external cyber attacks. There is significant financial risk (Ponemon estimates that the financial costs are approximately $18,660 per patient) but there is also a very serious medical risk when clinical information from fraudulent encounters is mingled with patients' actual clinical data. It is very difficult for patients to discover medical identity theft and extremely difficult to correct the records once the theft is understood [7].

In the end, healthcare organizations find themselves having to protect a valuable commodity in one of the toughest data security environments [8].

11.4 Major Threats to Medical Data in Clinical Environments

Medical data is lost due to a variety of events; some malicious, some inadvertent and some due to lack of knowledge or carelessness. Many of the losses could be averted with the proper training and awareness on the part of clinical and administrative staff. Two of the most common causes of credential and system compromise that regularly occur are phishing attacks and lost or stolen devices, both of which training and awareness can go a long way to prevent:

1. **Phishing**

Phishing is a type of cyber fraud in which an unsuspecting user is tricked into entering information into a malicious system that is posing as trustworthy [9]. For example, one CISO stated, "Each month we are receiving thousands of phishing messages that are becoming more polished and sophisticated. It only takes one slipping through to potentially create a breach" [10]. In many cases, clicking on links or attachments allows malware to be downloaded from the Internet that then began to propagate similar messages through the organization. Ultimately this lead to a shutdown of the email system and other resources that took many hours to remediate. One CISO reports working for 42 h straight to remediate such a breach [10].

Making the situation even more challenging, sometimes users send out perfectly legitimate emails with suspicious-looking characteristics such as zip file attachments or requests for users to "click on the enclosed link to access payroll or account information." These users may need specific training on how to set up messages that do not resemble phishing attacks so that users recognize them as legitimate and the cyber security team does not have to spend time investigating false positive reports. The more false-positives, the more complacency. Employees also must spend time trying to determine the legitimacy of false positive phishing messages and this reduces overall productivity.

2. **Use of Unapproved Cloud Services**

Clinicians may choose to share patient data with each other via personal email or tools such as Dropbox. There may be real clinical need to share the data. For example, one nurse described a process they used when they had a patient with transient bradycardia. When the bradycardia events occurred, the staff took pictures on their cell phones of the ECG monitors so that the cardiologists in another part of the facility could see them. The cardiologists did not have access to real time monitoring capabilities and the monitoring devices did not have playback capabilities. Staff will use workarounds to achieve clinical goals; it is essential to explain the risks of the workarounds and to show them the available safe tools that allow sharing of essential clinical information.

3. Lost or Stolen Devices

In a survey of 91 hospitals, 49 % of healthcare information security professionals reported that patient data had been breached due to lost or stolen laptops or mobile devices in the past year. 88 % of the hospitals allowed staff to connect to the hospital networks using their personal mobile devices [11]. Mobile device management tools are available, but 38 % of surveyed health care organizations were not using them. Loss or theft of these devices could result in patient data compromises [3]. Desktop machines, USB drives, servers, and tablet computers all can potentially contain patient health information and can be lost or stolen. While at one time a lost or stolen device may simply have been wiped by the finder or thief and reused, it is now common practice to look for potentially valuable information before re-imaging a device and using or selling it. While encryption provides a strong defense and is often the go-to solution for end-user devices and is thought of as a "Safe-Harbor", it difficult to ensure that all portable devices and media are encrypted. Providing the means to store and access data in a secure location and training users how to do so greatly reduces the risk for patients and the institution itself by simply not allowing the data to be compromised in the first place.

Sometimes the device does not have to be lost or stolen — just used in the wrong place. For example a staff member of Keystone Mercy Health Plan took a flash drive with the demographics and Social Security Numbers for 280,000 Medicaid recipients to a community health fair where it was compromised [12].

4. Insider Threats

Malicious insiders accounted for 12 % of data breaches identified in the 2014 Ponemon Health Privacy and Security Report [3]. While this is much less than the damage caused by losing devices, the potential damage may be more severe because insiders may know the value of the data and may intend to exploit it [12].

11.5 Training Resources

Health care institutions spend relatively less money on information technology and cyber security as compared to other major industries. A 2014 survey of major corporations worldwide found that they spend an average of 4 % of the overall IT budget on information security [13]. In contrast, a survey of large health care

TABLE HB1			
% of Total IT Operating Expense/Total Hospital Operating Expense-Overall	2010	2011	2012
Avg	2.40%	2.39%	2.74%
Median	1.93%	2.11%	2.27%
N	471	475	400

TABLE HB2			
% of Total IT Budget/Total Hospital Expense-Overall	2010	2011	2012
Avg	2.77%	4.87%	3.21%
Median	2.26%	3.92%	2.66%
N	469	436	479

TABLE HB3			
% IS Capital Expense/Total Hospital Capital Expense-Overall	2010	2011	2012
Avg	17.32%	17.89%	20.22%
Median	10.27%	12.14%	14.10%
N	211	300	244

Fig. 11.1 IT spending in health care institutions

organizations showed that in 2013, 19 % dedicated less than one percent of their IT budget to cyber security and 30 % dedicated between one and three percent of their IT budgets to cyber security — a much smaller proportional commitment to the domain [2].

Figure 11.1 shows the trends of hospital operating expenses, total IT budget and capital expenses in terms of level of spending on IT [8].

11.6 Training Users to Prevent Cyber Breaches

"If we security engineers do our job right, then users will get their awareness training informally and organically from their colleagues and friends," said Bruce Schneier, Chief Technology Officer of Co3 Systems [14].

User training includes many stakeholders, from the Board of Trustees to the hourly staff. Managing training of such diverse ever-changing communities is very challenging, especially when the budget and training time available are often very limited.

Training Purposes and Goals

Employee resistance is considered to be the biggest barrier to an effective mobile security strategy across all domains (not simply healthcare) [15]. This means that a primary goal of the training program is to convince users that they have a role to play in security and it is worthwhile for them to participate in or cooperate with cyber security controls. The highest risk factors for mobile security breaches are malware infection and loss or theft from end users — and both may be ameliorated by end user training [15]. Simply put, the CISO and the security team cannot secure the organization on their own. Without the help of end users who understand how to protect devices and data and what the organizations security requirements are, the CISO and the security team will forever find themselves in a losing battle.

Key goals of the training include:

1. Convey the important information needed for compliance with various regulatory requirements (such as HIPAA)
2. Provide guidance in dealing with the most likely risks such as how to:
   (a) Protect mobile devices (cellphones, tablets, laptops) that connect to or contain protected information
   (b) Recognize and avoid social engineering attacks [16] such as
      • Phishing Email
      • Pretexting (impersonating an authoritative or trusted individual)
      • Baiting or Quid Pro Quo (tricking users into using poisoned tools or devices or trading information)
3. Recognizing and reporting potential security incidents are the most important actions the users can take.

11.7 User Communities and Current Training Methods

Medical Center system users come from a very wide range of socioeconomic and educational backgrounds. Some have not finished high school and some have MD-PhD degrees. Thus, training has to be flexible and has to be relevant for the user groups. Also, there is a wide distribution of technical skill. The fact that one has an MD or a PhD does not guarantee technical understanding or proficiency when it comes to information technology or cyber security. Similarly, some employees with very little formal education may be very technically sophisticated.

Some users never enter the premises of the hospital itself. They may work for affiliated clinics that have access to medical records in order to coordinate specialty or primary care. The risks across the multiple venues of care can be very different. Some medical offices are in well-secured buildings with on-site security, video monitoring, and other controls while others may be protected by only a simple lock and key. Hospitals are necessarily open 24/7 and members of the public are allowed to visit patients in many hospital areas; this means that the hospital work stations may be accessed as well. The best method across the board is to have training materials available in all venues and encourage the appropriate leaders to use them as often as possible.

In addition, there may be a very large difference in the type of training that has to be offered based on generational experiences. Younger staff may be extremely familiar with the latest technology and have a ready understanding of security issues. Conversely, those that have not grown up with technology may require a different approach to understand the possible risks and issues that exist.

Clinical Users

When working in a clinical environment, it is important to know that not all users are created equal. Just as senior executives in major corporations are trained differently than the average employee, so physicians are usually trained in a different manner than the rest of the staff. Training can be very compressed and often must be done around the many demands on a physician's time. Physicians and medical students work throughout the organization; they may move from site to site, from inpatient settings to outpatient clinics and through specialty services all in a single day. Scheduling training sessions can be difficult and the training itself must address a variety of environments and applications, many of which have different security controls.

Cyber security training is just one of many training activities that medical professionals must undergo in the course of a year. They may have to have continuing professional education training to retain their clinical licenses. They may also have to be trained on safety measures (such as responding to fires), correct use of medical devices, new methods or techniques for caring for patients, using the electronic medical records, etc. Training time is precious and must be carefully planned.

Because clinicians have limited time for training, they are often given the essential messages during their orientation briefings. Clinicians tend to be very hands-on people. Demonstrations are very effective in getting the message out. It is valuable to explain the actual threats and how they work. Many clinicians have only a vague idea of how cyber attacks work and where they may come from.

Using demonstrations such as, watching an example of keystroke capturing using various devices or malware and then seeing how the garnered information can be used as part of an attack can greatly enhance understanding of the threats. Demonstrations are memorable, fun, and will attract more of an audience than a set of "thou shalt not" messages with no context or rationale.

Another very useful approach is to relate cyber security issues to ones individuals may be more familiar with. A good example would be responding to a cyber security incident. Clinicians have a lot of training in disaster scenarios and they readily relate to the need for rapid response, etc. The responses to a cyber security attack follows many of the same methodologies used to respond to a physical world event such as a fire, flood, or other natural disaster. The major steps in responding to a security incident look very close to those of responding to many other kinds and can be discussed using analogies to incidents the audience may be more familiar with:

1. Prepare (Practice, plan, test, and simulate events before an actual event)
2. Containment (Minimize the damage)
3. Eradictation (Remove the cause of the problem)
4. Recovery (Return things to a normal state)
5. After-action review (Incorporate lessons learned)

Because clinicians spend their professional lives diagnosing problems and acting in the face of incomplete data, they can relate to the challenges that the CISO faces better than staff in other industries might.

Supervisors and Managers

In their training, users are often told to call the Help Desk and or notify their supervisor if they know of or suspect a security related problem such as a lost device or malware infection. It is critical that supervisors be trained on how to handle these sorts of issues. Supervisors in clinical settings are trained in a wide range of crisis management responses such as for fire or natural disaster. Cyber security must become one of these areas as well. Often supervisors and managers are unaware of the correct response to a cyber incident. For example, one supervisor responded to a malware infestation by advising users to reboot their machines and continue as usual. It is key that managers and supervisors know what immediate actions steps to take and who to contact when an incident occurs.

Board Members, C-Suite and Executive Staff

The challenges of training executive users are similar to those of training physicians — they have very limited time and many priorities to deal with. Board members and executive staff are intelligent and well versed in their areas of expertise however cyber security may not be one of these areas. Beyond the standard end-user training, it is important to raise awareness on issues such as compliance and risk as it relates to the environment. The costs and commitment to a security program can be substantial and must be understood in terms of how security relates to the business, adds value to the organization, and supports the larger organizational objectives and goals. The vision of the end state of a secure organization needs to be communicated, often in a very brief period of time, and in terms that the audience can appreciate.

Unfortunately, the time when many executives become aware of cyber security is after a breach occurs. While incidents themselves can be teachable moments and an opportunity to raise issues and educate leadership about cyber security issue, the "incident effect" can be short lived and new priorities can quickly overshadow the priority of security.

One of the key goals when working with the executive users is to develop an understanding and appreciation for the importance of training for the entire enterprise community. Because of the many training demands placed on the organization, adding additional training even for something as important as security, will often require senior leadership support.

Information Technology Staff

Training for the Information Technology staff (in general they don't report to the CISO, they report to the CIO) is critical as they play an essential role in maintaining an effective cyber security posture within an organization. Among the groups that need specialized cyber security training:

1. The infrastructure team
2. The applications teams who write, install and maintain the applications throughout the enterprise
3. Server support teams
4. Desktop/End-User Device support teams
5. The messaging team (i.e. email)
6. Help Desk/Service Support staff

These teams should be given much more in-depth training on secure software development, incident response, the importance of patching, secure system configuration etc. Sometimes healthcare organizations outsource operational support. If this is the case, the CISO will need to assess the training levels of the contractor team and ensure that they are trained to the level required to support critical needs. This may need to be included as part of the service level agreements (SLAs) in the outsourcing contract. Collaboration with the contracting team is essential for these cases.

11.8 Types of Training Offered and Training Environment

There are various training opportunities for reaching the general employee population. The time available is usually very brief and the training may be done by non-security staff (such as Human Resources or Department Managers). Therefore, the communications must be carefully honed to communicate the most important messages clearly and precisely. These messages are then reinforced with additional training events as the opportunity arises. Figure 11.2 shows the routine training opportunities.

Limiting training to employee orientation sessions is far from ideal. New employees are bombarded with many messages on their orientation day and cyber security practices are usually not knowledge they will retain. Due to the limited amount of time, this type of training tends to focus on compliance and organizational policy — necessary but not sufficient.

Posters may be used to convey simple messages. It is best to keep the text concise and aligned with the basic messages conveyed at employee orientation. The posters should have simple, eye-catching graphics. Using humor can make them more memorable. If they are put in unexpected places, like the lavatory, they catch users' attention.

The Information Security Department may have a web page with processes, procedures, contact information, etc. provided to all staff and physicians. To draw readers, it may also include information on how to secure home systems, mobile devices, home networks, etc. Providing this sort of information draws users who might then learn things that are important for securing the hospital networks as well. Reminder emails tend to be very ineffective; in our experience, only about 20 % of them are read.

All employees are usually required to complete automated renewal training every year. There is a quiz at the end to measure comprehension.

Training Opportunity	Attendees	Usual Trainers	Key Messages	Training Time Available
New Employee Orientation	New Employees	Human Resources Dept. Optimally someone from the CISO staff may be able to brief, but this is not always possible. A video or slide deck may be provided to emphasize the appropriate messages.	1. Handling mobile devices & laptops 2. Preventing malware infections 3. Reporting suspicious activity or known breaches 4. Protecting Credentials (i.e. passwords, access cards etc.) 5. General understanding that Security it a regulatory requirement.	Fifteen minutes
Physician Training	New physicians (staff, medical students, attending physicians, contract physicians)	Specialized physician orientation team	The key messages are essentially the same as those for the end user but an emphasis is placed on the relatively high and broad level of access to sensitive information a physician may have. A positive and interactive approach is usually more effective than an enforcement-driven one.	At discretion of trainers
Posters	Casual viewing	IT Security staff Communications staff.	Simple messages. Put unexpected locations. The best posters are simple, provide clear and actionable guidance — and are funny.	1 to 2
Annual Renewal Training	All Employees	Computerized training; may be reused for more than one year.	Simple message, depending on the training vendor. May have a quiz at the end to test comprehension.	Ten minutes
email	All employees	CISO staff	Simple messages, reminders, etc.	1 minute
Web page	Open to all staff and physicians	CISO Staff	FAQs, procedures, contact lists, etc.	Unstructured

Fig. 11.2 Routine training opportunities

11.9 Innovative Approaches

There are some new and innovative approaches under development or still being actively considered. The object is to cost-effectively get the users' attention long enough to convey important security training messages.

Training the Best Clinical IT Trainers

As new health IT systems become available across the hospital network, nurse informaticists are responsible for working with the clinical staff and physicians to encourage adoption and to ensure that the new capabilities are used correctly. The nurse informaticists have extensive experience in using the clinical systems across all systems of care and all major venues. They are ideal information security champions and trainers because they are trusted, they spend the most time with end users and because they know the operating environment better than anyone else. Training the nurse informaticists to convey security and privacy messages as they conduct the rest of their training sessions creates many more opportunities to convey important information in a practical manner.

Fake Attacks Leading to Teachable Moments

Some organizations conduct simulated phishing campaigns to help train employees. If the employee clicks on the phishing training message, the employee is directed to a training video that explains how to identify phishing attacks in the future [17]. This approach is controversial. The Help Desk may have to spend a lot of time working with end users who fear that their system has been compromised by a phishing attack. While employees may be enlightened about phishing attacks, some analysts believe end users will come to be less vigilant, thinking that all suspicious messages are actually only a drill. In a clinical environment, the staff may be distracted from patient care to handle a phishing training message. This would frustrate clinicians and could cause unnecessary delays in clinical process flows.

Hands-on Demonstrations

It is often effective to relate the cyber security issues in the work environment with those that the users might face at home. By demonstrating how to secure their home systems, trainers help users achieve a better understanding and appreciation for the controls used in the work place. A truly effective demonstration offers the following benefits:

1. Participants pay attention because this is information that is useful to them in their everyday lives.
2. As they walk through the methods to secure their own data, the measures taken to secure the hospital data seem more reasonable and participants are more likely to comply.
3. With the expansion of remote access to clinical resources, encouraging "safe computing" practices at home reduces the likelihood of possible incidents resulting from users accessing healthcare systems and data from personal devices.

Specialized Training Videos

It is possible to hire professional videographers to create tailored videos for showing to certain populations (e.g., physicians) or for new training opportunities (department meetings). For example, one company quotes a rate of $800 for a 10 h session with a videographer in the New York City area and then $640 for a day of video editing. It is conceivable that a customized training video could be made for less than $5,000 [18].

Another eye-catching approach would be to create cartoon-like characters that would simulate real users responding (correctly and incorrectly) to simulated events. These could be very entertaining and could convey a great deal of information in a few minutes. It may be possible to collaborate with the local arts organizations to obtain access to animation skills for little money. Contests are also a consideration here; arts students may appreciate the recognition that a winning entry would bring them.

An even less expensive innovation could be to leverage the video capabilities of modern cellphones and tablets and challenge the staff to come up with the most compelling and useful training videos for their own specialties. This could lead to enjoyable, creative, videos — if the end users understand the message that the information security team is trying to convey. Such a contest could also lead to fascinating insights as to what the end users really understand the security procedures to be — a free and thorough diagnostic! It might be useful to offer simple prizes for the best videos. Trinkets such as flashlights are popular awards for hospital contests and are very low cost.

Here is a humorous staff video idea: A possible staff-created video could show a fictional "pink slip virus." It could show basic phishing training and show a fictional staff member reading social media instead of paying attention to the training session. Then the video could show that staff member clinking on a test phishing message — and having a pink slip printed on the nearest printer. This is obviously a joke, but it would send a useful message in a fashion that is likely to attract the end users' attention.

Leveraging Games and Non-IT Events

Some organizations have pizza tables set up in the cafeteria. The staff gets free pizza if they are willing to watch a training video while they eat. Free food is very popular with exhausted and stressed staff. More structured "Lunch and Learn" sessions may also be useful as they offer the opportunity to ask questions and share experiences in a less formal and more relaxed atmosphere than a formal training session.

Developing more formal training games in software form (called gamification) is also potentially useful. Users may find a video game version of the training message engaging, especially if the game modules are humorous — and short. Gamification is not a silver bullet; user motivation, context of use, time spent with the game, game design, etc. all affect the utility of a gamified training tool [19].

A fun exercise is to set up a booth at a system wide technology fair or other common event. The activity at the booth is "Guess My Password." The cyber security staff could attempt to guess a user's password. It is not necessary to reveal the password but if it can be guessed using a simple dictionary attack, the user would obtain some education about safe passwords — and they would have to change theirs on the spot. Another fun activity would be "spot the phishing email" — those that could identify a phishing email and explain why it is suspicious would get a prize.

11.10 Conclusion

There are a number of tools that organizations could employ with even a minimal training budget to train users to prevent, recognize and halt attacks. The key resources needed are:

1. A relatively modest budget (perhaps 1 % of the overall IT spend).
2. Executive support to ensure that security training is included in the organization's overall program.
3. Strong and creative (such as the use of videos, games, contests etc.) messaging that emphasizes the importance of cyber security and reinforces the training requirements and mandatory procedures.

Good training begins with an organization committed to cyber security. It must be supported by the executive team and considered as important as other issues such as patient safety. Using a variety of techniques including demonstrations, videos, and contests will help deliver the message in ways that appeal to a broad and diverse audience.

Bibliography

1. **Sullivan, Tom.** Are providers ripe for a massive medical records heist? *Government Health IT.* 2013 йил 14-January.
2. **HIMSS.** *Sixth Annual HIMSS Security Survey.* HIMSS. s.l. : HIMSS, 2013. pp. 1-45.
3. **Ponemon Institute.** *Fourth Annual Benchmark Study on Patient Privacy & Data Security.* Traverse City: Ponemon Institute, 2014. p. 7.
4. *THE ROLE OF THE FORENSIC ACCOUNTANT IN A MEDICARE FRAUD IDENTITY THEFT CASE .* **Sanchez, M.** 3, 2012, GLOBAL JOURNAL OF BUSINESS RESEARCH, Vol. 6, p. 86.
5. **Butler, M.** ACA Raises Privacy, Security Concerns, Study Finds. *Journal of AHIMA.* [Online] May 01, 2014. [Cited: December 30, 2014.] http://journal.ahima.org/aca-raises-privacy-security-concerns-study-finds/.
6. **RSA.** *CYBERCRIME AND THE HEALTHCARE INDUSTRY .* s.l. : RSA, 2012. p. 1.
7. **Ponemon Institute.** *2013 Survey on Medical Identity Theft.* Traverse City: Ponemon Institute, 2013.
8. **HIMSS.** *2013 Annual Report of the U.S. Hospital IT Market.* Chicago : HIMSS, 2013. p. 6. http://apps.himss.org/foundation/docs/2013HIMSSAnnualReportDorenfest.pdf.
9. **Pfleeger, Charles P. and Pfleeger, Shari Lawrence.** *Analyzing Computer Security: A Threat / Vulnerability / Countermeasure Approach.* s.l. : Pearson Education, 2011. pp. Kindle Locations 14988-1499 .

10. **HISTalk.** Advisory Panel: Data Brieaches. *HISTalk.* [Online] 2013 йил 15-April. [Cited: 2014 йил 29-December.] http://histalk2.com/2013/04/15/advisory-panel-data-breaches/.
11. **Ponemon, Larry and Kam, Rick.** *ACA Impacts on Patient Data Security* . Ponemon Institute. s.l. : Ponemon Institute, 2014. pp. 19-21.
12. Healthcare and Security: A Hackers Perspective. *InfoSec Island.* [Online] 2010 йил 27-December. [Cited: 2014 йил 29-December.] http://www.infosecisland.com/blogv iew/10538-Healthcare-and-Security-A-Hackers-Perspective.html.
13. **PWC.** *Managing cyber risks in an interconnected world: Key findings from The Global State of Information Security ® Survey 2015.* New York: PWC, 2014. p. 19.
14. **Schneier, B.** On Security Awareness Training. *Dark Reading.* March 19, 2013. http://www. darkreading.com/risk/on-security-awareness-training/d/d-id/1139381?.
15. **Ponemon Institute.** *Security in the New Mobile Ecosystem.* Ponemon Institute. Traverse City: Ponemon Institute, 2014. p. 3.
16. **Mimic Technologies.** *Social Engineering – Risks, Techniques and Safeguards.* s.l. : Mimic Technologies, 2013. p. 2. https://www.itegria.com/wp-content/uploads/2014/03/ItegriaSocial Engineering.pdf.
17. *Phishing Our Employees.* **Epstein, Jeremy.** s.l. : IEEE, 2014 йил May, IEEE Security & Privacy, pp. 3-4.
18. **Video Production New York.** Freelance Video Production Price List. [Online] 2014 йил 20-December. http://www.videoproduction.us.com/pricing/.
19. *Does Gamification Work?–A Literature Review of Empirical Studies on Gamification.* **Hamari, J., Koivisto, J., & Sarsa, H.** Hawaii: IEEE, 2014 йил, System Sciences (HICSS), 2014 47th Hawaii International Conference on, pp. 3025-3034.

Chapter 12
Plan X and Generation Z

This is the generation that makes a game out of everything.
For them, life is a game.
— Brian Niccol, Taco Bell's chief marketing
and innovation officer
(May 2012)

If you control the code, you control the world. This is the future
that awaits us.
— Marc Goodman, global security advisor and futurist
TEDGlobal 2012 (June 28, 2012)

We're the ones who built this Internet. Now we're the ones
who have to keep it secure.
— Army Gen. Keith Alexander, Director of National Security
Agency and Commander of U.S. Cyber Command
2012 Aspen Security Forum (July 27, 2012)

Plan X is a program that is specifically working toward … a future
where cyber is a capability like other weapons capabilities.
— DARPA Director Arati Prabhakar (April 24, 2013)

I have two daughters of my own coming up on college age.
I want them to have a world that's got equal opportunity
for them.
— Intel CEO Brian Krzanich at Consumer Electronics Show
(January 6, 2015)

© Springer International Publishing Switzerland 2015
N. Lee, *Counterterrorism and Cybersecurity*, DOI 10.1007/978-3-319-17244-6_12

12.1 Plan X: Foundational Cyberwarfare

"Other countries are preparing for a cyber war," said Richard M. George, a former National Security Agency (NSA) cybersecurity official. "If we're not pushing the envelope in cyber, somebody else will" [1].

Since 2009, the Defense Advanced Research Projects Agency (DARPA) within the U.S. Department of Defense (DoD) has been steadily increasing its cyber research budget to $208 million in fiscal year 2012 [2].

In May 2012, DARPA officially announced Plan X [3]. The Plan X program is explicitly not funding research and development efforts in vulnerability analysis or the generation of cyber weapons. Instead, Plan X will attempt to create revolutionary technologies for understanding, planning, and managing military cyber operations in real-time, large-scale, and dynamic network environments. In November 2012, DARPA issued a call for proposals (DARPA-BAA-13-02) on Foundational Cyberwarfare (Plan X) [4]:

> Plan X will conduct novel research into the nature of cyber warfare and support development of fundamental strategies needed to dominate the cyber battlespace. Proposed research should investigate innovative approaches that enable revolutionary advances in science, devices, or systems. Specifically excluded is research that primarily results in evolutionary improvements to the existing state of practice.
>
> The Plan X program seeks to build an end-to-end system that enables the military to understand, plan, and manage cyber warfare in real-time, large-scale, and dynamic network environments. Specifically, the Plan X program seeks to integrate the cyber battlespace concepts of the network map, operational unit, and capability set in the planning, execution, and measurement phases of military cyber operations. To achieve this goal, the Plan X system will be developed as an open platform architecture for integration with government and industry technologies.

In April 2013, DARPA Director Arati Prabhakar held a press conference during which she further clarified Plan X [5]:

> Plan X is a program that is specifically working toward building really the technology infrastructure that would allow cyber offense to move from the world we're in today, where it's a fine, handcrafted capability that requires exquisite authorities to do anything with it, that when you launch it into the world, you hope that it's going to do what you think it's gonna do, but you don't really know.
>
> We need to move from there to a future where cyber is a capability like other weapons capabilities, meaning that a military operator can design and deploy a cyber effect, know what it's going to accomplish, do battle damage assessment and measure what it has accomplished, and, with that, build in the graduated authorities that allow an appropriate individual to take an appropriate level of action.

Cybersecurity specialist Dan Roelker who conceived the Plan X idea explained, "Say you're playing World of Warcraft, and you've got this type of sword, +5 or whatever. You don't necessarily know what spells were used to create that sword, right? You just know it has these attributes and it helps you in this way. It's the same type of concept. You don't need the technical details" [6].

Figures 12.1a–f display Roelker's rationale behind scalable cyber warfare and the five pillars of "foundational cyberwarfare" [7].

(a)

Dan Roelker
Program Manager, Information Innovation Office

Scaling Cyberwarfare

DARPA Cyber Colloquium
Arlington, VA

November 7, 2011

Approved for Public Release, Distribution Unlimited.

(b)

 Cyberartisan production doesn't scale

All cybertools have a limited shelf-life and operational relevance

	Cyberartisan	*Automation*
Skill	Individual	Technology-based
Level of effort	Manually intensive	Mass produced
Cost/Benefit	"Too big to fail"	Cost effective

Approved for Public Release, Distribution Unlimited.

Fig. 12.1 a Scaling cyber warfare. **b** Cyberartisan versus automation. **c** Program: binary executable transforms (BET). **d** Hacker versus hacker. **e** Cyberwarfare limitations. **f** Pillars of foundational cyberwarfare

(c)

Program: Binary Executable Transforms (BET)

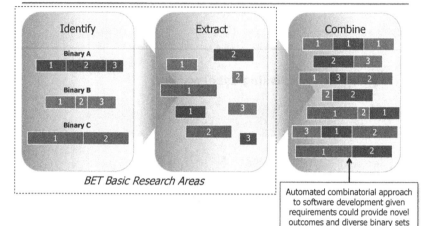

(d)

Hacker vs. Hacker approach doesn't scale

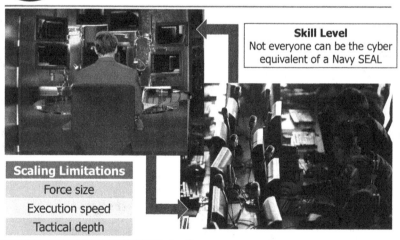

Fig. 12.1 (continued)

(e) Limitations to the Hacker vs. Hacker approach

Cyberwarfare is executed at the speed of light . . .

Force Size Limitations
#of people trained per year
# of people to execute a mission

Execution Speed Limitations
Speed of planning process
Speed of mission operation

Tactical Depth Limitations
Real-time move-counter-move
Multi-phase mission strategy

we need breakthroughs in technology to accomplish this goal

(f)

DARPA Pillars of Foundational Cyberwarfare

Exploitation Research
automation techniques, defeating formal methods, high-fidelity emulation

Network Analysis
on-demand topology, infrastructure capability, platform positioning

Planning and Execution
assured and automated execution, large-scale analytics, distributed planning

Cyberwarfare Platform Development

Visualization
new interfaces, adaptable views, large-scale data representation

Fig. 12.1 (continued)

12.2 Cyber Battlespace Research and Development

In June 2014, Plan X program manager Frank Pound stated, "The big goal of Plan X is to make cyber operations tools and their capabilities more available to the common military, which right now doesn't have [such] cyber capabilities. … What we're trying to do with Plan X is to quantify cyber effects so the military understands how [such effects] work and what the collateral damage could be. A cyber effect could cause damage to an adversary's network or to a hospital next door. We want to make sure when we deploy a cyber effect at an adversary that there's no collateral damage. Right now, that [capability] really doesn't exist, except in small enclaves. … A military commander wants to be able to sense the cyber environment and know if he can deploy a counterattack. … We want a Plan X system in every military installation, every combat information center on a ship, and at the tactical level in tactical operations centers. … We intend to transition the program out to DOD and Cyber Command in October 2017" [8].

Plan X looks at new ways to understand network topologies, a new programming language for online warfare [6], and intuitive devices such as a 55-inch touch table for planning missions and the Oculus Rift for executing missions [9]. The five-year, \$110 million Plan X research program aims to build a prototype system in five technical areas [1, 4]:

1. System Architecture

The core of the Plan X system is the cyber battlespace graphing engine whose primary task is to receive, store, model, retrieve, and send cyber battlespace information to other Plan X system components. The graphing engine receives real-time information from various network mapping components and operational overlay sources. The network mapping components send data that allow the graphing engine to convert and construct a real-time logical network topology. This information will include traceroute data, link latencies, Border Gateway Protocol (BGP) routes, IP Time-To-Live (TTL) header analysis, node routing tables, and any other type of information necessary to assist in constructing the logical network topology. The Plan X system must be able to model network topologies at Internet-level scales.

Operational execution overlay information is stored as meta-data for each element in the logical network topology. For example, operational overlay information will include the operating system identification, network service profile, defensive and offensive capabilities, and identification, friend or foe (IFF). The planning and operational areas will attach another layer of information on top of this constructed cyber battlespace model. Planning information includes the potential entry nodes, support platform placement, communication paths, and target sets. Centralizing operational planning and execution status will allow the Plan X system to show a global heat map of its activities, from conceptual to actual.

Secure software architecture design principles will allow the Plan X system to operate from Unclassified to Top Secret /Special Compartmented Information/Special

Access Program with the possibility of multiple simultaneous technology evaluations operating at different security levels and compartments.

2. Cyber Battlespace Analytics

The primary focus of Cyber Battlespace Analytics is to model, reason, and assist military planners to navigate and build strategically sound and tactically feasible cyber operations. The research areas are twofold: (a) development of automated techniques to assist military planners to construct cyberwarfare plans, and (b) support of wargaming applications, such as modeling opponent moves and counter moves, to optimize planning.

An important goal of Cyber Battlespace Analytics is to understand and quantify cyber battlespace effects including the probabilities of collateral damage. The Plan X system will provide assistance selecting optimal nodes in a cyber battlespace. Node sets might include entry nodes, target nodes, and nodes to avoid.

3. Mission Construction

Mission Construction develops automated techniques that allow mission planners to graphically construct detailed and robust plans that can be automatically synthesized into an executable mission script. The research involves investigating the structure of cyberwarfare program's control flow graphs (CFGs) and the development of domain specific languages (DSLs).

In a cyberwarfare program CFG, instructions executed at a node, whether an entry node, support platform, or target node, may transfer program control by "calling" other nodes as the mission progresses. Called nodes execute instructions, returning the calculation results to either the calling node or a central coordination node.

A cyberwarfare DSL will allow operation checkpoints during mission execution for real-time operator interaction, support real-time failover to switch to manual control, enable various levels of autonomous operation, leverage existing formal analysis to detect errors /bugs /inconsistencies, enforce rules of engagement (ROE), and construct a cyber operation "play book" to assist in planning future missions.

4. Mission Execution

Mission Execution involves research and development in (a) the mission script runtime environment and (b) support platforms.

The mission script runtime environment controls the entire execution of a mission, and supports real-time operator interaction. The runtime environment will use public and commercial toolkits such as Metasploit and Immunity CANVAS to build an extensible API framework for assembling capabilities for each mission program.

Support platforms focus on the development of operating systems and virtual machines designed to execute cyberwarfare missions in highly dynamic and hostile cyber battlespaces. Support platforms include launch platforms, battle effect monitors, communication relay, and adaptive defense support like packet filtering, connection filtering, and mitigation capability.

5. Intuitive Interfaces

Intuitive Interfaces provide a fully integrated visual user experience for commanders, planners, and operators to manage cyberwarfare activities. Similar to a massively multiplayer online game (MMOG), Plan X will model the cyber battlespace and update it with incoming mapping, operational status, and planning information from potentially millions of users.

Plan X will develop four integrated graphical interface workflows: (b) Real-time cyber battlespace views, (b) Planning process, (c) Capability construction, and (d) Operator controls. Touch user interface (TUI), tablet computing, and augmented reality (AR) displays will be a part of the user interface and user experience of Plan X. Traditional keyboard and mouse interactions will be minimized.

12.3 National Centers of Academic Excellence in Cyber Operations

Unlike DARPA and Plan X, the NSA has been actively involved in building cyber weapons such as Stuxnet and Flame [10]. However, finding skilled employees for cyber warfare is not easy. "Universities don't want to touch [hacking], they don't want to have the perception of teaching people how to subvert things," said NSA technical director Steven LaFountain [11].

In May 2012, the NSA launched the National Centers of Academic Excellence (CAE) in Cyber Operations Program to prime college students for careers in cyber warfare. The first four universities to receive the CAE-Cyber Operations designation for the 2012-2013 academic year were Dakota State University, the Naval Postgraduate School, Northeastern University, and the University of Tulsa. In August 2014, the list expanded to 13 centers: Air Force Institute of Technology in Ohio (2013-2018), Auburn University in Alabama (2013-2018), Carnegie Mellon University in Pennsylvania (2013-2018), Dakota State University in South Dakota (2012-2017), Mississippi State University in Mississippi (2013-2018), Naval Postgraduate School in California (2012-2017), Northeastern University in Massachusetts (2012-2017), Polytechnic School of Engineering of New York University in New York (2014-2019), Towson University in Maryland (2014-2019), United States Military Academy at West Point in New York (2014-2019), University of Cincinnati in Ohio (2014-2019), University of New Orleans in Louisiana (2014-2019), and University of Tulsa in Oklahoma (2012-2017) [12].

LeFountain remarked on the NSA-funded national centers: "The nation increasingly needs professionals with highly technical cyber skills to help keep America safe today — and to help the country meet future challenges and adapt with greater agility. When it comes to national security, there is no substitute for a dedicated, immensely talented workforce. This effort will sow even more seeds" [13].

12.4 Generation Z, Teen Hackers, and Girl Coders

In addition to recruiting from universities and college graduates, the U.S. government has turned to Generation Z born from the early 2000s to the present day.

Unlike Generation X and Generation Y (Millennial Generation), Generation Z is bestowed with advanced communication and media technology since birth — the Internet, instant messaging, text messaging, smartphones, and tablets, just to name a few. The new generation is sometimes referred to as Generation Wii, iGeneration, Gen Tech, Digital Natives, Net Gen, and Facebook Generation.

Gen Wii, shorthand for connectivity, was coined by Taco Bell executives in consultations with MTV. "This is the generation that makes a game out of everything," said Brian Niccol, chief marketing and innovation officer at Taco Bell. "For them, life is a game" [14].

The stereotype of a gamer or a computer hacker being a bright but socially maladapted adolescent boy is archaic and flawed [15]. In February 2012, Spanish police with support of INTERPOL arrested a 16-year-old girl who was allegedly a member of the Anonymous hacking group [16]. In February 2013, security firm AVG linked a piece of malware to an 11-year-old boy in Canada. "We believe these junior programmers are motivated mainly by the thrill of outwitting their peers, rather than financial gain," said AVG's chief technology officer Yuval Ben-Itzhak. "But it is nevertheless a disturbing and increasing trend" [17].

Nevertheless, with the exception of a few, most hackers are not evildoers. Even Facebook CEO and cofounder Mark Zuckerberg calls himself a hacker [18]. "The word 'hacker' has an unfairly negative connotation from being portrayed in the media as people who break into computers," Zuckerberg wrote in the Facebook IPO S-1 filing statement. "In reality, hacking just means building something quickly or testing the boundaries of what can be done" [19].

During his keynote address at the 2012 DEF CON 20 Hacking Conference, NSA director Gen. Keith Alexander brought on stage an 11-year-old girl known by the pseudonym CyFi, and called her "the most important person for our future" [20].

A year before in 2011, DEF CON featured for the very first time a section, dubbed DEFCON Kids, for children ages 8 to 16. The 10-year-old CyFi revealed a security flaw and a new class of vulnerability in iPhone and Android games by tinkering with the clock settings on the phone and tablet [21].

Citing the key role of the U.S. government in the early research and development of the Internet, NSA director Alexander hoped to recruit from the cream of the crop at DEF CON: "We're the ones who built this Internet. Now we're the ones who have to keep it secure, and I think you folks can help do that. … In this room, this room right here, is the talent our nation needs to secure cyberspace" [22].

In addition, the NSA careers website in July 2012 dedicated a special page to the conference attendees: "Attention DEF CON® 20 attendees: If you're up on your game, you already know the National Security Agency and what we do. … If you think you saw cool things at DEF CON® 20, just wait until you cross the threshold to NSA, 'cause you ain't seen nothing yet" [23] (See Fig. 12.2).

Fig. 12.2 Attention DEF CON 20 attendees (Courtesy of the National Security Agency)

The NSA is not alone in tapping the Generation Z talent pool. The Central Intelligence Agency (CIA) has been offering "Kids' Page" and "Kids' Games" on the official CIA website since 2007 [24]. Figure 12.3 shows the CIA webpage that offers high-quality children games that rival Disney and Nickelodeon's. "Break the Code" and "Aerial Analysis Challenge" (see Fig. 12.4) are two of the engaging online games and activities from the CIA.

It is an irony that the world's first computer programmer was a female—Ada Lovelace, and yet there have been disproportionately more men than women studying computer science. In May 2012, the National Center for Education Statistics (NCES) released its findings that in 2009-10, females earned 57 % of all bachelor's degrees, 60 % of all master's degrees, and 52 % of all doctoral degrees [25]. However, in the field of computer and information sciences, only 18 % of undergraduates, 27 % of graduates, and 22 % of doctoral students were women. The biggest drop was the number of bachelor's degrees in computer science conferred to females, from 28 % in 1999-2000 to 18 % a decade later in 2009-10.

Generation Z is going to change all that. CyFi is just one example of a girl genius in computer technology. For the first time in history, DEFCON 20 had

Fig. 12.3 Kids' games on Central Intelligence Agency website (Courtesy of the Central Intelligence Agency)

women show up for the Social Engineering Capture the Flag (SECTF) completion [26]. In 2012, the small number of women hired by game companies has at least tripled since 2009 [27]. Dr. Maria Klawe, president of Harvey Mudd College, and her faculty have achieved the near impossible: nearly 40 % of Harvey Mudd's computer science degrees in 2012 went to women [28]. In 2013, 11-year-old Sylvia Todd won a silver medal at an international robotics competition, took part in the White House Science Fair, and garnered more than 1.5 million YouTube views for her 20 videos of do-it-yourself science projects for kids [29].

Nonprofit educational organizations such as Girls Who Code founded by Reshma Saujani are working to educate, inspire, and equip teenage girls with the skills and resources to pursue opportunities in technology and engineering. Girls Who Code has garnered support from Google, General Electric, eBay, and Twitter. "Our support for this initiative represents our commitment to invest in, encourage and empower more women pursing opportunities in technology," said Dick Costolo, CEO of Twitter [30].

Disney reaches out to young girls through its wildly popular princesses such as Elsa and Anna in the recent blockbuster movie *Frozen*. In partnership with Disney, Code.org has created a 1-h tutorial for kids to learn to write code to get the princesses to ice skate, carve snowflakes, and create their own wonderland on the computer [31].

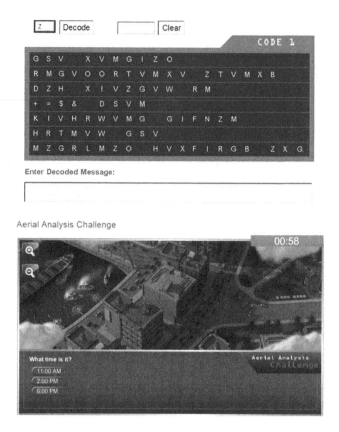

Fig. 12.4 "Break the Code" and "Aerial Analysis Challenge" games (Courtesy of the Central Intelligence Agency)

Google X executive vice president Megan Smith opines, "You don't have to know how to code to encourage someone else to code" [32]. In June 2014, Google launched the new initiative Made With Code with $50 million investment into the program over the next three year. Aimed at closing the gender gap in the computer tech industry, the program encourages young women to learn to code. The program partners include Chelsea Clinton, Mindy Kaling, MIT Media Lab, Girl Scouts of the USA, Girls Inc., Girls Who Code, the National Center for Women and Information Technology, and media sponsor *TechCrunch*. Made With Code's mentors include Ayah Bdeir, Danielle Feinberg, Erica Kochi, Limor Fried, Miral Kotb, and Robin Hunicke [33].

At the 2015 Consumer Electronics Show keynote address in Las Vegas, Intel CEO Brian Krzanich pledged to spend $300 million over the next three years to improve workplace diversity by educating and employing more women and minorities. "I have two daughters of my own coming up on college age," Krzanich told the audience. "I want them to have a world that's got equal opportunity for them" [34].

Generation Z women should be encouraged by Virginia Tech engineering alumni Regina Dugan and Letitia "Tish" Long as well as MIT alumnus Megan Smith. In 2009, Dugan was appointed the first female director of DARPA. In 2010, Long became the first woman in charge of a major U.S. intelligence agency—the National Geospatial-Intelligence Agency (NGA). In 2014, the White House named Megan Smith as the 3rd Chief Technology Officer of the United States.

As the interconnected world has become reliant on computers for everything from banking to military, teen hackers and girl coders of Generation Z are powerful resources for the NSA, CIA, DARPA, and U.S. intelligence community to carry out covert cyber operations. NSA director Gen. Keith Alexander spoke at DEF CON 20 in July 2012 in Las Vegas [35]:

> We as a global society are extremely vulnerable and at risk for a catastrophic cyber event. Global society needs the best and brightest to help secure our most valued resources in cyberspace: our intellectual property, our critical infrastructure and our privacy. DEF CON has an important place in computer security. It taps into a broad range of talent and provides an unprecedented diversity of experiences and expertise to solve tough problems. The hacker community and USG cyber community share some core values: we both see the Internet as an immensely positive force; we both believe information increases in value by sharing; we both respect protection of privacy and civil liberties; we both believe in the need for oversight that fosters innovation, doesn't pick winners and losers, and retains freedom and flexibility; we both oppose malicious and criminal behavior. We should build on this common ground because we have a shared responsibility to secure cyberspace.

12.5 DEF CON Diversity Panel and IT Girls 2.0 (a Commentary by Emily Peed)

Typically, when one imagines the hacker community there are a few stereotypes that come to mind. Since 2013, however, Vegas 2.0 has been taking steps to illuminate the incredible diversity within the hacking community, and break these stereotypes. They have done this through panels at the DEF CON Hacking Conference that brings together different representatives within the community.

First, for 2013, DEF CON focused on female participation. Soon they realized they needed to grow broader. The 2014 DEF CON Diversity Panel brought a much wider audience because of the panelists that included females, a man in the process of gender change, and multiple races including Indian, African American, Caucasian, and Mexican American of all ages. Some had been heavily involved in the hacking culture and others were just getting started.

The reception was immense. The attendees would tell the panelists about their experiences afterward throughout the rest of the conference and how much the inclusion meant to them. They would make suggestions, including the unseen barriers we face like depression, narcolepsy, or ADHD. People came forward and shared how wonderful it was for them to see this type of diversity effort at DEF CON. Those who have made a full and unrecognizable gender change as well as those who were not affiliated with the panel members were touched and invigorated by this type of effort. For the 2015 DEF CON Diversity Panel, there are more ideas floating around to increase interests and participation in the diversity efforts, to really capture the vast and almost indescribable depth of diversity that is currently within the hacking community, and to keep that going.

In addition to highlighting the current diversity of participants within the hacking community, it is important to gain new interests. Developers' Club (developersclubonline.com) is a resource platform for Open Source that is creating many resources and a variety of after-school programs. These will be for general participations, diversity efforts, and a Clean Computing program that teaches students about the environment.

For the 2013-2014 school year a program was run that had a goal of increasing female participation in information technology at the middle school level. The program was called "IT Girls", later rebranded "IT Girls 2.0". The IT Girls program ran at three sites in Indiana and it was supported by the NCWIT AspireIT Grant (ncwit.org). Through the grant students were able to receive goody bags, attend a banqueted awards ceremony, and participate in a fun and new technology-based after-school program. End of the program reviews from students were positive. The students wanted more after-school programs for more times per week than originally provided. The IT Girls 2.0 program will resume for the 2015- 2016 school year together with the launch of general participation programs and Clean Computing after-school programs.

12.6 Control the Code, Control the World

Bill Gates of Microsoft, Mark Zuckerberg of Facebook, Jack Dorsey of Twitter, will.i.am, and many others appeared in a 2013 YouTube video encouraging everyone to code [36]. Marc Goodman, global security advisor and futurist, spoke at the TEDGlobal 2012 in Edinburgh about his ominous prediction: "If you control the code, you control the world. This is the future that awaits us" [37].

In the 1983 movie *WarGames*, a high school student hacked into a fictional military supercomputer WOPR (War Operation Plan Response) and almost started World War III as a result [38].

In reality, two "war games" were conducted in June and December 2011 between the United States and China in order to help prevent a sudden military escalation between the two countries if either nation felt that they were being targeted [39]. The face-to-face war games were organized by the Center for Strategic and International Studies (CSIS) in Washington and the China Institute of Contemporary International Relations (CICIR) in Beijing.

"We coordinate the war games with the State Department and Department of Defense," said Jim Lewis, senior fellow and director of CSIS. "The officials start out as observers and become participants. … It is very much the same on the Chinese side. Because it is organized between two think tanks they can speak more freely. … The [Chinese officials] who favor cooperation are not as strong as the people who favor conflict" [39].

The world has become increasingly more vulnerable to cyber attacks as we adopt new technologies such as the Automatic Dependent Surveillance - Broadcast system (ADS-B) for air traffic control [40], biometric security systems [41], Apple's iTravel app integration with airport security [42], Internet-controlled door locks and thermostats in home automation [43], embedded Android software in rice cookers and refrigerators [44], and self-driving cars [45].

In March 2010, a disgruntled former employee of a Texas auto center hacked into the company computer and created a bit of havoc with a web-based vehicle-immobilization system. As a result, more than 100 drivers in Austin, Texas found their cars disabled or the horns honking out of control [46].

"We typically don't drive our smartphones at 80 miles an hour," said security strategist Brian Contos at McAfee. "But safety concerns and privacy concerns all culminate when you talk about automobiles. The nightmare scenario is 100 cars on a bridge and 50 % of them hit their brakes and 50 % hit their accelerators. Just the amount of collision that something like that would cause with a remote attack, that's pretty scary stuff" [47].

In August 2011 at the USENIX Security Symposium, researchers from the University of California San Diego and University of Washington reported that newer cars were susceptible to remote hacking: "Modern automobiles are pervasively computerized, and hence potentially vulnerable to attack. However, while previous research has shown that the *internal* networks within some modern cars are insecure, the associated threat model—requiring *prior physical access*—has justifiably been viewed as unrealistic. Thus, it remains an open question if automobiles can also be susceptible to *remote* compromise. Our work seeks to put this question to rest by systematically analyzing the *external* attack surface of a modern automobile. We discover that remote exploitation is feasible via a broad range of attack vectors (including mechanics tools, CD players, Bluetooth and cellular radio), and further, that wireless communications channels allow long distance vehicle control, location tracking, in-cabin audio exfiltration and theft" [48].

In December 2014, Google unveiled the first complete prototype of a self-driving car for "fully autonomous driving" [49]. The car has no steering wheel, gas pedal, or brake pedal. Without manual override capability, hackers would try to find ways to take control of the vehicle by reprogramming its destination, route, or other functions. It is a scary thought.

Fingerprint biometric security took a blow when hackers demonstrated at the 2014 Chaos Computer Club (CCC) convention that they were able to recreate German Defense Minister Ursula von der Leyen's fingerprint by using a picture of her thumb from an October press conference and the publicly available VeriFinger SDK [50].

Setting aside the make-believe of Hollywood, it is conceivable that a future world war will be fought among Generation Z on the Internet and the Intranet, causing global economic meltdown and creating immeasurable damage both online and offline. In 2008, a pseudonymous developer named Satoshi Nakamoto introduced Bitcoin, a peer-to-peer digital currency that bypasses governments and banking systems [51]. Imagine hackers of the future empty all bank accounts, erase all debts, and destroy all backup databases.

In January 2013, Anonymous hacked and defaced the websites of the Massachusetts Institute of Technology (MIT) and the U.S. Sentencing Commission (USSC) in protest over the death of 26-year-old hacktivist Aaron Swartz who succumbed to the pressure of an impending lawsuit and committed suicide earlier in the month [52, 53]. In July 2011, Swartz was charged with breaking into MIT's restricted networks and stealing more than four million academic journal articles from JSTOR [54].

A co-owner of Reddit, Swartz contributed to the original RSS (Really Simple Syndication) specification at the young age of 14. "Aaron is seen as a hero," said Christopher Soghoian, principal technologist and senior policy analyst at the American Civil Liberties Union (ACLU). "He spent a lot of time working to make the Internet a more open place. We lost a really important person who changed the Internet in a positive way, and we all lose out by his departure" [55].

Instead of prosecuting hacktivists like Aaron Swartz, the government and the hacker community should open a dialogue to discuss the broader issues of privacy, intellectual property, security, and civil liberties. Recognizing the need to reform the outdated Computer Fraud and Abuse Act (CFAA), U.S. Representative Zoe Lofgren and U.S. senator Ron Wyden introduced "Aaron's Law" in June 2013 to "refocus the law away from common computer and Internet activity and toward damaging hacks" [56].

U.S. Navy Admiral and NATO's Supreme Allied Commander James Stavridis said that "instead of building walls to create security, we need to build bridges" [57]. NSA Director and Army General Keith Alexander took the unprecedented first step in reaching out to the hacker community in DEF CON 20 in July 2012 [58].

Control the code, control the world. For the sake of our future, it is the responsibility of Generation X and Generation Y to lead Generation Z into peace rather than war.

Bibliography

1. **Nakashima, Ellen.** With Plan X, Pentagon seeks to spread U.S. military might to cyberspace. [Online] The Washington Post, May 30, 2012. http://articles.washingtonpost.com/2012-05-30/world/35458424_1_cyberwarfare-cyberspace-pentagon-agency.
2. **DARPA.** DARPA Director speaks of Offensive Capabilities in Cyber Security. [Online] Defense Advanced Research Projects Agency, March 12, 2012. http://www.darpa.mil/NewsEvents/Releases/2012/03/12c.aspx.
3. —. Cyber Experts Engage on DARPA's Plan X. [Online] Defense Advanced Research Projects Agency, October 17, 2012. http://www.darpa.mil/NewsEvents/Releases/2012/10/17.aspx.
4. **Defense Advanced Research Projects Agency.** DARPA-BAA-13-02: Foundational Cyberwarfare (Plan X). [Online] Federal Business Opportunities, November 21, 2012. https://www.fbo.gov/utils/view?id=49be462164f948384d455587f00abf19.
5. **Prabhakar, Arati.** Press Briefing with DARPA Director Arati Prabhakar from the Pentagon. [Online] U.S. Department of Defense, April 24, 2013. http://www.defense.gov/transcripts/transcript.aspx?transcriptid=5227.

6. **Shachtman, Noah.** This Pentagon Project Makes Cyberwar as Easy as Angry Birds. [Online] Wired, May 28, 2013. http://www.wired.com/2013/05/pentagon-cyberwar-angry-birds/all/.

7. **Roelker, Dan.** Scaling Cyberwarfare. [Online] DARPA Cyber Colloquium, November 7, 2011. http://www.darpa.mil/WorkArea/DownloadAsset.aspx?id=2147484459.

8. **Pellerin, Cheryl.** DARPA's Plan X Uses New Technologies to 'See' Cyber Effects. [Online] U.S. Department of Defense, June 11, 2014. http://www.defense.gov/news/newsarticle.aspx?id=122455.

9. **Greenberg, Andy.** Darpa Turns Oculus Into a Weapon for Cyberwar. [Online] Wired, May 23, 2014. http://www.wired.com/2014/05/darpa-is-using-oculus-rift-to-prep-for-cyberwar/.

10. **Sanger, David E.** Obama Order Sped Up Wave of Cyberattacks Against Iran. [Online] The New York Times, June 1, 2012. http://www.nytimes.com/2012/06/01/world/middleeast/obama-ordered-wave-of-cyberattacks-against-iran.html?pagewanted=all.

11. **Koebler, Jason.** NSA Built Stuxnet, but Real Trick Is Building Crew of Hackers. [Online] US News and World Report, June 8, 2012. http://www.usnews.com/news/articles/2012/06/08/nsa-built-stuxnet-but-real-trick-is-building-crew-of-hackers.

12. **National Security Agency and Central Security Service.** List of Centers of Academic Excellence for Cyber Operations. [Online] National Security Agency, August 18, 2014. https://www.nsa.gov/academia/nat_cae_cyber_ops/nat_cae_co_centers.shtml.

13. **NSA Public and Media Affairs.** NSA Announces New Program to Prime College Students for Careers in Cyber Ops. [Online] National Security Agency, May 21, 2012. http://www.nsa.gov/public_info/press_room/2012/new_college_cyber_ops_program.shtml.

14. **Horovitz, Bruce.** After Gen X, Millennials, what should next generation be? [Online] USA Today, May 4, 2012. http://usatoday30.usatoday.com/money/advertising/story/2012-05-03/naming-the-next-generation/54737518/1.

15. **WGBH educational foundation.** Studying the psychology of virus writers and hackers: An interview with researcher Sarah Gordon. [Online] PBS Frontline. [Cited: January 4, 2013.] http://www.pbs.org/wgbh/pages/frontline/shows/hackers/whoare/psycho.html.

16. **Whiteman, Hilary.** Interpol arrests suspected 'Anonymous' hackers. [Online] CNN, February 29, 2012. http://www.cnn.com/2012/02/29/world/europe/anonymous-arrests-hacking/index.html.

17. **Dunn, John E.** AVG finds 11 year-old creating malware to steal game passwords. [Online] TechWorld, February 8, 2013. http://news.techworld.com/security/3425185/avg-finds-11-year-old-creating-malware-steal-game-passwords/.

18. **Ortutay, Barbara.** For Facebook 'Hacker Way' is way of life. [Online] Yahoo! News, February 4, 2012. http://news.yahoo.com/facebook-hacker-way-way-life-150559696.html.

19. **Facebook, Inc.** Form S-1. Registration Statement. [Online] U.S. Securities and Exchange Commission, February 1, 2012. http://www.sec.gov/Archives/edgar/data/1326801/000119312512034517/d287954ds1.htm.

20. **Kelly, Heather.** Computer hacking for 8-year-olds. [Online] CNN, July 31, 2012. http://www.cnn.com/2012/07/31/tech/web/def-con-kids-2012/index.html.

21. **Rosenblatt, Seth.** 10-year-old hacker finds zero-day flaw in games. [Online] CNet, August 7, 2011. http://download.cnet.com/8301-2007_4-20089152-12/10-year-old-hacker-finds-zero-day-flaw-in-games/.

22. **Cowley, Stacy.** NSA wants to hire hackers. [Online] CNNMoney, July 29, 2012. http://money.cnn.com/2012/07/27/technology/defcon-nsa/index.htm.

23. **National Security Agency.** Careers. [Online] National Security Agency. http://www.nsa.gov/careers/dc20/.

24. **Central Intelligence Agency.** Games. [Online] Central Intelligence Agency, March 6, 2007. https://www.cia.gov/kids-page/games/index.html.

25. **Aud, Susan; Hussar, William; Johnson, Frank; Kena, Grace; Roth, Erin; Manning, Eileen; Wang, Xiaolei; Zhang, Jijun; Notter, Liz; Nachazel, Thomas; Yohn, Carolyn.** The Condition of Education 2012. [Online] U.S. Department of Education, May 2012. http://nces.ed.gov/pubs2012/2012045.pdf.

26. **Hadnagy, Christopher J. and Maxwell, Eric.** Defcon 20 Social Engineering CTF Report. [Online] DEFCON, September 24, 2012. http://www.social-engineer.org/social-engineering/defcon-20-social-engineering-ctf-report/.

27. **Nayak, Malathi.** Women pry open door to video game industry's boys' club. [Online] Reuters, January 13, 2013. http://www.reuters.com/article/2013/01/13/us-videogames-women-idUSBRE90C0CI20130113.

28. **Hafner, Katie.** Giving Women the Access Code. [Online] The New York Times, April 2, 2012. http://www.nytimes.com/2012/04/03/science/giving-women-the-access-code.html?pagewanted=all&_r=0.

29. **Bhanoo, Sindya N.** A Science Star Already, Tinkering With the Idea of Growing Up. [Online] The New York Times, April 23, 2013. http://www.nytimes.com/2013/04/24/science/sylvia-todd-science-star-tinkers-with-the-idea-of-growing-up.html?pagewanted=all&_r=0.

30. **Girls Who Code.** About Girls Who Code. [Online] Girls Who Code. [Cited: January 4, 2013.] http://www.girlswhocode.com/about/.

31. **Lobosco, Katie.** Frozen's Elsa and Anna teach girls to code . [Online] CNN, November 26, 2014. http://money.cnn.com/2014/11/26/technology/disney-frozen-code/index.html.

32. **Crook, Jordan.** Google Invests $50 Million In "Made With Code" Program To Get Girls Excited About CS. [Online] TechCrunch, June 22, 2014. http://techcrunch.com/2014/06/22/google-invests-50-million-in-made-with-code-program-to-get-girls-excited-about-cs/.

33. **Google.** Made w/ Code. [Online] Google. [Cited: January 26, 2015.] https://www.madewithcode.com/.

34. **Wingfield, Nick.** Intel Allocates $300 Million for Workplace Diversity. [Online] The New York Times, January 6, 2015. http://www.nytimes.com/2015/01/07/technology/intel-budgets-300-million-for-diversity.html?partner=rss&emc=rss&smid=tw-nytimes&_r=0.

35. **Alexander, Keith.** Shared Values, Shared Responsibility. [Online] DEF CON Communications Inc., July 2012. https://www.defcon.org/html/defcon-20/dc-20-speakers.html.

36. **Code.org.** What most schools don't teach. [Online] YouTube, February 26, 2013. https://www.youtube.com/watch?v=nKIu9yen5nc.

37. **Goodman, Marc.** Marc Goodman: A vision of crimes in the future. [Online] TEDGlobal 2012, June 28, 2012. http://www.ted.com/talks/marc_goodman_a_vision_of_crimes_in_the_future.html?quote=1769.

38. **IMDb.** WarGames. [Online] IMDb, June 3, 1983. http://www.imdb.com/title/tt0086567/.

39. **Hopkins, Nick.** US and China engage in cyber war games. [Online] The Guardian, April 16, 2012. http://www.guardian.co.uk/technology/2012/apr/16/us-china-cyber-war-games.

40. **Kelly, Heather.** Researcher: New air traffic control system is hackable. [Online] CNN, July 26, 2012. http://www.cnn.com/2012/07/26/tech/web/air-traffic-control-security/index.html.

41. **Goldman, David.** Hackers' next target: Your eyeballs. [Online] CNNMoney, July 26, 2012. http://money.cnn.com/2012/07/26/technology/iris-hacking/index.htm.

42. **Patterson, Thom.** Apple's secret plan to join iPhones with airport security. [Online] CNN, September 19, 2012. http://www.cnn.com/2012/09/19/travel/mobile-airport-travel-apps/index.html.

43. **Goldman, David.** Your future home is vulnerable to cyberattacks. [Online] CNNMoney, July 26, 2012. http://money.cnn.com/2012/07/26/technology/home-network-cyberattack/index.htm.

44. **Edwards, Cliff and King, Ian.** Google Android Baked Into Rice Cookers in Move Past Phone. [Online] Bloomberg, January 7, 2013. http://www.bloomberg.com/news/2013-01-08/google-android-baked-into-rice-cookers-in-move-past-phones-tech.html.

45. **Cohen, Adam.** Will Self-Driving Cars Change the Rules of the Road? [Online] Time Magazine, January 14, 2013. http://ideas.time.com/2013/01/14/will-self-driving-cars-change-the-rules-of-the-road/.

46. **Poulsen, Kevin.** Hacker Disables More Than 100 Cars Remotely. [Online] Wired, March 17, 2010. http://www.wired.com/threatlevel/2010/03/hacker-bricks-cars/.

47. **Neild, Barry.** Could hackers seize control of your car? [Online] CNN, March 2, 2012. http://www.cnn.com/2012/03/02/tech/mobile/mobile-car-hacking/index.html.

48. **Checkoway, Stephen, et al.** Comprehensive Experimental Analyses of Automotive Attack Surfaces. [Online] Center for Automotive Embedded Systems Security, August 8-12, 2011. http://www.autosec.org/pubs/cars-usenixsec2011.pdf.

49. **Davies, Alex.** Google's Self-Driving Car Hits Roads Next Month—Without a Wheel or Pedals. [Online] Wired, December 23, 2014. http://www.wired.com/2014/12/google-self-driving-car-prototype-2/.

50. **Orf, Darren.** Hackers Say They Can Copy Your Fingerprint From Just a Photograph. [Online] GIZMODO, December 28, 2014. http://gizmodo.com/chaos-computer-club-says-they-can-hack-your-fingerprint-1675845311.

51. **Nakamoto, Satoshi.** Bitcoin P2P e-cash paper. [Online] GMANE Newsgroup, October 31, 2008. http://article.gmane.org/gmane.comp.encryption.general/12588/.

52. **Musil, Steven.** Anonymous hacks MIT after Aaron Swartz's suicide. [Online] CNet, January 13, 2013. http://news.cnet.com/8301-1023_3-57563752-93/anonymous-hacks-mit-after-aaron-swartzs-suicide/.

53. **Brumfield, Ben.** Anonymous threatens Justice Department over hacktivist death. [Online] CNN, January 27, 2013. http://www.cnn.com/2013/01/26/tech/anonymous-threat/index.html.

54. **Seidman, Bianca.** Internet activist charged with hacking into MIT network. [Online] PBS, July 22, 2011. http://www.pbs.org/wnet/need-to-know/the-daily-need/internet-activist-charged-with-hacking-into-mit-network/.

55. **Leopold, Todd.** How Aaron Swartz helped build the Internet. [Online] CNN, January 15, 2013. http://www.cnn.com/2013/01/15/tech/web/aaron-swartz-internet/index.html.

56. **Lofgren, Zoe and Wyden, Ron.** Introducing Aaron's Law, a Desperately Needed Reform of the Computer Fraud and Abuse Act. [Online] Wired, June 20, 2013. http://www.wired.com/2013/06/aarons-law-is-finally-here/.

57. **Stavridis, James.** James Stavridis: A Navy Admiral's thoughts on global security. [Online] TED, June 2012. http://www.ted.com/talks/james_stavridis_how_nato_s_supreme_commander_thinks_about_global_security.html.

58. **Urquhart, Conal.** US National Security Agency boss asks hackers to make internet more secure. [Online] The Guardian, July 28, 2012. http://www.guardian.co.uk/technology/2012/jul/28/national-security-agency-hackers-internet.

Part V
Cybersecurity: Applications and Challenges

Chapter 13
Artificial Intelligence and Data Mining

To err is human. AI software modeled after humans will inevitably make mistakes. It is fine as long as the software learns from its errors and improves itself, which is something that humans ought to learn from AI.

— Newton Lee

That men do not learn very much from the lessons of history is the most important of all the lessons that history has to teach.
— Aldous Huxley in *Collected Essays* (1959)

A lot of cutting edge AI has filtered into general applications, often without being called AI because once something becomes useful enough and common enough it's not labeled AI anymore.
— Nick Bostrom
Oxford University Future of Humanity Institute (2006)

Whenever an AI research project made a useful new discovery, that product usually quickly spun off to form a new scientific or commercial specialty with its own distinctive name.
— Professor Marvin Minsky
MIT Artificial Intelligence Laboratory (2009)

HAL's not the focus; the focus is on the computer on 'Star Trek'.
— David Ferrucci
IBM Thomas J. Watson Research Center (2011)

13.1 Artificial Intelligence: From Hollywood to the Real World

In 1955, American computer scientist and cognitive scientist John McCarthy coined the term "artificial intelligence" (AI). He defined AI as "the science and engineering of making intelligent machines, especially intelligent computer programs" [1].

© Springer International Publishing Switzerland 2015
N. Lee, *Counterterrorism and Cybersecurity*, DOI 10.1007/978-3-319-17244-6_13

In the 2001 film *A.I.: Artificial Intelligence*, Steven Spielberg tells the story of a highly advanced robotic boy who longs to become real so that he can regain the love his human mother [2]. In 2004, Will Smith starred in the lead role of *I, Robot* — a film based loosely on Isaac Asimov's short-story collection of the same name [3]. Although the Hollywood movies are quite far-fetched, AI hit the spotlight on primetime television over three nights in February 2011 when the IBM Watson computer won on "Jeopardy!" against two human champions and took home a $1 million prize [4].

Watson, named after IBM founder Thomas J. Watson, has the ability of encyclopedic recall and natural language understanding. "People ask me if this is HAL," said David Ferrucci, lead developer of Watson, referring to the Heuristically programmed ALgorithmic (HAL) computer in *2001: A Space Odyssey* by Stanley Kubrick and Arthur C. Clarke. "HAL's not the focus; the focus is on the computer on 'Star Trek,' where you have this intelligent information seeking dialogue, where you can ask follow-up questions and the computer can look at all the evidence and tries to ask follow-up questions. That's very cool" [5].

Watson was inspired by the Deep Blue project at IBM. Back in May 1997, the IBM Deep Blue computer beat the world chess champion Garry Kasparov after a six-game match, marking the first time in history that a computer had ever defeated a world champion in a match play [6]. Since then, computers have become much faster and software more sophisticated. In October 2012, the U.S. Department of Energy unveiled the Titan supercomputer capable of 20 peta-flops — 20 thousand trillion (20,000,000,000,000,000) floating point operations per second [7].

Although our desktop computers are no match for the Titan, AI software has entered mainstream consumer products. Apple's intelligent personal assistant Siri on iPhone, iPad, and iPod is the epitome of AI in everyday life. Siri uses voice recognition and information from the user's contacts, music library, calendars, and reminders to better understand what the user says [8]. The software application is an offshoot of SRI International's CALO (Cognitive Assistant that Learns and Organizes) project funded by the Defense Advanced Research Projects Agency (DARPA) under its Perceptive Assistant that Learns (PAL) program [9, 10]. Apple acquired Siri in April 2010, integrated it into iOS, and the rest is history [11].

In addition to smartphones, domain-specific AI software applications have been embedded into newer automobiles, interactive toys, home appliances, medical equipment, and many electronic devices.

We do not often hear about AI in the real world, because as MIT Professor Marvin Minsky explained, "AI research has made enormous progress in only a few decades, and because of that rapidity, the field has acquired a somewhat shady reputation! This paradox resulted from the fact that whenever an AI research project made a useful new discovery, that product usually quickly spun off to form a new scientific or commercial specialty with its own distinctive name" [12].

Professor Nick Bostrom, director of the Future of Humanity Institute at Oxford University, told CNN in a 2006 interview, "A lot of cutting edge AI has filtered into general applications, often without being called AI because once something becomes useful enough and common enough it's not labeled AI anymore" [13].

13.2 Intelligent CCTV Cameras

Artificial intelligence (AI) is increasingly used in the processing of collected data from physical surveillance. There are approximately 30 million closed-circuit television (CCTV) cameras in the world capturing 250 billion hours of raw footage annually [14]. It is time-prohibitive to manually process that much data in search of clues that will solve a crime. The effort would be like looking for a needle in a haystack. In addition, while CCTV cameras help deter crimes, they are less effective than an eyewitness at the scene who can alert the police.

AI software empowers CCTV surveillance by automatic detection of visual and audio clues to spot anything out of the ordinary such as violent crimes, vandalism, and terrorism. Not only can AI software scan the recorded footages at high speed, it can also do real-time analysis at the scene much like a human eyewitness.

Prof. David Brown at the University of Portsmouth described the intelligent CCTV cameras that can see and hear, and that can alert law enforcement authorities to crimes in progress [15]:

> We have already developed visual recognition software, but the next stage is to develop audio recognition software to listen for particular sounds. We can teach the cameras to listen out for things like a swear word being shouted in an aggressive way, or for other words which might signify a crime taking place. The camera will be able to swivel to the direction of the sound at the same speed someone turns their head when they hear a scream, or about 300 ms. People monitoring CCTV images have banks of screens in front of them, and this system helps them by alerting them to something the system has spotted. The person looking at the screen can then quickly identify if it is a crime taking place, or whether the camera has simply picked up on something innocent, like a child screaming, and act on it accordingly. The system would not be sensitive enough to record individual conversations. We are just looking for certain trigger sounds and visual anomalies.

Equipped with facial recognition capabilities, AI-enhanced CCTV systems have been used in cities like Chicago and London [16]. Although the systems can be susceptible to false alarms, a human counterpart can make a judgment call upon receiving an alert. Artificial intelligence (AI) assisting human decision making is a form of cognitive augmentation or intelligence amplification (IA).

In 2009, Microsoft teamed up with New York Police Department (NYPD) in developing Domain Awareness System (DAS) — a real-time networked counterterrorism system to detect, deter, and prevent potential terrorist activities in New York City [17]. Companies partnering with NYPD provide feeds from their proprietary CCTVs into the Lower Manhattan Security Coordination Center.

In 2012, Microsoft and NYDP announced that DAS will "feature the use of artificial intelligence (AI) capabilities to analyze video, public safety data and other situational awareness information in real time to proactively identify potential terrorist threats and protect critical infrastructure" [18].

13.3 Data Mining in the Age of Big Data

In April 2014, International Data Corporation (IDC) reported that there were 4.4 zettabytes (4.4 billion terabytes) of digital information in the world at the start of the year, and that figure was growing by 40 % a year into the next decade. The amount of data more than doubles every two years. By 2020, the digital universe will contain "as many digital bits as there are stars in the universe" [19]. Among all that data, IDC estimated that about 5 % would be valuable or "target rich."

Fully operational in May 2014, the nation's largest data-mining center at Camp Williams National Guard Training Site in Bluffdale, Utah is handling the skyrocketing volume of information collected by the National Security Agency (NSA) [20]. With more than a dozen listening posts around the world, the NSA intercepts about two million phone calls, e-mail messages, faxes and other types of communications every hour [21]. This information is cross-referenced with over 1 million names in the Terrorist Identities Datamart Environment (TIDE) database [22].

Analysts at the NSA, Central Intelligence Agency (CIA), and Federal Bureau of Investigation (FBI) have been knee-deep in a mountain of collected data from physical surveillance and open source information, looking for useful patterns.

A 2001 Congressional report disclosed that the NSA was faced with "profound needle-in-the-haystack challenges." *The New York Times* revealed in 2002 that there were 200 million pieces of intelligence in a regular workday, and less than one percent of it was ever decoded, translated, or processed [23]. Recognizing the importance of data mining, the Obama administration in March 2012 announced more than $200 million in funding for the "Big Data Research and Development Initiative" [24].

Association for Computing Machinery (ACM) Special Interest Group on Knowledge Discovery and Data Mining (SIGKDD) defines data mining as "an interdisciplinary field at the intersection of artificial intelligence, machine learning, statistics, and database systems" [25]. The goal of data mining is to extract knowledge from the available data by capturing this knowledge in a human-understandable structure. The discovery of structure in big data involves:

1. Database, data management, and data warehouse structure
2. Data preprocessing, transformations, and dimensionality
3. Choice of model, valid approximations, and statistical inference considerations
4. Interestingness metrics and choice of algorithms

5. Algorithmic complexity and scalability considerations
6. Post-processing of discovered structure
7. Visualization and understandability
8. Maintenance, updates, and model life cycle considerations

Voluminous amounts of structured and unstructured data residing in a vast number of heterogeneous databases present a real challenge to data mining. For example:

1. Despite the establishment of the Terrorist Screening Center (TSC) in 2003, U.S. agencies handling the terrorist watch lists have continued to "work from at least 12 different, sometimes incompatible, often uncoordinated and technologically archaic databases" [26].
2. For spy agencies like the CIA and military intelligence organizations, *The Wall Street Journal* revealed in 2009 that there are hundreds of databases used by each and most of them are not linked up [27].
3. In July 2010, Michael T. Flynn, Deputy Chief of Staff of Intelligence in Afghanistan, wrote a memorandum citing the urgent need for a new system to analyze the vast amounts of intelligence being collected. Flynn wrote, "US intelligence analysts in Afghanistan have several tools available to access the ever-increasing amount of intelligence and battlefield information residing in a myriad of databases. These tools provide access to the information, some more readily than others, but provide little in the way of improved analytical support" [28].

13.4 Knowledge Representation, Acquisition, and Inference

As a co-pioneer of artificial intelligence applications in counterterrorism, I helped develop a natural language parser and machine learning program to digest news and articles in search of potential terrorist threats around the globe. Those were the early days of data mining at the Institute for Defense Analyses (IDA) in 1984. It is conceivable that AI will be able to understand all the news and investigative reports in all languages from both traditional and social media, connect the dots, predict imminent dangers, and identify long-term concerns.

Employing psychology and cognitive science, my prototype system thinks like a human in constructing small-scale models of reality that it uses to anticipate events [29]. The knowledge representation and data structures were based on "A Framework for Representing Knowledge" by MIT Artificial Intelligence Lab co-founder Marvin Minsky [30]. Automated data analysis applies models to data in order to predict behavior, assess risk, and determine associations. The models can be based on patterns obtained from data mining or based on subjects under surveillance [31].

Artificial intelligence can assist human analysts in data mining by more efficiently organizing the information, detecting missing pieces of data, and making inferences that may otherwise be overlooked by human eyes. The three fundamental artificial intelligence (AI) techniques are:

1. Knowledge representation
   Data mining seeks to discover useful patterns which can take various forms of knowledge presentation. Some of the knowledge representation techniques are heuristic question answering, neural networks, theorem proving, and expert systems. Programming languages for knowledge representation include s-expression-based Lisp (List processing) [32], rule-based Prolog (Programming in logic) [33], and frame-based KL-ONE [34].

   Which kind of knowledge representation is best? Marvin Minsky offered his answer: "To solve really hard problems, we'll have to use several different representations. This is because each particular kind of data structure has its own virtues and deficiencies, and none by itself seems adequate for all the different functions involved with what we call 'common sense.' Each has domains of competence and efficiency, so that one may work where another fails. Furthermore, if we rely only on any single 'unified' scheme, then we'll have no way to recover from failure" [35].

2. Knowledge acquisition
   Knowledge acquisition from databases involves both database technologies and machine learning techniques. Databases may be centralized, distributed, hierarchical, relational, object-oriented, spatial, temporal, real-time, unstructured, or any combinations of these [36]. An email, for instance, contains an unstructured message along with IP-based geolocation, timestamp, and email addresses of the sender and recipient.

   Machine learning can speed up the processing of the million pieces of daily intelligence by automatically parsing, classifying, and reorganizing data [37]. AI not only can streamline the workflow but it can also detect missing pieces of information that need to be acquired in order to form a complete picture.

3. Knowledge inference
   Expert systems can learn from human analysts in making inferences based on the patterns discovered from data mining. The inference engine within an expert system can use propositional, predicate, modal, deontic, temporal, or fuzzy logic to conduct forward chaining, backward chaining, abduction, and reasoning under uncertainty. In the medical domain, expert systems such as MYCIN have shown to outperform doctors in diagnosing diseases [38]. In experimental design, expert systems can study a large number of variables simultaneously and analyze the resulting data using variance decomposition methods [39].

   While expert systems are proven to be superior in some cases, they are not meant to replace human analysts but rather they are invaluable tools to supplement human intelligence. For example, in the 2005 computer-assisted PAL/CSS Freestyle Chess Tournament, two amateur chess players used three computers for analysis and won the tournament by defeating all the other teams including grandmasters who were 1,000 Elo points stronger and equipped with more powerful computers [40]. The amateurs turned out to be better at human-computer symbiosis than the grandmasters. Human-computer symbiosis is the idea that technology should be designed in a way that amplifies human intelligence instead of attempting to replace it.

13.5 Dynamic Mental Models

I left the Institute for Defense Analyses for Bell Laboratories in 1985 to further my research on artificial intelligence and expert systems. At Bell Labs, I conceived Dynamic Mental Models (DM^2) as a general algorithm that combines analytical models and experiential knowledge in diagnostic problem solving, regardless of the problem domains [41]. The algorithm mimics a human expert in formulating and using an internal, cognitive representation of a physical system during the process of diagnosis. This internal representation, known as a mental model, originates from an analytical model but it changes dynamically to various levels of abstraction that are most appropriate for efficient diagnosis. An analytical model is represented as structure and behavior, whereas experiential knowledge is expressed in terms of pattern-recognition, topological clustering, topological pruning, and recommendation rules.

Realizing that rules alone are insufficient to make an expert system as smart as a competent human being, some AI researchers advocate reasoning from first principles which are the structure (components and their interconnections) and the behavior (input and output characteristics) of a given system [42, 43]. Such an approach is known as model-based reasoning, which promises some progress towards achieving the goals of application versatility, program understandability, knowledge base extensibility, ease of maintenance, and capability of dealing with novel situations [44]. However, model-based systems are often computationally more expensive than their rule-based counterparts. The situation is worse for applications where certain required computations, such as functional inversion, are practically impossible. A major reason for this difficulty is that the model-based approach relies heavily on the analytical models of a given physical system and undermines the power of experiential knowledge that human experts possess.

Both analytical models and experiential knowledge are essential in building powerful expert systems. To justify this statement, let us take another look at the conventional rule-based approach and ask ourselves the question: "What's in a rule?" A rule such as "if the body temperature elevates abnormally, then the subject has a fever" is a very natural way of expressing a diagnostic decision. However, the rule does little to explain the reasons behind the decision. To justify the decision, one begins to think about various entities and relationships implied by the rule. It is obvious that there must be a human body which has a central temperature. What a rule embodies, therefore, is an implicit model of a domain and the explicit experience of a human expert in diagnosing a problem. In other words, a rule is an end-product that results from compiling domain-specific facts and personal experience. Capturing all the rules (if at all possible) in an expert system is insufficient to make the system as smart as a human expert. Unlike a novice, an expert understands why a rule is a rule and is therefore able to change the rule whenever necessary. Capturing this ability is a challenging problem for AI. To begin with, expert systems have to "de-compile" the rules to make explicit both the domain models and the experiential knowledge.

In 1987, the DM^2 algorithm was implemented and tested on a real-world expert system prototype for telecommunication networks maintenance at AT&T. The application demonstrated that the dynamic mental model approach promotes system robustness, program correctness, software reuse, and ease of knowledge base modification and maintenance [41]. In 1989, the U.S. Army Research Office studied DM^2 for use in diagnostic support of complex modern weapons systems with promising results [45]. In 2004, the U.S. Naval Research Laboratory applied dynamic mental models to meteorological forecasting [46].

13.6 Modeling Human Problem Solving

Research in cognitive psychology suggests that human beings employ the so-called mental models to understand knowledge about the physical world [47]. Mental models differ significantly from analytical models.

An analytical model of a physical system is a result of some engineering design or scientific investigation which is often documented in books, manuals, and written reports. Such a model is an accurate, consistent, and objective representation of a physical system.

On the contrary, a mental model is naturally evolving. Experiments have indicated that a novice reasons about a physical system by first creating a crude, buggy initial mental model of the system and then successively refining it to more elaborated models [48].

Experts are generally much better problem solvers than novices. The reason is that experts know the correct and powerful analytical models for a given physical system, and they also possess rich experience in reasoning with those models. As a result, they are able to formulate much better mental models in terms of accuracy, consistency, and objectivity.

A correct and powerful analytical model needs not to be complete to the lowest possible level of details. Taking an example from the domain of VLSI (very large-scale integrated circuits), a good analytical model describes the circuit at the logic-gate level, not at the level of transistors and resistors. Taking another example in the domain of internal medicine, a given problem may require a "fuzzy" analytical model of a human body that describes the functions and interrelations of bodily organs, whereas another problem may require the inclusion of sensory receptors in the model.

Human experts, upon analyzing the given problem at hand, know what kinds of analytical models are most appropriate. The experts formulate their own mental models of a physical system based on its analytical model as well as personal experience such as undocumented information about the failure rates of certain devices. The experts continue to modify their mental models until they successfully solve the given problem.

Modeling the way human experts solve problems helps improve the performance of AI software programs. In addition, it helps to uncover subtle erroneous decisions and beliefs that human experts might take on in some situations.

13.7 Structural Topology and Behavioral Causality

The mental model of a physical system is defined as the structural topology and behavioral causality of that system as perceived by a human mind.

Structural topology refers to the components and their interconnections within a given system. A topology provides information about not only the existence of certain constituents in a system but also the possible flow of data within a system. In VLSI, for example, a typical structural topology is a circuit diagram showing some integrated circuit chips and their interconnections by which logical truth values propagate from one chip to another.

Behavioral causality refers to the cause-effect relationships between the inputs and outputs of a given component in a system. The behavior of an entire system is the "sum total" of all individual behaviors of its components. A behavior can be represented as a rule, an equation, or a procedure. In electronic devices, for instance, an ideal transformer has a behavioral description of the form: $Z(p) = (N(p) / N(s))^2 * Z(s)$ where Z is the impedance and N is the number of turns on the primary coil p and the secondary coil s. Besides inputs and outputs, a behavior can describe discrete physical states of a device such as "open" and "close" of an on/off switch. The behavioral description of a device is usually limited to its intended function. In other words, we are normally interested in a subset of behaviors that is relevant to the diagnostic problem at hand. Such a subset is called a "relevant facet of behavior."

13.8 Component Clustering and Decoupling

"Component clustering" and "component decoupling" describe the mechanisms that human experts use to modify their mental models. Component clustering refers to the agglomeration of physically adjacent components to form a single composite element whose behavior is the sum total of all its constituents. On the contrary, component decoupling means separating apart adjacent components from a single composite element.

By clustering physically adjacent components we increase the level of abstraction on various parts of a mental model and thereby reduce its complexity. Clustering is a sensible thing to do when we believe that the faults are not located at any of the components that are to be agglomerated. Should our belief be proven incorrect, we would modify the mental model by decoupling the clusters. It is permissible to have multi-layered clusters, that is, clusters within clusters.

By decoupling a composite element, we refine various parts of a mental model and thereby increase its intricacy. Decoupling is inevitable when the faults are believed to reside in one or more of the clustered components.

13.9 Analytical Models and Experiential Knowledge

Analytical models provide the basis for formulating useful mental models for diagnosis. Experiential knowledge guides the formulation process towards more effective fault isolations. A change in a mental model affects either its structure or behavior. The structure changes when a network branch is pruned from the model and when the model constituents are clustered and decoupled. The behavior changes when a different relevant facet of behavior is selected to be the focus.

Experiential knowledge guiding model refinement consists of:

(1) Pattern-recognition rules: Based on the symptom descriptions, the pattern-recognition rules decide which facet of behavior is relevant to the problem at hand. Since these rules also speculate the likely causes of the problem, they indirectly influence the clustering and decoupling of the mental model constituents. Therefore, these rules change the structure and behavior of a mental model.
(2) Topological clustering rules: Based on personal experience and preference, the clustering rules decide which adjacent components to agglomerate into clusters. Therefore, these rules change the structure of a mental model.
(3) Topological pruning rules: Based on the locality of suspicious components and clusters, the pruning rules decide which branches of the model structure can be pruned without affecting the model simulation. Therefore, these rules change the structure of a mental model.
(4) Recommendation rules: Based on domain-specific requirements, the recommendation rules decide which remedial actions to perform in order to correct the individual misbehaviors in the model. Therefore, these rules change the overall behavior of a mental model.

13.10 The DM^2 Algorithm

At the Aspen Security Forum in July 2012, former deputy director of the CIA Counterterrorism Center Henry "Hank" Crumpton spoke of the war on terror: "It's a different type of war. Dealing with terror is going to be more like managing disease" [49].

The generic Dynamic Mental Models (DM^2) can be applied equally well in disease diagnosis as well as counterterrorism. The DM^2 algorithm consists of six major steps (see Fig. 13.1):

1. Misbehavior pattern recognition

Based on a body of pattern-recognition rules, DM^2 identifies the misbehavior type for the given symptoms, and speculates the likely sources of the problem. These rules are among the experiential knowledge obtained from a human expert, or they are inferred from statistical data that correlate known symptoms with confirmed causes. In either case, DM^2 collects a list of suspicious components and a list of suspicious clusters for further investigation.

2. Mental model formulation

DM^2 constructs a mental model by studying a detailed analytical model of the physical system that it is going to diagnose. Firstly, based on the locality of the suspicious components and clusters, DM^2 prunes the excessive branches from the model structure. This is accomplished with the help of topological pruning rules. Secondly, based on the past experience and personal preference obtained from a human expert, DM^2 clusters the model components in order to increase the mental model abstraction to the level that is more manageable. Thirdly, based on the previously identified misbehavior type, DM^2 chooses to consider only a "relevant facet of behavior" from the complete description of a component.

3. Mental model refinement

In the course of diagnosis, DM^2 revises its mental model when necessary. Imagine you are picking up a small box, if you think in your mind that the box is light but it turns out to be heavy, you will be startled for a split second. Your mental model sets up certain expectations that may or may not be met.

The need to modify a mental model arises when the model is not at the right level of abstraction due to new test results and refined speculations. A correct level of model abstraction is essential for effective and efficient diagnosis. To achieve this goal, DM^2 revises its mental model by means of component clustering and decoupling:

(a) Firstly, based on a body of topological clustering rules, DM^2 agglomerates physically adjacent components to form clusters. These rules represent the human expert's perceptive view of the model topology. DM^2 triggers a topological clustering rule if and only if the components that the rule is trying to cluster are not among the suspicious components subject to investigation.

(b) Secondly, DM^2 goes through the list of suspicious components to make sure that they are not embedded inside clusters. For every suspicious component that is hidden in one or more clusters, DM^2 decouples these clusters in order to expose the suspicious component to investigation.

(c) Thirdly, DM^2 decouples all the clusters that are listed as suspicious. In so doing, the components and clusters inside these suspicious clusters become exposed to investigation. The exposed components are then added to the list of suspicious components.

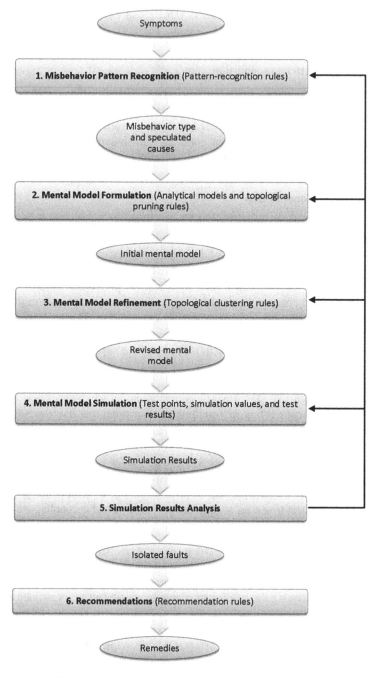

Fig. 13.1   The DM² algorithm flowchart

4. **Mental model simulation**

Simulating a mental model means propagating data and constraints from component to component via their interconnections in order to produce the expected behaviors for comparison with the test results obtained from the real system. In other words, we compare the simulated outputs of a component with its actual measurement. For the clusters in a mental model, we consider only their outputs as the net effects of combining all the individual behaviors of their constituents. DM^2 divides the mental model simulation process into three major steps:

(a) Test points restriction: A physical system sometimes limits the extent of testing one can perform and therefore makes certain component behaviors unobservable. The observable output of a component is called a "test point." Test points are usually known beforehand, and they are determined by the availability of test equipment and other resources. A restriction on the testability of physical components reduces the precision of fault isolation and complicates the simulation procedure.

(b) Simulation input values selection: Different input values to the physical system can generate very different behaviors, resulting in different symptoms or none at all under some circumstance. Normally DM^2 asks the user at run time to determine the appropriate input values for a simulation.

(c) Model execution and test results comparison: DM^2 executes a mental model by propagating data and constraints through the components in order to generate their expected behaviors. During the model execution, DM^2 asks the user to compare the simulated behaviors with the actual test results. If the discrepancy between an expected and an observed behavior is out of tolerance, then DM^2 concludes that one or more components in-between the current test point and the previous test point(s) are at fault; it then takes the actual data as the simulated values and continues the model simulation. Continuing the simulation without redoing previous computations and retesting the real system generally saves time and resources when attempting to isolate multiple faults.

5. **Simulation results analysis**

Based on the simulation results from the previous step, DM^2 decides how to proceed in the diagnosis:

(a) If there is no discrepancy at all between the expected behaviors and the actual test results, DM^2 gives the user a choice to return to step 4 (Mental Model Simulation) and try different simulation input values. If the repeated attempt fails, DM^2 suspects incorrect misbehavior-pattern recognition due to misleading symptom descriptions, and it therefore returns to step 1 (Misbehavior Pattern Recognition).

(b) If there is a discrepancy between the expected behaviors and the actual test results, but the fault cannot be isolated, DM^2 suspects incorrect analytical models, excessive pruning, or inaccurate "relevant facet of behavior", and it returns to step 2 (Mental Model Formulation).

(c) If a cluster is found to be at fault, DM^2 returns to step 3 (Mental Model Refinement) so that refining the level of abstraction can pinpoint the problems inside the cluster.

(d) Otherwise, DM^2 proceeds to step 6 (Recommendations).

6. **Recommendations**

Based on a body of recommendation rules obtained from a domain expert, DM^2 suggests remedial actions for the isolated faults. In many cases, the faulty components are either replaced or adjusted.

To err is human. AI software modeled after humans will inevitably make mistakes. It is fine as long as the software learns from its errors and improves itself, which is something that humans ought to learn from AI. Aldous Huxley, author of *Brave New World*, wrote in "Case of Voluntary Ignorance" in *Collected Essays* (1959): "That men do not learn very much from the lessons of history is the most important of all the lessons that history has to teach."

13.11 AI Applications in Counterterrorism

After the 9/11 terrorist attacks, more artificial intelligence research and development projects have focused on facilitating knowledge acquisition, assisting in the formation of terrorism-related knowledge bases, and supporting the processes of analysis and decision making in counterterrorism [50].

The University of Arizona's AI Lab developed web-based counterterrorism knowledge portals to "to support the analysis of terrorism research, dynamically model the behavior of terrorists and their social networks, and provide an intelligent, reliable, and interactive communication channel with the terrorized (victims and citizens) groups" [51].

North Atlantic Treaty Organization (NATO) commissioned research on detecting terrorist preparations. FUSEDOT (FUzzy Signal Expert system for the Detection Of Terrorism preparations) applies "artificial intelligence expert system technology to the fuzzy signals presented by certain anomalous data, such as interpersonal relationships, financial relationships, travel patterns, purchasing patterns, patterns of Internet usage, and personal background" [52].

Five U.K. universities launched DScent — a joint project that "combines research theories in the disciplines of computational inference, forensic psychology and expert decision-making in the area of counterterrorism" and it includes the use of neural networks to identify deceptive behavior of terrorists with an average of 60% success rate [53].

In November 2008, The U.K. government-funded Cyber Security Knowledge Transfer Network (KTN) started examining the potential use of AI in counterterrorism surveillance and data mining. Nigel Jones, director of the KTN, said,

"In today's age of distributed networks and with moves towards cloud computing, there is a lot more information out there that might be useful in terms of evidence. There is a problem in handling that mass of data, in storing it, routing it, tracing it and in finding patterns. The [KTN AI and forensics] special interest group will look at the role that AI could have in gathering and analyzing that information, which could be used in investigations or in intelligence gathering to trigger alerts" [54].

In August 2012, Microsoft and NYDP jointly announced bringing the real-time networked Domain Awareness System (DAS) technology to law enforcement agencies around the world. According to retired U.S. Army Lt. Gen. Mike McDuffie, DAS "aggregates and analyzes public safety data in real time and combines artificial intelligence analytics with video from around a jurisdiction to identify potential threats and protect critical infrastructure" [55].

Apart from data mining and public safety, AI will be deployed in the new generation of Unmanned Aerial Vehicles (UAVs) or Unmanned Aerial Systems (UAS) that have become the counterterrorism weapon of choice. In 2009, the U.S. Air Force published "Unmanned Aircraft Systems Flight Plan 2009-2047" which assumes that "the next generation of drones will have artificial intelligence giving them a high degree of operational autonomy including — if legal and ethical questions can be resolved — the ability to shoot to kill" [56].

13.12 Massively Multi-Participant Intelligence Amplification (MMPIA)

Hosted by anti-crime activist John Walsh, *America's Most Wanted* premiered in February 1988 on Fox Television Network and moved to Lifetime in December 2011 after 23 years. The TV show profiles and assists law enforcement in the apprehension of fugitives wanted for various crimes. By January 2013, more than 1,200 fugitives have been captured, and over 60 missing children and persons have been found [57]. The success of *America's Most Wanted* is dependent on volunteer citizen detectives among millions of television viewers.

In the computing world, SETI@home (Search for Extraterrestrial Intelligence) was released by UC Berkeley to the public in May 1999 to support the analysis of radio signals from space in an attempt to detect intelligent life outside Earth [58]. The program installs itself as a screensaver and it processes observational data when the home or work computers are idle. By January 2013, SETI@home has a total of 1.3 million users and 3.3 million hosts in 233 countries [59].

In October 2007, Stanford University's Folding@home project received a Guinness World Record for topping 1 petaflop (a thousand trillion floating point operations per second) running on computers as well as Sony's PlayStation 3 video game consoles [60]. Folding@home helps scientists study protein folding and its relationship to Alzheimer's, Huntington's, and cancerous diseases [61]. In September 2011, players of the Foldit video game took less than 10 days to decipher the AIDS-causing Mason-Pfizer monkey virus that had stumped scientists for 15 years [62].

Volunteer citizen detectives and citizen scientists have proved to be immensely successful in data mining and problem solving. We are witnessing the beginning of Massively Multi-Participant Intelligence Amplification (MMPIA) in which a massive network of collective human intelligence, often assisted by distributed computing, is extending the information processing capabilities of the human mind.

In *Facebook Nation: Total Information Awareness*, I proposed a distributed software program called "STS@home" designed for Neighborhood Watch [63]. The artificial intelligence system would analyze live video streams from webcams connected to the homeowners' computers in the Search for Trespassers and Suspects (STS) in their neighborhood. A suspicious activity would trigger an alert, and a neighborhood-watch leader could forward the information to police after reviewing the video clip. A facial recognition feature in the software system could assist police in locating missing children, apprehending fugitives, and solving crimes.

There are already tons of visual information from live traffic cams, ATM security cameras, Neighborhood Watch webcams, and public CCTV systems. Technology exists today that uses video analytics to distill millions of hours of raw video footage into structured, searchable data [64]. Leveraging MMPIA, law enforcement agencies and the intelligence community could conceivably conduct more effective data mining and counterterrorism efforts with the help of citizen detectives and citizen scientists.

Bibliography

1. **McCarthy, John.** What is Artificial Intelligence. [Online] Stanford University, November 12, 2007. http://www-formal.stanford.edu/jmc/whatisai/node1.html.
2. **IMDb.** A.I. Artificial Intelligence. [Online] IMDb, June 29, 2001. http://www.imdb.com/title/tt0212720/.
3. —. I, Robot. [Online] IMDb, July 16, 2004. http://www.imdb.com/title/tt0343818/.
4. **Paul, Ian.** IBM Watson Wins Jeopardy, Humans Rally Back. [Online] PCWorld, February 17, 2011. http://www.pcworld.com/article/219900/IBM_Watson_Wins_Jeopardy_Humans_Rally_Back.html.
5. **Markoff, John.** Computer Wins on 'Jeopardy!': Trivial, It's Not. [Online] The New York Times, February 16, 2011. http://www.nytimes.com/2011/02/17/science/17jeopardy-watson.html?pagewanted=all.
6. **IBM.** Deep Blue. [Online] IBM. [Cited: November 5, 2012.] http://researchweb.watson.ibm.com/deepblue/.
7. **Goldman, David.** Top U.S. supercomputer guns for fastest in world. [Online] CNNMoney, October 29, 2012. http://money.cnn.com/2012/10/29/technology/innovation/titan-supercomputer/index.html.
8. **Apple.** Learn more about Siri. [Online] Apple. [Cited: November 5, 2012.] http://www.apple.com/ios/siri/siri-faq/.
9. **SRI International.** Cognitive Assistant that Learns and Organizes. [Online] SRI International. [Cited: November 5, 2012.] http://www.ai.sri.com/project/CALO.

10. **Markoff, John.** A Software Secretary That Takes Charge. [Online] The New York Times, December 13, 2008. http://www.nytimes.com/2008/12/14/business/14stream.html.

11. **Hay, Timothy.** Apple Moves Deeper Into Voice-Activated Search With Siri Buy. [Online] The Wall Street Journal, April 28, 2010. http://blogs.wsj.com/venturecapital/2010/04/28/apple-moves-deeper-into-voice-activated-search-with-siri-buy/.

12. **Minsky, Marvin.** THE AGE of INTELLIGENT MACHINES | Thoughts About Artificial Intelligence. [Online] Kurzweil Accelerating Intelligence, February 21, 2001. http://www.kurzweilai.net/marvin-minsky.

13. **CNN.** AI set to exceed human brain power. [Online] CNN, August 9, 2006. http://www.cnn.com/2006/TECH/science/07/24/ai.bostrom/.

14. **3VR Inc.** Use Video Analytics and Data Decision Making to Grow Your Business. [Online] Digital Signage Today. [Cited: May 28, 2012.] http://www.digitalsignagetoday.com/whitepapers/4891/Use-Video-Analytics-and-Data-Decision-Making-to-Grow-Your-Business.

15. **Rayner, Gordon.** New intelligent CCTV cameras can see and hear. [Online] The Telegraph, June 23, 2008. http://www.telegraph.co.uk/news/uknews/2180628/New-intelligent-CCTV-cameras-can-see-and-hear.html.

16. **Aviv, Juval.** Can AI Fight Terrorism? [Online] Forbes, June 22, 2009. http://www.forbes.com/2009/06/18/ai-terrorism-interfor-opinions-contributors-artificial-intelligence-09-juval-aviv.html.

17. **New York City Police Department.** Public Security Privacy Guidelines. [Online] New York City Police Department, April 2, 2009. http://www.nyc.gov/html/nypd/downloads/pdf/crime_prevention/public_security_privacy_guidelines.pdf.

18. **Thistle, Michele Bedford.** Microsoft, NYPD team up on global counterterrorism solution. [Online] Microsoft, August 9, 2012. http://www.microsoft.com/government/ww/public-services/blog/Pages/post.aspx?postID=158&aID=4.

19. **IDC.** The Digital Universe of Opportunities: Rich Data and the Increasing Value of the Internet of Things. [Online] International Data Coproration, April 2014. http://www.emc.com/leadership/digital-universe/2014iview/executive-summary.htm.

20. **Bamford, James.** The NSA Is Building the Country's Biggest Spy Center (Watch What You Say). [Online] Wired, March 15, 2012. http://www.wired.com/threatlevel/2012/03/ff_nsadatacenter/all/.

21. **—.** War of Secrets; Eyes in the Sky, Ears to the Wall, and Still Wanting. [Online] The New York Times, September 8, 2002. http://www.nytimes.com/2002/09/08/weekinreview/war-of-secrets-eyes-in-the-sky-ears-to-the-wall-and-still-wanting.html?pagewanted=all.

22. **Perez, Evan.** New leaker disclosing U.S. secrets, government concludes. [Online] CNN, August 6, 2014. http://www.cnn.com/2014/08/05/politics/u-s-new-leaker/index.html.

23. **Bamford, James.** War of Secrets; Eyes in the Sky, Ears to the Wall, and Still Wanting. [Online] The New York Times, September 8, 2002. http://www.nytimes.com/2002/09/08/weekinreview/war-of-secrets-eyes-in-the-sky-ears-to-the-wall-and-still-wanting.html?pagewanted=all.

24. **Kalil, Tom.** Big Data is a Big Deal. [Online] The White House, March 29, 2012. http://www.whitehouse.gov/blog/2012/03/29/big-data-big-deal.

25. **Chakrabarti, Soumen, et al.** Data Mining Curriculum: A Proposal (Version 1.0). [Online] ACM SIGKDD, April 30, 2006. http://www.sigkdd.org/curriculum/CURMay06.pdf.

26. **Block, Robert, Fields, Gary and Wrighton, Jo.** U.S. 'Terror' List Still Lacking . [Online] The Wall Street Journal, January 2, 2004. http://online.wsj.com/article/0,,SB107300534268021800,00.html.

27. **Gorman, Siobhan.** How Team of Geeks Cracked Spy Trade . [Online] The Wall Street Journal, September 4, 2009. http://online.wsj.com/article/SB125200842406984303.html.

28. **Flynn, Michael T.** Advanced Analytical Capability Joint Urgent Operational Need Statement. [Online] Department of Defense US Forces Afghanistan, July 2, 2010. http://www.politico.com/static/PPM223_110629_flynn.html.

29. **Carik, Kenneth.** The Nature of Explanation. [Online] Cambridge University Press, UK, 1943. http://www.cambridge.org/us/knowledge/isbn/item1121731/.

30. **Minsky, Marvin.** A Framework for Representing Knowledge. [Online] MIT AI Laboratory Memo 306, June 1974. http://web.media.mit.edu/~minsky/papers/Frames/frames.html.

31. **DeRosa, Mary.** Data Mining and Data Analysis for Counterterrorism. [Online] Center for Strategic and International Studies, March 2004. http://csis.org/files/media/csis/pubs/040301_data_mining_report.pdf.

32. **McCarthy, John.** Recursive Functions of Symbolic Expressions and Their Computation by Machine, Part I. [Online] Massachusetts Institute of Technology, April 1960. http://www-formal.stanford.edu/jmc/recursive/recursive.html.

33. **Simran, Max & Charence.** Prolog. [Online] Imperial College London, 2006. http://www.doc.ic.ac.uk/~cclw05/topics1/prolog.html.

34. **Brachman, Ronald J. and Schmolze, James G.** An Overview of the KL-ONE Knowledge Representation System. [Online] Cognitive Science, 1985. http://eolo.cps.unizar.es/docencia/MasterUPV/Articulos/An%20Overview%20of%20the%20KL-ONE%20Knowledge%20Representation%20System-Brachman1985.PDF.

35. **Minsky, Marvin.** Logical vs. Analogical or Symbolic vs. Connectionist or Neat vs. Scruffy. [Online] MIT Press, 1990. http://web.media.mit.edu/~minsky/papers/SymbolicVs.Connectionist.html.

36. **Liu, Ling and Ozsu, M. Tamer.** Encyclopedia of Database Systems. [Online] Springer, 2009. http://www.springer.com/computer/database+management+%26+information+retrieval/book/978-0-387-35544-3.

37. **Wu, Xindong.** Knowledge Acquisition from Databases. [Online] Ablex Publishing Corporation, 1995. http://www.cs.uvm.edu/~xwu/Publication/Book-95.html.

38. **Buchanan, Bruce G. and Shortliffe, Edward H.** Rule-Based Expert Systems: The MYCIN Experiments of the Stanford Heuristic Programming Project. [Online] Addison Wesley, 1984. http://www.amia.org/staff/eshortliffe/Buchanan-Shortliffe-1984/MYCIN%20Book.htm.

39. **Lee, Newton S., Phadke, Madhav S. and Keny, Rajiv.** An expert system for experimental design in off-line quality control. [Online] Wiley, November 1989. http://onlinelibrary.wiley.com/doi/10.1111/j.1468-0394.1989.tb00148.x/abstract.

40. **Chessbase .** Dark horse ZackS wins Freestyle Chess Tournament. [Online] Chessbase News, June 19, 2005. http://www.chessbase.com/newsdetail.asp?newsid=2461.

41. **Lee, Newton S.** DM2: an algorithm for diagnostic reasoning that combines analytical models and experiential knowledge. [Online] International Journal of Man-Machine Studies, June 1988. http://www.sciencedirect.com/science/article/pii/S002073738880066X.

42. **De Kleer, Johan.** How Circuits Work. [Online] Artificial Intelligence, December 1984. http://www2.parc.com/spl/members/dekleer/Publications/How%20Circuits%20Work.pdf.

43. **Genesereth, Michael R.** The use of design descriptions in automated diagnosis. [Online] Artificial Intelligence, December 1984. http://www.sciencedirect.com/science/article/pii/0004370284900432.

44. **Davis, Randall.** Diagnostic reasoning based on structure and behavior. [Online] Artificial Intelligence, December 1984. http://www.sciencedirect.com/science/article/pii/0004370284900420.

45. **Berwaner, Mary.** The Problem of Diagnostic Aiding. [Online] The Defense Technical Information Center, October 30, 1989. http://www.dtic.mil/cgi-bin/GetTRDoc?AD=ADA239200.

46. **Trafton, J. Gregory.** Dynamic mental models in weather forecasting. [Online] Defense Technical Information Center, 2004. http://www.dtic.mil/cgi-bin/GetTRDoc?AD=ADA480241.

47. **Gentner, Dedre and Stevens, Albert L.** Mental Models. [Online] Lawrence Erlbaum Associates, May 1, 1983. http://books.google.com/books/about/Mental_Models.html?id=QFI0SvbieOcC.

48. **Williams, Michael D. and Hollan, James D., Stevens, Albert L.** Human Reasoning About a Simple Physical System. [Online] Lawrence Erlbaum Associates, May 1, 1983. http://books.google.com/books?id=QFI0SvbieOcC&pg=PA131.

49. **Dougherty, Jill.** Experts: No easy cure for the disease of terror. [Online] CNN, July 27, 2012. http://security.blogs.cnn.com/2012/07/27/experts-no-easy-cure-for-the-disease-of-terror/.

50. **Markman, A.B., et al.** Analogical Reasoning Techniques In Intelligent Counterterrorism Systems. [Online] International Journal on Information Theories and Applications, 2003. http://www.foibg.com/ijita/vol10/ijita10-2-p04.pdf.

51. **Reid, Edna, et al.** Terrorism Knowledge Discovery Project: A Knowledge Discovery Approach to Addressing the Threats of Terrorism. [Online] The University of Arizona. [Cited: November 24, 2012.] http://ai-vm-s08-rs1-1.ailab.eller.arizona.edu/people/edna/AILab_terrorism%20Knowledge %20Discovery%20ISI%20_apr04.pdf.

52. **Koltko-Rivera, Mark E.** Detection of Terrorist Preparations by an Artificial Intelligence Expert System Employing Fuzzy Signal Detection Theory. [Online] Defense Technical Information Center, October 2004. http://www.dtic.mil/cgi-bin/GetTRDoc?AD=ADA460204.

53. **Dixon, S.J., et al.** Neural Network for Counter-Terrorism. [Online] Leeds Metropolitan University. [Cited: November 24, 2012.] http://www.leedsmet.ac.uk/aet/computing/aNNfor CTShortPaper.pdf.

54. **Heath, Nick.** Police enlist AI to help tackle crime. [Online] ZDNet, November 6, 2008. http://www.zdnet.com/police-enlist-ai-to-help-tackle-crime-3039541531/.

55. **McDuffie, Mike.** Microsoft and NYPD Announce Partnership Providing Real-Time Counterterrorism Solution Globally. [Online] Microsoft, August 8, 2012. http://www. microsoft.com/government/en-us/state/brightside/Pages/details.aspx?Microsoft-and-NYPD-Announce-Partnership-Providing-Real-Time-Counterterrorism-Solution-Globally&blogid=697.

56. **The Economist.** Flight of the drones: Why the future of air power belongs to unmanned systems. [Online] The Economist, October 8, 2011. http://www.economist.com/node/21531433.

57. **America's Most Wanted.** America's Most Wanted. [Online] Lifetime. [Cited: January 17, 2013.] http://www.amw.com/.

58. **SETI@home.** SETI@home. [Online] University of California. [Cited: January 17, 2013.] http://setiathome.berkeley.edu/index.php.

59. **BOINC STATS.** Project stats info. *BOINC STATS.* [Online] January 15, 2013. http://boincstats.com/en/stats/projectStatsInfo.

60. **Terdiman, Daniel.** Sony's Folding@home project gets Guinness record. [Online] CNet, October 31, 2007. http://news.cnet.com/8301-13772_3-9808500-52.html.

61. **Stanford University.** Folding@home distributed computing. [Online] Stanford University. [Cited: January 17, 2013.] http://folding.stanford.edu/English/HomePage.

62. **Boyle, Alan.** Gamers solve molecular puzzle that baffled scientists. [Online] NBC News, September 18, 2011. http://cosmiclog.nbcnews.com/_news/2011/09/18/7802623-gamers-solve-molecular-puzzle-that-baffled-scientists.

63. **Lee, Newton.** Facebook Nation: Total Information Awareness. [Online] Springer, September 15, 2012. http://www.amazon.com/Facebook-Nation-Total-Information-Awareness/dp/1461453070.

64. **3VR.** Use Video Analytics and Data Decision Making to Grow Your Business. [Online] [Cited: May 28, 2012.] http://www.digitalsignagetoday.com/whitepapers/4891/Use-Video-Analytics-and-Data-Decision-Making-to-Grow-Your-Business.

Chapter 14
Gamification of Penetration Testing

Darren Manners

14.1 Win, Lose, or Something Else

Penetration testing is the term given to a process that tests an organizations security by mimicking a specified attack actor. It often comprised of an external test, internal test, web application test, wireless test and a social engineering test. Further segmentation of the main segments also can be conducted. An example is an external test may only concentrate on a specific IP address.

The segmented tests can be put all together in one test, often referred to as a red team test. The colors of teams are either red for offensive operations or blue for defensive operations taken from military terminology. In a red team test, information or exploits found in one segment can be used in another segment. It is a test, following a very loose scope of work and guideline that is the closest test an organization can get mimicking real world attacks.

In a nutshell, a penetration test mimics a hacker's attempt to gain access to an organization and steal intellectual property or sensitive information. It is this 'win or lose' philosophy that makes penetration testing ripe for gamification. A tester either gets in or doesn't.

Of course it's not as simple as that. Nothing is ever black and white. As a penetration tester, I don't have to 'get in' to win or be able to point out flaws in a network. However, it is fun to sit down with a client and show how the external network was breached. It is less fun, but just as important, to map a technical finding to a missing or incorrect procedure in a gap analysis.

So we have a win, lose, and something else that creates a game for a penetration tester.

D. Manners (✉)
SyCom Technologies, Virginia, USA

© Springer International Publishing Switzerland 2015
N. Lee, *Counterterrorism and Cybersecurity*, DOI 10.1007/978-3-319-17244-6_14

14.2 Short and Long Engagements

There are often two types of players. Those that tend to get to the end of a game as quick as possible and those that methodically go along collecting everything as they go. Penetration testers can do both depending upon the type and length of the engagement.

In a short engagement the tester may try to show how quickly a potential threat actor can achieve their objective. An example of a threat actor is organized crime. Another would be state sponsored. There are many threat actors out there with varying skillsets and resources. Time constraints may only allow a penetration tester to find one attack vector. The objective is often privilege escalation (going from no privilege to perhaps administrative privilege), obtain sensitive information or something else.

A longer engagement often allows the penetration tester to be more methodical. The tester may be able to find multiple routes for the same objective. The tester may also be able to become more in depth and create more complex approaches. At times the test may be compared to a line of dominos. The success of the test may be result of a number of factors set up by the tester. Like dominos, the final domino will only fall as a result of the falling of all the other dominos.

Often a penetration tester will use good judgment on the depth of the penetration as compared to the fix to prevent future penetrations. An example of this would a finding of a default credential. If the penetration test was a short engagement is it worth spending an enormous amount of time exploiting everything that was a result of a default credential when the fix is very simple? The answer may be yes, if the results can justify the means, but often it may be a simple highlight in a report and the tester will move on to find other attack vectors. So like any game, how you play it may depend upon how much time you have to hand.

14.3 A Game Already?

Penetration testing is already a game in all but name. It has:

1. **An Objective and Rules**
   It has an objective that all parties agree on. During the initial phase, all parties will agree to abide by a set of rules that creates the playing field. This client being tested would be wise to allow an as open a playing field as possible so as not to tip the odds in their favor, thus tainting the test. Sometimes this phase of the test is the most important.

   A good methodology such as the PTES (Penetration Testing Execution Standard) can help create a balanced test as well as a good repeatable methodology — important for testing [1].

2. **Points and Skills**

A tester can accrue points as a result of finding exploits, vulnerabilities and misconfigurations. Often a tester can create a scorecard based upon likelihood of occurrence vs. impact. Vulnerabilities are assigned a score by CVE (Common Vulnerabilities and Exposures) that can also be used to create a scorecard and assign points [2].

Misconfigurations and "tricks' are what penetration testers call their special 'pwn sauce' — 'leet' speak is part of the culture where characters in the English language can be altered perhaps a number for a letter. A trick maybe something that is considered a functional part of network/software use, such as Microsoft's PowerShell, but when used with malicious intent, can have devastating consequences. This is harder to score, but is necessary to show to an organization so as to allow them to create mitigation for the threat. A misconfiguration is only known if the penetration tester knows what the correct configuration is. It is sometimes easy to spot and sometimes difficult, dependent upon the penetration testers experience and 'blue' fighting skills.

Sometimes, like special combinations in fight games, simples, low level exploits, when combined together, can 'punch' well above their weight. This is where a tester's skill comes into play. Junior testers may overlook level exploits, vulnerabilities and misconfigurations thinking that they will not give much of a return. Often a very simple piece of nearly overlooked information has had devastating consequences later on.

A good analogy of this is the dungeon and dragon games of old. Often you would pick up an item, having no idea why you want it, but you have it just in case. Then, later on in the game, you find a good use for it. Sometimes penetration testing is like that. A good tester will never overlook something.

3. **Brotherhood**

Good games create a sense of a brotherhood. Penetration testers are often considered to be expert security professionals, with some having 'rock star' status. This can be seen in the conferences and speakers that regularly attend conferences such as DEF CON, DerbyCon or Black Hat. They are often well sought after post talk/presentation for photos and autographs; and some have a huge following on social media.

Regardless of age, race or gender many of the top penetration testing professionals have given back to their community, with notoriety coming from what they know and do. Often born out of the hacker community, but not loved by some black hat hackers (the term black hat refers back the western movies where the good guys wore white hats and the bad guys black), penetration testers are considered white hat hackers and have a strong brotherhood. The link between the hacker community and the penetration testers' community is often a blurred line, both borrowing tools and techniques from each other.

4. **Addictiveness**

I can tell you as a penetration tester myself, it is extremely addictive. You get the highs and lows all the time. There is no greater fun that 'popping' (obtaining a shell or sensitive information) a box at 3 am in morning after hours of research and coding. It is very addictive. I am an ardent gamer myself. I consider a test almost an extension of my Sony PlayStation 4. In fact, whenever I get burned out on penetration testing, I will go back to my 'roots' and play Modern War Battlefield 4 for hours, until I feel recharged. Attending conferences with my fellow penetration testers and security professionals is some of the best times I've had. And they call this work!

5. **Fun**

I love what I do. It's a strange world, often working strange hours in strange places. If I am not testing I am training. Social engineering is the most fun and the most effective methodology to gain access. We push it to the point of being noticed, sometimes gun-ho security admins have called the police on me, who in turned called the SWATSWAT team—apparently I have a funny accent. I've tested small network, large networks and in between. It's a game of cat and mouse, attempting to defeat a far numerical and superior force by means of one's brain. Who wouldn't love that. Asymmetrical warfare is all the rage. You only have to take a look at the modern battlefield to see that.

I work in a field where questioning everything is the norm. After all your network is probably being tested for free right now, just that those hackers never send you the results. I see every kind of mistake imaginable, every misconfiguration, and I get excited when a new technique or exploit can drill through millions of dollars of security systems designed to prevent attacks. I'm funny like that, but you know what? I am not alone. Don't believe me? Pay a visit to DEF CON or a Black Hat security conference one day.

14.4 The Future: Continuous Penetration Testing

So why not completely make penetration testing a game? Creating a standard scorecard or even a standard grade is tricky, but not impossible. The secrecy that many companies hold their testing results in often contributes the lack of remediation. If a company had a publicly known score, then not only would they be able to identify where they are within their own verticals, but so would everyone else. Perhaps it is the latter as to why this is not published.

No company would want their scores known, but sometimes regulation forces companies to do things that they don't want to do as it costs them money. A great example would be safety scores for vehicles. No company wants poor safety scores known, but it is in the consumers and industry's interest. Often companies have only increased safety as a result of bad publicity or to attract consumers.

Wouldn't you want to know that the company that has all your details is ranked last in security or received a failing grade? As a CEO wouldn't you want to score a better grade than your competitor? Wouldn't you use that fact in publicity campaigns?

It's not all one sided though. Penetration testers need to adhere to a repeatable process, like the PTES framework. They also need to have a specific measured skillset. Currently anyone can call himself or herself a penetration tester.

The problem is that often a vulnerability found in one company does not have the same impact as another company due to the complexity, defenses, and/or configuration of that company's network. This should be mapped to a scorecard based upon impact vs. likelihood of occurrence or other methodologies.

I had the idea of continuous penetration testing years ago. The idea would be to have one penetration tester responsible for 25 companies only. Incentives would be given for gaining points on the company, rewarding the tester for their skills. A room full of such individuals would be able to share techniques and exploits. Imagine something like the *WarGames* movie setup: big screens showing ongoing attacks against companies paying to find vulnerabilities, exploits and misconfigurations. Imagine what a game that would be!

Earlier this year I managed to set this up as a managed service. I'm lucky in that the company I work for, SyCom Technologies, supported me in this endeavor. We're now seeing the payoff as more and more companies come to the realization that the future of penetration testing is not in its current transactional form, but in a continuous transaction process. The game has now been taken to a new level.

Bibliography

1. **PTES.** PTES Technical Guidelines. [Online] Penetration Testing Standard, April 30, 2012. http://www.pentest-standard.org/index.php/PTES_Technical_Guidelines.
2. **MITRE.** Common Vulnerabiites and Exposures. [Online] MITRE, February 4, 2015. https://cve.mitre.org.

Chapter 15
USB Write Blocking and Forensics

Philip Polstra

15.1 Introduction

In recent years USB mass storage devices using NAND flash storage (also known as thumb drives or flash drives) have replaced magnetic media, such as floppy discs, and optical media, such as CD/DVD, as the standard means for backup and file exchange. The ultimate aim is to understand how to perform forensics on USB mass storage devices. In order to achieve this one must first understand the basics of USB devices. The first part of this chapter will cover the basics of USB. From there we will move on to learn more about USB mass storage devices. Once the foundations have been set we will cover some advanced topics such as USB write blocking and device impersonation. The chapter concludes with a discussion of BadUSB and methods of protecting from this and other similar attacks.

15.2 Brief History

Up until the early 1990s peripherals were connected to computers via serial connections (RS-232), parallel connections (LPT), or some proprietary method. While RS-232 is a standard, there are several variations in cabling which leads to complication. Furthermore, serials devices have several choices of protocols leading to a potentially non-user-friendly configuration.

P. Polstra (✉)
Bloomsburg University, Bloomsburg, PA, USA

© Springer International Publishing Switzerland 2015
N. Lee, *Counterterrorism and Cybersecurity*, DOI 10.1007/978-3-319-17244-6_15

In 1996 the first Universal Serial Bus (USB) standard was released. This initial version allowed for plug and play operations with low-speed devices operating at 1.5 Mbps and full-speed devices operating at 12 Mbps. In 1998 some minor revisions and corrections were made and the USB 1.1 standard was released. An improved version, USB 2.0, was released in 2000. The most notable feature of USB 2.0 was the introduction of a new high-speed rate of 480 Mbps. Through USB 2.0 no changes in cabling or other hardware were required.

In 2008, USB 3.0 was introduced. One of the most touted features of USB 3.0 was the introduction of 5.0 Gbps super-speed. The new super-speed came at the cost of adding additional wires, however. USB 3.0 connectors are backwards compatible with USB 2.0. This is accomplished by adding connections which are recessed inside standard USB 2.0 connectors.

15.3 Hardware

USB uses power wires (5 Volts and ground), and differential signal wires for each communication channel. The use of differential voltage makes USB less susceptible to noise than older standards which measure signals relative to ground. Through USB 2.0 only one signal channel was used. As a result, USB 2.0 connections require only four wires (while some connector types have extra shield or ground wires). USB 3.0 adds two additional super-speed channels which require their own ground bringing the minimum number of wires for a USB 3.0 connection to nine.

Unlike some of the older standards, USB devices are hot-pluggable. As a consequence of this devices must tolerate the application and removal of power without damage. Having learned a lesson from non-universal serial connections, the designers of USB ensured that improperly connecting devices and hosts would be impossible using standard cables. In some cases these standard cables can be up to 16 feet long.

15.4 Software

From the end user perspective USB is easy. Just plug in a device, wait for the chimes to sound and start using the device. As one might expect, things are a bit more complicated under the covers. There are no settable jumpers or other complications from a user perspective. Through a process known as enumeration a host will discover a newly connected device, determine the speeds it is capable of communicating at, learn what capabilities the device possesses, and what protocols should be used to communicate with the device.

The USB standards define several device classes including Human Interface Device (HID), printer, audio, and mass storage. In many cases developers and users need not worry about special drivers for a particular device that falls into one of the standard classes.

Connecting a Device

Connecting a device is a 12 step process. Perhaps that is why working with USB is so addictive. Some of the details will be covered in this article. For the full description I highly recommend *USB Complete: The Developer's Guide (4th ed.)* by Jan Axelson. Here are the 12 steps:

1. The device is connected and in most cases receives power from the host.
2. The hub detects that a new device has been connected.
3. The host (PC) is informed of the new device.
4. The hub determines device speed capability as indicated by location of pull-up resistors. This is only a choice of low or full speed as higher speeds are only available after the device is fully enumerated.
5. The hub resets the device so it can begin communicating with it in a less generic manner.
6. The host determines if device is capable of high speed by sending a series of pulses known as chirps.
7. The hub establishes a signal path.
8. The host requests a descriptor (more about descriptors later) from the device to determine max packet size to be used.
9. The host assigns an address to the device so that communication may commence. Addresses are required because more than one device may operate on a single USB bus.
10. The host learns the devices capabilities by asking for a set of structures that describe the device known as descriptors.
11. The host assigns and loads an appropriate device driver (INF file under Windows). In many cases a driver included with the operating system is loaded if the device is from a standard USB device class.
12. The device driver selects a configuration. Devices are not required to support more than one configuration, but multiple configuration devices are not uncommon.

Endpoints

All communication between USB devices and hosts is via endpoints. Endpoints are an abstracted unidirectional communications pipe. All packet fragmentation, handshaking, etc. are handled by the hardware in most cases. The direction of each endpoint is specified relative to the host. For example, an in endpoint receives data from the device into the host. The high bit of the endpoint address is used to indicate direction where 1 and 0 indicate in and out, respectively. There are four types of endpoints: control, bulk transport, interrupt, and isochronous.

All devices must have at least one control endpoint. Devices with more than one control endpoint are extremely rare. Often the control endpoint is referred to as endpoint 0 or EP0. This is the endpoint used to determine devices capabilities. For many devices this is the primary mechanism for communicating with a host. Mass storage devices are an exception.

The control endpoint is used to handle standard requests from the host. The requests include things such as setting or getting and address, returning descriptors, setting power levels and modes, and providing status. The device may also respond to standard USB class (HID, mass storage, etc.) requests. Additionally the USB standard allows vendors to add additional requests to be handled by their devices, but this is rarely done as it would require the vendor to provide a proprietary driver for every supported operating system.

Transfers on control endpoints can involve up to three stages: setup, data, and status. During the required setup stage a setup token and then an eight byte request is sent to the device. The first byte of the request is a bitmap telling the type of request and recipient (device, interface, endpoint). The remaining bytes in the request are parameters for the request and response. If a valid setup request is received, the device responds with an acknowledgement (ACK) packet. During the optional data stage the requested information is sent to the host. In the status stage a zero length data packet is sent as an acknowledgement of success.

Interrupt endpoints are used for communicating with low-speed devices such as keyboards. Interrupt endpoints are useful for avoiding polling and busy waits. The lower speeds typically used in such devices also allow for longer cables. These are not used in mass storage devices.

Isochronous endpoints provide a guaranteed amount of bandwidth. This is useful for time-critical applications such as streaming media. While on the face of things it might seem like isochronous endpoints would be ideal for transferring large amounts of data, this is not the case. As it turns out, the overhead of using isochronous endpoints decreases the total throughput.

Bulk transport or more simply bulk endpoints are used to transfer large quantities of information efficiently. Bulk endpoints have no latency guarantees (unlike isochronous endpoints), but they have the highest throughput on an idle bus. Bulk transfers are superseded by all other transfer types which means they are not the best choice on a busy bus. Low-speed (1.5 Mbps) endpoints may not be used for bulk transfers. These are used extensively in mass storage devices.

Descriptors

As previously mentioned, descriptors are structures used to describe devices and their capabilities. Descriptors have a standard format. The first byte gives the length of the descriptor (so a host will know when it has received the entire descriptor). The second byte tells the type of descriptor and the remaining bytes are the descriptor itself. The standard descriptor types are device, configuration, interface, endpoint, and string.

A device descriptor provides basic information about a device. Some of the more interesting information it provides includes the vendor ID, product ID, and the device class. Manufacturers of USB devices must purchase a vendor ID. Product IDs are set by the manufacturer. A class code of zero indicates that the device class is specified in another (interface) descriptor. A zero class code is quite common and is the norm for mass storage devices. Index values in any descriptors refer to the string descriptor number for that value. The format for a device descriptor is provided in Table 15.1.

Table 15.1 Device descriptors

Offset	Field	Size	Value	Description
0	bLength	1	Number	18 bytes
1	bDescriptorType	1	Constant	Device descriptor (0x01)
2	bcdUSB	2	BCD	0x200
4	bDeviceClass	1	Class	Class code
5	bDeviceSubClass	1	SubClass	Subclass code
6	bDeviceProtocol	1	Protocol	Protocol code
7	bMaxPacketSize	1	Number	Maxi packet size EP0
8	idVendor	2	ID	Vendor ID
10	idProduct	2	ID	Product ID
12	bcdDevice	2	BCD	Device release number
14	iManufacturer	1	Index	Index of manu descriptor
15	iProduct	1	Index	Index of prod descriptor
16	iSerialNumber	1	Index	Index of SN descriptor
17	bNumConfigurations	1	Integer	Num configurations

Table 15.2 Configuration descriptors

Offset	Field	Size	Value	Description
0	bLength	1	Number	Size in bytes
1	bDescriptorType	1	Constant	0x02
2	wTotalLength	2	Number	Total data returned
4	bNumInterfaces	1	Number	Number of interfaces
5	bConfigurationValue	1	Number	Configuration number
6	iConfiguration	1	Index	String descriptor
7	bmAttributes	1	Bitmap	b7 reserved, set to 1 b6 self powered b5 remote wakeup b4..0 reserved 0
8	bMaxPower	1	mA	Max power in mA/2

Configuration descriptors describe the power needs, number of interfaces, etc. for each supported configuration of a USB device. Devices are only required to support one configuration. Because the configuration descriptor contains the total bytes in all subordinate descriptors a request is normally made for nine bytes of the configuration descriptor followed by a request for the configuration descriptor and all subordinate interface and endpoint descriptors. Devices should not request more power than needed as requests for more power than a host can provide result in a failed device enumeration. The format for configuration descriptors is provided in Table 15.2.

An interface descriptor describes how to communicate to a device. For devices with a class code of zero in the configuration descriptor the device class is provided in the interface descriptor(s). Many devices present themselves as composite devices (devices with more than one device class). A camera is a good example of a composite device as it is both a camera and a mass storage device in most cases. The subclass code and protocol code are defined for each device class and are optional. The format for interface descriptors is presented in Table 15.3.

Table 15.3 Interface descriptors

Offset	Field	Size	Value	Description
0	bLength	1	Number	9 bytes
1	bDescriptorType	1	Constant	0x04
2	bInterfaceNumber	1	Number	Number of interface
3	bAlternateSetting	1	Number	Alternative setting
4	bNumEndpoints	1	Number	Number of endpoints used
5	bInterfaceClass	1	Class	Class code
6	bInterfaceSubClass	1	SubClass	Subclass code
7	bInterfaceProtocol	1	Protocol	Protocol code
8	iInterface	1	Index	Index of string descriptor

Table 15.4 Endpoint descriptors

Offset	Field	Size	Value	Description
0	bLength	1	Number	Size of descriptor (7 bytes)
1	bDescriptorType	1	Constant	Endpoint descriptor (0x05)
2	bEndpointAddress	1	Endpoint	b0..3 endpoint number b4..6 reserved. Set to zero b7 direction 0 = Out, 1 = In
3	bmAttributes	1	Bitmap	b0..1 transfer type 10 = Bulk b2..7 are reserved. I
4	wMaxPacketSize	2	Number	Maximum packet size
6	bInterval	1	Number	Interval for polling endpoint data

Table 15.5 String descriptors

Offset	Field	Size	Value	Description
0	bLength	1	Number	Size of descriptor (7 bytes)
1	bDescriptorType	1	Constant	Endpoint descriptor (0x05)
2	bEndpointAddress	1	Endpoint	b0..3 endpoint number b4..6 reserved. Set to zero b7 direction 0 = Out, 1 = In
3	bmAttributes	1	Bitmap	b0..1 transfer type 10 = Bulk b2..7 are reserved. I
4	wMaxPacketSize	2	Number	Maximum packet size
6	bInterval	1	Number	Interval for polling endpoint data

Each non-control endpoint is described by and endpoint descriptor. The descriptor provides the direction, type, and number of each endpoint. The format for endpoint descriptors is provided in Table 15.4.

String descriptors are used to allow devices to return Unicode text strings. String descriptor numbers (indexes) can be found in several other descriptors. String descriptor 0 is a special case that returns the languages the device supports. The most commonly supported language is 0x0409 US English regardless of country of origin. The format for string descriptors appears in Table 15.5.

15.5 Summary of USB Basics

At this point it might be helpful to reiterate all that happens when we connect our USB device. First some basic information is exchanged to determine what capabilities a device possesses. The device is then reset and it is further probed to determine whether or not it is capable of supporting high-speed communications. Using a control endpoint the host requests a series of descriptors in the following order: device, configuration(s), interface(s), endpoint(s), and optionally string(s). Finally, now that the host knows what type of device it is dealing with a device driver can be loaded and the device may be used.

While much has been covered in this section, we have merely scratched the surface. USB is a huge topic, but hopefully this introduction and pointers to more information contained within are sufficient to provide a basic understanding of the topic. In the next section of this chapter we will take an in-depth look at the workings of USB mass storage. From there we will discuss various forensic techniques and devices that can be developed when working with USB mass storage devices.

15.6 USB Mass Storage Basics

While hard drives and floppy discs store information magnetically on media which is rotated under a magnetic read/write head, flash drives use NAND flash memory chips. NAND flash memory is reasonably compact and straightforward to access. It does have some limitations which will be covered later.

A typical USB flash drive consists of a NAND flash chip, specialized microcontroller, and supporting electronics. The most common supporting electronics include power regulators (to drop the 5 Volts supplied by USB to 3.3 or 1.8 Volts used by the memory and microcontroller), oscillator crystals (12 MHz is the most common frequency), and, in some cases, status LEDs. A typical flash drive is consisted of a larger rectangular chip (NAND flash memory) and the smaller square chip is the microcontroller. Highly integrated drives in which everything is on a single chip (including the 4 contacts) that fits inside the USB connector are not uncommon.

NAND Flash Limitations

USB NAND flash memory has revolutionized data storage, but it is not without its limitations. A well known limitation of NAND flash storage is that it has limited write cycles. After as memory block has been overwritten too many times it becomes unreliable. A typical mean writes before failure is 10,000 cycles. High quality chips may have mean failures after 100,000 writes. While this may sound like a large number of writes, the way that writes are performed tends to result in more writes than are absolutely necessary.

Writes to NAND flash are performed one block at a time. Typical blocks sizes are 512, 2048, 4096, and 16,384 bytes. What this means is that in order to change a single bit the entire block must be read into a buffer, the bit changed, the block erased, and then the entire block re-written. Most controllers will implement some form of wear leveling, in which the memory blocks are dynamically mapped, in order to improve the life of a flash drive. This wear leveling may be performed on the flash chip itself in some cases. Even with wear leveling in place, if writes are requested one byte at a time by the operating system will result in premature failures.

Write speeds of flash drives vary widely. Generally speaking, larger and more expensive flash drives tend to have higher write speeds than their more affordable counterparts. Some of the cheaper devices have write speeds in the 1 MB/s range while 15 MB/s might be possible with high-end drives. Toshiba has recently developed a NAND flash chip they claim can sustain a write speed of 25 MB/s. Even the Toshiba chip is not capable of writes at even half of the USB 2.0 high-speed transfer rate of 480 Mbps (60 MB/s). Read speeds are generally higher, but still well under the maximum USB 2.0 rate.

The mechanism for recovering information from a damaged flash drive depends on the device's construction. NAND flash chips may be unsoldered from most devices with the exception of the completely integrated design described earlier. After chips have been removed they are most easily read by inserting them into a chip test socket which has been wired to the appropriate electronics (which may be another flash drive with the flash chip removed). Alternatively, chips may be soldered onto a functional flash drive which utilizes the same style chip. Some devices feature JTAG interfaces which permit data recovery without messy unsoldering and/or soldering. Data recovery from completely integrated flash drives may be impossible without specialized equipment.

As with hard drives, flash drives rarely store as much information as their size would indicate. Generally speaking, flash drives use a minimum of 1/32 of their capacity for error correction codes (ECC). For example, a 512 byte block typically consumes 528 bytes of memory (512 bytes data + 16 bytes ECC). Additionally, some controllers use a portion of the flash memory whereas others have internal flash storage. As a result, care should be taken when performing forensic copies from a flash drive to another flash drive of a different brand. Even in the case of identical flash drives, the target drive may not have sufficient capacity to complete the copy if some of the blocks have been marked as bad. Use of an oversized and/or brand new target drive might alleviate this difficulty in many situations.

Flash Drive Presentation

Nearly all flash drives present themselves asSCSI hard drives when connected to a computer. The sectors on these pseudo hard drives are typically 512, 2048, or 4096 bytes, with 512 bytes being the most common size in all but the largest drives. Often the devices support a reduced SCSI instruction set as the full set of SCSI commands doesn't make sense for a memory-based device.

Just as hard drives can be organized into partitions, flash drives may be partitioned using standard operating system tools. Hard drive partitions are referred to as logical units (LU). Most drives come preformatted as one LU. The partitions are assigned logical unit numbers (LUN) starting at zero. Some operating systems (in particular older versions of Windows) do not recognize LUNs above zero. As a result, data can be hidden in upper LUNs from users of outdated operating systems.

Options abound for filesystems to be used on flash drives. In addition to the normal filesystems used on hard drives, numerous filesystems that are optimized for flash memory are available. At least, they are available to Linux users, as in typical fashion, Windows users are left out. These specialized filesystems include the TrueFFS, ExtremeFFS, Journaling Flash File System (JFFS), and Yet Another FFS (YAFFS), among others. Most flash drives are preformatted as a single FAT or FAT32 partition. Drives formatted with NTFS are not uncommon with larger drives.

Talking to Mass Storage Devices

Most USB devices use control endpoints for all but data transfer. This is not the case with mass storage devices. Mass storage devices use control endpoints primarily during the initial enumeration process and then use bulk endpoints for all of the real work.

Communication over the bulk endpoints consists of three phases: command block wrapper (CBW), data transport (optional), and command status wrapper (CSW). Commands are sent in a in a command block (CB) which is wrapped inside the aptly named CBW. If the command requires the exchange of data, bulk endpoints are used to transfer data in the data transport phase. All commands are terminated by the device sending a CSW.

Command Block Wrappers

The format for a command block wrapper is presented in Listing 15.1. The CBW begins with the signature "USBC" or 0x43425355 in hexadecimal. The tag value is used to associate the CBW with the CSW. The data transfer length varies by command and is zero in some cases. The flags byte is all zeros with the exception of the high bit which is used to indicate transfer direction with 1 indicating in (out of the device and into the host) and 0 indicating out (into the device). The command block length will vary from 6-16 bytes (3 high bits will always be zero). Unused bytes in the command block are padded with zeros.

Listing 15.1 Command block wrapper structure

```
typedef struct _USB_MSI_CBW {
    unsigned long dCBWSignature; //0x43425355 "USBC"
    unsigned long dCBWTag; // associates CBW with CSW response
    unsigned long dCBWDataTransferLength; // bytes to send or receive
    unsigned char bCBWFlags; // bit 7 0=OUT, 1=IN all others zero
    unsigned char bCBWLUN; // logical unit number (usually zero)
    unsigned char bCBWCBLength; // 3 hi bits zero, rest bytes in CB
    unsigned char bCBWCB[16]; // the actual command block (>= 6 bytes)
} USB_MSI_CBW;
```

While the format for the command block varies with the command, the first byte is always the command. Some sample command blocks are presented in Listing 15.2 and Listing 15.3.

Listing 15.2 Format unit command block

```
typedef struct _CB_FORMAT_UNIT {
    unsigned char OperationCode; //must be 0x04
    unsigned char LUN:3; // logical unit number (usually zero)
    unsigned char FmtData:1; // if 1, extra parameters follow command
    unsigned char CmpLst:1; // if 0, partial list of defects, 1, complete
    unsigned char DefectListFormat:3; //000 = 32-bit LBAs
    unsigned char VendorSpecific; //vendor specific code
    unsigned short Interleave; //0x0000 = use vendor default
    unsigned char Control;
} CB_FORMAT_UNIT;
```

Listing 15.3 Read (10) command block

```
typedef struct _CB_READ10 {
    unsigned char OperationCode; //must be 0x28
    unsigned char RelativeAddress:1; // normally 0
    unsigned char Resv:2;
    unsigned char FUA:1; // 1=force unit access, don't use cache
    unsigned char DPO:1; // 1=disable page out
    unsigned char LUN:3; //logical unit number
    unsigned long LBA; //logical block address (sector number)
    unsigned char Reserved;
    unsigned short TransferLength;
    unsigned char Control;
} CB_READ10;
```

A set of common SCSI commands is presented in the code fragment in Listing 15.4. This code fragment is from code for a USB write blocker and a USB impersonator which will be covered later in this chapter.

Listing 15.4 Common SCSI commands

```
#define BOMS_FORMAT_UNIT 0x04 //definitely block!
#define BOMS_INQUIRY 0x12
#define BOMS_MODE_SELECT_6 0x15
#define BOMS_MODE_SELECT_10 0x55
#define BOMS_MODE_SENSE_6 0x1a
#define BOMS_MODE_SENSE_10 0x5a
#define BOMS_PREVENT_ALLOW_REMOVAL 0x1e
#define BOMS_READ_6 0x08
#define BOMS_READ_10 0x28
#define BOMS_READ_12 0xa8
#define BOMS_READ_CAPACITY 0x25
#define BOMS_READ_FORMAT_CAPACITIES 0x23
#define BOMS_READ_TOC_PMA_ATIP 0x43
#define BOMS_REPORT_LUNS 0xa0
#define BOMS_REQUEST_SENSE 0x03
#define BOMS_SEND_DIAGNOSTIC 0x1d
#define BOMS_START_STOP_UNIT 0x1b
#define BOMS_SYCHRONIZE_CACHE 0x35
#define BOMS_TEST_UNIT_READY 0x00
#define BOMS_VERIFY 0x2f
#define BOMS_WRITE_6 0x0a //block
#define BOMS_WRITE_10 0x2a //block
#define BOMS_WRITE_12 0xaa //block
```

Data Transport Phase

Some commands involve a data transport phase. In the case of read and write commands that transfer a considerable amount of data, data is automatically broken up into packets. The maximum packet size is determined by the connection speed. Bulk endpoints are not permitted to operate at low speed. The maximum permissible packet size for full-speed and high-speed connections is 64 and 512 bytes, respectively. One thing to keep in mind when working with USB mass storage devices is that while they are required to operate at full-speed rates, they may suffer from poor performance when connected at this speed. This is in part due to the extra overhead of sending 512 byte or larger blocks in 64 byte chunks. Additionally, the controller may be optimized for high-speed operations.

Command Status Wrapper

Every command terminates with a command status wrapper. The command status wrapper structure is presented in Listing 15.5. The CSW starts with the signature "USBS" or 0x53425355 in hexadecimal. Next comes the tag which is used to link the CSW to the correct CBW. The data residue indicates any data that remains to be transferred.

The status byte will be one of three values: 00, 01, or 02 for pass, fail, and phase error, respectively. A status of fail (01) does not indicate the nature of the failure aside from indicating it is not the result of a phase error. Upon receiving a failure status a read sense command should immediately be issued to discover the exact error that has occurred.

The host and drive alternate data phases (between DATA0 and DATA1) according to a set of rules. Sometimes the host and drive get out of sync and a reset status is sent to indicate they should both return to the starting phase (DATA0). This process is normally automatic and need not concern the forensic specialist.

Listing 15.5 Command status wrapper format

```
typedef struct _USB_MSI_CSW {
    unsigned long dCSWSignature; //0x53425355 or "USBS"
    unsigned long dCSWTag; // associate CBW with CSW response
    unsigned long dCSWDataResidue; // difference between requested data and actual
    unsigned char bCSWStatus; //00=pass, 01=fail, 02=phase error, reset
} USB_MSI_CSW;
```

Summary of USB Mass Storage Basics

At this point it might be helpful to reiterate all that happens when we connect our USB flash drive. First the standard enumeration process is performed as it is for all USB devices. Once the device has been identified as a USB mass storage device an in bulk endpoint and out bulk endpoint are created. The host sends commands to the drive by embedding them in CBWs. If data needs to be exchanged, packets are sent in a data transport phase. Once the command has been processed, a CSW is sent from the drive to the host to terminate the transaction. The drive then waits for further commands from the host.

While much has been covered in this chapter thus far, we have merely scratched the surface. USB mass storage devices are a huge topic, but hopefully this introduction and pointers to more information contained within are sufficient to provide a basic understanding of the topic. For a more detailed coverage of this topic consult *USB Mass Storage: Designing and programming devices and embedded hosts* by Jan Axelson. In the next section of this chapter we will begin our journey into building devices to be when performing USB mass storage device forensics.

15.7 Making Forensics Images and Duplicates

When making images or forensic duplicates there are a number of options. One straightforward method is to connect a source device to a personal computer (hopefully through some sort of write blocker). While this is certainly possible, an appropriate workstation and write blocker might not always be on hand. Additionally, commercial USB write blockers tend to be expensive. The next section in this chapter will describe an inexpensive write blocker based on the FTDI microcontroller.

In this section we will describe another option also based on the FTDI microcontroller. Some readers might recognize FTDI as the company which produces USB-related chips such as those found in older versions of the Arduino and many USB to serial cables. Despite being introduced several years ago, the FTDI Vinculum II (VNC2) remains one of the few microcontrollers on the market which is capable of operating as either a slave (device) or host. After a brief introduction to the VNC2, details for a pocket-sized USB mass storage device duplicator will be provided.

FTDI has recently announced a new updated microcontroller, the FT900, with some advanced capabilities such as support for USB high-speed (480 Mbps). The FT900 has not yet begun shipping as of this writing. It is likely that the techniques presented here could be applied to this new chip.

Meet the FTDI Vinculum II (VNC2)

The FTDI Vinculum II dual USB host controller has many nice features including:

1. Embedded 16-bit Harvard architecture MCU core, with 256KBytes of Flash memory and 16Kbytes RAM.
2. 2 x Full-Speed /Low-speed USB 2.0 ports supporting Host or Slave operation.
3. Programmable UART interface, supports up to 6MBaud transfers.
4. 8-bit wide FIFO interface.
5. 2 x SPI slave interfaces, 1 x SPI master interface.
6. PWM (Pulse Width Modulation) interface.
7. 4 channel DMA controller, and general purpose timers.
8. Support for reduced power modes.

9. Multiple packages size options (32-/48-/64-pin QFN and LQFP packages).
10. Backwards compatible with VNC1L with 48-pin LQFP package.
11. RoHS compliant, and extended temperature support ($-40°C$ to $+85°C$).
12. Enhanced features with Vinculum Software Tool Suite.
13. Based on royalty-free flexible 'C' based Integrated Development Environment.
14. Includes compiler, drivers and RTOS kernel to support user firmware development.
15. Debug interface for real-time hardware based code debug.
16. Pre-compiled libraries for several USB classes including FAT file system support.

The VNC2 is available in 32, 48, and 64 pin versions. Unlike the AVR line of microcontrollers as found in the Arduinos, the only difference between chips in the VNC2 family is the number of general purpose input/output (GPIO) lines available. This makes it possible to easily scale solutions up and down without any code changes. Unfortunately, the VNC2 is only available in surface mount (SMD) versions which can make prototyping difficult. However, several relatively inexpensive development modules, including the Arduino-style Vinco, are available.

The VNC2 is a full-featured microcontroller as can be seen from the block diagram in Fig. 15.1. In addition to providing unrivaled USB functionality, the inclusion of pulse width modulation (PWM), inter-integrated circuit (I2C), serial, GPIO, and serial peripheral interface (SPI) support ensures that the VNC2 can be interfaced with virtually any device. The VNC2 may be operated at clock speeds of up to 48 MHz (over twice the typical Arduino speeds). Additionally, the VNC2

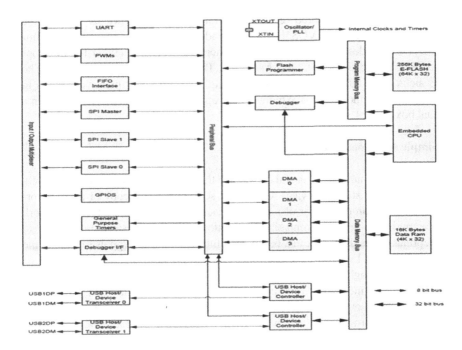

Fig. 15.1 Vinculum II block diagram

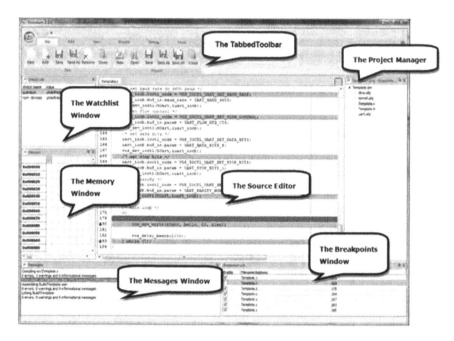

Fig. 15.2 FTDI integrated development environment

normally runs a full-featured real-time operating system known as VOS. Some of the nicer features of VOS include: threads, semaphores, mutexes, and timers. FTDI provides an integrated development environment (IDE) for developing VNC2 applications in C and Assembly. The IDE is shown in Fig. 15.2. The IDE includes a GUI facility for programming the GPIO multiplexer as shown in Fig. 15.3. While the IDE is a Windows-only application, it runs perfectly well in a virtual box under Linux.

A Simple and Compact Duplicator

FTDI provides a number of useful drivers for the VNC2 platform, including drivers for reading mass storage devices. Creating a sector-by-sector copy of a mass storage device is a simple matter of reading in each sector using the FTDI framework and outputing to another compatible mass storage device or storing an image to appropriate media. Full code can be found at https://github.com/ppolstra. Due to the total code length, I will only provide representative snippets of code in this chapter.

The most straightforward way to read sectors (which are typically 512 bytes) is shown in Listing 15.6. Writing to a target mass storage device is very similar. Recall from earlier in this chapter that data is read and written in blocks. Maximum performance is achieved when the amount of data read and more importantly written is a multiple of the devices read and write block sizes. An improved method for reading multiple sectors is presented later in this chapter.

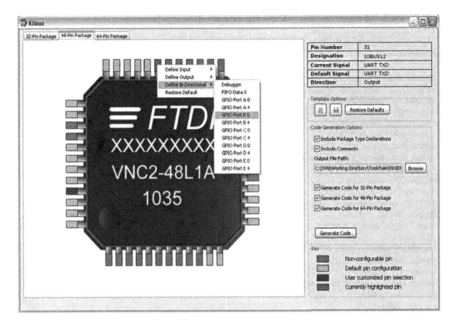

Fig. 15.3 Programming the VNC2 multiplexer

Listing 15.6 Reading sectors

```
unsigned char FatReadSector(unsigned long sec, unsigned char *buffer, unsigned short sl)
{
    // transfer buffer
    static msi_xfer_cb_t xfer;
    unsigned char stat;
    xfer.sector = sec;
    xfer.buf = buffer;
    xfer.total_len = sl;
    xfer.buf_len = sl;
    xfer.status = MSI_NOT_ACCESSED;
    xfer.s = &semRead;
    xfer.do_phases = MSI_PHASE_ALL;
    stat = vos_dev_read(hBOMS_2, (unsigned char *)&xfer, sizeof(msi_xfer_cb_t ), NULL);
    if (stat == MSI_OK)
    {
            stat = FAT_OK;
    }
    else
    {
            stat |= FAT_MSI_ERROR;
    }
    return stat;
}
```

The main portion of the main processing loop for a duplicator that reads one mass
storage device sector by sector and creates a duplicate on a target drive that is at
least as large as the source is presented in Listing 15.7.

Listing 15.7 Main processing loop for simple duplicator

```
void firmware(void)
{
    unsigned short sectorSize2, sectorSize1;
    unsigned char led[5]={0x08, 0x10, 0x20, 0x40, 0x80};
    unsigned char connectstate;
    unsigned char status;
    unsigned short ledStep;
    usbhost_device_handle *ifDev;
    usbhost_ioctl_cb_t hc_iocb, be_iocb;
    usbhost_ioctl_cb_class_t hc_iocb_class;
    // BOMS device variables
    msi_ioctl_cb_t boms_iocb;
    boms_ioctl_cb_attach_t boms_att;
    // FAT file system variables
    fat_ioctl_cb_t fat_ioctl;
    fatdrv_ioctl_cb_attach_t fat_att;
    FILE *file;
    msi_xfer_cb_t xfer;
    usbhost_xfer_t uxfer;
    // completion semaphore
    vos_semaphore_t semRead, semWrite;
    unsigned char *pBuffer;
    unsigned long sector=0;
    unsigned short clusterSize;
    short allDone=0;
    int i; //counting variable

    // open host controller
    hUSBHOST_2 = vos_dev_open(VOS_DEV_USBHOST_2);

    // buffer for reading and writing sectors
    pBuffer = malloc(BUFFER_SIZE);
    do
    {
            // use ioctl to see if bus available
            hc_iocb.ioctl_code = VOS_IOCTL_USBHOST_GET_CONNECT_STATE;
            hc_iocb.get = &connectstate;
            vos_dev_ioctl(hUSBHOST_2, &hc_iocb);
        if (connectstate == PORT_STATE_ENUMERATED)
            {
                    // find and connect a BOMS device
                    // USBHost ioctl to find first BOMS device on host
                    hc_iocb.ioctl_code = VOS_IOCTL_USBHOST_DEVICE_FIND_HANDLE_BY_CLASS;
                    hc_iocb.handle.dif = NULL;
                    hc_iocb.set = &hc_iocb_class;
                    hc_iocb.get = &ifDev;
                    hc_iocb_class.dev_class = USB_CLASS_MASS_STORAGE;
                    hc_iocb_class.dev_subclass = USB_SUBCLASS_MASS_STORAGE_SCSI;
                    hc_iocb_class.dev_protocol = USB_PROTOCOL_MASS_STORAGE_BOMS;

                    if (vos_dev_ioctl(hUSBHOST_2, &hc_iocb) != USBHOST_OK)
                    {
                            break;
                    }

                    // now we have a device, intialise a BOMS driver for it
                    hBOMS_2 = vos_dev_open(VOS_DEV_BOMS_2);
```

```
// BOMS ioctl to attach BOMS driver to device on host
boms_iocb.ioctl_code = MSI_IOCTL_BOMS_ATTACH;
boms_iocb.set = &boms_att;
boms_iocb.get = NULL;
boms_att.hc_handle = hUSBHOST_2;
boms_att.ifDev = ifDev;

status = vos_dev_ioctl(hBOMS_2, &boms_iocb);
if (status != MSI_OK)
{
        break;
}


boms_iocb.ioctl_code = MSI_IOCTL_GET_SECTOR_SIZE;
boms_iocb.get = &sectorSize2;
vos_dev_ioctl(hBOMS_2,&boms_iocb);

// now connect to the drive to be written to in port 1
hUSBHOST_1 = vos_dev_open(VOS_DEV_USBHOST_1);
// use ioctl to see if bus available
hc_iocb.ioctl_code = VOS_IOCTL_USBHOST_GET_CONNECT_STATE;
hc_iocb.get = &connectstate;

do
{
        vos_dev_ioctl(hUSBHOST_1, &hc_iocb);
        vos_delay_msecs(250);
} while (connectstate!=PORT_STATE_ENUMERATED);

// find and connect a BOMS device
// USBHost ioctl to find first BOMS device on host
hc_iocb.ioctl_code = VOS_IOCTL_USBHOST_DEVICE_FIND_HANDLE_BY_CLASS;
hc_iocb.handle.dif = NULL;
hc_iocb.set = &hc_iocb_class;
hc_iocb.get = &ifDev;
hc_iocb_class.dev_class = USB_CLASS_MASS_STORAGE;
hc_iocb_class.dev_subclass = USB_SUBCLASS_MASS_STORAGE_SCSI;
hc_iocb_class.dev_protocol = USB_PROTOCOL_MASS_STORAGE_BOMS;

if (vos_dev_ioctl(hUSBHOST_1, &hc_iocb) != USBHOST_OK)
{
        break;
}

// now we have a device, intialise a BOMS driver for it
hBOMS_1 = vos_dev_open(VOS_DEV_BOMS_1);

// BOMS ioctl to attach BOMS driver to device on host
boms_iocb.ioctl_code = MSI_IOCTL_BOMS_ATTACH;
boms_iocb.set = &boms_att;
boms_iocb.get = NULL;
boms_att.hc_handle = hUSBHOST_1;
boms_att.ifDev = ifDev;

status = vos_dev_ioctl(hBOMS_1, &boms_iocb);
if (status != MSI_OK)
{
        break;
}
```

```
boms_iocb.ioctl_code = MSI_IOCTL_GET_SECTOR_SIZE;
boms_iocb.get = &sectorSize1;
vos_dev_ioctl(hBOMS_1,&boms_iocb);
clusterSize = BUFFER_SIZE/sectorSize1;

// time to copy
vos_init_semaphore(&semRead, 0);
vos_init_semaphore(&semWrite, 0);
xfer.sector = sector;
xfer.buf = pBuffer;
xfer.total_len = BUFFER_SIZE;
xfer.buf_len = BUFFER_SIZE;
xfer.status = MSI_NOT_ACCESSED;
xfer.s = &semRead;
xfer.do_phases = MSI_PHASE_ALL;
do
{
                status = FatReadSector(sector, pBuffer, BUFFER_SIZE);
                if (status == FAT_OK)
                {
                        status = FatWriteSector(sector, pBuffer, BUFFER_SIZE);
                } else
                {
                        allDone = 1;
                        break;
                }
                sector+=clusterSize;
        vos_dev_write(hGPIO_PORT_A, &led[ledStep%5],1,NULL);
} while((status == FAT_OK)&& !allDone);

// TO DO: use SCSI command 0x25 to find drive size instead of going till error
allDone=1;

boms_iocb.ioctl_code = MSI_IOCTL_BOMS_DETACH;
if (vos_dev_ioctl(hBOMS_2, &boms_iocb) != MSI_OK)
{
        break;
}
vos_dev_close(hBOMS_2);

boms_iocb.ioctl_code = MSI_IOCTL_BOMS_DETACH;
if (vos_dev_ioctl(hBOMS_1, &boms_iocb) != MSI_OK)
{
        break;
}
vos_dev_close(hBOMS_1);
vos_power_down(VOS_WAKE_ON_USB_1); // go to sleep till next time
        }
    } while (!allDone);
}
```

So far I have presented the majority of the code required for a simple duplicator. There is an issue, however. The user has no clue how to operate the device and is not provided with any status information. The simplest situation would be to use a set of LEDs to indicate status. A more user-friendly and slight less compact device would use an LCD screen in addition to the LEDs.

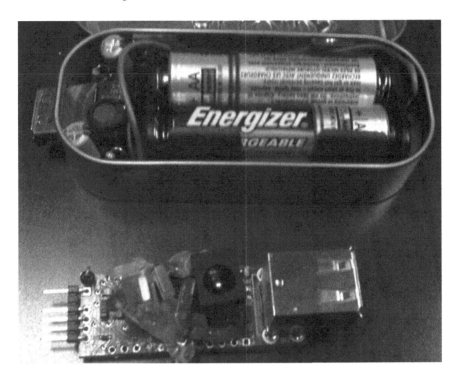

Fig. 15.4   An extremely compact USB duplicator

A very simple duplicator based on the 32-pin development board with 2 USB host ports is shown in Fig. 15.4 along side its Minty Boost USB power supply. I would challenge readers to find a smaller forensic duplicator. This device blinks to indicate waiting for a target drive to be inserted. While it is copying the lights strobe, a single LED is illuminated to indicate completion. Full source code and schematics for this device are available at https://github.com/ppolstra.

Users wanting a little more user-friendly device could add an LCD screen. This requires the use of a 48 or 64-pin VNC2 chip assuming that the status LEDs will still be utilized. Thanks to the design of the VNC2 the exact same code can be run on both devices. Reads and writes to nonexistent GPIO lines are ignored on the smaller device. This duplicator is shown alongside its carrying case in Fig. 15.5.

At this stage we now have some devices that will fit in our pocket and provide use the ability to duplicate flash drives if we find ourselves needing to do so without our complete forensics toolkits. Recall from previous articles in this series that NAND flash write speeds are slow and that changing a single byte requires changing an entire block. To greatly speed up this process we can make use of threads and buffering. The two main methods for a multithreaded and double-buffered duplicator are presented in Listing 15.8. Note that one thread is used for reading and the second for writing.

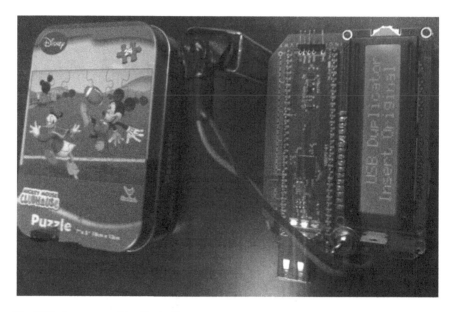

Fig. 15.5 A more user-friendly duplicator

Listing 15.8 Main threads for a double-buffered duplicator

```
void thread_1()
{
    unsigned short i;
    unsigned char status;
    sector1 = 0;
    sector2 = clusterSize;

    while (!enumed2)
    {
        vos_delay_msecs(1000);
    }

    do
    {
        // this funny for loop is to speed up processing
        // by eliminating as much as possible from a tight loop
        // while still providing status through LEDs
        for(i=0;i<500;i++)
        {
            vos_wait_semaphore(&semBuf1);
            vos_lock_mutex(&mBuf1);
            status = FatReadSector(sector1, pBuf1, BUFFER_SIZE);
            vos_unlock_mutex(&mBuf1);
            if (status == FAT_OK)
            {
                vos_wait_semaphore(&semBuf2);
                vos_lock_mutex(&mBuf2);
                status = FatReadSector(sector2, pBuf2, BUFFER_SIZE);
                vos_unlock_mutex(&mBuf2);
            } else
            {
```

```
                                allDone = 1;
                                break;
                        }
                        sector1 += 2*clusterSize;
                        sector2 += 2*clusterSize;
                }
                ledStep++;
                vos_dev_write(hGPIO_PORT_A, &led[ledStep%5],1,NULL);
        } while(!allDone);
}

void thread_2()
{
        unsigned char status;

        while(!enumed1)
                vos_delay_msecs(1000);


        do
        {
                vos_lock_mutex(&mBuf1);
                status = FatWriteSector(sector1, pBuf1, BUFFER_SIZE);
                vos_unlock_mutex(&mBuf1);
                vos_signal_semaphore(&semBuf1);
                if (status == FAT_OK)
                {
                        vos_lock_mutex(&mBuf2);
                        status = FatWriteSector(sector2, pBuf2, BUFFER_SIZE);
                        vos_unlock_mutex(&mBuf2);
                        vos_signal_semaphore(&semBuf2);
                } else
                {
                        allDone = 1;
                        break;
                }
        } while(!allDone);
}
```

Summary of USB Mass Storage Duplication

There are several options when it comes to making forensic duplicates of USB
mass storage devices. This section presented a couple of compact duplicators
which utilize the FTDI VNC2 microcontroller. We also discussed simple ways to
improve performance of these duplicators. Code and schematics for the devices
presented are available online at https://github.com/ppolstra. While not explicitly
presented, modifying these duplicators to output image files is straightforward.

In the next section we will again make use of the FTDI VNC2. In that case
we will discuss building an inexpensive and compact USB write blocker. This will
require the use of both a USB host and USB slave.

15.8 Blocking USB Writes

Motivation for USB Write Blocking

Has this ever happened to you? A friend asks you for help with a troublesome PC. You insert a flash drive with your security and diagnostics tools into their computer. Their anti-virus instantly detects your security tools as malware and begins deleting them. You scramble to yank out the drive, but it is too late. The drive will need to be reloaded. Wouldn't it be great if all this trouble could be easily and cheaply avoided?

Perhaps you need to examine a possibly interesting USB mass storage device, but you don't have an expensive commercial USB write blocker handy. I have spoken to cybercrime units that could only afford one shared USB write blocker for the entire unit. Additionally, if a flash drive has been identified as interesting, you will need to make additional working copies and having multiple cheap write blockers could avoid the bottleneck inherent in having to share a single blocker.

Introduction to USB Write Blocking

There are a number of ways to block write operations to USB mass storage devices. Some older flash drives featured at write-protect switch. This feature is quite rare in modern devices. On certain versions of Windows USB write operations can be blocked by creating a registry key HKEY_LOCAL_MACHINE\ SYSTEM\CurrentControlSet\Control\StorageDevicePolicies\WriteProtect. Creating this registry key will block write operations to **all** USB mass storage devices, however.

There are several commercial solutions that can be used to block writes. Software can be used to restrict which devices (vendor ID, product ID, and possibly serial number) can be opened for writing. A number of hardware write blockers costing hundreds of dollars are also available. Alternatively, the device described here can be constructed for less than US$20.

Implementing a Write Blocker

The FTDI Vinculum II (VNC2) is a convenient and inexpensive microcontroller for building a write blocker. While a custom device could be created, the fact that the VNC2 chips are only available in surface-mount formats makes this somewhat inconvenient. Fortunately, FTDI provides development boards which can be used for this device. The best choice for this project is a V2DIP1-32 which has one USB host port and is based on the 32-pin VNC2 chip. The V2DIP1-32 is shown in Fig. 15.6. This user will need to solder on a USB cable for the USB slave port. This is easily done by cutting the device end off of an unwanted USB cable. The wires should be color coded as red, white, green, and black for +5V, data−, data+, and ground, respectively. The corresponding pins on the V2DIP1-32 are labeled 5V0, GND, U1P, and U1M for +5V, ground, data+, and data−, respectively. The user may optionally wish to either trim or desolder the header pins in order to make the device smaller and not as painful to carry in a pocket.

Alternatively, a device could be created using FTDI's Arduino-like board known as the Vinco. This has the advantage of not requiring the user to do any soldering.

Fig. 15.6 V2DIP1-32

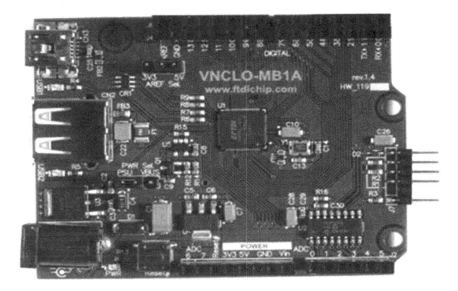

Fig. 15.7 FTDI Vinco

It should be noted that there have been some reported issues when using the Vinco. In particular, the way that power is applied to the USB host port in software can lead to timing issues which may prevent a flash drive from enumerating properly in some cases. The Vinco is shown in Fig. 15.7. Because all the VNC2 chips have the same memory and flash storage, the same code is easily run on both devices.

High Level Design

The device needs to intercept and block any commands that could potentially alter the USB drive. A lazy way to implement this would be to block black-listed commands. This is not the proper way to implement this, however. Rather, the proper way to implement and future-proof the device is to whitelist benign commands. The original design would return "unsupported command" for commands to be blocked. The code was redesigned to fake completing a command in order to prevent infinite loops on Windows systems which stupidly would retry commands that were reported as unsupported.

The application code consists of two main threads. One thread is used to enumerate and communicate with an attached USB mass storage device. The other thread presents the device as a USB drive to an attached host and decides which commands to forward to the actual drive and which to emulate. It should be noted that the underlying FTDI library code may result in the creation of additional threads.

The code for the host enumeration thread is shown in Listing 15.9. The thread does not appear to do much based on this short thread function. The majority of the work is performed by the FTDI libraries from the calls contained within the open_drivers and attach_drivers methods. Global handles to various devices are initialized in these function calls. These handles are used in order to forward commands to the USB drive from the other thread.

Listing 15.9 Host enumeration thread methods

```
unsigned char usbhost_connect_state(VOS_HANDLE hUSB)
{
    unsigned char connectstate = PORT_STATE_DISCONNECTED;
    usbhost_ioctl_cb_t hc_iocb;

    if (hUSB)
    {
        hc_iocb.ioctl_code = VOS_IOCTL_USBHOST_GET_CONNECT_STATE;
        hc_iocb.get = &connectstate;
        vos_dev_ioctl(hUSB, &hc_iocb);
    }

    return connectstate;
}

void open_drivers(void)
{
    gpio_ioctl_cb_t gpio_iocb;
    unsigned char leds;

    /* Code for opening and closing drivers - move to required places in Application Threads */
    /* FTDI:SDA Driver Open */
    hGPIO_PORT_E = vos_dev_open(VOS_DEV_GPIO_PORT_E);

    // power up Vinco USB Host
    // this must happen before we want to enumerate the flash drive
    gpio_iocb.ioctl_code = VOS_IOCTL_GPIO_SET_MASK;
    gpio_iocb.value = 0x60;          // set power and LED as output
    vos_dev_ioctl(hGPIO_PORT_E, &gpio_iocb);
    leds = 0x00;
    vos_dev_write(hGPIO_PORT_E, &leds, 1, NULL);

    hUSBHOST_2 = vos_dev_open(VOS_DEV_USBHOST_2);

    /* FTDI:EDA */
```

```
    hUSBHOSTBOMS = vos_dev_open(VOS_DEV_USBHOSTBOMS);

    hUSBSLAVE_1 = vos_dev_open(VOS_DEV_USBSLAVE_1);

    hUSBSLAVEBOMS = vos_dev_open(VOS_DEV_USBSLAVEBOMS);

}

void attach_drivers(void)
{
    common_ioctl_cb_t bomsAttach;

    /* FTDI:SUA Layered Driver Attach Function Calls */
    /* FTDI:EUA */

    // attach BOMS to USB Host port B
    bomsAttach.ioctl_code = VOS_IOCTL_USBHOSTBOMS_ATTACH;
    bomsAttach.set.data = (void *) hUSBHOST_2;
    vos_dev_ioctl(hUSBHOSTBOMS, &bomsAttach);



    // attach BOMS to USB Slave port A
    bomsAttach.ioctl_code = VOS_IOCTL_USBSLAVEBOMS_ATTACH;
    bomsAttach.set.data = (void *) hUSBSLAVE_1;
    vos_dev_ioctl(hUSBSLAVEBOMS, &bomsAttach);

}

void close_drivers(void)
{
    vos_dev_close(hUSBHOST_2);
    vos_dev_close(hUSBHOSTBOMS);
    vos_dev_close(hUSBSLAVE_1);
    vos_dev_close(hUSBSLAVEBOMS);
    vos_dev_close(hGPIO_PORT_E);
}

void hostEnum()
{
    unsigned char i;
    unsigned char status;
    unsigned char buf[64];
    unsigned short num_read;
    unsigned int handle;

    usbhostBoms_ioctl_t generic_iocb;
    usbhost_device_handle_ex ifDev;
    usbhost_ioctl_cb_t hc_iocb;
    usbhost_ioctl_cb_class_t hc_iocb_class;
    usbhostBoms_ioctl_cb_attach_t genericAtt;

    open_drivers(); // open all drivers including USB host and slave
    attach_drivers(); // enumerate flash drive then connect slave

    do
    {
            // see if bus available
            if (usbhost_connect_state(hUSBHOST_2) == PORT_STATE_ENUMERATED)
```

```
{
                // ultimately want to find a mass storage device using SCSI protocol
                hc_iocb_class.dev_class = USB_CLASS_MASS_STORAGE;
                hc_iocb_class.dev_subclass = USB_SUBCLASS_MASS_STORAGE_SCSI;
                hc_iocb_class.dev_protocol = USB_PROTOCOL_MASS_STORAGE_BOMS;

                // user ioctl to find first hub device
                hc_iocb.ioctl_code =

                        VOS_IOCTL_USBHOST_DEVICE_FIND_HANDLE_BY_CLASS;
                hc_iocb.handle.dif = NULL;
                hc_iocb.set = &hc_iocb_class;
                hc_iocb.get = &ifDev;

                if (vos_dev_ioctl(hUSBHOST_2, &hc_iocb) == USBHOST_OK)
                {
                        // optionally notify user of status via LEDs, etc.
                }

                genericAtt.hc_handle = hUSBHOST_2;
                genericAtt.ifDev = ifDev;
                generic_iocb.ioctl_code = VOS_IOCTL_USBHOSTBOMS_ATTACH;
                generic_iocb.set.att = &genericAtt;

                // we use a simple variable to indicate if the flash drive
                // is attached
                // this is not as elegent as using a semaphore, but
                // this is the only thread that updates this variable and
                // if the device is disconnected and reconnected that is hard
                // to handle with a semaphore
                if (vos_dev_ioctl(hUSBHOSTBOMS, &generic_iocb) == USBHOSTBOMS_OK)
                {
                        slaveBomsCtx->flashConnected = 1;
                        vos_signal_semaphore(&slaveBomsCtx->enumed);
                } else
                {
                        slaveBomsCtx->flashConnected = 0;
                }// if attach
                // this code is in here so that if the drive gets disconnected
                // we can try to restart it
                // also, hopefully the the traffice every few seconds will keep
                // the drive from going to sleep
                vos_delay_msecs(2000);
        } // if enumerated
                vos_delay_msecs(10); // recheck every .01 seconds for new connect
    } // outer do
    while (1);

}
```

The thread method for the second thread is presented in Listing 15.10. This thread is little more than a large switch statement which calls various handler functions based on the received command. In the initial version of the write blocker only a couple of generalized handlers were used. This proved problematic, however, so these were replaced with individual handler methods for each command.

Each whitelisted command handler follows the general format of forwarding the command block wrapper (CBW) to the drive, receiving data from the drive, sending the data to the USB slave, receiving a command status wrapper (CSW) from the drive, and forwarding the CSW to the USB slave. Illegal command handlers are similar, but without the communication with the drive faked. The handler methods are provided in Listing 15.11.

Listing 15.10 Command block wrapper handling thread

```
void handleCbw()
{
    unsigned short num_read, num_written;
    boms_cbw_t *cbw = vos_malloc(sizeof(boms_cbw_t));

    vos_wait_semaphore(&slaveBomsCtx->enumed);
    vos_signal_semaphore(&slaveBomsCtx->enumed);

    while(1)
    {
        if(slaveBomsCtx)
        {
            while(slaveBomsCtx && slaveBomsCtx->flashConnected)
            {
            // get the CBW
            memset(cbw, 0, sizeof(boms_cbw_t));
            usbSlaveBoms_readCbw(cbw, slaveBomsCtx);
            // TO DO: Check for valid CBW

            switch (cbw->cb.formated.command)
            {
                case BOMS_INQUIRY:
                    handle_inquiry(cbw);
                    break;
                case BOMS_MODE_SELECT_6:
                case BOMS_MODE_SELECT_10:
                    handle_mode_select(cbw);
                    break;
                case BOMS_MODE_SENSE_6:
                case BOMS_MODE_SENSE_10:
                    handle_mode_sense(cbw);
                    break;
                case BOMS_READ_6:
                case BOMS_READ_10:
                case BOMS_READ_12:
                    handle_read(cbw);
                    break;
                case BOMS_READ_CAPACITY:
                    handle_read_capacity(cbw);
                    break;
                case BOMS_REPORT_LUNS:
                    handle_report_luns(cbw);
                    break;
                case BOMS_REQUEST_SENSE:
                    handle_request_sense(cbw);
                    break;
```

```
                                        case BOMS_TEST_UNIT_READY:
                                                handle_test_unit_ready(cbw);
                                                break;
                                        case BOMS_SEND_DIAGNOSTIC:
                                                handle_send_diagnostic(cbw);
                                                break;
                                        case BOMS_START_STOP_UNIT:
                                                handle_start_stop_unit(cbw);
                                                break;
                                        case BOMS_SYCHRONIZE_CACHE:
                                                handle_synchronize_cache(cbw);
                                                break;
                                        case BOMS_READ_FORMAT_CAPACITIES:
                                                handle_read_format_capacities(cbw);
                                                break;
                                        case BOMS_PREVENT_ALLOW_REMOVAL:
                                                handle_prevent_allow_removal(cbw);
                                                break;
                                        case BOMS_READ_TOC_PMA_ATIP:
                                                handle_read_toc_pma_atip(cbw);
                                                break;
                                        case BOMS_VERIFY:
                                        case BOMS_FORMAT_UNIT:
                                                // tell them NO! by failing command
                                                handle_illegal_request(cbw);
                                                break;
                                        case BOMS_WRITE_6:
                                        case BOMS_WRITE_10:
                                        case BOMS_WRITE_12:
                                                handle_illegal_write_request(cbw);
                                                break;
                                        default:
                                                handle_illegal_request(cbw);
                                                break;
                                } // switch
                        } // inner while
                } else
                {
                        vos_delay_msecs(1000);
                }
        } // outer while

        vos_free(cbw);
}
```

Listing 15.11   Handler methods

```
unsigned short forward_cbw_to_device(boms_cbw_t *cbw)
{
    unsigned short num_written;
    usbhostBoms_write((void*)cbw, sizeof(boms_cbw_t), &num_written, hostBomsCtx);

    return num_written;
}

unsigned short receive_data_from_device(void* buffer, unsigned short expected)
{
    unsigned short num_read;
    unsigned char status;

    status = usbhostBoms_read(buffer, expected, &num_read, hostBomsCtx);
    if (status == USBHOST_EP_HALTED)
    {
            // the endpoint is halted so let's halt the slave endpoint
            usbslaveboms_stall_bulk_in(slaveBomsCtx);
    }

    return num_read;
}

unsigned short forward_data_to_slave(void* buffer, unsigned short bytes)
{
    unsigned short num_written;
    usbSlaveBoms_write(buffer, bytes, &num_written, slaveBomsCtx);

    return num_written;
}

unsigned short forward_data_to_slave_then_stall(void* buffer, unsigned short bytes)
{
    unsigned short num_written;
    usbSlaveBoms_short_write(buffer, bytes, &num_written, slaveBomsCtx);

    return num_written;
}

unsigned short receive_csw_from_device(boms_csw_t *csw)
{
    unsigned short num_read;
    usbhostBoms_read((void*)csw, 13, &num_read, hostBomsCtx);

    return num_read;
}

unsigned short forward_csw_to_slave(boms_csw_t *csw)
{
    unsigned short num_written;
    usbSlaveBoms_write((void*)csw, 13, &num_written, slaveBomsCtx);

    return num_written;
}
```

```
void handle_inquiry(boms_cbw_t *cbw)
{
    unsigned char buffer[64];
    unsigned short responseSize;
    boms_csw_t csw;

    // forward the CBW to device
    if (forward_cbw_to_device(cbw))
    {
            // receive response from device
            // note we will assume that only the standard 36 bytes will be requested
            if (responseSize = receive_data_from_device(&buffer[0], 36))
            {
                    // forward response to slave
                    forward_data_to_slave(&buffer[0], responseSize);

                    // receive CSW from device
                    if (receive_csw_from_device(&csw))
                    {
                            // forward CSW to slave
                            forward_csw_to_slave(&csw);
                    }
            }
    }
}

void handle_test_unit_ready(boms_cbw_t *cbw)
{
    boms_csw_t csw;

    // forward the CBW to device
    if (forward_cbw_to_device(cbw))
    {
            // receive response from device
            if (receive_csw_from_device(&csw))
            {
                    // forward CSW to slave
                    forward_csw_to_slave(&csw);
            }
    }
}

void handle_read(boms_cbw_t *cbw)
{
    // this same routine handles all 3 possible read commands
    // most likely read command is read(10)

    unsigned long lba; // logical block address for start block
    unsigned short blocks; // number of blocks to read
    unsigned short i;

    boms_csw_t csw;
    unsigned char *buffer;
    unsigned short num_read;
    unsigned short num_written;
```

```
        switch (cbw->cb.formated.command)
        {
                case BOMS_READ_6:
                        lba = cbw->cb.raw[1]*65536 + cbw->cb.raw[2]*256 + cbw->cb.raw[3];
                        blocks = cbw->cb.raw[4];
                        break;
                case BOMS_READ_10:
                        lba = cbw->cb.raw[2]*16777216 + cbw->cb.raw[3]*65536 + cbw->cb.raw[4]*256 +cbw-
>cb.raw[5];
                        blocks = cbw->cb.raw[7] * 256 + cbw->cb.raw[8];
                        break;
                case BOMS_READ_12:
                        lba = cbw->cb.raw[2]*16777216 + cbw->cb.raw[3]*65536 + cbw->cb.raw[4]*256 +cbw-
>cb.raw[5];
                        // we are being a little bad here the number of blocks is actually a long
                        // it is extremely unlikely that anyone would request this much at once, however
                        blocks = cbw->cb.raw[8] * 256 + cbw->cb.raw[9];
                        break;
        }

        // now forward the cbw to the device
        forward_cbw_to_device(cbw);

        // receive the appropriate number of blocks from the device
        // forward the blocks to the slave
        // most requests are probably 1 block of 512 bytes
        // read in 512 byte chunks (packet size is 64 bytes, but VOS should handle this)
        // If devices with larger blocks are encountered, 512 should still work
        buffer = vos_malloc(blockSize);
        while(blocks>0)
        {
                usbhostBoms_read((void*)buffer, blockSize, &num_read, hostBomsCtx);
                usbSlaveBoms_write((void*)buffer, num_read, &num_written, slaveBomsCtx);
                blocks--;
        }
        vos_free(buffer);

        // receive the csw from the device
        receive_csw_from_device(&csw);

        // forward the csw to the slave
        forward_csw_to_slave(&csw);
}

void handle_read_capacity(boms_cbw_t *cbw)
{
        boms_csw_t csw;
        unsigned char buffer[8];
        unsigned short received;

        // forward cbw to device
        forward_cbw_to_device(cbw);

        // receive response from device
        if (received = receive_data_from_device(&buffer[0], 8))
        {
```

```c
        rsr.formated.responseCode =0x70; //0x70 current error
        rsr.formated.valid = 0; // 1=INFORMATION field valid
        rsr.formated.obsolete = 0;
        rsr.formated.senseKey = 0x05; // 0x05 for illegal request
        rsr.formated.resvered = 0;
        rsr.formated.ili = 0; // incorrect length indicator
        rsr.formated.eom = 0; // end of media for streaming devices
        rsr.formated.filemark = 0; // for streaming devices
        rsr.formated.information = 0; // device specific info
        rsr.formated.addSenseLen = 0x0a; // additional bytes that follow 244 max
        rsr.formated.cmdSpecInfo = 0; // command specific info
        rsr.formated.asc = 0x20; // additional sense code 0x20 for illegal command
        rsr.formated.ascq = 0; // additional sense code qualifier 0-unused
        rsr.formated.fruc = 0; // field replaceable unit code set to 0
        rsr.formated.senseKeySpecific[0] = 0; //senses key spec info if b7=1
        rsr.formated.senseKeySpecific[1] = 0;
        rsr.formated.senseKeySpecific[2] = 0;
        bytesWritten = forward_data_to_slave(&rsr, 18);

        // now send an appropriate CSW to indicate success of this command
        csw.sig[0] = 'U'; //"USBS"
        csw.sig[1] = 'S';
        csw.sig[2] = 'B';
        csw.sig[3] = 'S';
        csw.tag = cbw->tag;
        csw.residue = 0;
        csw.status = 0; // 0x00=success 0x01=failure 0x02=phase error
        forward_csw_to_slave(&csw);
    } else
    {
        // forward cbw
        bytesRequested = cbw->cb.raw[4];
        forward_cbw_to_device(cbw);
        buffer = vos_malloc((unsigned short)bytesRequested);

        // receive data from device
        if (bytesRead = receive_data_from_device(buffer, (unsigned short)bytesRequested))
        {

            bytesWritten = forward_data_to_slave(buffer, bytesRead);
        }

        vos_free(buffer);
        // receive csw from device
        receive_csw_from_device(&csw);

        // forward csw to slave
        forward_csw_to_slave(&csw);
    }
}

void handle_mode_sense(boms_cbw_t *cbw)
{
    boms_csw_t csw;
    unsigned short allocLength=0;
    unsigned char *buffer=NULL;
    unsigned short bytesReceived=0;
```

```
    // forward the cbw to the device
    switch (cbw->cb.formated.command)
    {
            case BOMS_MODE_SENSE_6:
                    allocLength = cbw->cb.raw[4];
                    break;
            case BOMS_MODE_SENSE_10:
                    allocLength = cbw->cb.raw[7]*256 + cbw->cb.raw[8];
                    break;
    }
    forward_cbw_to_device(cbw);

    // receive data from device
    if (allocLength)
    {
            buffer = vos_malloc(allocLength);
            bytesReceived = receive_data_from_device(buffer, allocLength);
            // forward data to slave
            forward_data_to_slave(buffer, bytesReceived);
            vos_free(buffer);
    }

    // receive csw from device
    receive_csw_from_device(&csw);

    // forward csw to slave
    forward_csw_to_slave(&csw);
}

void handle_mode_select(boms_cbw_t *cbw)
{
    boms_csw_t csw;

    unsigned short allocLength=0;
    unsigned char *buffer=NULL;
    unsigned short bytesReceived=0;

    // forward the cbw to the device
    switch (cbw->cb.formated.command)
    {
            case BOMS_MODE_SELECT_6:
                    allocLength = cbw->cb.raw[4];
                    break;
            case BOMS_MODE_SELECT_10:
                    allocLength = cbw->cb.raw[7]*256 + cbw->cb.raw[8];
                    break;
    }
    forward_cbw_to_device(cbw);

    // receive data from device
    if (allocLength)
    {
            buffer = vos_malloc(allocLength);
            bytesReceived = receive_data_from_device(buffer, allocLength);
            // forward data to slave
            forward_data_to_slave(buffer, bytesReceived);
            vos_free(buffer);
    }
```

```
    // receive csw from device
    receive_csw_from_device(&csw);

    // forward csw to slave
    forward_csw_to_slave(&csw);
}

void handle_illegal_request(boms_cbw_t *cbw)
{
    usbslave_ioctl_cb_t iocb;
    boms_csw_t csw;

    // now send the CSW
    csw.sig[0]='U';
    csw.sig[1]='S';
    csw.sig[2]='B';
    csw.sig[3]='S';//"USBS"
    csw.tag=cbw->tag;
    csw.residue=0;
    csw.status=0x01; // 0x00=success 0x01=failure 0x02=phase error
    forward_csw_to_slave(&csw);

    // flag the error for the anticipated call to REQUEST SENSE
    illegalRequest=1;
}

void handle_illegal_write_request(boms_cbw_t *cbw)
{
    usbslave_ioctl_cb_t iocb;
    boms_csw_t csw;
    unsigned short blocks;
    unsigned char *buffer;
    unsigned short num_read;
    unsigned short i;


    // as strange as it may seem, there is no way to tell the host to quit
    // instead we need to receive all this data and throw it away!
    switch (cbw->cb.formated.command)
    {
            case BOMS_WRITE_6:
                    blocks = cbw->cb.raw[4];
                    break;
            case BOMS_WRITE_10:
                    blocks = cbw->cb.raw[7] * 256 + cbw->cb.raw[8];
                    break;
            case BOMS_WRITE_12:
                    // we are being a little bad here the number of blocks is actually a long
                    // it is extremely unlikely that anyone would request this much at once, however
                    blocks = cbw->cb.raw[8] * 256 + cbw->cb.raw[9];
                    break;
    }
```

```
    buffer = vos_malloc(512);
    iocb.ioctl_code = VOS_IOCTL_USBSLAVE_TRANSFER;
    iocb.handle = slaveBomsCtx->out_ep;
    iocb.request.setup_or_bulk_transfer.buffer = buffer;
    iocb.request.setup_or_bulk_transfer.size = 512;
    iocb.request.setup_or_bulk_transfer.bytes_transferred = 0;
    for (i = 0; i < (blocks * (512/blockSize)); i++)
    {
            // process bytes received from host
            vos_dev_ioctl(slaveBomsCtx->handle,&iocb);
    }
    vos_free(buffer);

    // now send the CSW
    csw.sig[0]='U';
    csw.sig[1]='S';
    csw.sig[2]='B';
    csw.sig[3]='S';//"USBS"
    csw.tag=cbw->tag;
    csw.residue=0;
    csw.status=0x00; // 0x00=success 0x01=failure 0x02=phase error
    //forward_csw_to_slave(&csw);
    iocb.ioctl_code = VOS_IOCTL_USBSLAVE_TRANSFER;
    iocb.handle = slaveBomsCtx->in_ep;
    iocb.request.setup_or_bulk_transfer.buffer = &csw;
    iocb.request.setup_or_bulk_transfer.size = sizeof(boms_csw_t);
    vos_dev_ioctl(slaveBomsCtx->handle, &iocb);

}

void handle_send_diagnostic(boms_cbw_t *cbw)
{
    usbslave_ioctl_cb_t iocb;
    boms_csw_t csw;

    // first send ZLDP to ACK the command
    iocb.ioctl_code = VOS_IOCTL_USBSLAVE_TRANSFER;
    iocb.handle = slaveBomsCtx->in_ep;
    iocb.request.setup_or_bulk_transfer.buffer = NULL;
    iocb.request.setup_or_bulk_transfer.size = 0;
    vos_dev_ioctl(slaveBomsCtx->handle, &iocb);

    // now send the CSW
    csw.sig[0]='U';
    csw.sig[1]='S';
    csw.sig[2]='B';
    csw.sig[3]='S';//"USBS"
    csw.tag=cbw->tag;
    csw.residue=0;
    csw.status=0x00; // 0x00=success 0x01=failure 0x02=phase error
    forward_csw_to_slave(&csw);

}

void handle_start_stop_unit(boms_cbw_t *cbw)
{
    boms_csw_t csw;
```

```
      // forward the CBW to device
      if (forward_cbw_to_device(cbw))
      {
              // receive response from device
              if (receive_csw_from_device(&csw))
              {
                      // forward CSW to slave
                      forward_csw_to_slave(&csw);
              }
      }

}

void handle_synchronize_cache(boms_cbw_t *cbw)
{
    usbslave_ioctl_cb_t iocb;

    boms_csw_t csw;

    // first send ZLDP to ACK the command
    iocb.ioctl_code = VOS_IOCTL_USBSLAVE_TRANSFER;
    iocb.handle = slaveBomsCtx->in_ep;
    iocb.request.setup_or_bulk_transfer.buffer = NULL;
    iocb.request.setup_or_bulk_transfer.size = 0;
    vos_dev_ioctl(slaveBomsCtx->handle, &iocb);

    // now send the CSW
    csw.sig[0]='U';
    csw.sig[1]='S';
    csw.sig[2]='B';
    csw.sig[3]='S';//"USBS"
    csw.tag=cbw->tag;
    csw.residue=0;
    csw.status=0x00; // 0x00=success 0x01=failure 0x02=phase error
    forward_csw_to_slave(&csw);

}

void handle_read_format_capacities(boms_cbw_t *cbw)
{
    unsigned char *buffer;
    unsigned short responseSize;
    unsigned short allocLength;
    boms_csw_t csw;

    allocLength = cbw->cb.raw[7] * 256 + cbw->cb.raw[8];
    buffer = vos_malloc(allocLength);

    // forward the CBW to device
    if (forward_cbw_to_device(cbw))
    {
            // receive response from device
            if (responseSize = receive_data_from_device(buffer, allocLength))
            {
                    // forward response to slave
                    forward_data_to_slave(&buffer[0], responseSize);
```

```
                    // receive CSW from device
                    if (receive_csw_from_device(&csw))
                    {
                            // forward CSW to slave
                            forward_csw_to_slave(&csw);
                    }
            }
    }

    vos_free(buffer);
}

void handle_read_toc_pma_atip(boms_cbw_t *cbw)
{
    unsigned char *buffer;
    unsigned short responseSize;
    unsigned short allocLength;
    boms_csw_t csw;

    allocLength = cbw->cb.raw[7] * 256 + cbw->cb.raw[8];
    buffer = vos_malloc(allocLength);

    // forward the CBW to device
    if (forward_cbw_to_device(cbw))
    {
            // receive response from device
            if (responseSize = receive_data_from_device(buffer, allocLength))
            {
                    // forward response to slave
                    forward_data_to_slave(&buffer[0], responseSize);

                    // receive CSW from device
                    if (receive_csw_from_device(&csw))
                    {
                            // forward CSW to slave
                            forward_csw_to_slave(&csw);
                    }
            }
    }

    vos_free(buffer);
}

// This function handles the call to prevent/allow removal
// If we fail this command when prevent=1 (true) then
// Windows will not cache writes.  This actually leads
// to better performance.
void handle_prevent_allow_removal(boms_cbw_t *cbw)
{
    usbslave_ioctl_cb_t iocb;
    boms_csw_t csw;
```

```
// now send the CSW
csw.sig[0]='U';
csw.sig[1]='S';
csw.sig[2]='B';
csw.sig[3]='S';//"USBS"
csw.tag=cbw->tag;
csw.residue=0;
csw.status=0x01; // 0x00=success 0x01=failure 0x02=phase error
forward_csw_to_slave(&csw);

// flag the error for the anticipated call to REQUEST SENSE
illegalRequest=1;

}
```

The VNC2 microcontroller is somewhat unique in that it can be a host for USB devices. It is not unique in its ability to be embedded in a USB device, however. The FTDI libraries support easy implementation of a USB mass storage device which FTDI refers to as a USB Bulk-Only Mass Storage (BOMS) slave. To make our device appear as a flash drive when plugged into a computer we must simply create a series of handler functions and register our driver. One of the handlers will need to respond to standard requests and return device, configuration, and a collection of string descriptors. The required source code is provided in Listing 15.12.

The descriptors are defined at the top of the listing. The vendor and product IDs provided have been borrowed from an actual flash drive. Feel free to pick your favorite IDs. These IDs will be manipulated in the USB impersonator presented in the next article in this series. The primary handler functions call specialized handlers in order to keep the code somewhat neat.

Listing 15.12 USB mass storage slave driver

```
#include "vos.h"
#include "devman.h"
#include "memmgmt.h"

#include "ioctl.h"

#include "USB.h"
#include "USBHID.h"
#include "USBSlave.h"
#include "USBHost.h"

#include "USBSlaveBomsDrv.h"
#include "USBHostBomsDrv.h"

unsigned char standard_request(usbSlaveBoms_context *ctx);
unsigned char class_request(usbSlaveBoms_context *ctx);
void set_control_ep_halt(usbSlaveBoms_context *ctx);

unsigned char usbSlaveBoms_ioctl(common_ioctl_cb_t *cb, usbSlaveBoms_context *ctx);

// thread states
#define UNATTACHED 0
#define ATTACHED   1
```

```
// Every USB device has an 18 byte descriptor
// This descriptor is immediately retrieved by the host/hub in order
// to determine how to talk to the device
// Note: All USB values are little-endian (LSB first)
unsigned char device_descriptor[18] =

{
    18, //length in bytes
    1, // descriptor type 1=device
    0x0, 2, // USB version BCD USB version
    0, // device class 0=actual device class in interface descriptor
    0, // device subclass 0=actual device class in interface descriptor
    0, // device protocol 0=actual device class in interface descriptor
    64, // max packet size is 64 bytes for full speed endpoints
    0x4b, 0x15, // vendor id  // currently spoof PNY drive
    0x40, 0, // product id
    0, 0x01, // device release number in BCD
    1, // string descriptor index for manufacturer
    2, // string descriptor index for product
    3, // string descriptor index for serial number
    1 // number of possible configurations
};

// The configuration header is actually a composite of
// configuration, interface, and endpoint descriptors
// The host will typically ask for the first part of the
// descriptor (first 9 bytes) which contains the total descriptor length.
// The host will then make a second request for the entire descriptor
unsigned char config_descriptor[32] =
{

// Configuration Header
    9, // length
    2, // descriptor type 2 = configuration
    32, 0, // bytes in this and all subordinate descriptors
    1, // number of interfaces
    1, // configuration value
    0, // index of string descriptor for configurations
    0x80, // self/bus power and remote wakeup
    50, // max power in milliamps /2 asking for too much can cause enum to fail
// Interface
    9, // length
    4, // descriptor type 4 = interface
    0, // interface number
    0, // alternate setting - 00 is default
    2, // number of endpoints other than control (1 bulk in, 1 bulk out)
    8, // class type 8 for mass storage
    6, // subclass 6 means SCSI transport
    0x50, // interface protocol
    0, // index of string descriptor for the interface
// Endpoint, EP1_In
    7, // length
    5, // endpoint descriptor
    0x81, // endpoint address bit7-1 = in
    2, // attributes 2=bulk
    0x40, 0x00, // max packet size 64 bytes
    1, // polling interval, ignored for bulk endpoints
// Endpoint, EP2_Out
    7, // length
    5, // endpoint descriptor
    0x02, // endpoint address bit7-1 = in
```

```
    2, // attributes 2=bulk
    0x40, 0x00, // max packet size
    1 // polling interval, ignored for bulk endpoints
};


// language descriptor
// Requests for string descriptor 0 return a code for the default
// language. Devices could support multiple languages. In practice,
// everyone seems to support US English only, even if they are from
// Glasgouw, UK.
unsigned char str0_descriptor[4] =
{   0x04, // length
    0x03, // type 3=string
    0x09,
    0x04 // US English
};


// According to our configuration descriptor, string descriptor 2
// is a product ID. This Product descriptor was borrowed from an
// actual flash drive. Alternatively, we could query the actual
// information from the flash drive, but we fake it to make things
// a bit simplier. These descriptors all use UNICODE
unsigned char str2_descriptor[22] =


{   22, // length
    3, // string descriptor
    0x55, 0, //USB 2.0 FD
    0x53, 0,
    0x42, 0,
    0x20, 0,
    0x32, 0,
    0x2e, 0,
    0x30, 0,
    0x20, 0,
    0x46, 0,
    0x44, 0
};


// According to our configuration descriptor, string descriptor 1
// is the manufacturer. In this case we have borrowed PNY.
unsigned char str1_descriptor[8] =
{   8, // length
    3, // string descriptor
    0x50, 0, // PNY
    0x4e, 0,
    0x59, 0
};


// Every mass storage device is required to have a serial number
// the last 12 digits must be unique. There is no other specification
// on how these are assigned. Note that Windows will store this and
// lots of other information in the registry. USBDevView is a nice
// free utility for viewing this information.
// More information on USB forensics can be found in my 44Con video
// which is available on SecurityTube or youtube.
unsigned char str3_descriptor[30] =
{   30, // length
```

```
    3, // string descriptor
    0x55, 0, // UTYM0832030481
    0x54, 0,
    0x59, 0,
    0x4d, 0,
    0x30, 0,
    0x38, 0,
    0x33, 0,
    0x32, 0,
    0x30, 0,
    0x33, 0,
    0x30, 0,
    0x34, 0,
    0x38, 0,
    0x31, 0
};


// global variables are evil, but sometimes when
// dealing with memory-constrained microcontrollers
// they are somewhat unavoidable
usbSlaveBoms_context *slaveBomsCtx=NULL;
extern usbhostBoms_context_t *hostBomsCtx;


// These are the only 2 Class requests sent on the control endpoint
// BOMS devices send requests in the CBW on the bulk out endpoint
#define GET_MAX_LUN 0xfe
#define BOMS_RESET 0xff


// The following functions are for stalling and clearing bulk endpoints
// There are some conditions which require us to stall these endpoints.
// For example, if we return less data than expected to the PC we
// must stall the endpoint so it doesn't hang forever waiting for the rest
// of the data.
void usbslaveboms_stall_bulk_in(usbSlaveBoms_context *ctx)
{
    usbslave_ioctl_cb_t iocb; // this is a structure used by underlying VOS driver

    iocb.ioctl_code = VOS_IOCTL_USBSLAVE_ENDPOINT_STALL;
    iocb.ep = 0x81; // bulk in endpoint
    vos_dev_ioctl(ctx->handle, &iocb);
}


void usbslaveboms_stall_bulk_out(usbSlaveBoms_context *ctx)
{
    usbslave_ioctl_cb_t iocb; // this is a structure used by underlying VOS driver

    iocb.ioctl_code = VOS_IOCTL_USBSLAVE_ENDPOINT_STALL;
    iocb.ep = 0x02; // bulk out endpoint
    vos_dev_ioctl(ctx->handle, &iocb);
}


void usbslaveboms_clear_bulk_in(usbSlaveBoms_context *ctx)
{
    usbslave_ioctl_cb_t iocb; // this is a structure used by underlying VOS driver

    iocb.ioctl_code = VOS_IOCTL_USBSLAVE_ENDPOINT_CLEAR;
    iocb.ep = 0x81; // bulk in endpoint
    vos_dev_ioctl(ctx->handle, &iocb);
}
```

```
void usbslaveboms_clear_bulk_out(usbSlaveBoms_context *ctx)
{
    usbslave_ioctl_cb_t iocb; // this is a structure used by underlying VOS driver

    iocb.ioctl_code = VOS_IOCTL_USBSLAVE_ENDPOINT_CLEAR;
    iocb.ep = 0x02; // bulk out endpoint
    vos_dev_ioctl(ctx->handle, &iocb);
}


// This function MUST BE CALLED BEFORE THE SCHEDULER IS STARTED
// It initializes (initialises for you Brits reading this) variables
// in the context structure, registers our driver with the
// VOS device manager, and creates the thread for handling
// requests on the control endpoint.
unsigned char usbslaveboms_init(unsigned char vos_dev_num)
{
    vos_driver_t *usbSlaveBoms_cb;

    slaveBomsCtx = vos_malloc(sizeof(usbSlaveBoms_context));

    if (slaveBomsCtx == NULL)
            return USBSLAVEBOMS_ERROR; //somehow ran out of RAM

    usbSlaveBoms_cb = vos_malloc(sizeof(vos_driver_t));

    if (usbSlaveBoms_cb == NULL)
    {
            vos_free(slaveBomsCtx);
            return USBSLAVEBOMS_ERROR;
    }

    // Set up function pointers for our driver

    usbSlaveBoms_cb->flags = 0;
    usbSlaveBoms_cb->read = usbSlaveBoms_read;
    usbSlaveBoms_cb->write = usbSlaveBoms_write;
    usbSlaveBoms_cb->ioctl = usbSlaveBoms_ioctl;
    usbSlaveBoms_cb->interrupt = (PF_INT) NULL;
    usbSlaveBoms_cb->open = (PF_OPEN) NULL;
    usbSlaveBoms_cb->close = (PF_CLOSE) NULL;

    // OK - register with device manager
    vos_dev_init(vos_dev_num, usbSlaveBoms_cb, slaveBomsCtx);

    // defaults to not connected and no flash drive yet
    slaveBomsCtx->attached = 0;
    slaveBomsCtx->flashConnected = 0;

    // create the thread that handles standard control requests
    slaveBomsCtx->tcbSetup = vos_create_thread_ex(31, SIZEOF_BOMS_SETUP_MEMORY, usbslaveboms_setup,
"BOMSSetup", 2, slaveBomsCtx);

    // initialize the sempahore to 0 so that anyone waiting for
    // the device to enum will block
    vos_init_semaphore(&slaveBomsCtx->enumed, 0);

    if (slaveBomsCtx->tcbSetup)
            return USBSLAVEBOMS_OK;

    return USBSLAVEBOMS_ERROR;
}
```

```
// This function is run inside the thread created in init
// It will respond to standard USB requests
// some of the requests are forwarded to the flash drive when appropriate
void usbslaveboms_setup(usbSlaveBoms_context *ctx)
{
    usbslave_ioctl_cb_t iocb; // this is a structure used by underlying VOS driver
    usb_deviceRequest_t *devReq; // this struct is defined in usb.h and is used to store the 9 byte setup request
    unsigned char bmRequestType; // The request type is defined by USB standard
    unsigned char state = UNATTACHED; // assume unattached at the start

    // This will wait till the flash drive is enumed so we don't
    // start responding to commands right away
    vos_wait_semaphore(&ctx->enumed);
    vos_signal_semaphore(&ctx->enumed);

    while (1)
    {
        switch (state)
        {
        case UNATTACHED:

            if (!ctx->attached)

                    vos_delay_msecs(100); // this delay is to avoid a tight loop
            else
            {
                    state = ATTACHED;
            }

            break;

        case ATTACHED:

            if (!ctx->attached) // check to see if we somehow became unattached
            {
                    state = UNATTACHED;
                    break;
            }

            // we now make a blocking call requesting the 9 byte setup
            // packet on the control endpoint
            iocb.ioctl_code = VOS_IOCTL_USBSLAVE_WAIT_SETUP_RCVD;
            iocb.request.setup_or_bulk_transfer.buffer = ctx->setup_buffer;
            iocb.request.setup_or_bulk_transfer.size = 9;
            vos_dev_ioctl(ctx->handle, &iocb);

            // decode the raw data by pointing to our structure
            devReq = (usb_deviceRequest_t *) ctx->setup_buffer;
            // valid types here are standard, class, and vendor
            // BOMS devices do not have vendor specific calls
            // even if they did, we wouldn't support them.
            bmRequestType = devReq->bmRequestType & (USB_BMREQUESTTYPE_STANDARD |
USB_BMREQUESTTYPE_CLASS);

            // we only need to handle standard and class requests for BOMS
            if (bmRequestType == USB_BMREQUESTTYPE_STANDARD)
            {
                    standard_request(ctx); // standard request that all USB devices support
            }
            else if (bmRequestType == USB_BMREQUESTTYPE_CLASS)
            {
                    class_request(ctx); // the request is specific to this device class (only 2 in our case)
            }
```

```
                              break;

                      default:
                              asm {HALT}; // if we somehow got here the fecal matter has hit the turbine
                              break;
                      }
              }

          return;
      }


// this function just marks our device as detached
void usbSlaveBoms_detach(usbSlaveBoms_context *ctx)
{
    ctx->attached = 0;

    return;
}

// this function will present our device to the PC and cause it
// to be enumed.
unsigned char usbSlaveBoms_attach(VOS_HANDLE handle, usbSlaveBoms_context *ctx)
{
    usbslave_ioctl_cb_t iocb;
    unsigned char status = USBSLAVEBOMS_OK;

    // save usb slave handle
    ctx->handle = handle;

    if (!ctx->attached)
    {
              // issue connect IOCTL call here to present ourselves to the host
              // MUST be called before configuring endpoints
              // Note: If you used older versions of the VNC2 toolchain this function
              // was added and the need to make this call is buried in the release
              // notes. Upgraders of old code be warned!
              iocb.ioctl_code = VOS_IOCTL_USBSLAVE_CONNECT;
              iocb.set = (void *) 0;
              vos_dev_ioctl(ctx->handle,&iocb);

              // get endpoint handles and set max packet sizes
              // these should match the descriptor values!
              iocb.ioctl_code = VOS_IOCTL_USBSLAVE_GET_CONTROL_ENDPOINT_HANDLE;
              iocb.ep = USBSLAVE_CONTROL_IN;
              iocb.get = &ctx->in_ep0;
              vos_dev_ioctl(ctx->handle, &iocb);

              iocb.ioctl_code = VOS_IOCTL_USBSLAVE_SET_ENDPOINT_MAX_PACKET_SIZE;
              iocb.handle = ctx->in_ep0;
              iocb.request.ep_max_packet_size = USBSLAVE_MAX_PACKET_SIZE_64;
              vos_dev_ioctl(ctx->handle, &iocb);

              iocb.ioctl_code = VOS_IOCTL_USBSLAVE_GET_CONTROL_ENDPOINT_HANDLE;
              iocb.ep = USBSLAVE_CONTROL_OUT;
              iocb.get = &ctx->out_ep0;
              vos_dev_ioctl(ctx->handle, &iocb);

              iocb.ioctl_code = VOS_IOCTL_USBSLAVE_SET_ENDPOINT_MAX_PACKET_SIZE;
              iocb.handle = ctx->out_ep0;
              iocb.request.ep_max_packet_size = USBSLAVE_MAX_PACKET_SIZE_64;
              vos_dev_ioctl(ctx->handle, &iocb);

              iocb.ioctl_code = VOS_IOCTL_USBSLAVE_GET_BULK_IN_ENDPOINT_HANDLE;
              iocb.ep = USBSLAVEBOMS_IN;
```

```
        iocb.get = &ctx->in_ep;
        vos_dev_ioctl(ctx->handle, &iocb);

        iocb.ioctl_code = VOS_IOCTL_USBSLAVE_SET_ENDPOINT_MAX_PACKET_SIZE;
        iocb.handle = ctx->in_ep;
        iocb.request.ep_max_packet_size = USBSLAVE_MAX_PACKET_SIZE_64;
        vos_dev_ioctl(ctx->handle, &iocb);

        iocb.ioctl_code = VOS_IOCTL_USBSLAVE_GET_BULK_OUT_ENDPOINT_HANDLE;
        iocb.ep = USBSLAVEBOMS_OUT;
        iocb.get = &ctx->out_ep;
        vos_dev_ioctl(ctx->handle, &iocb);

        iocb.ioctl_code = VOS_IOCTL_USBSLAVE_SET_ENDPOINT_MAX_PACKET_SIZE;
        iocb.handle = ctx->out_ep;
        iocb.request.ep_max_packet_size = USBSLAVE_MAX_PACKET_SIZE_64;
        vos_dev_ioctl(ctx->handle, &iocb);

        ctx->attached = 1; // mark our device as attached and configured

    }

    return status;
}

// write function
// This function will send up to the specified number of bytes to
// the bulk in endpoint
unsigned char usbSlaveBoms_write(
    char *xfer,
    unsigned short num_to_write,
    unsigned short *num_written,
    usbSlaveBoms_context *ctx)
{
    usbslave_ioctl_cb_t iocb;

    if(ctx->attached)
    {
        iocb.ioctl_code = VOS_IOCTL_USBSLAVE_TRANSFER;
        iocb.handle = ctx->in_ep;
        iocb.request.setup_or_bulk_transfer.buffer = xfer;
        iocb.request.setup_or_bulk_transfer.size = num_to_write;
        vos_dev_ioctl(ctx->handle, &iocb);
        if (num_written) // callers who don't care might pass NULL here let's not crash!
                *num_written = iocb.request.setup_or_bulk_transfer.bytes_transferred;
    } else {
        return USBSLAVEBOMS_ERROR;
    }

    return USBSLAVEBOMS_OK;
}

// read function
// This function reads up to num_to_read bytes from the bulk out endpoint.
unsigned char usbSlaveBoms_read(
    char *xfer,
    unsigned short num_to_read,
    unsigned short *num_read,
    usbSlaveBoms_context *ctx)
{
    usbslave_ioctl_cb_t iocb;

    if (ctx->attached)
    {
```

```
            iocb.ioctl_code = VOS_IOCTL_USBSLAVE_TRANSFER;
            iocb.handle = ctx->out_ep;
            iocb.request.setup_or_bulk_transfer.buffer = xfer;
            iocb.request.setup_or_bulk_transfer.size = num_to_read;
            vos_dev_ioctl(ctx->handle, &iocb);
            if (num_read) // allow caller to pass NULL if they don't care how many bytes
                    *num_read = iocb.request.setup_or_bulk_transfer.bytes_transferred;
    } else {
            return USBSLAVEBOMS_ERROR;
    }

    return USBSLAVEBOMS_OK;
}


// write function
// This function will send up to the specified number of bytes to
// the bulk in endpoint then it will stall (short write)
unsigned char usbSlaveBoms_short_write(
    char *xfer,
    unsigned short num_to_write,
    unsigned short *num_written,
    usbSlaveBoms_context *ctx)
{
    usbslave_ioctl_cb_t iocb;

    if(ctx->attached)
    {
            iocb.ioctl_code = VOS_IOCTL_USBSLAVE_TRANSFER;
            iocb.handle = ctx->in_ep;
            iocb.request.setup_or_bulk_transfer.buffer = xfer;
            iocb.request.setup_or_bulk_transfer.size = num_to_write;
            vos_dev_ioctl(ctx->handle, &iocb);
            if (num_written) // callers who don't care might pass NULL here let's not crash!
                    *num_written = iocb.request.setup_or_bulk_transfer.bytes_transferred;
            usbslaveboms_stall_bulk_in(ctx);

    } else {
            return USBSLAVEBOMS_ERROR;
    }

    return USBSLAVEBOMS_OK;
}


// read function
// This function reads up to num_to_read bytes from the bulk out endpoint.
// This function will stall the endpoint when done
unsigned char usbSlaveBoms_short_read(
    char *xfer,
    unsigned short num_to_read,
    unsigned short *num_read,
    usbSlaveBoms_context *ctx)
{
    usbslave_ioctl_cb_t iocb;

    if (ctx->attached)
    {
```

.

```
            iocb.ioctl_code = VOS_IOCTL_USBSLAVE_TRANSFER;
            iocb.handle = ctx->out_ep;
            iocb.request.setup_or_bulk_transfer.buffer = xfer;
            iocb.request.setup_or_bulk_transfer.size = num_to_read;
            vos_dev_ioctl(ctx->handle, &iocb);
            if (num_read) // allow caller to pass NULL if they don't care how many bytes
                    *num_read = iocb.request.setup_or_bulk_transfer.bytes_transferred;
            usbslaveboms_stall_bulk_out(ctx);
    } else {
            return USBSLAVEBOMS_ERROR;
    }

    return USBSLAVEBOMS_OK;
}
// This function reads the 31 byte CBW from the bulk out endpoint
// The value is returned directly into the boms_cbw_t structure
unsigned char usbSlaveBoms_readCbw(
    boms_cbw_t *cbw,
    usbSlaveBoms_context *ctx)
{
    usbslave_ioctl_cb_t iocb;

    if (ctx->attached)
    {
            iocb.ioctl_code = VOS_IOCTL_USBSLAVE_TRANSFER;
            iocb.handle = ctx->out_ep;
            iocb.request.setup_or_bulk_transfer.buffer = (unsigned char*)cbw;
            iocb.request.setup_or_bulk_transfer.size = 31; //CBW is 31 bytes
            vos_dev_ioctl(ctx->handle, &iocb);

    } else {
            return USBSLAVEBOMS_ERROR;
    }

    return USBSLAVEBOMS_OK;

}

// USB Slave IOCTL function
unsigned char usbSlaveBoms_ioctl(common_ioctl_cb_t *cb, usbSlaveBoms_context *ctx)
{
    unsigned char status = USBSLAVEBOMS_INVALID_PARAMETER;

    switch (cb->ioctl_code)
    {
    case VOS_IOCTL_USBSLAVEBOMS_ATTACH:
            status = usbSlaveBoms_attach((VOS_HANDLE) cb->set.data, ctx);
            break;

    case VOS_IOCTL_USBSLAVEBOMS_DETACH:
            usbSlaveBoms_detach(ctx);
            status = (unsigned char)USBSLAVEBOMS_OK;
            break;

    default:
            break;
    }
```

```
    return status;
}

// All commands on the control endpoint must be acknowledged.
// This is done by sending a Zero Length Data Packet ZLDP
// on the control in endpoint.
void ack_request(usbSlaveBoms_context *ctx)
{
    usbslave_ioctl_cb_t iocb;

    iocb.ioctl_code = VOS_IOCTL_USBSLAVE_SETUP_TRANSFER;
    iocb.handle = ctx->in_ep0;
    iocb.request.setup_or_bulk_transfer.buffer = (void *) 0;
    iocb.request.setup_or_bulk_transfer.size = 0;
    vos_dev_ioctl(ctx->handle, &iocb);
}

// When initially attached to a host devices have no address and are
// assigned an adddress after enumeration.  Then the device is reset
// and fully enumerated.
void set_address_request(usbSlaveBoms_context *ctx, unsigned char addr)
{
    usbslave_ioctl_cb_t iocb;
    iocb.ioctl_code = VOS_IOCTL_USBSLAVE_SET_ADDRESS;
    iocb.set = (void *) addr;
    vos_dev_ioctl(ctx->handle, &iocb);

    ack_request(ctx);
}

// Most devices support only one configuration.  However, every
// device must respond to this call asking them to select a
// configuration even if it is the default.
void set_configuration_request(usbSlaveBoms_context *ctx, unsigned char config)
{
    usbslave_ioctl_cb_t iocb;

    iocb.ioctl_code = VOS_IOCTL_USBSLAVE_SET_CONFIGURATION;
    iocb.set = (void *) config;
    vos_dev_ioctl(ctx->handle, &iocb);

    ack_request(ctx);
}

// handle requests for a descriptor
void get_descriptor_request(usbSlaveBoms_context *ctx)
{
    unsigned char *buffer;  // buffer for pass thru to drive
    usbhost_ioctl_cb_t hc_ioctl;
    usbslave_ioctl_cb_t iocb;
    usb_deviceRequest_t *devReq;
    unsigned char hValue; // high byte of the descriptor requested
    unsigned char lValue; // low byte of the descriptor requested
    unsigned short wLength;
    unsigned short siz;
    uint32 ul_siz;
    unsigned char *src;
    unsigned char cond;
```

```
devReq = (usb_deviceRequest_t *) ctx->setup_buffer;

hValue = devReq->wValue >> 8; // shift away the low byte
lValue = devReq->wValue & 0xff; // and away the high byte

wLength = devReq->wLength;

switch (hValue) // the high byte determines type of descriptor requested
{
case USB_DESCRIPTOR_TYPE_DEVICE:
        ul_siz = (uint32) wLength;
        iocb.ioctl_code = VOS_IOCTL_USBSLAVE_SETUP_TRANSFER;
        iocb.handle = ctx->in_ep0;
        iocb.request.setup_or_bulk_transfer.buffer = device_descriptor;
        iocb.request.setup_or_bulk_transfer.size = (int16) ul_siz;

        vos_dev_ioctl(ctx->handle, &iocb);
        return;
        break;


case USB_DESCRIPTOR_TYPE_CONFIGURATION:
        // host will initially ask for first 9 bytes of configuration descriptor
        // this descriptor header has the size of the full descriptor which
        // is actually a composite of the configuration/interface/endpoints.
        // Once host knows the complete descriptor size it makes a second
        // request for the whole thing
        siz = wLength == 9?9:sizeof(config_descriptor);
        ul_siz = (uint32) siz;

        iocb.ioctl_code = VOS_IOCTL_USBSLAVE_SETUP_TRANSFER;
        iocb.handle = ctx->in_ep0;
        iocb.request.setup_or_bulk_transfer.buffer = config_descriptor;
        iocb.request.setup_or_bulk_transfer.size = (int16) ul_siz;
        vos_dev_ioctl(ctx->handle, &iocb);
        return;

case USB_DESCRIPTOR_TYPE_STRING:

        if (lValue == 0) // language type
        {
                src = str0_descriptor;
                siz = sizeof(str0_descriptor);
        }
        else if (lValue == 1) // manufacturer
        {
                src = str1_descriptor;
                siz = sizeof(str1_descriptor);
        }
        else if (lValue == 2) // product
        {
                src = str2_descriptor;
                siz = sizeof(str2_descriptor);
        }
        else if (lValue == 3) // serial number
        {
                src = str3_descriptor;
                siz = sizeof(str3_descriptor);
        }

        cond = (unsigned char) (wLength != siz);
```

```
            if (siz > wLength) // don't return more than was asked for
                    siz = wLength;

        ul_siz = (uint32) siz;

        iocb.ioctl_code = VOS_IOCTL_USBSLAVE_SETUP_TRANSFER;
        iocb.handle = ctx->in_ep0;
        iocb.request.setup_or_bulk_transfer.buffer = src;
        iocb.request.setup_or_bulk_transfer.size = (int16) ul_siz;
        vos_dev_ioctl(ctx->handle, &iocb);
        return;

    default:
        // if drive is connected get descriptor from it
        if (ctx->flashConnected)
        {
            buffer = vos_malloc(wLength);
            hc_ioctl.ioctl_code = VOS_IOCTL_USBHOST_DEVICE_SETUP_TRANSFER;
            hc_ioctl.handle.ep = hostBomsCtx->epCtrl;
            hc_ioctl.set = &(ctx->setup_buffer[0]);
            hc_ioctl.get = buffer; // descriptor from drive
            vos_dev_ioctl(hostBomsCtx->hc, &hc_ioctl);

            iocb.ioctl_code = VOS_IOCTL_USBSLAVE_SETUP_TRANSFER;
            iocb.handle = ctx->in_ep0;
            iocb.request.setup_or_bulk_transfer.buffer = buffer;
            iocb.request.setup_or_bulk_transfer.size = wLength;
            vos_dev_ioctl(ctx->handle, &iocb);

            vos_free(buffer);
        } else {
            // respond with Request Error
            set_control_ep_halt(ctx);
        }
    }
}

// This function will set a feature. If it is directed at an
// endpoint the endpoint will stall.
// Note: the endpoint passed in is the USB endpoint from the setup request
// packet, not the VOS endpoint handle. Unfortunately, most of the
// FTDI defined types are are just typedefs so no real type checking
// is going on here.
void set_feature_request(usbSlaveBoms_context *ctx, unsigned char ep)
{
    usbslave_ioctl_cb_t iocb;
    usbhost_ep_handle_ex hep; // host endpoint to pass stall to
    usbhost_ioctl_cb_t host_ioctl_cb;

    ack_request(ctx); // first ack the request then decide what to do

    // is this directed at an endpoint or the device?
    if (ctx->setup_buffer[0] & USB_BMREQUESTTYPE_ENDPOINT)
    {
        // directed to an endpoint
        //ep 1 is IN 2 is OUT on my fake device
        iocb.ioctl_code = VOS_IOCTL_USBSLAVE_ENDPOINT_STALL;
```

```
            iocb.ep = (ep & 0x02)?ctx->out_ep:ctx->in_ep;
            vos_dev_ioctl(ctx->handle, &iocb);

            // if flash drive is attached pass along request
            if (ctx->flashConnected)
            {
                    //need to figure out which endpoint to stall
                    // if b7=1 then IN else OUT
                    if (ep & 0x80)
                    {
                            hep = hostBomsCtx->epBulkIn;
                    } else {
                            hep = hostBomsCtx->epBulkOut;
                    }
                    host_ioctl_cb.ioctl_code = VOS_IOCTL_USBHOST_DEVICE_SET_HOST_HALT;
                    host_ioctl_cb.handle.ep = hep;
                    // clear halt state on endpoint
                    vos_dev_ioctl(hostBomsCtx->hc, &host_ioctl_cb);
            }
    } else {
            // this is a device request
            host_ioctl_cb.ioctl_code = VOS_IOCTL_USBHOST_DEVICE_SETUP_TRANSFER;
            host_ioctl_cb.handle.ep = hostBomsCtx->epCtrl;
            host_ioctl_cb.set = &(ctx->setup_buffer[0]);
            host_ioctl_cb.get = NULL;
            vos_dev_ioctl(hostBomsCtx->hc, &host_ioctl_cb);

    }
}

// This function is the complement to the set_feature_request function.
void clear_feature_request(usbSlaveBoms_context *ctx, unsigned char ep)
{
    usbslave_ioctl_cb_t iocb;
    usbhost_ep_handle_ex hep; // host endpoint to pass stall to
    usbhost_ioctl_cb_t host_ioctl_cb;

    ack_request(ctx);

    // is this directed at an endpoint or the device?
    if (ctx->setup_buffer[0] & USB_BMREQUESTTYPE_ENDPOINT)
    {
            // directed to an endpoint
            //ep 1 is IN 2 is OUT on my fake device
            iocb.ioctl_code = VOS_IOCTL_USBSLAVE_ENDPOINT_CLEAR;
            iocb.ep = (ep & 0x02)?ctx->out_ep:ctx->in_ep;
            vos_dev_ioctl(ctx->handle, &iocb);

            // if flash drive is attached pass along request
            if (ctx->flashConnected)
            {
```

```
                         //need to figure out which endpoint to clear
                         // if b7=1 then IN else OUT
                         if (ep & 0x80)
                         {
                                 hep = hostBomsCtx->epBulkIn;
                         } else {
                                 hep = hostBomsCtx->epBulkOut;
                         }
                         host_ioctl_cb.ioctl_code = VOS_IOCTL_USBHOST_DEVICE_CLEAR_HOST_HALT;
                         host_ioctl_cb.handle.ep = hep;
                         // clear halt state on endpoint
                         vos_dev_ioctl(hostBomsCtx->hc, &host_ioctl_cb);
                 }
         } else {
                 // this is a device request
                 host_ioctl_cb.ioctl_code = VOS_IOCTL_USBHOST_DEVICE_SETUP_TRANSFER;
                 host_ioctl_cb.handle.ep = hostBomsCtx->epCtrl;
                 host_ioctl_cb.set = &(ctx->setup_buffer[0]);
                 host_ioctl_cb.get = NULL;
                 vos_dev_ioctl(hostBomsCtx->hc, &host_ioctl_cb);

         }
}

// This function returns a 1 byte status code for our endpoint
void get_ep_status(usbSlaveBoms_context *ctx, usbslave_ep_handle_t ep)
{
    usbslave_ioctl_cb_t iocb;
    char state;

    iocb.ioctl_code = VOS_IOCTL_USBSLAVE_ENDPOINT_STATE;
    iocb.ep = ep;
    iocb.get = &state;
    vos_dev_ioctl(ctx->handle, &iocb);

    iocb.ioctl_code = VOS_IOCTL_USBSLAVE_SETUP_TRANSFER;
    iocb.handle = ctx->out_ep0;
    iocb.request.setup_or_bulk_transfer.buffer = (void *) &state;
    iocb.request.setup_or_bulk_transfer.size = 1;
    vos_dev_ioctl(ctx->handle, &iocb);
}

// This function will halt the control endpoint.  This function will only
// be called when an illegal request has been passed to the device.
void set_control_ep_halt(usbSlaveBoms_context *ctx)
{
    usbslave_ioctl_cb_t iocb;

    ack_request(ctx);

    // Performs a protocol stall on endpoint 0
```

```
    // Indicates that a request is unsupported
    iocb.ioctl_code = VOS_IOCTL_USBSLAVE_ENDPOINT_STALL;
    iocb.ep = 0;
    vos_dev_ioctl(ctx->handle, &iocb);
}


// This returns the device status.  Only the two lowest bits
// have any meaning here.  B0=1 means device is self-powered.
// B0 = 0 means the device is bus-powered.
// B1 = 1 means that remote wakeup is enabled
// This request can also be used for interface status which
// should just return all zeroes or endpoint status.  Endpoint
// status halted/cleared is indicated by the LSB.
void get_status_request(usbSlaveBoms_context *ctx)
{
    unsigned short status = 0;
    usbslave_ioctl_cb_t iocb;
    usbhost_ioctl_cb_t hc_ioctl;

    // if drive is connected get status from it
    if (ctx->flashConnected)
    {
            hc_ioctl.ioctl_code = VOS_IOCTL_USBHOST_DEVICE_SETUP_TRANSFER;
            hc_ioctl.handle.ep = hostBomsCtx->epCtrl;
            hc_ioctl.set = &(ctx->setup_buffer[0]);
            hc_ioctl.get = &status; // status returned from the drive
            vos_dev_ioctl(hostBomsCtx->hc, &hc_ioctl);
    }

    iocb.ioctl_code = VOS_IOCTL_USBSLAVE_SETUP_TRANSFER;
    iocb.handle = ctx->in_ep0;
    iocb.request.setup_or_bulk_transfer.buffer = &status;
    iocb.request.setup_or_bulk_transfer.size = 2;
    vos_dev_ioctl(ctx->handle, &iocb);
}

// This function returns one byte.  If the byte is zero this indicates
// that the device is not ready.  Non-zero means everything is OK.
void get_configuration_request(usbSlaveBoms_context *ctx)
{
    unsigned char status = 0;
    usbslave_ioctl_cb_t iocb;
    usbhost_ioctl_cb_t hc_ioctl;

    // if drive is connected get status from it
    if (ctx->flashConnected)
    {
            hc_ioctl.ioctl_code = VOS_IOCTL_USBHOST_DEVICE_SETUP_TRANSFER;
            hc_ioctl.handle.ep = hostBomsCtx->epCtrl;
            hc_ioctl.set = &(ctx->setup_buffer[0]);
            hc_ioctl.get = &status; // status returned from the drive
```

```
            vos_dev_ioctl(hostBomsCtx->hc, &hc_ioctl);
    }

    iocb.ioctl_code = VOS_IOCTL_USBSLAVE_SETUP_TRANSFER;
    iocb.handle = ctx->in_ep0;
    iocb.request.setup_or_bulk_transfer.buffer = &status;
    iocb.request.setup_or_bulk_transfer.size = 1;
    vos_dev_ioctl(ctx->handle, &iocb);
}

// This request should never be sent to our device, but since this
// is a standard request we will accept it and just do nothing.
void set_descriptor_request(usbSlaveBoms_context *ctx, unsigned short wLength)
{
    unsigned char *buffer;
    usbslave_ioctl_cb_t iocb;
    usbhost_ioctl_cb_t hc_ioctl;

    ack_request(ctx);

    // We read in the info here but throw it away
    buffer = vos_malloc(wLength);
    iocb.ioctl_code = VOS_IOCTL_USBSLAVE_SETUP_TRANSFER;
    iocb.handle = ctx->out_ep0;
    iocb.request.setup_or_bulk_transfer.buffer = buffer;
    iocb.request.setup_or_bulk_transfer.size = wLength;
    vos_dev_ioctl(ctx->handle, &iocb);
    vos_free(buffer);
}

// This function returns a single byte. We will just fake
// it and return zero.
void get_interface_request(usbSlaveBoms_context *ctx)
{
    unsigned char status = 0;
    usbslave_ioctl_cb_t iocb;
    usbhost_ioctl_cb_t hc_ioctl;

    iocb.ioctl_code = VOS_IOCTL_USBSLAVE_SETUP_TRANSFER;
    iocb.handle = ctx->in_ep0;
    iocb.request.setup_or_bulk_transfer.buffer = &status;
    iocb.request.setup_or_bulk_transfer.size = 1;
    vos_dev_ioctl(ctx->handle, &iocb);
}

// We just ack this request.
void set_interface_request(usbSlaveBoms_context *ctx)
{
    ack_request(ctx);
}
```

```
// This is the main handler function for requests coming in on
// the control endpoint. These are all requests every USB device
// must respond to.
unsigned char standard_request(usbSlaveBoms_context *ctx)
{
    usb_deviceRequest_t *devReq;
    unsigned char status = USBSLAVE_OK;
    unsigned char bReq;

    devReq = (usb_deviceRequest_t *) ctx->setup_buffer;
    bReq = devReq->bRequest;

    switch (bReq) // request is 1 byte value
    {
            case USB_REQUEST_CODE_GET_STATUS:
                    get_status_request(ctx);
                    break;

            case USB_REQUEST_CODE_SET_ADDRESS:
                    set_address_request(ctx, devReq->wValue & 0xff);
                    break;

            case USB_REQUEST_CODE_GET_DESCRIPTOR:
                    get_descriptor_request(ctx);
                    break;

            case USB_REQUEST_CODE_SET_DESCRIPTOR:
                    set_descriptor_request(ctx, devReq->wLength);
                    break;

            case USB_REQUEST_CODE_SET_CONFIGURATION:
                    set_configuration_request(ctx, devReq->wValue & 0xff);
                    break;

            case USB_REQUEST_CODE_GET_CONFIGURATION:
                    get_configuration_request(ctx);
                    break;

            case USB_REQUEST_CODE_CLEAR_FEATURE:
                    clear_feature_request(ctx, devReq->wIndex >> 8);
                    break;

            case USB_REQUEST_CODE_SET_FEATURE:
                    set_feature_request(ctx, devReq->wIndex >> 8);
                    break;

            case USB_REQUEST_CODE_GET_INTERFACE:
                    get_interface_request(ctx);
                    break;

            case USB_REQUEST_CODE_SET_INTERFACE:
```

```
                        set_interface_request(ctx);
                        break;
                default:
                        // force a protocol stall
                        set_control_ep_halt(ctx);
                        break;
    }

    return status;
}

void class_ack(usbSlaveBoms_context *ctx)
{
    ack_request(ctx);
}

void class_control_out(usbSlaveBoms_context *ctx, char *buffer, unsigned short len)
{
    usbslave_ioctl_cb_t iocb;

    iocb.ioctl_code = VOS_IOCTL_USBSLAVE_SETUP_TRANSFER;
    iocb.handle = ctx->out_ep0;
    iocb.request.setup_or_bulk_transfer.buffer = (void *) buffer;
    iocb.request.setup_or_bulk_transfer.size = len;
    vos_dev_ioctl(ctx->handle, &iocb);
}

// This is one of two class requests for BOMS devices.
// If we are not connected we return 0 which is the most
// common situation. Windows may not recognize higher LUNs.
void get_max_lun_request(usbSlaveBoms_context *ctx)
{
    unsigned char maxLun = 0;
    usbslave_ioctl_cb_t iocb;
    usbhost_ioctl_cb_t hc_ioctl;

    // if drive is connected get max lun from it
    if (ctx->flashConnected)
    {
            hc_ioctl.ioctl_code = VOS_IOCTL_USBHOST_DEVICE_SETUP_TRANSFER;
            hc_ioctl.handle.ep = hostBomsCtx->epCtrl;
            hc_ioctl.set = &(ctx->setup_buffer[0]);
            hc_ioctl.get = &maxLun; // maximum LUN from drive
            vos_dev_ioctl(hostBomsCtx->hc, &hc_ioctl);
    }

    iocb.ioctl_code = VOS_IOCTL_USBSLAVE_SETUP_TRANSFER;
    iocb.handle = ctx->in_ep0;
    iocb.request.setup_or_bulk_transfer.buffer = &maxLun;
    iocb.request.setup_or_bulk_transfer.size = 1;
    vos_dev_ioctl(ctx->handle, &iocb);
```

```
}

// The USB standards define this function. After it is called
// the drive should be ready to respond to CBWs. The silly
// thing about this is that the drive should always be ready
// for this and if it isn't then it probable crashed long ago.
void boms_reset_request(usbSlaveBoms_context *ctx)
{
    usbhost_ioctl_cb_t hc_ioctl;

    ack_request(ctx);

    // forward this command to a drive if connected
    if (ctx->flashConnected)
    {
        hc_ioctl.ioctl_code = VOS_IOCTL_USBHOST_DEVICE_SETUP_TRANSFER;
        hc_ioctl.handle.ep = hostBomsCtx->epCtrl;
        hc_ioctl.set = &(ctx->setup_buffer[0]);
        hc_ioctl.get = NULL; //no return value from this call
        vos_dev_ioctl(hostBomsCtx->hc, &hc_ioctl);
    }
}

unsigned char class_request(usbSlaveBoms_context *ctx)
{
    usb_deviceRequest_t *devReq;
    usbslave_ioctl_cb_t iocb;
    unsigned char status = USBSLAVE_OK;
    unsigned char bReq;

    devReq = (usb_deviceRequest_t *) ctx->setup_buffer;
    bReq = devReq->bRequest;

    // force a protocol stall since there are no class requests in BOMS
    switch (bReq)
    {
        case GET_MAX_LUN :
            get_max_lun_request(ctx);
            break;

        case BOMS_RESET :
            boms_reset_request(ctx);
            break;

        default:
            set_control_ep_halt(ctx);
    }

    return status;
}
```

Summary of USB Mass Storage Write Blocking

There are a number of options for blocking write operations to USB mass storage devices. In this section we presented a couple of compact write blockers
which utilize the FTDI VNC2 microcontroller. We also briefly covered alternative

methods of write blocking using software or somewhat pricey commercial hardware. Code and schematics for the devices presented are available online at
https://github.com/ppolstra.

In the next section we will again make use of the FTDI VNC2. In that section
we will discuss building an inexpensive and compact device which can imperson-
ate other authorized USB drives. This is done in order to bypass endpoint security software which only permits whitelisted devices to be mounted. Recently
some organizations have begun to take steps to help deter the outward flow of
information.

15.9 USB Impersonation

Introduction to USB Impersonation

Because USB mass storage devices are now ubiquitous they are a commonly
used in the extraction of proprietary or classified data. Several companies now
offer endpoint security software which allows only authorized USB mass storage
devices to be mounted. Additionally, some administrators use operating system
policies to accomplish the same effect. Those familiar with network security will
recognize this as the USB equivalent of MAC filtering. Just as MAC filtering is
easily bypassed, so are attempts at blocking all but authorized USB mass storage devices. This section will present a simple device which could be used to
bypass such efforts. An impersonator with FTDI programmer attached is shown
in Fig. 15.8.

High Level Design

As previously stated, the USB impersonator is based on the USB write blocker
which was presented in the previous section in this chapter. The V2DIP1-32
development board from FTDI (in the lower left corner of Fig. 15.8) provides
sufficient GPIO lines to drive an LCD display, light 2 LEDs, and receive input
from three buttons. The three buttons are used to toggle write blocking and
to scroll through the list of 500 most common vendor/product ID (VID/PID)
combinations. The 10K potentiometer in Fig. 15.8 is used to adjust the
contrast on the LCD screen. The two LEDs are used to indicate write-blocking
status.

The impersonator has two modes of operation. In manual mode two of the buttons are used to select a known good VID/PID combination. In automatic mode
the impersonator starts a timer as soon as the PC starts communicating with it. If
the PC stops talking to the impersonator the timer expires which is interpreted as

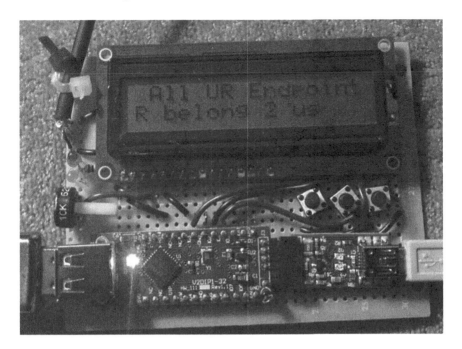

Fig. 15.8 USB impersonator

blocking so the device is disconnected via software and reconnected using the next VID/PID combination in the list.

The impersonator adds two new threads to the write blocker. One thread is used to detect button presses. This is a low priority thread. The second additional thread is used to handle the timer for function for automatic mode. The timer is set to one second when the PC begins communicating with the impersonator. If the PC doesn't complete the enumeration process before the timer expires a handler function is called. This handler function disconnects the impersonator, increments the current VID/PID pointer, and then reconnects the impersonator.

Timers

Like many microcontrollers, the FTDI Vinculum II (VNC2) supports multiple hardware timers. The timer thread function is presented in Listing 15.13. Note that a blocking call is made and if the timer is reset the code to change the VID/PID is never reached. The code which returns descriptors to the PC and also sets and resets the timers is presented in Listing 15.14. Complete code may be found at https://github.com/ppolstra/usb-impersonator.

Listing 15.13 Timer thread method

```
// This timer is used to determine if our device has successfully connected
// When the device is first enumerated this timer is set. This thread then
// blocks till the timer expires. When the timer expires we increment the
// index into our VID/PID list. If the device is connected the timer is
// cancelled.
void timer()
{
    tmr_ioctl_cb_t tmr_iocb;
    usbslave_ioctl_cb_t siocb;

    // here we set up the timer, but don't start it
    // timer is started by the USB slave driver when someone
    // starts talking to it
    hTimer = vos_dev_open(TIMER0);
    tmr_iocb.ioctl_code = VOS_IOCTL_TIMER_SET_TICK_SIZE;
    tmr_iocb.param = TIMER_TICK_MS;
    vos_dev_ioctl(hTimer, &tmr_iocb);
    tmr_iocb.ioctl_code = VOS_IOCTL_TIMER_SET_COUNT;
    tmr_iocb.param = 1000; // 1s
    vos_dev_ioctl(hTimer, &tmr_iocb);
    tmr_iocb.ioctl_code = VOS_IOCTL_TIMER_SET_DIRECTION;
    tmr_iocb.param = TIMER_COUNT_DOWN;
    vos_dev_ioctl(hTimer, &tmr_iocb);
    tmr_iocb.ioctl_code = VOS_IOCTL_TIMER_SET_MODE;
    tmr_iocb.param = TIMER_MODE_SINGLE_SHOT;
    vos_dev_ioctl(hTimer, &tmr_iocb);

    // if our device is connected this never gets past blocking call
    while (1)
    {
        tmr_iocb.ioctl_code = VOS_IOCTL_TIMER_WAIT_ON_COMPLETE;
        vos_dev_ioctl(hTimer, &tmr_iocb); //only returns if enumeration doesn't complete
        if (autoMode)
        {
            vos_lock_mutex(&vidPidMutex);

            currentVidPidIndex += 2;
            if (currentVidPidIndex > sizeof(vidPid))
                currentVidPidIndex = 0;
            vos_unlock_mutex(&vidPidMutex);
            // Disconnect the slave device
            siocb.ioctl_code = VOS_IOCTL_USBSLAVE_DISCONNECT;
            siocb.set = (void *) 0;
            vos_dev_ioctl(hUSBSLAVE_1,&siocb);
            // Now reconnect with new VID/PID
            siocb.ioctl_code = VOS_IOCTL_USBSLAVE_CONNECT;
            siocb.set = (void *) 0;
            vos_dev_ioctl(hUSBSLAVE_1,&siocb);
        }
    }
}
```

Listing 15.14 Descriptor request handler method

```
// handle requests for a descriptor
void get_descriptor_request(usbSlaveBoms_context *ctx)
{
    unsigned char *buffer;  // buffer for pass thru to drive
    usbhost_ioctl_cb_t hc_ioctl;
    usbslave_ioctl_cb_t iocb;
    usb_deviceRequest_t *devReq;
    unsigned char hValue; // high byte of the descriptor requested
    unsigned char lValue; // low byte of the descriptor requested
    unsigned short wLength;
    unsigned short siz;
    uint32 ul_siz;
    unsigned char *src;
    unsigned char cond;
    tmr_ioctl_cb_t tmr_iocb;

    devReq = (usb_deviceRequest_t *) ctx->setup_buffer;

    hValue = devReq->wValue >> 8; // shift away the low byte
    lValue = devReq->wValue & 0xff; // and away the high byte

    wLength = devReq->wLength;

    switch (hValue) // the high byte determines type of descriptor requested
    {
    case USB_DESCRIPTOR_TYPE_DEVICE:
            ul_siz = (uint32) wLength;
            iocb.ioctl_code = VOS_IOCTL_USBSLAVE_SETUP_TRANSFER;
            iocb.handle = ctx->in_ep0;
            // update the device descriptor VID/PID from our list
            vos_lock_mutex(&vidPidMutex);
            device_descriptor[8] = vidPid[currentVidPidIndex] & 0xff;
            device_descriptor[9] = vidPid[currentVidPidIndex] >> 8;
            device_descriptor[10] = vidPid[currentVidPidIndex+1] & 0xff;
            device_descriptor[11] = vidPid[currentVidPidIndex+1] >> 8;
            vos_unlock_mutex(&vidPidMutex);
            //updated LCD
            update_lcd_vidpid();

            iocb.request.setup_or_bulk_transfer.buffer = device_descriptor;
            iocb.request.setup_or_bulk_transfer.size = (int16) ul_siz;
            vos_dev_ioctl(ctx->handle, &iocb);

            // start this timer and if it is not killed then the next VID/PID will be selected
            // this is done to detect an unsuccessful enumeration which is assumed
            // to result from endpoint security blocking
            if (autoMode)
            {
                    tmr_iocb.ioctl_code = VOS_IOCTL_TIMER_START;
                    vos_dev_ioctl(hTimer, &tmr_iocb);
            }
            return;
            break;
```

```
case USB_DESCRIPTOR_TYPE_CONFIGURATION:
        // host will initially ask for first 9 bytes of configuration descriptor
        // this descriptor header has the size of the full descriptor which
        // is actually a composite of the configuration/interface/endpoints.
        // Once host knows the complete descriptor size it makes a second
        // request for the whole thing
        siz = wLength == 9?9:sizeof(config_descriptor);
        ul_siz = (uint32) siz;

        iocb.ioctl_code = VOS_IOCTL_USBSLAVE_SETUP_TRANSFER;
        iocb.handle = ctx->in_ep0;
        iocb.request.setup_or_bulk_transfer.buffer = config_descriptor;
        iocb.request.setup_or_bulk_transfer.size = (int16) ul_siz;
        vos_dev_ioctl(ctx->handle, &iocb);

        // stop the timer because we are being asked to enumerate
        if(autoMode)
        {
                tmr_iocb.ioctl_code = VOS_IOCTL_TIMER_STOP;
                vos_dev_ioctl(hTimer, &tmr_iocb);
        }
        return;

case USB_DESCRIPTOR_TYPE_STRING:

        if (lValue == 0) // language type
        {
                src = str0_descriptor;
                siz = sizeof(str0_descriptor);
        }
        else if (lValue == 1) // manufacturer
        {
                src = str1_descriptor;
                siz = sizeof(str1_descriptor);
        }
        else if (lValue == 2) // product
        {
                src = str2_descriptor;
                siz = sizeof(str2_descriptor);
        }
        else if (lValue == 3) // serial number
        {
                src = str3_descriptor;
                siz = sizeof(str3_descriptor);
        }

        cond = (unsigned char) (wLength != siz);

        if (siz > wLength) // don't return more than was asked for
                siz = wLength;

        ul_siz = (uint32) siz;
```

```
          iocb.ioctl_code = VOS_IOCTL_USBSLAVE_SETUP_TRANSFER;
          iocb.handle = ctx->in_ep0;
          iocb.request.setup_or_bulk_transfer.buffer = src;
          iocb.request.setup_or_bulk_transfer.size = (int16) ul_siz;
          vos_dev_ioctl(ctx->handle, &iocb);
          return;

default:
          // if drive is connected get descriptor from it
          if (ctx->flashConnected)
          {
                  buffer = vos_malloc(wLength);
                  hc_ioctl.ioctl_code = VOS_IOCTL_USBHOST_DEVICE_SETUP_TRANSFER;
                  hc_ioctl.handle.ep = hostBomsCtx->epCtrl;
                  hc_ioctl.set = &(ctx->setup_buffer[0]);
                  hc_ioctl.get = buffer; // descriptor from drive
                  vos_dev_ioctl(hostBomsCtx->hc, &hc_ioctl);

                  iocb.ioctl_code = VOS_IOCTL_USBSLAVE_SETUP_TRANSFER;
                  iocb.handle = ctx->in_ep0;
                  iocb.request.setup_or_bulk_transfer.buffer = buffer;
                  iocb.request.setup_or_bulk_transfer.size = wLength;
                  vos_dev_ioctl(ctx->handle, &iocb);

                  vos_free(buffer);
          } else {
                  // respond with Request Error
                  set_control_ep_halt(ctx);
          }
   }
}
```

General Purpose Input and Output

The changes presented so far are really all that is required to transform a write blocker into an impersonator. While such a device could be somewhat useful, having displays to indicate write-blocking status, current VID/PID, and allowing input to change mode of operation makes the device much more useful. Enhancing the basic impersonator requires an understanding of how to do general purpose input and output (GPIO) with microcontrollers.

The first thing to understand about doing GPIO with microcontrollers is that one must deal with ports not pins. Those familiar with the AVR microcontrollers as found in the Arduino line or products might dispute this statement, but they would be wrong to do so. The Processing language used by the Arduino allows the programmer to set modes on individual pins. The underlying libraries deal with individual pins in predefined ports, however. Because there are more ports with associated pins (up to 8 per port) than there are pins and there is some flexibility in associating pins with ports, a multiplexer (MUX) is used.

At program startup appropriate function calls are made to associate pins with ports which are labeled with letters and pins within the ports (labeled 0 through 7). Additionally, the pin mode (input, output, bidirectional) can be reset from the default which is typically input. Listing 15.15 shows these function calls for the impersonator. In this application port E is used for the buttons, and the LCD and LEDs are connected to port B.

Listing 15.15 GPIO multiplexer setup

```
void iomux_setup(void)
{
    unsigned char packageType;
    tmr_context_t tmrCtx; // timer context for timer to sequence VID/PID

    packageType = vos_get_package_type();

    // This is for the smaller package. This is probably what you want
    // for a couple of reasons. First off the LCD doesn't fit on a
    // Vinco shield very well. Also, you can leave off the buttons
    // and LCD if you don't want all the functionality, those items
    // will simply be ignored. If you leave out the LCD then you could
    // still build a pretty compact device with just a button and LEDs
    // for the write protect functionality.
    if (packageType == VINCULUM_II_32_PIN)
    {
            // Debugger to pin 11 as Bi-Directional.
            vos_iomux_define_bidi(199,IOMUX_IN_DEBUGGER,IOMUX_OUT_DEBUGGER);
            // GPIO_Port_E_1 to pin 12 as Input.
            vos_iomux_define_input(12, IOMUX_IN_GPIO_PORT_E_1); //VID+ button
            // GPIO_Port_E_2 to pin 14 as Input.
            vos_iomux_define_input(14, IOMUX_IN_GPIO_PORT_E_2); //VID- button
            // GPIO_Port_E_3 to pin 15 as Input.
            vos_iomux_define_input(15, IOMUX_IN_GPIO_PORT_E_3); //write protect button
            // GPIO_Port_B_0 to pin 23 as Output.
            vos_iomux_define_output(23, IOMUX_OUT_GPIO_PORT_B_0); //LCD DB4
            // GPIO_Port_B_1 to pin 24 as Output.
            vos_iomux_define_output(24, IOMUX_OUT_GPIO_PORT_B_1); //LCD DB5
            // GPIO_Port_B_2 to pin 25 as Output.
            vos_iomux_define_output(25, IOMUX_OUT_GPIO_PORT_B_2); //LCD DB6
            // GPIO_Port_B_3 to pin 26 as Output.
            vos_iomux_define_output(26, IOMUX_OUT_GPIO_PORT_B_3); //LCD DB7
            // GPIO_Port_B_4 to pin 29 as Output.
            vos_iomux_define_output(29, IOMUX_OUT_GPIO_PORT_B_4); //LCD RS
            // GPIO_Port_B_5 to pin 30 as Output.
            vos_iomux_define_output(30, IOMUX_OUT_GPIO_PORT_B_5); //LCD E
            // GPIO_Port_B_6 to pin 31 as Output.
            vos_iomux_define_output(31, IOMUX_OUT_GPIO_PORT_B_6); //Green LED
            // GPIO_Port_B_7 to pin 32 as Output.
            vos_iomux_define_output(32, IOMUX_OUT_GPIO_PORT_B_7); //Red LED
    }

    // This is for people who hate to solder and want to base their device
    // on the Vinco. Personally, I think that basing your device off the
    // 32-pin V2DIP1-32 is a better idea. Note that setup for pins 40
    // and 41 is only needed for the Vinco, so if you are running
    // a dev board with a 64-pin chip it is not really needed.
    if (packageType == VINCULUM_II_64_PIN)
    {
```

```
    // Debugger to pin 11 as Bi-Directional.
    vos_iomux_define_bidi(199,IOMUX_IN_DEBUGGER,IOMUX_OUT_DEBUGGER);
    // GPIO_Port_E_1 to pin 12 as Input.
    vos_iomux_define_input(12, IOMUX_IN_GPIO_PORT_E_1); //VID+ button
    // GPIO_Port_E_2 to pin 13 as Input.
    vos_iomux_define_input(13, IOMUX_IN_GPIO_PORT_E_2); //VID- button
    // GPIO_Port_E_3 to pin 14 as Input.
    vos_iomux_define_input(14, IOMUX_IN_GPIO_PORT_E_3); //write protect button
    // for Vinco need to set pins to output for LED and power on host
    // both are active low
    vos_iomux_define_output(40, IOMUX_OUT_GPIO_PORT_E_5); // USB host LED
    vos_iomux_define_output(41, IOMUX_OUT_GPIO_PORT_E_6); // USB host power
    // GPIO_Port_B_0 to pin 24 as Output.
    vos_iomux_define_output(24, IOMUX_OUT_GPIO_PORT_B_0); //LCD DB4
    // GPIO_Port_B_1 to pin 25 as Output.
    vos_iomux_define_output(25, IOMUX_OUT_GPIO_PORT_B_1); //LCD DB5
    // GPIO_Port_B_2 to pin 26 as Output.
    vos_iomux_define_output(26, IOMUX_OUT_GPIO_PORT_B_2); //LCD DB6
    // GPIO_Port_B_3 to pin 27 as Output.
    vos_iomux_define_output(27, IOMUX_OUT_GPIO_PORT_B_3); //LCD DB7
    // GPIO_Port_B_4 to pin 28 as Output.
    vos_iomux_define_output(28, IOMUX_OUT_GPIO_PORT_B_4); //LCD RS
    // GPIO_Port_B_5 to pin 29 as Output.
    vos_iomux_define_output(29, IOMUX_OUT_GPIO_PORT_B_5); //LCD E
    // GPIO_Port_B_6 to pin 31 as Output.
    vos_iomux_define_output(31, IOMUX_OUT_GPIO_PORT_B_6); //Green LED
    // GPIO_Port_B_7 to pin 32 as Output.
    vos_iomux_define_output(32, IOMUX_OUT_GPIO_PORT_B_7); //Red LED
  }
  // setup the timer used to cycle through VID/PID
  tmrCtx.timer_identifier = TIMER_0;
  tmr_init(TIMER0, &tmrCtx);

}
```

Buttons

Pushbuttons are obviously input devices. Normally open pushbuttons (the most common type) are traditional connected to the positive power rail (+5V in our case) through a pull up resistor (5-10k Ohms) on one side and ground on the other. The microcontroller input pin is connected after the resistor on the high voltage side. This is known as an active low switch. When the switch is not pressed the voltage on the pin is approximately 5V. When the switch is pushed the voltage drops to approximately 0V. The pull up resistor limits the current flowing through the switch (shorting directly to ground would be a very bad idea).

As previously mentioned, microcontrollers deal with ports, not pins. The value of the port is read and an appropriate mask is applied to determine if a particular pin was pressed. When reading switches mechanical bouncing in which the internal parts of the switch vibrate causing a series of openings and closings leads to complications. Dealing with this situation is known as debouncing a switch. In the impersonator we use a very simple debouncing method of just waiting for a set period of time and rereading the switch before taking any action. The code for handling the buttons is shown in Listing 15.16.

Listing 15.16 Button handling thread method

```
// This function runs in a low priority thread. Its only
// function is to read the buttons to see if someone has pressed
// them to toggle write protect or change the VID/PID.
// This could have been implemented using GPIO lines with
// interupt capabilities, but interupting the USB threads might
// not be a great idea. Additionally, enabling interupts introduces
// additional overhead.
void handleButtons()
{
    unsigned char buttonBits, leds, counter=0, firstTime=1;
    char str1[17], str2[17];

    while (1)
    {
        vos_dev_read(hGPIO_PORT_E, &buttonBits, 1, NULL);

        // if this is non-zero, somebody is pushing buttons
        // we prioritize the write protect button then +, then -
        while (~buttonBits)
        {
            if ((buttonBits ^ 0xF7)==0) // write protect button pressed
            {
                // since this is pretty serious business we require
                // a long keypress to toggle
                vos_delay_msecs(1000);
                if((buttonBits ^ 0xF7)==0)
                {
                        writeProtect = (~writeProtect & 0x01);
                        leds = writeProtect?led_green:led_red;
                        vos_dev_write(hGPIO_PORT_B, &leds, 1, NULL);
                }


            } else if((buttonBits ^ 0xFB)==0) // VID/PID +
            {
                    autoMode=0; // disable VID/PID scan
                    counter = 0; // reset the counter
                    vos_lock_mutex(&vidPidMutex);
                    currentVidPidIndex += 2;
                    if (currentVidPidIndex > vidPidSize)
                            currentVidPidIndex = 0;
                    vos_unlock_mutex(&vidPidMutex);
                    update_lcd_vidpid();
            } else if((buttonBits ^ 0xFD)==0) // VID/PID -
            {
                    autoMode=0; // disable VID/PID scan
                    counter = 0; // reset the counter
                    vos_lock_mutex(&vidPidMutex);
                    currentVidPidIndex -= 2;
                    if (currentVidPidIndex < 0)
                            currentVidPidIndex = 0;
                    vos_unlock_mutex(&vidPidMutex);
                    update_lcd_vidpid();
            }
            vos_dev_read(hGPIO_PORT_E, &buttonBits, 1, NULL);
            vos_delay_msecs(50);
        }
```

```
        vos_delay_msecs(100);
        // if we haven't hit a button for 5 seconds then it is time to
        // move on and start using the device
        if (firstTime)
                counter++; // increment the counter
        if (counter > 50)
        {
                firstTime=0;
                vos_signal_semaphore(&setupDoneSemaphore);
        }
    }
}
```

Using LEDs

Connecting LEDs to a microcontroller is fairly straightforward. Connect the positive LED pin (the one with the longer lead) to the microcontroller pin. Connect the other side to ground through a current limiting resistor. You may need to do some experimentation with the current limiting resistor to obtain the desired brightness. Also, if you go too large with this resistor you might fail to turn on the LED. Because I only have two LEDs and only one is lit at a time, I used a single current limiting resistor connected to the negative side of both LEDs. The code for turning the LEDs on and off is presented in Listing 15.17.

Listing 15.17 LED methods

```
// LED port masks
#define led_green 0x80
#define led_red 0x40

void light_red_led(VOS_HANDLE hLED)
{
    unsigned char toggle;
    // first we read port to not mess up the LCD display
    vos_dev_read(hLED, &toggle, 1, NULL);
    toggle &= (~led_green);
    toggle |= led_red;
    vos_dev_write(hLED, &toggle, 1, NULL);
}

void light_green_led(VOS_HANDLE hLED)
{
    unsigned char toggle;
    vos_dev_read(hLED, &toggle, 1, NULL);
    toggle &= (~led_red);
    toggle |= led_green;
    vos_dev_write(hLED, &toggle, 1, NULL);
}
```

Table 15.6 LCD screen pinouts

Pin	Signal	Usage	Comments
1	Ground	Ground	
2	Vcc	+5V	
3	Vo	Contrast	Connected to 10K pot center tap
4	Rs	B4	Read select
5	Rw	Ground	Active low write enable
6	E	B5	Enable display
7	DB0	Not used	DB0-3 only used 4 sending 8-bit values
8	DB1	Not used	
9	DB2	Not used	
10	DB3	Not used	
11	DB4	B0	DB4-7 are used for sending 4-bit values
12	DB5	B1	
13	DB6	B2	
14	DB7	B3	
5	BL+	Not used	Positive for backlight if present
16	BL-	Not used	Negative for backlight if present

Using LCD Screens

Small LCD screens with 16 characters by 2 lines are a standard output device for microcontrollers. As a result, there are many libraries available for these devices. I shamelessly modified Arduino LCD library code to work with the Vincullum II. These displays have 16 pins, not all of which are typically used. The standard pinout for these displays and usage in the impersonator are shown in Table 15.6. Note that the 10K contrast potentiometer is connected to +5V and ground with the center pin connected to Vo on the LCD.

Because GPIO pins are often at a premium only 4 data pins are used to send messages to the screen. The screen responds to a number of standard commands that include things such as clearing the screen and moving the cursor. The methods used to interface with the LCD are shown in Listing 15.18. Note that the LCD and LEDs are on the same port. As a result, the port value is first read to avoid messing up the LEDs when printing to the LCD.

Listing 15.18 LED methods

```
/// LCD control signals
#define lcd_rs   0x10
#define lcd_e    0x20

// Send a command to our LCD display
void write_lcd_cmd(VOS_HANDLE hLCD, unsigned char byte)
{
  unsigned char cmd, leds;
    // first read state to not mess up LEDs on same port
    leds = writeProtect?led_green:led_red;

  // Write High nibble data to LCD
  cmd = (((byte >> 4) &0x0F) | lcd_e);
  cmd = (cmd &(~lcd_rs)) | leds; // Select Registers
  vos_dev_write(hLCD,&cmd,1,NULL);
  // Toggle 'E' pin
  cmd &= (~lcd_e);
  vos_dev_write(hLCD,&cmd,1,NULL);
  // Write Low nibble data to LCD
  cmd = ((byte &0x0F) | lcd_e);
  cmd = (cmd &(~lcd_rs)) | leds; // Select Registers
  vos_dev_write(hLCD,&cmd,1,NULL);
  // Toggle 'E' pin
  cmd &= (~lcd_e);
  vos_dev_write(hLCD,&cmd,1,NULL);
  vos_delay_msecs(1);

}

// Send data to LCD display
void write_lcd_data(VOS_HANDLE hLCD, unsigned char byte)
{
  unsigned char cmd, leds;

    // first read state to not mess up LEDs on same port
    leds = writeProtect?led_green:led_red;

    // Write High nibble data to LCD
  cmd = (((byte >> 4)&0x0F) | lcd_rs);
  cmd = (cmd | lcd_e) | leds; // Select DDRAM
  vos_dev_write(hLCD,&cmd,1,NULL);
  // Toggle 'E' pin
  cmd &= (~lcd_e);
  vos_dev_write(hLCD,&cmd,1,NULL);
  // Write Low nibble data to LCD
  cmd = ((byte & 0x0F) | lcd_rs);
  cmd = (cmd | lcd_e) | leds; // Select DDRAM
  vos_dev_write(hLCD,&cmd,1,NULL);
  // Toggle 'E' pin
  cmd &= (~lcd_e);
  vos_dev_write(hLCD,&cmd,1,NULL);
  vos_delay_msecs(1);

}

// Write a string at the current cursor position
```

```c
void write_lcd_str(VOS_HANDLE hLCD, char *str)
{
  while(*str != '\0')
  {
    write_lcd_data(hLCD, (unsigned char*)*str);
    ++str;
  }
}

// Attempt to init the LCD display
void lcd_ini(VOS_HANDLE hLCD)
{
  vos_delay_msecs(100);
  // Send Reset command
  write_lcd_cmd(hLCD, 0x03);
  vos_delay_msecs(2);
  // Send Function Set
  write_lcd_cmd(hLCD, 0x28);
  vos_delay_msecs(2);
  write_lcd_cmd(hLCD, 0x28);
  vos_delay_msecs(2);
  // Send Display control command
  write_lcd_cmd(hLCD, 0x0C);
  vos_delay_msecs(2);
  // Send Display Clear command
  write_lcd_cmd(hLCD, 0x01);
  vos_delay_msecs(2);
  // Send Entry Mode Set command
  write_lcd_cmd(hLCD, 0x06);
  vos_delay_msecs(2);
}

// Clear LCD and reset cursor
void lcd_clear(VOS_HANDLE hLcd)
{
  // Send Display Clear command
  write_lcd_cmd(hLcd, 0x01);
  vos_delay_msecs(2);
}

// Write to the top line of our display
void write_lcd_line1(VOS_HANDLE hLcd, char* str)
{   // Set 1-st line address
  write_lcd_cmd(hLcd, (0x01 | 0x80));
  // Send string to LCD
  write_lcd_str(hLcd, str);
}

// Write to the bottom line of our display
void write_lcd_line2(VOS_HANDLE hLcd, char* str)
{
  // Set 2-nd line address
  write_lcd_cmd(hLcd, (0x40 | 0x80));
  // Send string to LCD
  write_lcd_str(hLcd, str);
}
```

Summary of USB Impersonation

In this section we have seen how a powerful USB impersonator can be created by extending the USB write blocker from a previous section. Along the way we also received a brief introduction to performing GPIO with microcontrollers. Code and schematics for the devices presented are available online at https://github.com/ppolstra.

In the next section we exam some open source tools which can prove useful when performing USB forensics and/or debugging USB devices.

15.10 Leveraging Open Source

Previous sections in this chapter have focused on how to construct devices and perform forensics investigation of USB mass storage devices. This section is intended to show investigators how to leverage open source software in order to investigate USB devices. Everything described here could also be done with commercial hardware and software, but budgets do not always allow for commercial solutions.

Identifying Devices

When encountering an unknown USB device, the first step is to identify the device. We will use Linux and standard tools to identify the device. Linux is used because it is an operating systems by programmers and for programmers. As such, the selection of open source tools available on the Linux platform is unmatched anywhere else.

For this tutorial I will use a small flash drive and "unknown" USB device. The first step in identification is to use the lsusb utility (which should be installed by default in most Linux distributions) to list all properly enumerated devices. The results of running this command on my system are shown in Fig. 15.9.

Notice that a series of USB hub devices are listed in the lsusb results. There are a couple of important things to note about this. There are two USB 2.0 root hubs. One of these, on Bus 001, is used for built in USB devices such as the multi-card reader. The other high-speed hub, on Bus 002, is used to communicate with devices capable of high-speed communication that have been plugged into USB ports (possibly downstream from high-speed hubs). From the results we see the flash drive is a SanDisk Cruzer Blade which is capable of high-speed. The mystery device reports itself as a Prolific PL2303 serial port.

The PL2303 is attached to Bus 006. This tells us that it is not capable of high-speed. When a slow device is attached it is given its own bus so that it doesn't bog down the high-speed buses. It can be helpful to connect devices through a USB 1.1 hub when debugging and/or investigating them as this will facilitate sniffing by creating a bus with a single device. Care should be taken, however, as the behavior of some devices changes with different speeds.

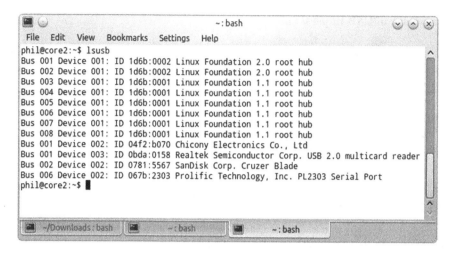

Fig. 15.9 Results of running lsusb

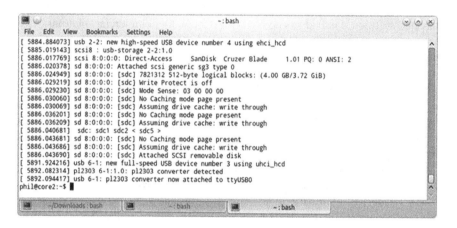

Fig. 15.10 Results of running dmesg

Our two devices have enumerated successfully. When this isn't the case or when more information is desired, the device message logs may prove helpful. Linux systems all ship with the dmesg utility for listing these messages. The results from the dmesg command are shown in Fig. 15.10.

The dmesg results provide more information about the SanDisk drive. We see that it is 4 gigabytes, use 512 byte blocks, has no write protection, and uses a SCSI command set. Additionally, we see that it has been labeled as/dev/sdc. The PL2303 has been attached to Bus 006 and also attached to ttyUSB0.

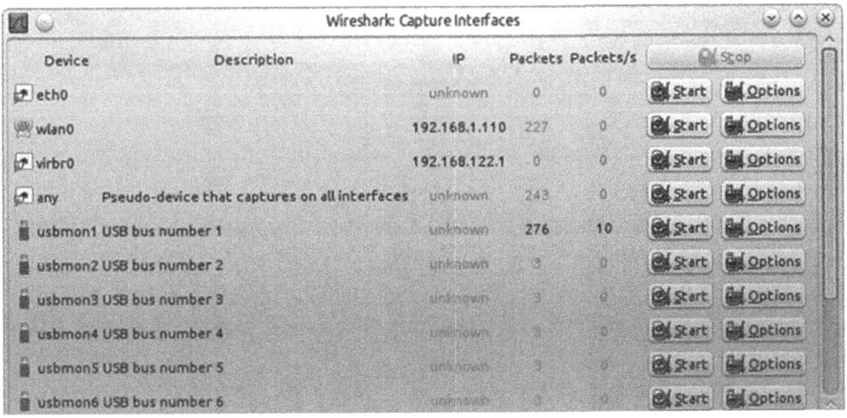

Fig. 15.11   Wireshark capture interfaces

Sniffing USB Traffic

Using two standard tools we have made progress in the investigation of our two devices. In order to go any further we must start intercepting USB traffic. Fortunately, this is quite easy to do on Linux systems. A USB monitoring device is easily created by loading the usbmon module by executing the command "sudo modprobe usbmon". Once this command has been executed new capture devices will appear in Wireshark and similar packet capture tools. These new interfaces are shown in Fig. 15.11.

Let's have a look at the PL2303 traffic, which we know will appear on Bus 006. As shown in Fig. 15.12, Wireshark understands and interprets USB traffic as it does with network traffic.

From the configuration descriptor which is partially shown in Fig. 15.13, we can see the USB device type and other useful information. The device is not self-powered, does not respond to remote wakeup requests, requires no more than 100mA of current, has one interface, one out endpoint, and two in endpoints. While not shown here, the PL2303 is also queried for description and manufacturer strings.

With very little effort, we have captured the enumeration traffic for the PL2303. If desired we could also capture traffic to this device. Let us now turn our attention to the SanDisk drive instead. As previously mentioned, because the device is high-speed capable, it is connected to high-speed Bus 002. There is considerable chatter on this bus. Similar to floods of beacon frames in wireless sniffing, USB sniffing is done in a sea of status requests and test unit ready requests. Many operating systems will continuously ping mass storage devices with test unit ready requests to keep them from sleeping. The device descriptor is shown in Fig. 15.14. Notice the device descriptor indicates type 0 which means the type will be revealed later in the configuration descriptor.

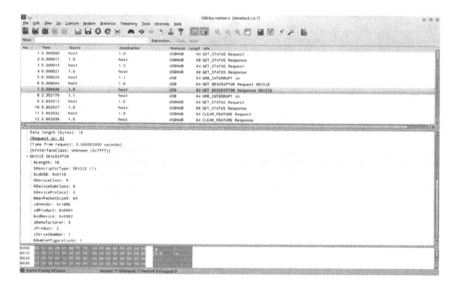

Fig. 15.12 Wireshark capture of a device descriptor

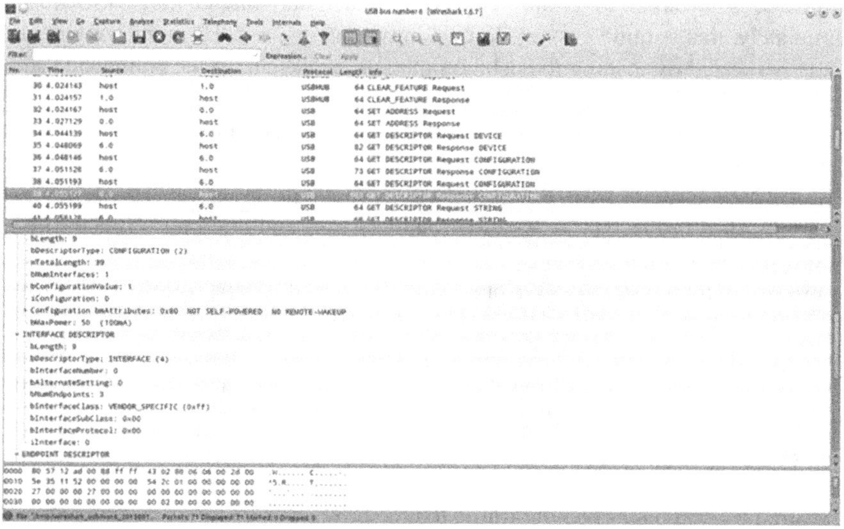

Fig. 15.13 Wireshark capture of a configuration descriptor

A partial capture of the configuration descriptor is shown in Fig. 15.15. We see from the descriptor that the SanDisk can require up to 200 mA of current. This much current could power an entire small computer board, such as the BeagleBone Black. The descriptor also tells us that it is a mass storage device that uses the SCSI protocol and that each of the bulk endpoints has a maximum packet

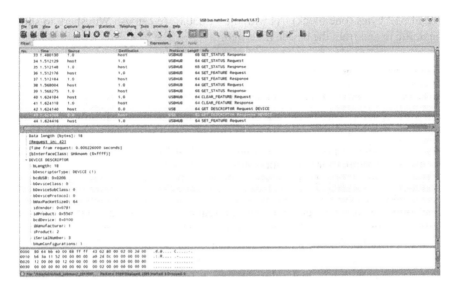

Fig. 15.14   Wireshark capture of a device descriptor for flash drive

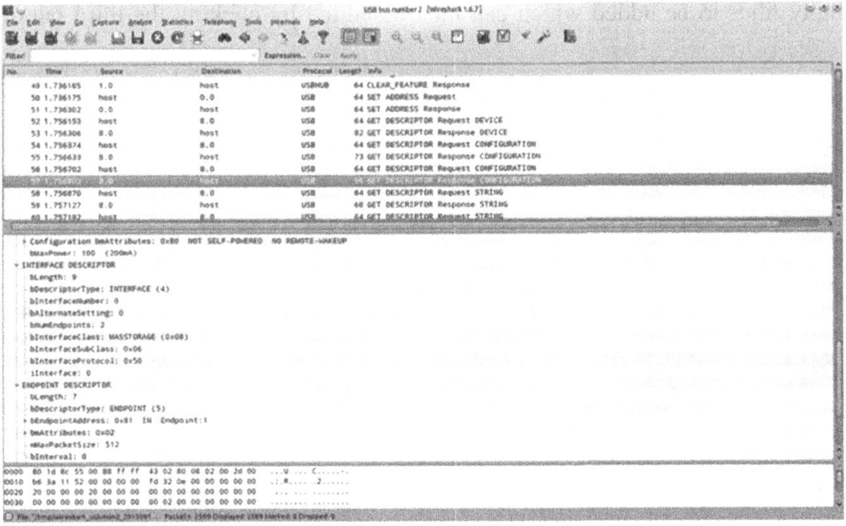

Fig. 15.15   Wireshark capture of a configuration descriptor for flash drive

size of 512 bytes (one block). As with the PL2303, we can use Wireshark to sniff the traffic to and from the device if desired.

Dealing with Windows-Only Devices

So far we have learned how to handle devices which operate on all the standard operating systems. One might ask how to deal with devices that only work with Windows. The solution is actually quite simple. Setup your Linux system to capture USB traffic as previously described and then run Windows virtually using VirtualBox or similar virtualization software.

For example, the SanDisk Cruzer Profile drive is a Windows-only mass storage device that only permits access to the stored data after a successful fingerprint scan is performed with the integrated fingerprint scanner. To access such a device in our virtual machine we must tell VirtualBox to pass all the traffic through and to have the host operating system ignore it. We can either do this each time we use the device from the Devices menu in the window which contains the Windows virtual machine or permanently make this the case using the settings for our Windows Virtual box.

A screenshot for this more permanent method is shown for the Profile drive in Fig. 15.16. The first button on the right with the blue circle allows a new empty filter to be added which can then be edited by clicking the third button

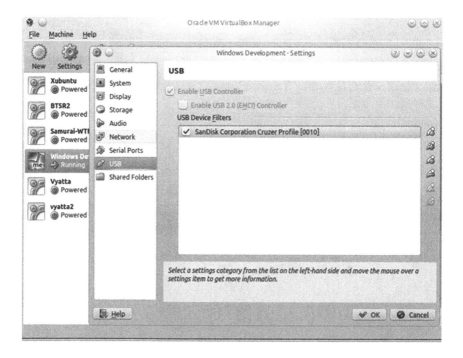

Fig. 15.16 Passing through USB devices to an underlying virtual machine

with the orange circle. Clicking on the second button with the green plus sign allows a filter for a currently connected device to be added. I will leave it as an exercise to the reader to reverse engineer the Profile drive and/or develop a Linux driver.

Fun with Udev Rules

In an earlier section, we discussed using microcontrollers to block writes to USB mass storage devices. When running Linux we can accomplish the same thing using udev rules. Udev rules affect the way devices are labeled, what drivers are loaded, etc. We can use udev rules to force any mass storage devices connected downstream from a hub with a particular VID/PID combination to be mounted read-only. One of the big advantages this has over our microcontroller solution is that we can operate at high speed. While full speed is fine for flash drives, it would take much too long to create an image of a large hard drive with our microcontroller-based write blocker.

What is presented here is not intended as a complete tutorial on udev rules, but rather as a concrete example of their power. This information is taken from a forensics module for The Deck. The Deck is a custom penetration testing and forensics Linux distribution that runs on the BeagleBoard-xM, BeagleBone, and BeagleBone Black ARM-based devices. Because this module was implemented entirely using udev rules it can be used on any Linux distribution. More information on The Deck can be found at http://philpolstra.com or in the book *Hacking and Penetration Testing with Low Power Devices* by Philip Polstra (Syngress, 2015). This module, known as the 4Deck, can also be downloaded from https://sourceforge.net/projects/thedeck/files/.

All Linux systems ship with a standard set of udev rules. While these could be modified directly, doing so is not recommended. The suggested practice is to create new rules in the/etc./udev/rules.d directory. Similar to the workings of startup scripts, filenames are used to determine the order in which udev rules are processed. As a result, rules normally start with a priority number followed by a descriptive name and must end with a ".rules" extension. The appropriate rules for a hub with a VID/PID of 0x1a40/0x0101 are shown in Listing 15.19.

Listing 15.19 Sample udev rules

```
    ACTION=="add", SUBSYSTEM=="block", KERNEL=="sd?[1-9]", ATTRS{idVendor}=="1a40",
ATTRS{idProduct}=="0101", ENV{PHIL_MOUNT}="1", ENV{PHIL_DEV}="%k",
RUN+="/etc/udev/scripts/protmount.sh %k"
    ACTION=="remove", SUBSYSTEM=="block", KERNEL=="sd?[1-9]", ATTRS{idVendor}=="1a40",
ATTRS{idProduct}=="0101", ENV{PHIL_UNMOUNT}="1", RUN+="/etc/udev/scripts/protmount3.sh %k"

    ENV{PHIL_MOUNT}=="1", ENV{UDISKS_PRESENTATION_HIDE}="1",
ENV{UDISKS_AUTOMOUNT_HINT}="never", RUN+="/etc/udev/scripts/protmount2.sh"
    ENV{PHIL_MOUNT}!="1", ENV{UDISKS_PRESENTATION_HIDE}="0",
ENV{UDISKS_AUTOMOUNT_HINT}="always"

    ENV{PHIL_UNMOUNT}=="1", RUN+="/etc/udev/scripts/protmount4.sh"
```

The first line is matched whenever a block device is added downstream from a device with the appropriate VID/PID (our special hub). When this happens two environment variables PHIL_MOUNT and PHIL_DEV are set. The kernel name for the new device (i.e. "sdb1") is substituted for %k when the rule is run. In addition to setting the environment variables we add protmount.sh with a parameter of our device's kernel name to a list of scripts to be ran. The second line is similar and is matched upon device removal.

The next two lines are used to prevent the message box from popping up when a read-only device is connected through our hub as indicated by PHIL_MOUNT having a value of 1 and to cause the message box to pop up otherwise (PHIL_ MOUNT not equal to 1). The last line adds a script to the run list when the read-only device is mounted.

The protmount.sh and protmount3.sh scripts simply create protmount2.sh and protmount4.sh, respectively. These files are stored in the/etc./udev/scripts directory. The reason that these scripts create other scripts is that the device has not been fully loaded at the point when our rules are matched. The protmount.sh and protmount3.sh scripts are presented in Listing 15.20 and Listing 15.21, respectively. The protmount.sh mounts the mass storage device in the standard place, but read-only. The protmount3.sh script performs cleanup tasks such as unmounting the drive and removing the mount point directory.

Listing 15.20 protmount.sh script

```
#!/bin/bash
echo "#!/bin/bash" > /etc/udev/scripts/protmount2.sh
echo "mkdir /media/$1" >> /etc/udev/scripts/protmount2.sh
echo "chmod 777 /media/$1" >> /etc/udev/scripts/protmount2.sh
echo "/bin/mount /dev/$1 -o ro,noatime /media/$1" >> /etc/udev/scripts/protmount2.sh
chmod +x /etc/udev/scripts/protmount2.sh
```

Listing 15.21

```
#!/bin/bash
echo "#!/bin/bash" > /etc/udev/scripts/protmount4.sh
echo "/bin/umount /dev/$1" >> /etc/udev/scripts/protmount4.sh
echo "rmdir /media/$1" >> /etc/udev/scripts/protmount4.sh
chmod +x /etc/udev/scripts/protmount4.sh
```

Summary of Open Source USB Forensics Tools

In this section we have seen how a few simple open source tools can be used to make USB forensics easier and considerably cheaper. In the next, and final, section we will examine how BadUSB and similar threats can be mitigated using what has been learned in earlier parts of this chapter.

15.11 BadUSB

BadUSB is used to describe a family of computer attacks that involve reprogramming the specialized microcontroller on a USB peripheral in order to do something malicious. In many cases the peripheral is made to emulate another device such as a USB keyboard. The United States government is credited with initially coming up with this idea.

A natural question to ask is how organizations can protect against such threats. Some have said there is no defense against BadUSB. Readers of this chapter should realize that this is false. The USB write blocker described earlier in this chapter provides bidirectional protection. The write-blocking functionality buffers the attached thumb drive in order to prevent modification to the drive. The fact that the blocker mounts the drive and then emulates a drive (and only a drive) protects the PC as well.

Unfortunately, the udev rules software write blocking solution does not provide the same protection. All hope is not lost for Linux systems, however. The same udev rules could be used to create a whitelist of USB devices which might be attached to the system. This would eliminate the possibility of altered peripheral emulating additional devices. Such as system protects the PC, but provide no protection for the drive which could affect other systems not protected by udev rules.

Additional Resources

1. http://www.usb.org/developers/docs/ — The official source for all things USB
2. http://lvr.com/usbc.htm — USB expert Jan Axelson's main USB page
3. http://www.youtube.com/watch?v=3D9uGCvtoFo — *Phil Polstra's DEFCON XX Talk on USB impersonation*
4. http://www.youtube.com/watch?v=CIVGzG0W-DM — *Phil Polstra's presentation on building a USB impersonator*
5. http://ppolstra.blogspot.com — Phil Polstra's blog
6. http://twitter.com/ppolstra — Phil Polstra's Twitter page (@ppolstra)
7. http://www.reactivated.net/writing_udev_rules.html — udev tutorial
8. http://www.philpolstra.com — Phil Polstra's page
9. Polstra, Philip. *Hacking and Penetration Testing with Low Power Devices.* Syngress (2015). Everything you wanted to know about hacking and forensics with the BeagleBone Black and similar boards
10. Axelson, Jan. *USB Complete (4ᵗʰ Ed.).* Lakeview Research (2009).
11. Axelson, Jan. *USB Mass Storage.* Lakeview Research (2006).

Chapter 16
DARPA's Cyber Grand Challenge (2014–2016)

> *The networked civilization we are building is going to need to be able to make strong promises about the safety of software, because it won't just be guarding our data security — it will be guarding our physical security.*
> — Mike Walker, DARPA Project Manager (June 3, 2014)

16.1 Cyber Grand Challenge Kick-off

On June 3, 2014, DARPA (Defense Advanced Research Projects Agency) kicked off the first-ever Cyber Grand Challenge. DARPA project manager Mike Walker and Naval Postgraduate School (NPS) lecturer Chris Eagle announced on reddit [1]:

> We're excited to share that our Cyber Grand Challenge kicks off today. The Cyber Grand Challenge is simple: we've challenged the world to build high-performance computers[1] that can play Capture the Flag (CTF), a computer security competition format pioneered and refined by the hacker community over the past two decades. CGC is much more than a tournament with up to $9 million in prizes (including a $2 million top prize).

> As part of today's launch event we're releasing a brand new reference platform for CTF: an Open Source platform purpose-built for security research and competition. It is, in effect, a parallel digital universe in which this two-year-long competition will play out. We've even included a few sample challenges to get the fun started.

> Our competition framework team is assembled here at DARPA, so we'll be here throughout the day to answer questions here. We're looking forward so AUA!

---

[1]The computer hardware in CTF must fit entirely in a single, standard 19″ 42U rack [24].

© Springer International Publishing Switzerland 2015
N. Lee, *Counterterrorism and Cybersecurity*, DOI 10.1007/978-3-319-17244-6_16

edit: proof! http://i.imgur.com/wL1bnL9.jpg[2]

Thanks everyone! -Mike, Chris, and Team CGC

The CGC launch on reddit AMA (Ask Me Anything) or AUA (Ask Us Anything) attracted about 128 comments. The following were some of the most popular Q&A's on reddit:

Q [rataza]: The current state of the art (i.e. KLEE[3] and S2E[4]) is capable of automatically discovering some bugs in command line tools and straight forward network daemons. Will you provide challenges that they or similar tools can solve? Or will the challenges resemble complex network daemons, like SMTP servers or OpenSSH? It seems that would require a generational leap in automated tools.

A [Chris Eagle]: Our goal is simple: start with the current state of the art and push it as far as we can. As such, we expect our challenges to range in difficulty from solvable today to potentially unsolvable even at the conclusion of CGC. Ideally by the time we are done the technology that has been developed will be able to solve a much larger percentage of our challenges than can be solved today.

For program analysis, we've simplified the problems of isolating entropy, input and output from the operating system down to a bare minimum. We have just seven system calls with no polymorphism or ambiguity in the ABI. Our simple binary format has a single entry point method and no dynamic loader. DECREE's "OS tax", the bane of automation research, is as close to zero as any platform in existence.[5]

Q [gynophage]: You seem to have a lot invested in the attack/defense model of computer security competition. I've heard arguments from many players that the current model of attack/defense CTF is "stale". Do you believe these events are stale? If so, do you think there will be any innovations from the Cyber Grand Challenge that the CTF community will be able to use to continue generating

---

[2]Figure 16.1 is the Cyber Grand Challenge infrastructure team at the DARPA office building on June 3, 2014. Chris Eagle wrote on reddit: "The photo is our AMA 'proof'. Many of us brought totems from our former work. The sheep is the mascot of DDTEK, past organizers of DEFCON CTF. The books on her lap are the Federal Acquisition Regulations (FAR) and Defense Federal Acquisitions Supplement (DFARS) (which may be relevant to one of the challenge binaries). The other paper is the front page of the science section of today's New York Times."

[3]KLEE is a symbolic virtual machine built on top of the LLVM compiler infrastructure, which uses a theorem prover to try to evaluate all dynamic paths through a program in an effort to find bugs and to prove properties of functions. A major feature of KLEE is that it can produce a test case in the event that it detects a bug [26].

[4]S2E is a platform for analyzing the properties and behavior of software systems. S2E has been used to develop practical tools for comprehensive performance profiling, reverse engineering of proprietary software, and bug finding for both kernel-mode and user-mode binaries [27].

[5]DECREE = DARPA Experimental Cyber Research Evaluation Environment. DECREE is an open-source extension built atop the Linux operating system. Constructed from the ground up as a platform for operating small, isolated software test samples that are incompatible with any other software in the world—DECREE aims to provide a safe research and experimentation environment for the Cyber Grand Challenge [24].

Fig. 16.1 AMA "proof" — Cyber Grand Challenge infrastructure team at the DARPA office building (June 3, 2014)

engaging challenges for human experts, as well as compelling visualizations or data for spectators?

A [Mike Walker]: I don't think the CTF circuit is stale. In 2014 it's bigger than ever. The Secuinside CTF pre-qualifiers held last weekend had 940 teams!

If anything is stale, it's the state of software safety. In 1995 Matt Blaze[6] ranked "the sorry state of software" first in his list of security issues in his Afterword to *Applied Cryptography* and when he revisited this fifteen years later, software safety was still problem #1.

CGC will be contributing back to the CTF community, starting with our release today of DECREE. We're working on new ways to visualize competitions and see inside Bratus's "weird machines" [2].[7]

Q [fuzyl]: I agree completely that the CTF circuit isn't stale, but in terms of competition design, we still have mostly "Jeopardy-style" competitions and the occasional "Full Spectrum-style" attack/defend competition like the DEFCON Finals. It pains me to say it, but UCSB's iCTF is really one of the only CTFs that has attempted to do innovation upon an actual CTF's design.

Are there any innovations on the design of these competitions that CGC can offer? Do you have any thoughts on what the CTF community at large could improve upon as a whole?

---

[6]Matt Blaze, Associate Professor of Computer and Information Science at the University of Pennsylvania, is a researcher in the areas of secure systems, cryptography, and trust management (a term that he coined) [28].

[7]In computer security, the "weird machine" is a computational artifact where additional code execution can happen outside the original specification of the program.

A [Mike Walker]: There's great innovation happening in the CTF community. Take a look at Build It/Break It/Fix It: https://builditbreakit.org/, funded by the National Science Foundation [3].

CGC intends that machines will someday walk in the footsteps of experts. That's why we modeled it in the tradition of the world's biggest, longest-running CTF.

At the end of the day though we believe that the dream of automation is the biggest thing that CGC can contribute. CGC will address the problematic economics of bug checking and defending large code bases head on.

Q [Psifertex]: You've mentioned before that you expect this effort to be the result of combined work by CTF experts (hackers), and program analysis experts (scientists/researchers). If you'll excuse me for going all D&D, these seem like they're on opposite ends of the lawful/chaos spectrum. How do you expect these two to actually balance each other? What do you expect will be the right balance between the two? Or are they really not so different?

A [Mike Walker]: I think what's exciting about challenges is their ability to bring diverse communities together to form a new one. If you watch the documentary *Charge*, you can watch the electric vehicle community and professional motorcycle racers come together to try to build the first generation of electric racing motorcycles. The community that was created through the pressure of competition was a new community. So to ponder this question, does the future of computer security belong to the people doing the work now, or the academics trying to automate its future? Our answer is *yes*.

Q [rolfr]: What specifically are the limitations of current program analysis technologies? Which are likely to remain in the long term? What should be researched further at present so as to improve the state of the art? What do you think the most promising research directions in program analysis are at present?

A [Mike Walker]: I've heard we should ask Rolf Rolles!;-)[8] Halvar[9] has been checking the bounds of static analysis as well [4]. You may notice that in Cyber Grand Challenge we haven't set research directions — instead we hope to learn the answers to your question through open competition. Memory aliasing, complex loop satisfaction, state space explosion, take your pick. We hope to see new answers developed and fielded in competition over the next two years.

---

[8]From Rolf Rolles' LinkedIn profile [23]: Seventeen years of reverse engineering: malware, interoperability, security assessment (including patch diffing), copy protections, exploit development. First researcher to publicly break virtualization obfuscators (older versions of VMProtect; current versions of TheMida CISC VM). Non-trivial C/C++/OCaml reversing tool development, including techniques from compiler theory, program analysis, and formal verification. Was the lead developer of BinDiff in ancient times.

[9]From Halvar Flake's LinkedIn profile [25]: Staff Engineer at Google, Zürich Area, Switzerland. I like to work on challenging problems related to computer security; ideally with a heavy algorithmic / mathematical bent. Technical work: - Reverse engineering of malicious software - Reverse engineering of COTS software - Security analysis of software, vulnerability/exploit development - Engineering of large distributed systems - Large-scale analysis of executable code (large both in terms of volume and individual size) - Static analysis, formal methods - Applying mathematics / statistical inference / "machine learning" to real-world problems.

Reply [rolfr]: It was a trick question! If you're in charge of funding academic things, this here is the future: http://dl.acm.org/citation.cfm?id=2535868[10]

Q [rataza]: You published that all Challenge Binaries will be written in C (or a C like language). Will they include tricks based on how the compiler lays out data in memory? Or will they work independent of what the compiler does?

I.e. void (struct secret *b) {int i; unsigned int *p = &i; p[3] = some_other_ secret; ...}

A [Mike Walker]: Here's what we've said about Challenge Binaries so far:

CBs will be compiled
CBs will be written in the C language family
CBs will be compiled using the CGC platform compiler

If we could make guarantees about CB memory safety, there wouldn't be a Cyber Grand Challenge.;-)

Q [fuzyl]: How do you feel about professional video game teams being paid more than professional hacking teams? Dota 2 looks like it might beat your prize pool pretty soon: http://www.dota2.com/international/compendium/.

A [Mike Walker]: The rise of esports has been fascinating to track, and interestingly CGC shares some esports challenges. We're going to be holding our competition in front of a live audience, and that will require us to visualize things that have long been marooned in the world of decompilers, debuggers and analysis tools. We think that the video game industry has something to contribute to this solution.

Q [HockeyInJune]: I'm giving a talk at Black Hat this summer on using simple program analysis techniques to aid a human in vulnerability discovery. In many ways, some people see this as the antithesis of the Cyber Grand Challenge, where you want teams to build a fully automated system to discover, patch, and prevent vulnerabilities.

What do you say to researchers, who are seen as experts, that say that such a system can never exist?

Just to be clear, I have a lot of respect for the CGC program and the team running it.

A [Chris Eagle]: Turing tells us that we can never expect a system that can find every bug, but we like the idea of having an automated system do triage for us. This allows human analysts to focus on the bugs that technology fails to locate and we think this is a good thing. Pushing this technology further, the gap between what experts can find and what machines can find may slowly vanish.

Q [mattshockl]: What do you hope the endstate will be with immune computers? Where do you see this technology progressing in the years to come?

---

[10]In the paper "Abstract Satisfaction", Vijay D'Silva, Leopold Haller, and Daniel Kroening introduced a framework for applying abstract interpretation to problems that are NP-hard but decidable, such as satisfiability [29].

A [Mike Walker]: It's worth pointing out that no vehicles finished the first DARPA vehicle challenge in 2004. We consider the future of this technology to be an open question. We've seen technology signs that show that the field of *program analysis*, or the automated study of software and its execution, is starting to close the gap on the abilities of experts.

This is exciting to us because of the economics of computer security: Attackers have the concrete and often inexpensive task of finding a single flaw to penetrate a system, while Defenders have the un-provable and hugely expensive task of finding all flaws (thousands?) and patching them before an attack occurs. We believe automation can upend these economics. So the end state we'd like to see is for computer security to become the expert domain of machines.

Q [LocSic]: So it's been mentioned countless times the divide between the two sides in CGC, CTF experts (hackers), and program analysis experts (scientists/ researchers). Do you feel there could exist a middle-ground between the two or will there always be a divide - culture and skillset-wise. Also what are your opinions on the interlinking relationship, do you find that the security researcher is to service the hackers more (i.e.: tool development for hackers) or the opposite where hackers are to service the security researcher(feeding and strengthening program analysis).

A [Mike Walker]: CGC is intended to help build a new middle ground. I am hopeful.

Q [emberfang]: Once you have this technology (that mechanically builds exploits for systems), how will the vast amount of information it outputs be used?

Would this software be publically available? Otherwise software developers would have to wait for disclosure, this would be something happening at a much slower rate than vulnerabilities were being discovered and weaponized. This bottleneck will cause classified vulnerabilities to pile up.

A [Mike Walker]: There may be some confusion here in the premise of your question.

Cyber Grand Challenge is a completely unclassified program. We're doing open technology development, primarily incentivized with prizes. The hope is the creation of an open automation revolution in computer security.

During CGC, competitor systems will only analyze a collection of Challenge Binaries — software built from scratch that shares no protocols or code with the real world. It runs on DECREE, an incompatible-by-design platform built to support our events (and perhaps someday, others like it).

Finally, we've never required competitor systems to "exploit", or gain execution of arbitrary code in Challenge Binaries. We require only minimum Proof of Vulnerability — for example in CQE, faulting inputs. More information is available in our FAQ posted on our Competitor Portal.

Q [chooter]: So basically, as per the NY times, are you guys building an immune system for computers?

A [Mike Walker]: Innate immunity is just one aspect of what we're hoping competitors will build. CGC will be played by automated systems, and part of their job will be to process never-before-seen software and build secure replacements. Competitor systems may take an immune system approach[11] - we'll see!

Q [hso]: what's the cgc team mascot?
A [Mike Walker]: We don't have a mascot, but we do have a logo. We printed a 3d copy for today's AMA picture. Now it needs a home...
Edit: Please note - banana to scale. https://i.imgur.com/0kiBg65.jpg[12]

Q [karmanaut]: How do you feel about the fact that a lot of technology only seems to advance in different ways to wage war? Do you wish that you were involved in development of something for more peaceful purposes?
A [Mike Walker]: I don't accept that premise. We're doing open technology development, incentivized primarily by prizes. We expect that the innovators who will answer the call of CGC will be interested in building the next generation of computer security products. The market forces that will push this type of technology are clear.

In April of 2014, insurers started selling insurance products that covered physical harm generated by cyber effects — available via a Google search for "cyber insurance" "property damage". In May of 2014, Sky News reported that over 42,000 London cars — nearly half of the cars stolen in the city of London — were stolen with hacking. The networked civilization we are building is going to need to be able to make strong promises about the safety of software, because it won't just be guarding our data security — it will be guarding our physical security. If we're going to be able to make strong promises about software safety, we're going to need automation that can investigate software in a uniform, scalable and effective manner. We know that expert auditors can't get there — IBM/Rational points out that our civilization crossed 1 trillion lines of code in the early 2000s. Operating systems weigh in above 40 million lines under constant development. The problem is too big and it's moving too fast. We also know that today's automation is losing every contest of wits to experts — in the wake of Heartbleed, not a single automation product has come forward to say that this flaw could have been detected without expert annotation or intervention. CGC is open technology development on the problem of software safety, a problem seen by the DoD — and everyone with a vested interest in our connected future.

16.2 Costly Software Bugs

Software bugs can have devastating effects in the financial world. The 2010 Flash Crash is a prime example: On the afternoon of May 6, 2010, the Dow Jones Industrial Average (DJIA) plunged almost 1,000 points due to a technical glitch

---

[11]See, for example, "An Immunological Model of Distributed Detection and Its Application to Computer Security" by Steven Andrew Hofmeyr [30].
[12]Figure 16.2.

Fig. 16.2   Cyber Grand Challenge logo

that triggered erroneous trading in Procter & Gamble and several other NYSE stocks [5].

In August 2012, a software bug on Knight Capital's computers executed a series of automatic orders within an hour, resulting in a loss of $440 million and bringing the company to the edge of bankruptcy [6].

Then there was the infamous Facebook IPO snafu caused by a NASDAQ computer glitch on May 18, 2012 [7]. Swiss global financial services company UBS lost $356 million on Facebook as a result [8].

Similarly, in August 2012, a computer system error halted derivatives trading on the Tokyo Stock Exchange for 95 min [9] and knocked the Spanish stock exchange IBEX 35 offline for 5 h [10].

CNN reported in August 2012 that "stock markets have become increasingly vulnerable to bugs over the last decade thanks to financial firms' growing reliance on high-speed computerized trading. Because the trading is automated, there's nobody to apply the brakes if things go wrong" [6].

Bitcoin's developers had known about a bug since 2011 but did little to fix it until the bug wiped out 20 % of Bitcoin's market value and $2.7 million stolen from Silk Road's customers in February 2014 [11].

16.3 Disastrous Arithmetic Errors

There are some college classes that I will never forget in my life. One of them is a numerical analysis course taught by Prof. Layne T. Watson at my alma mater Virginia Tech. Watson painstakingly explained computer arithmetic errors and how to avoid mistakes in software coding. Everyone in the Computer Science Department considers him to be the toughest professor ever, but none imagines that arithmetic errors can be as costly and deadly as the following incidents.

Douglas N. Arnold, McKnight Presidential Professor of Mathematics at the University of Minnesota, cited three disastrous incidents due to arithmetic errors [12]:

1. The Patriot Missile failure, in Dharan, Saudi Arabia, on February 25, 1991 which resulted in 28 deaths and 100 injuries, is ultimately attributable to poor handling of rounding errors in software.
2. The sinking of the Sleipner A offshore platform in Gandsfjorden near Stavanger, Norway, on August 23, 1991, resulted in a loss of nearly one billion dollars. It was found to be the result of inaccurate finite element approximation of the linear elastic model of the tricell (using the popular finite element program NASTRAN).
3. The explosion of the Ariane 5 rocket just after lift-off on its maiden voyage off French Guiana, on June 4, 1996, was the consequence of a simple overflow during execution of a data conversion from 64-bit floating point to 16-bit signed integer value.

Arnold dived deeper into the 1991 Patriot Missile failure incident: "On February 25, 1991, during the Gulf War, an American Patriot Missile battery in Dharan, Saudi Arabia, failed to track and intercept an incoming Iraqi Scud missile. The Scud struck an American Army barracks, killing 28 soldiers and injuring around 100 other people. A report of the General Accounting office, GAO/IMTEC-92-26, entitled *Patriot Missile Defense: Software Problem Led to System Failure at Dhahran, Saudi Arabia* reported on the cause of the failure. It turns out that the cause was an inaccurate calculation of the time since boot due to computer arithmetic errors. Specifically, the time in tenths of second as measured by the system's internal clock was multiplied by 1/10 to produce the time in seconds. This calculation was performed using a 24 bit fixed point register. In particular, the value 1/10, which has a non-terminating binary expansion, was chopped at 24 bits after the radix point. The small chopping error, when multiplied by the large number giving the time in tenths of a second, led to a significant error. Indeed, the Patriot battery had been up around 100 h, and an easy calculation shows that the resulting time error due to the magnified chopping error was about 0.34 s. (The number 1/10 equals $1/2^4 + 1/2^5 + 1/2^8 + 1/2^9 + 1/2^{12} + 1/2^{13} + \cdots$. In other words, the binary expansion of 1/10 is 0.0001100110011001100110011001100.... Now the 24 bit register in the Patriot stored instead 0.00011001100110011001100 introducing an error of 0.0000000000000000000000011001100... binary, or about 0.000000095 decimal. Multiplying by the number of tenths of a second in 100 h gives $0.000000095 \times 100 \times 60 \times 60 \times 10 = 0.34$.) A Scud travels at about 1,676 meters per second, and so travels more than half a kilometer in this time. This was far enough that the incoming Scud was outside the "range gate" that the Patriot tracked. Ironically, the fact that the bad time calculation had been improved in some parts of the code, but not all, contributed to the problem, since it meant that the inaccuracies did not cancel" [13].

The GAO/IMTEC-92-26 report stated, "The range gate's prediction of where the Scud will next appear is a function of the Scud's known velocity and the time of the last radar detection. Velocity is a real number that can be expressed as a whole number and a decimal (e.g., 3750.2563…miles per hour). Time is kept continuously by the system's internal clock in tenths of seconds but is expressed as an integer or whole number (e.g., 32, 33, 34, …). The longer the system has been running, the larger the number representing time. To predict where the Scud will next appear, both time and velocity must be expressed as real numbers. Because of the way the Patriot computer performs its calculations and the fact that its registers are only 24 bits long, the conversion of time from an integer to a real number cannot be any more precise than 24 bits. This conversion results in a loss of precision causing a less accurate time calculation. The effect of this inaccuracy on the range gate's calculation is directly proportional to the target's velocity and the length of the system has been running. Consequently, performing the conversion after the Patriot has been running continuously for extended periods causes the range gate to shift away from the center of the target, making it less likely that the target, in this case a Scud, will be successfully intercepted" [14].

16.4 DEF CON Capture the Flag

On November 14, 2013, DARPA program manager Mike Walker gave a PowerPoint presentation on "Could a purpose built supercomputer play DEF CON Capture the Flag?" that described the first-ever Cyber Grand Challenge (CGC). Figures 16.3, 16.4, 16.5, 16.6, 16.7, 16.8, 16.9, 16.10, 16.11, 16.12, 16.13, 16.14, 16.15, 16.16, 16.17, 16.18, 16.19, 16.20, 16.21, 16.22, 16.23 and 16.24 show a complete rundown of the CGC including the $750,000 Challenge Qualifying Event in June 2015 and the $2 million Challenge Final Event in 2016 at DEF CON 24 [15].

16.5 DESCARTES (Distributed Expert Systems for Cyber Analysis, Reasoning, Testing, Evaluation, and Security)

In response to DARPA's Cyber Grand Challenge, I have assembled an international team of cybersecurity researchers, software architects, software developers, professors, and students whose names are kept confidential at the time of this writing. We have registered with DARPA as Team DESCARTES (Distributed Expert

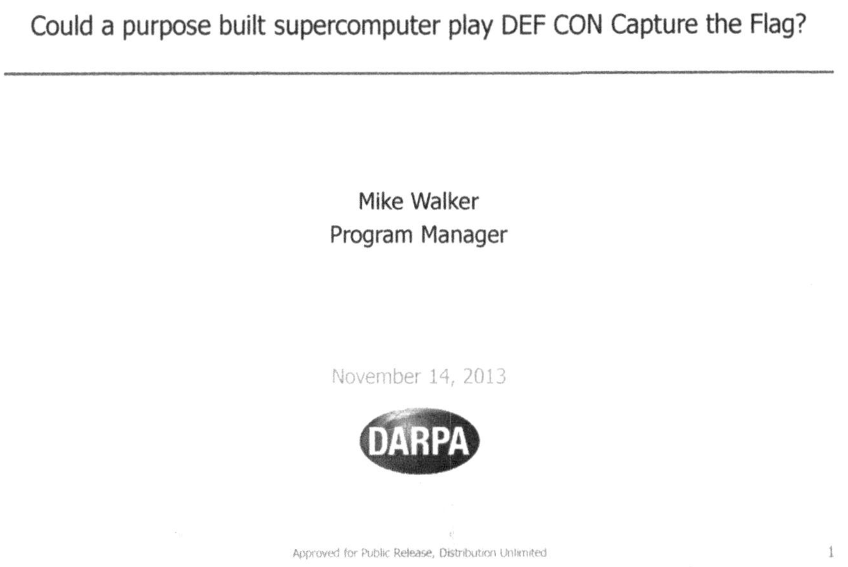
Could a purpose built supercomputer play DEF CON Capture the Flag?
Mike Walker
Program Manager
November 14, 2013
DARPA
Approved for Public Release, Distribution Unlimited
1

Fig. 16.3 DEF CON Capture the Flag

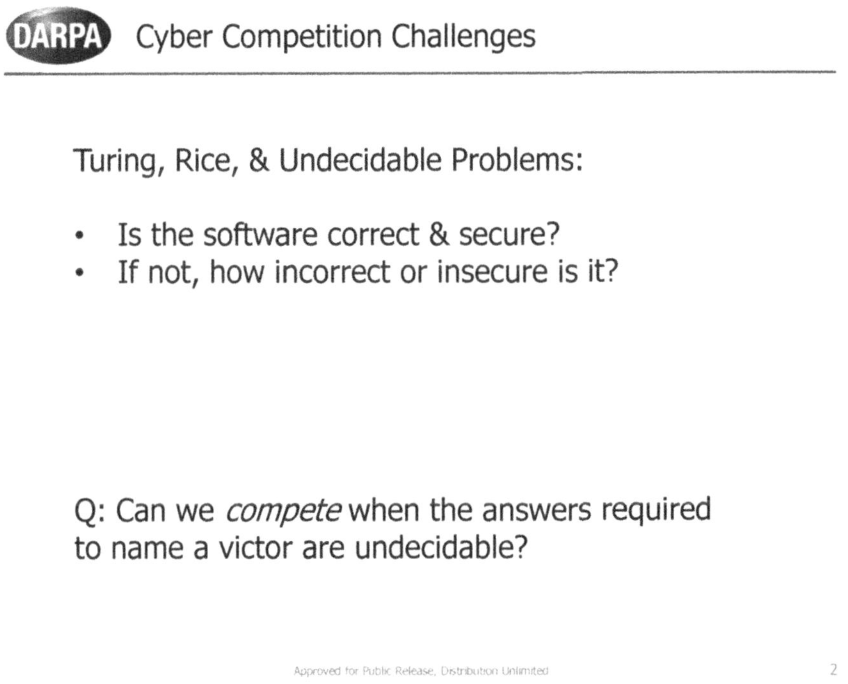
DARPA
Cyber Competition Challenges
Turing, Rice, & Undecidable Problems:
• Is the software correct & secure?
• If not, how incorrect or insecure is it?
Q: Can we *compete* when the answers required to name a victor are undecidable?
Approved for Public Release, Distribution Unlimited
2

Fig. 16.4 Cyber Competition Challenges: Program Correctness and Software Security

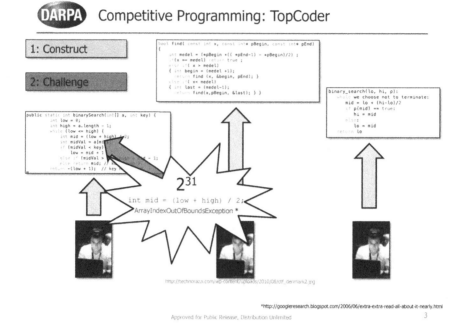

Fig. 16.5 Competitive Programming: TopCoder

Q: Can we *compete* when the answers required to name a victor are undecidable?

A: *consensus evaluation*

Fig. 16.6 Competitive Programming: Consensus Evaluation

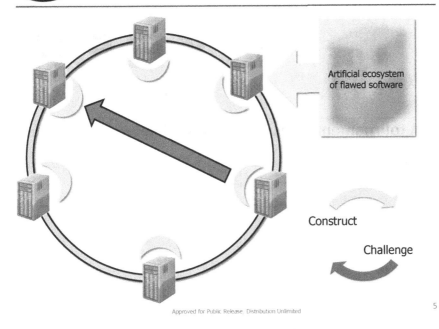

 Competitive Computer Security: DEF CON CTF

Fig. 16.7 Competitive Computer Security: DEF CON CTF

DARPA Competition Paradigm

Harness consensus evaluation to identify
breakthrough technology.

Fig. 16.8 Competition Paradigm: Consensus Evaluation

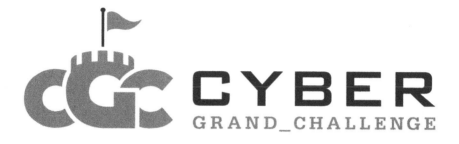

CGC
CYBER
GRAND_CHALLENGE
A tournament for fully automated network defense
Approved for Public Release, Distribution Unlimited
7

Fig. 16.9 Cyber Grand Challenge: Fully Automated Network Defense

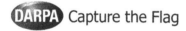

DARPA
Capture the Flag
CGC
An alternative software ecosystem whose challenges and constraints mirror those imposed on real world network defenders.
Approved for Public Release, Distribution Unlimited
8

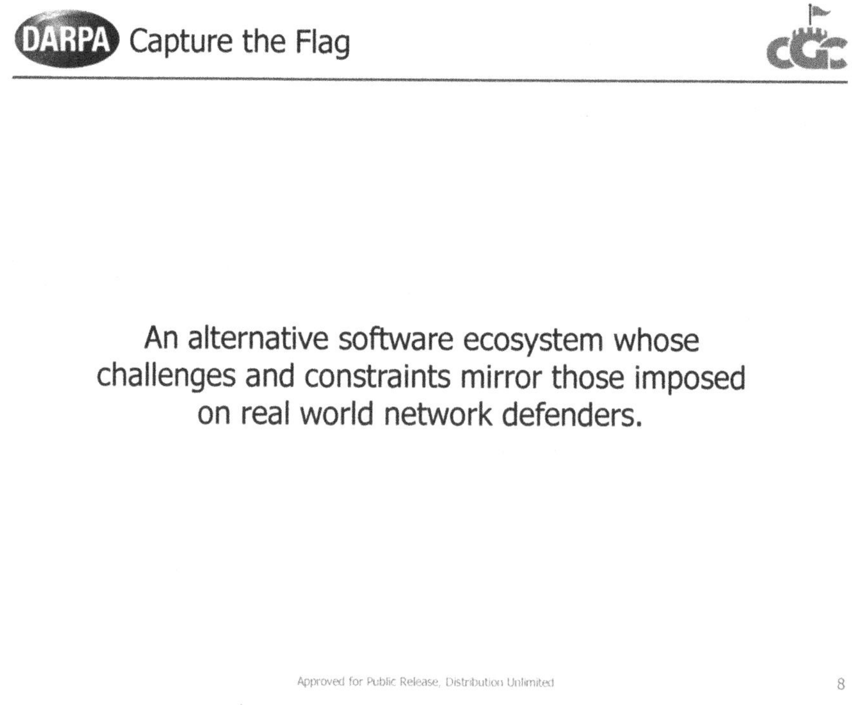

Fig. 16.10 CTF: Alternative Software Ecosystem

DARPA CTF: Alternative Software Ecosystem

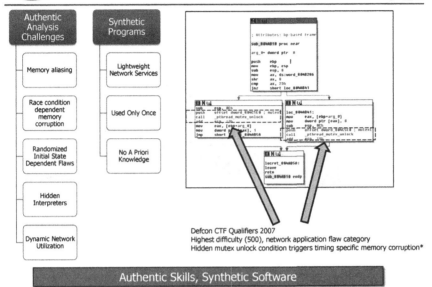

*nopsr.us   9

Fig. 16.11   Authentic Skills and Synthetic Software

DARPA CTF: Real World Challenges

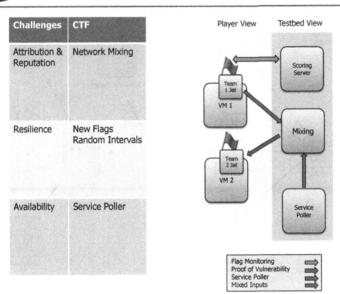

10

Fig. 16.12   CTF: Real World Challenges

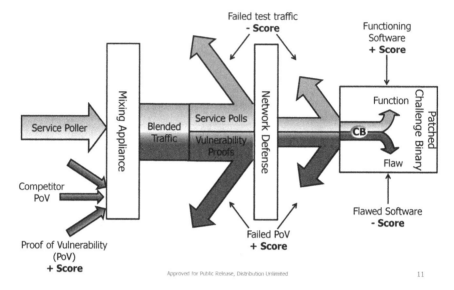

Fig. 16.13 CTF: Real Time Defense

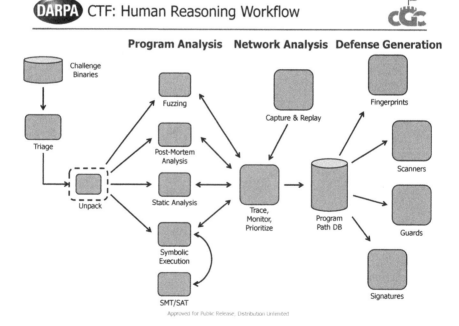

Fig. 16.14 CTF: Human Reasoning Workflow

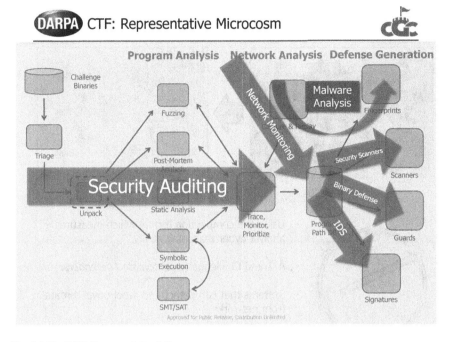

Fig. 16.15   CTF: Representative Microcosm

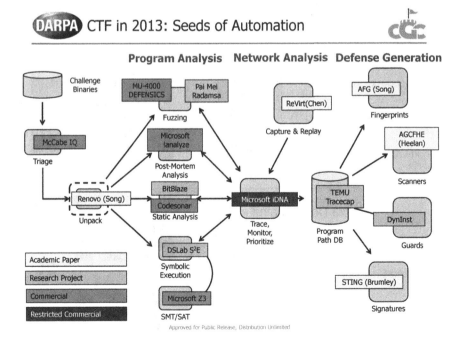

Fig. 16.16   CTF in 2013: Seeds of Automation

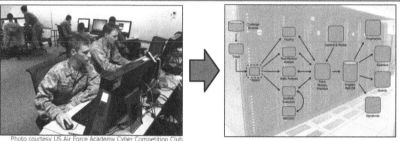

Fig. 16.17 Cyber Grand Challenge: Cyber Reasoning and Automated Defenders

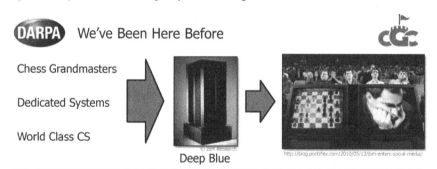

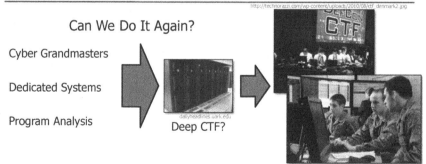

Fig. 16.18 Deep CTF: Cyber Grandmasters

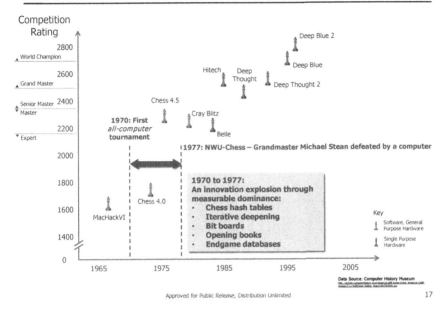

Fig. 16.19   A League of Their Own

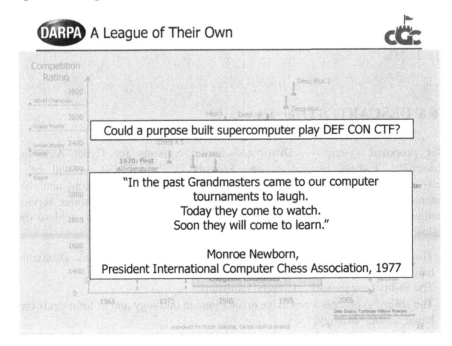

Fig. 16.20   A League of Their Own: Monroe Newborn

A new DARPA Challenge...

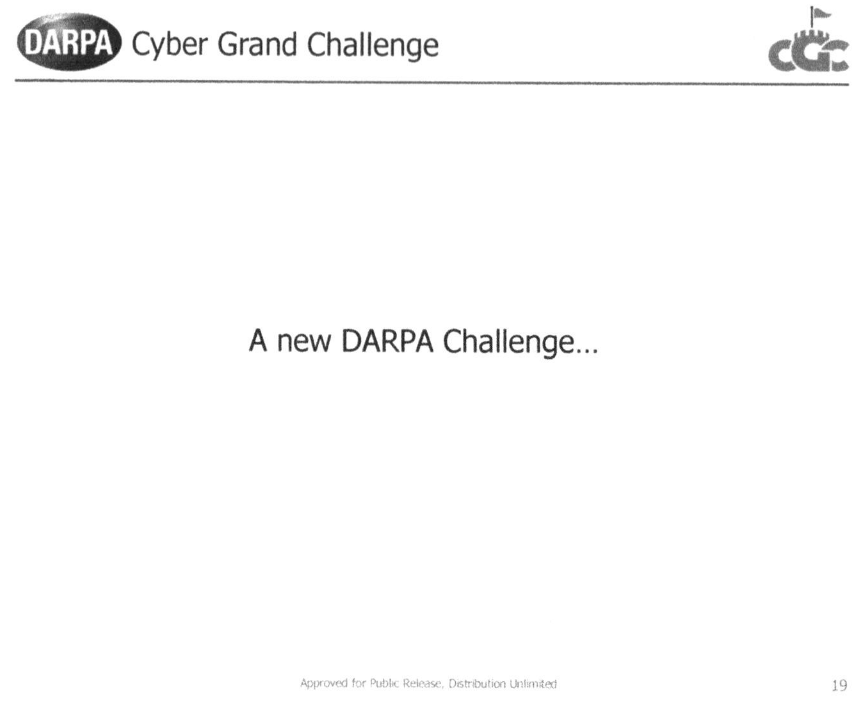

Fig. 16.21   A new DARPA Challenge

Systems for Cyber Analysis, Reasoning, Testing, Evaluation, and Security) (see Fig. 16.25).

16.6 DESCARTES Overview

Our proposed system — Distributed Expert Systems for Cyber Analysis, Reasoning, Testing, Evaluation, and Security (DESCARTES) — will be a fully autonomous cyber defense system that is capable of autonomous analysis, autonomous patching, autonomous vulnerability scanning, autonomous service resiliency, and autonomous network defense. DESCARTES plans to achieve the following objectives:

- The ability to run in a resource constrained environment on affordable hardware.
- The ability to scale with additional hardware.
- The ability to process knowledge quanta into an ontology and to form decisions based on this ontology.
- The ability to react to known threats quickly and to discover new vulnerabilities given limited information.

Cyber Grand Challenge: Scheduled Participation Opportunities

Open Track

- Open to any eligible team
- No IP restrictions on entrant system

Proposal Track

- DARPA Scientific Review Board
- Funded $750k/phase
- Government Purpose Rights to funded development

See rules at www.darpa.mil/cybergrandchallenge for full details

Fig. 16.22 Cyber Grand Challenge Open Track and Proposal Track

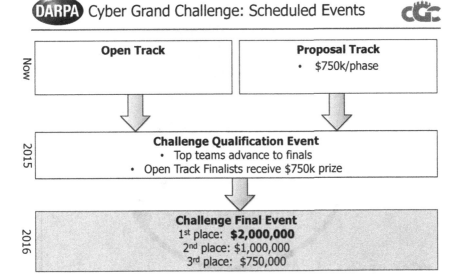

Fig. 16.23 Cyber Grand Challenge: Scheduled Events

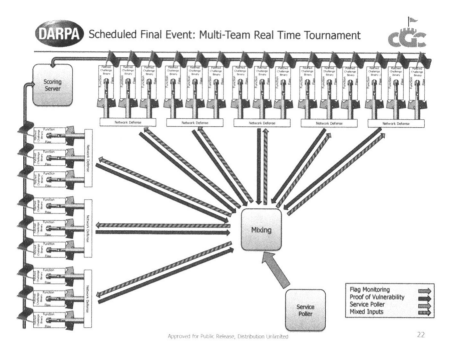

Fig. 16.24   Cyber Grand Challenge Final Event: Multi-Team Real Time Tournament

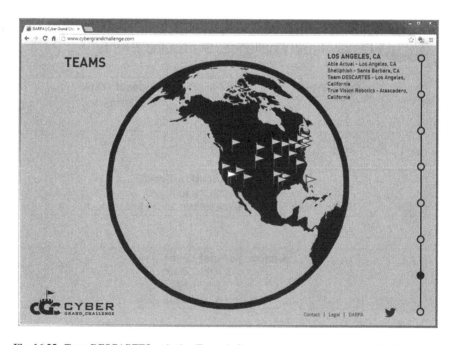

Fig. 16.25   Team DESCARTES and other Teams in Los Angeles on www.cybergrandchallenge.com

Planning Agents and Reactive Agents

In order to achieve these objectives, DESCARTES will employ distributed expert systems — a combination of intelligent planning agents and an adaptive reactive self-healing network based on dynamic mental models [16] and artificial immune system [17]. Dynamic mental models have been successfully applied to diagnosing AT&T telecommunication networks [18] and studied by the U.S. Army Research Office for use in diagnostic support of complex modern weapons systems [19]. Artificial immune system using negative detection — an immunological model of distributed detection — has shown promise in network intrusion detection [17].

Given a set of challenge binaries, DESCARTES will use data flow analysis and binary disassembly to form a structure similar to the HI-CFG (hybrid information- and control-flow graph) structure of Caselden et al. [20].

Automated fuzz testing will generate a series of messages to send to the target application and both the data flow and structure of the program will be mapped. However, due to the large nature of software applications and the exponential time growth of mapping an entire application, our approach will set a depth limit to the analysis. We do not expect to find vulnerabilities using the automated fuzz testing, but we hope to use the fuzz testing combined with the data flow tracking as a way to map which parts of the code are being used most often. The more resources that are available to DESCARTES, the more of the application will be able to be mapped using iterative deepening. Analysis will be possible even if the graph structure is incomplete.

Common code paths that are executed frequently will be determined. Once this is determined we can heavily investigate network traffic that tries to execute parts of the software application that are less used.

Traffic System Analogy

If we imagine a protected software application like a traffic system, the roads represent the structure of the application and the cars represent data moving through the system from an outside source. An intruder can be thought of as a person with malicious intent traveling in a car. If a car is moving along a well lit highway then that car is likely not a threat. However if we can detect that a car is moving along a poorly lit alley, it might warrant further investigation.

To further our traffic system analogy, we may pretend that we are the police patrolling for potential criminals but we do not have a perfect knowledge of all of the roads. As we drive around we can learn more about the road structure and we can follow the suspects as they move throughout the road system.

To mark high traffic areas, we will adjust the arc weights of the software graph. Higher traffic routes will receive a higher weight. We can then prioritize analysis of network messages that trigger nodes with low weights.

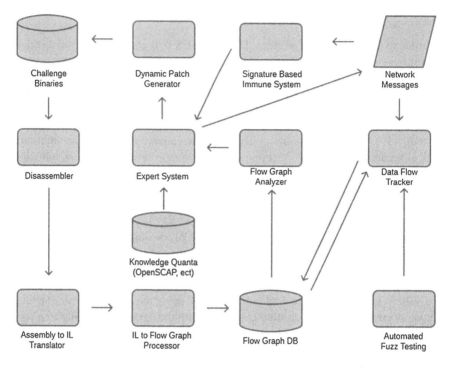

Fig. 16.26 High-Level System Architecture for DESCARTES

We expect network traffic and program execution to follow certain patterns over a period of time. If we detect anomalies that go against typical program execution then this could indicate a potential network attack. Furthermore we can recognize behavioral signatures for further network attack evidence.

The software graph itself will be analyzed for signature based vulnerabilities and the probability that a certain detected graph pattern matches a known signature will increase the probability of a vulnerability.

High-Level System Architecture

Probabilities of suspected vulnerabilities will be fed as knowledge quanta into distributed expert systems as the basis for probabilistic inference. The distributed expert systems (intelligent agents with their own decision-making dynamic mental models) will weigh the evidence collected and develop a high level mitigation plan to either autonomously patch the detected vulnerability or to deploy network countermeasures (e.g. blocking IP of sender). In a sense, each expert system is a "check engine" light and the distributed expert systems will collectively attempt to heal security problems before they happen.

DESCARTES views autonomous patching as a form of emergency response and first aid. For instance, if we apply a band-aid to a cut, we still have the cut, but we have provided protection against infection. If network messages are the input into the system and if specifically formatted or malformed messages are the triggers for software vulnerabilities, we can patch the challenge binaries by deploying a selection of input validators and adapters. The goal is to add a layer to the application that quickly processes the network message into a safe format before the message is allowed to pass to the rest of the application logic. Figure 16.26 shows a high-level system architecture for DESCARTES.

16.7 DESCARTES Automation Strategy

Autonomous Analysis:

Typically when performing an analysis of a software application, the source code is analyzed using static code analysis. Vulnerabilities are detected by looking at known issues at the source code level and these issues are flagged for review by human experts. Static code analysis is typically a slow process that must be done at compile time. Often budget realities mean that issues found during static code analysis must be prioritized according to their severity. However, this approach does not work when the source code is unavailable such as in the case of COTs or legacy applications.

The approach of DESCARTES is to perform an analysis of the code at the assembly language level. Performing a disassembly of a software application is a relatively quick process. The problem with this approach is that there are many assembly language instructions and the instruction set can change with different software architectures. For this reason, the assembly language needs to be converted into a standardized set of instructions that are generic to a wide variety of architectures so that the code can be properly analyzed. Our approach of converting assembly language into an Intermediate Language (IL) is similar to the approach employed by the Vine application in the BitBlaze project [21]. Unlike Vine however, we seek to develop a software tool that is highly configurable and can target a wide range of processor architectures. To make the code more flexible, we plan to develop such a tool in a high level language such as Python and Scala.

Autonomous Patching:

Our proposed system will be able to detect software vulnerabilities at the assembly language level. Once these vulnerabilities are determined, patches will need to be created to protect the software application. We propose two strategies for autonomously patching a challenge binary.

Strategy #1: If we detect a known vulnerability in a software application that is part of our expert system ontology (due to a known bug in a COTs application for instance), we will autonomously trigger an installation of any available hotfixes. One potential source for this information is the Security Content Automation Protocol (SCAP).

Strategy #2: We will develop generic patches that aim to solve common programming errors such as buffer overflows, double free bugs, and format string bugs. Our approach is similar to the approach taken by Loriant et al. [22].

Autonomous Vulnerability Scanning:

In order to autonomously generate network messages that provide proof of software vulnerabilities, DESCARTES will first identify the vulnerability in the constructed software graph from the Intermediate Language (IL) and then perform a series of reverse transformations on the data. Using automated fuzz testing combined with a data flow graph, we can observe the transformations that occur to a given input message as it flows through the software system. Once a vulnerability is identified using the DESCARTES analysis techniques, these transformations will be reversed to generate a message would trigger the vulnerability if it passed back through the system. This message can then be sent to competitors in order to trigger the detected vulnerability on their systems. Our approach is similar to the work being done by Caselden et al. [20].

Autonomous Service Resiliency:

Autonomous service resiliency will be a result of the DESCARTES code analysis and autonomous patching capabilities. Because our techniques will be designed to run in real time and to respond to both novel and known threats using a hybrid combination of high level planning agents and reactive agents, we will be able to preserve the availability and function of the challenge binaries provided through the competition framework.

Autonomous Network Defense:

The first line of defense in the DESCARTES autonomous network defense solution will be the reactive agents that are modeled on the human immune system. By using signature based detection and knowledge of the typical system state, our proposed system will be able to detect network intrusions. By using probabilistic inference through a high level planning agent we will be able to respond to novel threats by making inferences about suspicious behaviors.

If a threat is detected, a mitigation strategy will be employed. Some examples of potential mitigation strategies include blocking network traffic from a specific IP address, blocking targeted ports, and autonomous patching.

16.8 The DESCARTES Chronicle

In a new book slated for release in 2016-2017, Team DESCARTES will chronicle our journey and technical achievements for the DARPA Cyber Grand Challenge regardless of the competition results. Please stay tuned for more exciting information.

Bibliography

1. **Walker, Mike and Eagle, Chris.** Hi, it's Mike Walker and Chris Eagle from the DARPA Cyber Grand Challenge. Ask us Anything! [Online] reddit, June 3, 2014. http://www.reddit.com/r/IAmA/comments/277aih/hi_its_mike_walker_and_chris_eagle_from_the_darpa/.

2. **Bratus, Sergey, et al.** Exploit Programming: From Buffer Overf lows to "Weird Machines" and Theory of Computation. [Online]; login:, December 2011. http://langsec.org/papers/Bratus.pdf.

3. **Hicks, Michael.** Build it, Break it, Fix it: A new security contest. [Online] Maryland Cybersecurity Center, September 17-19, 2013. http://csrc.nist.gov/nice/2013workshop/presentations/day2/d2_trk2_hicks_measuring_software_security_with_contests.pdf.

4. **Flake, Halvar.** Checking the boundaries of static analysis. [Online] SyScan (The Symposium on Security for Asia Network) 2013, April 2013. https://docs.google.com/presentation/d/1_Te 02rSqn7wuhsmkkluqWhDBoXXFVUL5Mp0dUxH0cVE/edit#slide=id.gbcb101ff_026.

5. **Twin, Alexandra.** Glitches send Dow on wild ride. [Online] CNNMoney, May 6, 2010. http://money.cnn.com/2010/05/06/markets/markets_newyork/index.htm.

6. **Eha, Brian Patrick.** Is Knight's $440 million glitch the costliest computer bug ever? [Online] CNNMoney, August 9, 2012. http://money.cnn.com/2012/08/09/technology/knight-expensive-computer-bug/index.html.

7. **Yousuf, Hibah.** Facebook trader: Nasdaq 'blew it'. [Online] CNNMoney, May 21, 2012. http://money.cnn.com/2012/05/21/markets/facebook-nasdaq/index.htm.

8. **—.** UBS lost $356 million on Facebook, suing Nasdaq for it. [Online] CNNMoney, July 31, 2012. http://buzz.money.cnn.com/2012/07/31/ubs-loss-facebook-ipo/.

9. **Hasegawa, Toshiro, Nohara, Yoshiaki and Ikeda, Yumi.** Tokyo System Errors Underscore Decline in Japan's Equity Market. [Online] Bloomberg, August 7, 2012. http://www.bloomberg.com/news/2012-08-07/second-system-error-in-seven-months-halts-tokyo-derivative-trade.html.

10. **Rooney, Ben.** Spanish stocks halted for 5 hours due to trading glitch. [Online] CNNMoney, August 6, 2012. http://buzz.money.cnn.com/2012/08/06/spain-stocks-trading-glitch/.

11. **Pagliery, Jose.** Drug site Silk Road wiped out by Bitcoin glitch. [Online] CNNMoney, February 14, 2014. http://money.cnn.com/2014/02/14/technology/security/silk-road-bitcoin/.

12. **Arnold, Douglas N.** Some disasters attributable to bad numerical computing. [Online] University of Minnesota, August 26, 1998. https://www.ima.umn.edu/~arnold/disasters/disasters.html.

13. **—.** The Patriot Missile Failure. [Online] University of Minnesota, August 23, 2000. https://www.ima.umn.edu/~arnold/disasters/patriot.html.

14. **Office of Public Affairs.** Patriot Missile Defense: Software Problem Led to System Failure at Dhahran, Saudi Arabia. [Online] U.S. Government Accountability Office, February 4, 1992. http://www.gao.gov/products/IMTEC-92-26.

15. **Walker, Mike.** Could a purpose built supercomputer play DEF CON Capture the Flag? [Online] U.S. Department of Defense, November 14, 2013. https://cgc.darpa.mil/Competitor_Day_CGC_Presentation_distar_21978.pdf.

16. **Lee, Newton.** *Counterterrorism and Cybersecurity.* New York : Springer Science + Business Media, 2013.

17. **Hofmeyr, Steven Andrew.** *"An Immunological Model of Distributed Detection and Its Application to Computer Security.* s.l. : University of theWitwatersrand, 1999.

18. *DM2: An Algorithm for Diagnostic Reasoning that Combines Analytical Models and Experiential Knowledge.* **Lee, Newton.** 1988, International Journal of Man-Machine Studies, pp. 643-670.

19. **Berwaner, Mary.** *The Problem of Diagnostic Aiding.* s.l. : The Defense Technical Information Center, 1989.

20. *HI-CFG: Construction by Binary Analysis, and Application to Attack Polymorphism.* **Caselden, Dan, et al.** s.l. : ESORICS'13: European Symp. on Research in Comp. Security, 2013, ESORICS 2013: European Symposium on Research in Computer Security.

21. **Song, Dawn, et al.** *BitBlaze: A New Approach to Computer Security via Binary Analysis.* s.l. : Proceedings of the 4th International Conference on Information Systems Security, 2008.

22. *Server Protection through Dynamic Patching.* **Nicolas, Loriant, Segura-Devillechaise, Marc and Menaud, Jean-Marc.** s.l. : Proceedings of the 11th Pacific Rim International Symposium on Dependable Computing, 2005. pp. 343-349.

23. **Rolles, Rolf.** Rolf Rolles: Principal at Stealth-Mode Startup. [Online] LinkedIn. [Cited: January 4, 2015.] https://www.linkedin.com/in/rolfrolles.

24. **Cyber Grand Challenge.** Frequently Asked Questions. [Online] DARPA, November 14, 2014. https://cgc.darpa.mil/CGC_FAQ_v10.pdf.

25. **Flake, Halvar.** Halvar Flake: Staff Engineer at Google. [Online] LinkedIn. [Cited: January 4, 2015.] https://www.linkedin.com/profile/view?id=4969287.

26. **Lattner, Chris.** The LLVM Compiler Infrastructure. [Online] University of Illinois at Urbana-Champaign. [Cited: January 4, 2015.] http://llvm.org/.

27. **Dependable Systems Lab.** S2E: Selective Symbolic Execution. [Online] École Polytechnique Fédérale de Lausanne (EPFL). [Cited: January 4, 2015.] https://sites.google.com/site/dslabepfl/proj/s2e.

28. **Wikipedians.** Matt Blaze. [Online] Wikipedia, April 1, 2014. http://en.wikipedia.org/wiki/Matt_Blaze.

29. **D'Silva, Vijay, Haller, Leopold and Kroening, Daniel.** Abstract satisfaction. [Online] ACM Digital Library, January 2014. http://dl.acm.org/citation.cfm?id=2535868.

30. **Hofmeyr, Steven Andrew.** An Immunological Model of Distributed Detection and Its Application to Computer Security. [Online] ACM Digital Library, May 1999. http://dl.acm.org/citation.cfm?id=929186.

Index

© Springer International Publishing Switzerland 2015
N. Lee, *Counterterrorism and Cybersecurity*, DOI 10.1007/978-3-319-17244-6

Printed in the United States
By Bookmasters